MODELLING SMALL DEFORMATIONS OF POLYCRYSTALS

Professor Jean Mandel (1907–1982)

MODELLING SMALL DEFORMATIONS OF POLYCRYSTALS

Edited by

JOHN GITTUS

Director, Safety & Reliability Directorate, United Kingdom Atomic Energy Authority, Culcheth, Warrington, UK

and

JOSEPH ZARKA

Centre National de la Recherche Scientifique, Ecole Polytechnique, Palaiseau, France

ELSEVIER APPLIED SCIENCE PUBLISHERS
LONDON and NEW YORK

ELSEVIER APPLIED SCIENCE PUBLISHERS LTD
Crown House, Linton Road, Barking, Essex IG11 8JU, England

Sole Distributor in the USA and Canada
ELSEVIER SCIENCE PUBLISHING CO., INC.
52 Vanderbilt Avenue, New York, NY 10017, USA

WITH 5 TABLES AND 101 ILLUSTRATIONS

Softcover reprint of the hardcover 1st edition 1986

British Library Cataloguing in Publication Data

Modelling small deformations of polycrystals.
1. Crystals—Defects 2. Deformation
(Mechanics)
I. Gittus, John II. Zarka, Joseph
548'.842 QD921

Library of Congress Cataloging in Publication Data

Modelling small deformations of polycrystals.

English and French.
Bibliography: p.
Includes index.
1. Dislocations in crystals. 2. Deformations
(Mechanics) 3. Materials—Mechanical properties.
I. Gittus, John. II. Zarka, Joseph.
QD921.M674 1986 548'.842 85-29167

ISBN-13: 978-94-010-8357-7 e-ISBN-13: 978-94-009-4181-6

DOI: 10.1007/978-94-009-4181-6

Phototypesetting by Tech-Set, Gateshead, Tyne & Wear.

Avant-Propos

L'oeuvre scientifique de Jean Mandel a été exceptionnellement riche dans le domaine de la Mécanique des Solides; les sujets qu'il a traités sont véritablement très différents, que ce soit pour la théorie de la plasticité, le flambage, la Mécanique des Sols, la viscoélasticité, la théorie des modèles réduits, la thermodynamique ou la percolation dans les milieux poreux. Mais, à travers une oeuvre si complète, l'intérêt de Jean Mandel pour le rattachement des propriétés de la matière (résistance, déformabilité, viscosité) à celles de leurs constituants intimes ne s'est jamais démenti. Ce qu'on appelle parfois, pour les matériaux, le passage du microscopique au macroscopique fut une direction de recherches très constante chez lui et il n'a cessé de la promouvoir au sein du Laboratoire de Mécanique des Solides qu'il dirigeait.

Dans la plasticité des métaux, on sait que les déformations permanentes doivent être recherchées dans les glissements intracristallins et plus généralement dans les dislocations et les divers défauts microstructuraux. Avant d'aborder la déformation du polycristal, il faut connaître les mécanismes qui se produisent au sein du cristal, quels sont les différents systèmes de glissement qui peuvent être activés et aussi les mécanismes élémentaires du maclage. Jean Mandel a montré comment passer du comportement du monocristal à celui du polycristal et a donné des relations entre la déformation permanente globale du polycristal et la déformation plastique des monocristaux.

Le travail absorbé dans la déformation plastique est divisé en deux parties, l'une dissipée en chaleur, l'autre transformée en énergie bloquée dans les cristaux: une partie de l'écrouissage peut alors être due aux contraintes résiduelles exactement comme cela se produit dans une structure. Il s'agit là d'un exemple particulier rattaché à la théorie de la plasticité des métaux. Mais on peut en citer d'autres pour d'autres matériaux. Par exemple, pour les milieux constitués de plusieurs corps comme les matériaux composites ou le béton, Jean Mandel a donné des bornes du module d'élasticité global, plus serrées que celles qui avaient été proposées avant lui. Pour ces mêmes matériaux, et pour les sols, l'anisotropie élastique est fonction de l'arrangement des grains. Pour les milieux granulaires, elle est liée aussi à la forme des grains; ainsi, par exemple, un empilement régulier de grains ellipsoïdaux est plus déformable perpendiculairement que parallèlement au grand axe des ellipses.

Pour les roches, il est assez bien connu que le passage micro–macro se fait encore mal. Certes, il est expérimentalement montré que les granites ou les calcaires cristallins sont d'autant plus résistants que leurs cristaux élémentaires sont petits, mais l'anisotropie, la déformabilité, la fatigue oligocyclique, etc, se rattachent mal aux propriétés des grains. L'expérience et le calcul montrent que les éléments microscopiques qui gouvernent le comportement des matériaux de type rocheux sont les microfissures intracristallines ou de joints de grains, de la même façon que le comportement physique ou mécanique d'un massif rocheux est lié aux diaclases, aux failles et à la stratification.

L'étude de la fissure, dont le premier modèle est celui de Griffith, a été reprise par Jean Mandel, notamment pour la singularité en bout de fissure et pour les conditions de propagation. Ces travaux n'ont peut-être pas apporté à la Mécanique des Roches autant que nous l'avions espéré; par contre, ils ont été très évidemment à la base de certains développements faits au Laboratoire de Mécanique des Solides dans le domaine de la Mécanique de la rupture ou de la fatigue et cela montre une fois de plus que s'il est très utile de choisir des orientations de travail, il est important aussi de savoir les quitter.

Le sujet choisi pour ce livre est un thème d'une grande actualité; c'est le meilleur hommage que nous pouvions rendre à Jean Mandel que de rassembler ici des contributions de qualité.

P. HABIB
Directeur du Laboratoire de Mécanique des Solides,
Ecole Polytechnique
Palaiseau, France

Foreword

The scientific work of Jean Mandel has been exceptionally rich in the area of the mechanics of solids; the subjects which he has treated have been extremely diverse, from the theory of plasticity, buckling, soil mechanics, visco-elasticity, the theory of reduced models, and thermodynamics, to percolation in porous media. But throughout this comprehensive work Jean Mandel has always maintained his interest in forming connections between the properties of materials (strength, deformability, viscosity) and the properties of their basic constituents. What is sometimes referred to in materials science as the transition from the microscopic to the macroscopic has for him been a very constant direction of research, which he never ceased to encourage in the Laboratoire de Mécanique des Solides of which he was the director.

It is known that in the plasticity of metals permanent deformations must be sought in intercrystalline slip and more generally in dislocations and the various microstructural defects. Before deformation of polycrystals is tackled, it is necessary to understand the mechanisms which take place within the crystal: the different systems of slip which may be activated and also the elementary mechanisms of twinning. Jean Mandel has shown how to make the transition from the behaviour of the single crystal to that of the polycrystal and has given the relationships between the overall permanent deformation of the polycrystal and the plastic deformation of the single crystal.

The work absorbed in plastic deformation is divided into two parts, one being dissipated as heat, and the other being transformed into energy locked within the crystal: a part of the work hardening may then be due to residual stresses, exactly as occurs in a structure. In that case we are dealing with a special example of the theory of plasticity of metals. But other examples can be quoted for other materials. For example, for media with several constituents, such as composite materials or concrete, Jean Mandel has given closer limits for the overall modulus of elasticity than have been proposed before. For these same materials, and for soils, elastic anisotropy is a function of the arrangement of the grains, so that a regular stack of ellipsoidal grains, for example, is more deformable perpendicular to the major axis of the ellipses than parallel to it.

For rocks, it is fairly well known that the micro–macro transition is still poor. It is true that it has been shown experimentally that granites and crystalline limestones are stronger the smaller their elementary crystals, but anisotropy, deformability, low cycle fatigue, etc., are not well related to the properties of the grains. Experiments and calculation show that the microscopic elements which govern the behaviour of rocky materials are intercrystalline micro-cracking and grain boundaries, in the same way that the physical or mechanical behaviour of rock masses is related to diaclases, faults and stratification.

The study of cracks, the first model of which was that of Griffith, was taken up by Jean Mandel, mainly with regard to the singularity at the crack tip, and the conditions of propagation. It has not been possible to apply this work to the mechanics of rocks to the extent to which it had been hoped; on the other hand, it has very clearly been the basis of certain developments made at the Laboratoire de Mécanique des Solides in the area of fracture mechanics and fatigue — which shows once more, that while it is very useful to choose the directions to be taken in one's work, it is also important to know when to abandon them.

The subject chosen for this book is a topic of great current interest; to bring together contributions of high quality is the greatest homage we could pay to Jean Mandel.

P. Habib
Directeur du Laboratoire de Mécanique des Solides,
Ecole Polytechnique
Palaiseau, France

Preface

In solid mechanics, materials are taken to be continuous and we define, at any chosen point in the body, the stress tensor and the strain tensor. The chosen point represents in fact a *volume element.* Consider now the very simple structure shown in Fig. 1. This is a system of six rods articulated at A, B, C and D and we wish to examine the relation between Q, the applied force at A, and q, the displacement at this point. Now:

(i) This relation can be directly obtained, at the *macroscopic scale,* by a phenomenological, global approach. The system is put in a 'black box'; the coefficients and functions in the relation may

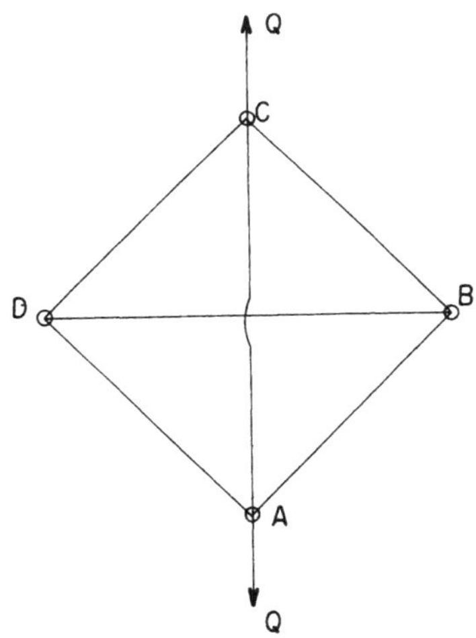

Fig. 1.

then be easily determined by experiment since the loading is one-dimensional. However, it will not be possible to use the relation so-derived for another structure of different geometry; we should have to undertake another series of experiments.

(ii) Alternatively, one can proceed as follows: On the *microscopic scale* by first determining the behaviour of *each rod,* after which we examine how the rods are put together as a system and finally calculate the behaviour of this (hyperstatic) system to obtain the relation (Q, q).

In the *microscopic* approach, the data for the behaviour of the individual rods could be used in any other structure; however, in order to determine the global behaviour of the new structure, we shall have to repeat the calculations. When these calculations are written out in an explicit form, the advantage of the microscopic approach over the macroscopic approach becomes obvious.

Under the *macroscopic* approach, several theories have been developed in order to describe the behaviour of materials. For elasticity, they were (and they still are) very useful. For elastoplasticity and elastovisco-plasticity, they are not yet able to represent all the known effects: it would be necessary to carry out an infinity of experiments in order to achieve this purpose.

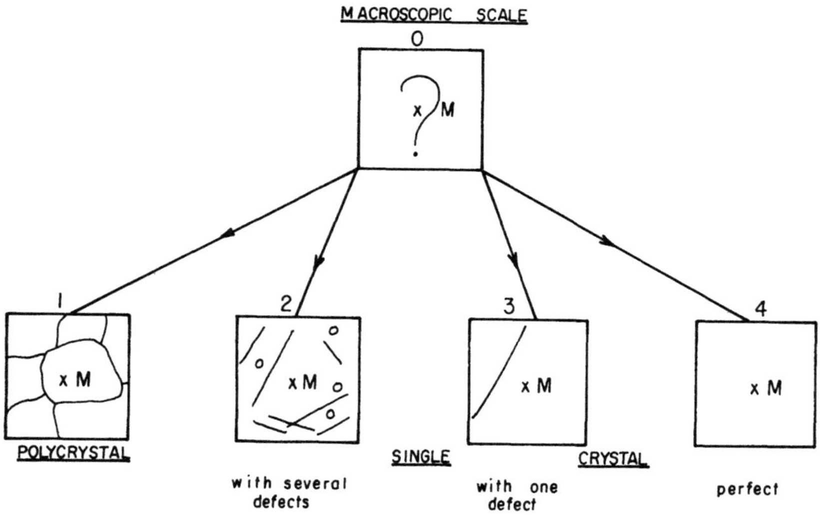

Fig. 2

For a metal, it is possible to use different physical scales for the volume element around the point. Special constitutive laws are associated with each of these scales. It is easily seen that by choosing smaller and smaller scales (microscopic scales) for the volume element, it is easier and easier to express the constitutive relations. The *atomic* scale is not available for continuous media; just above the atomic scale we find the scale of the perfect single crystal that exhibits classical elastic behaviour.

The smaller the microscopic scale, the more difficult it becomes to return to the macroscopic scale (since there are more and more points in the medium and by consequence more and more unknowns) (Fig. 2). So we have to find an *optimal* microscopic scale.

An International Colloquium dedicated to the corresponding problem of modelling large strains (30 September–2 October 1985) will be the subject of a further book to be published by Elsevier Applied Science Publishers.† Our main purpose in the present book is to review the various elements which are now available for the purpose of modelling small strains.

J. GITTUS
J. ZARKA

† *Large Deformations of Solids: Physical Basis and Mathematical Modelling*, Edited by J. Gittus, J. Zarka and S. Nemat-Nasser, Elsevier Applied Science Publishers, London, 1986.

Contents

PART II: PHYSICAL MODELLING

PART III: PHENOMENOLOGICAL MODELLING

PART IV: SYNTHESIS

List of Contributors

A. S. ARGON

Department of Mechanical Engineering, Massachusetts Institute of Technology, Cambridge, Massachusetts 02139, USA

J. GITTUS

Director, Safety & Reliability Directorate, UKAEA, Wigshaw Lane, Culcheth, Warrington WA3 4NE, UK

F. KRÖNER

Institut für Theoretische und Angewandte Physik, Universität Stuttgart, 7 Stuttgart 80, Pfaffenwaldring 57/VI, Federal Republic of Germany

Z. MRÓZ

Polish Academy of Sciences, Institute of Fundamental Technological Research, ul. Swietokrzyska 21, 00–049, Warsaw, Poland

S. NEMAT-NASSER

Department of Applied Mechanics and Engineering Sciences, Mail Code B-010, University of California, San Diego, La Jolla, California 92093, USA

Q. S. NGUYEN

Maître de Recherches au CNRS, Maître de Conférences à l'Ecole Polytechnique, Laboratoire de Mécanique des Solides, Ecole Polytechnique, 91128 Palaiseau Cedex, France

C. STOLZ
Laboratoire de Mécanique des Solides, Ecole Polytechnique, 91128 Palaiseau Cedex, France

A. ZAOUI
Laboratoire des Propriétés Mécaniques et Thermodynamiques des Matériaux, CNRS, Université Paris-Nord, Avenue J. B. Clément, 93430 Villetaneuse, France

J. ZARKA
Directeur de Recherches au CNRS, Conseiller Scientifique du CETIM, Laboratoire de Mécanique des Solides, Ecole Polytechnique, 91128 Palaiseau Cedex, France

PART I

GENERAL ELEMENTS

CHAPTER 1

Physical Basis of Modelling Inelastic Behaviour of Solids

A. S. ARGON
Department of Mechanical Engineering, Massachusetts Institute of Technology, Cambridge, USA

ABSTRACT

In this chapter, some basic approaches to modelling the inelastic response of solids with microstructure are discussed; the difficulty of the task where a strictly continuum approach may be tractable but not quite valid, and a full dislocation mechanics approach may be proper but intractable, is considered. In addition, a brief survey of kinematical aspects of strain production and mechanisms governing plastic resistance is given.

1. INTRODUCTION

The study of the mechanical behaviour of materials has traditionally evolved through phenomenological approaches. Although from time to time some mechanistic models were suggested by the more inquisitive investigator, the basic approach in engineering consisted of performing tests on the materials of interest in the ranges of stress, temperature, and strain rate of interest, and fitting this information to some phenomenological functional form. In the best of such early phenomenological approaches, the functional forms were chosen to satisfy operational symmetry or objectivity conditions, and were normalized to have properties for correct scaling. Often in these parameterization exercises, certain underlying physical processes such as thermal activation of some rate-controlling process of deformation emerged and were noted.

On the whole, however, knowledge of underlying mechanisms was not particularly valued, as often this knowledge remained only qualitative and did not enhance operational reliability. The phenomenological approach had the advantage that microstructural information was not required and materials were treated as 'black box' continua.

Prior to the more definitive developments of dislocation theory of crystal plasticity in the late 1940s and early 50s, the phenomenological approach has led to the statement of equations of state for inelastically deforming media, in analogy to the equations of state of ideal gases. However, in recognition of the fact that a deformed solid may exist at different states as a result of previous deformation history, the total accumulated equivalent plastic strain suffered by the solid was considered as a necessary parameter of state in addition to the externally controllable parameters of the deviatoric and mean normal components of the stress tensor and the temperature. Apart from the difficulty that the total integrated equivalent plastic strain suffered by a solid could not be determined for a bar of steel sitting on the shelf, many discrepancies and much lack of precision was found in such equations of state considering the equivalent plastic strain as a state parameter. Thus, in the more modern revival of the equation of state approach to inelastic behaviour (e.g. by Hart *et al.* (1975)), the accumulated wisdom of dislocation theory of crystal plasticity is utilized by introducing the current isotropic plastic resistance of the solid, as determined, e.g., from a hardness test, into the equation of state as the appropriate state variable, transforming it into a constitutive relation. This approach is now widely used by the practising structural engineer not only for property determination, but also as input in computational approaches to structural design.

Before the developments of modern dislocation theory, the practising metallurgist was quite aware of microstructure, and used the art of metallography as an effective tool for quality control of alloys, associating certain microstructures with desirable hardness and ductility of the alloy. Although this provided very effective visual features of the microstructure, it provided no operationally reliable mechanistic understanding of the phenomenological information on inelastic behaviour gathered by the structural engineer. Such micro-structural monitoring of mechanical behaviour by modern tools of both light and electron microscopy is also widely used in industry in conjunction with the phenomenological, state variable approach for property determinations.

However operationally valuable these traditional approaches might appear, they are inadequate because they cannot deal with the evolution of the state variables with processing that alters one or more components of the microstructure, and may result even in altered forms of constitutive relations in addition to a mere change in the state variable. This has resulted in a quest for the identification and understanding of more fundamental unit mechanisms that govern the deformation resistance together with the evolutionary changes that alter it. These changes may also introduce microstructural damage that can both affect the deformation itself and also initiate fracture. This alternative approach must be based on the correct physical mechanisms to be of predictive value. It relies heavily on periodic measurements of microstructural features of phase forms, defect density and the state of damage at the appropriate scale to monitor the evolution of the state variables in the deforming solid. The main ingredient of the approach is mechanistic modelling of the evolution of the microstructure and its relation to the current deformation resistance, through the use of mechanistic constitutive relations. A complicating feature of mechanistic modelling is that it may have to be performed at one of two different levels by two very different approaches, made necessary by the non-continuum nature of crystal plasticity. It is this complexity in relation to the scale of the microstructure that needs to be appreciated before any modelling is attempted. We deal with this problem of scale below, after we have considered microstructure, and before we discuss mechanisms of deformation that are preferred in different ranges of stress and temperature.

2. MICROSTRUCTURE

2.1. Pre-Smoothing, Post-Smoothing

The notion of a continuum for a crystalline solid is in most cases an abstraction assumed by the applied mechanician for operational convenience. In the sense that the notion of a continuum might be used as a medium that is continuous with uniform properties on a scale slightly larger than atomic dimensions, it is applicable only to simple, non-polymeric fluids. In many cases, however, the medium is known to be heterogeneous and the notion of continuum is used to describe it on a scale very much larger than the scale of the heterogeneities. When this approach of 'pre-smoothing' is taken *a priori,* without any knowledge of the distribution and aggregation of specific microstructure, information on all internal detail, on the distribution of internal stresses between differently deforming phases, is forfeited. Since this information is necessary to understand the overall deformation resistance of the solid and the evolution of internal damage, this becomes impossible. The material is treated as a 'black box', and the investigator is forever subjected to surprises.

Most solids have microstructure in the form of grains and phase domains of different sizes, shapes and orientation. The sujbect of when this microstructure is of such a scale as to require special considerations for its interaction with glide dislocations is treated in Section 3. In many instances, the scale of the microstructure is coarse enough to be out of the range of such specific considerations of dislocation theory, and the individual component phases can be considered as continua with well-defined plastic resistances and hardening behaviour. In this instance, the individual component phases of the overall solid deform collectively as a heterogeneous assembly of continua, interacting with each other only through the boundary conditions applicable at their various interfaces. The deformations of such coarse heterogeneous assemblies are best considered in full detail, preserving the information of the internal variations of effective deformation resistances in individual component phases and the associated internal stresses. This can then be followed by averaging or 'post-smoothing' these variations to obtain the behaviour of the overall assembly. The preserved information of local variations of internal stress in the 'post-smoothing' approach permits the monitoring of the evolution of internal damage in addition to the overall deformation resistances, and removes the surprises and mysteries that are otherwise associated with the 'black-box' point of view.

2.2. Classification of Microstructure

The smallest scale of microstructure is at the level of lattice defects, which are best classified as intrinsic and extrinsic types.

Intrinsic defects are present even in super-pure solids, while extrinsic ones result from additional compositional constituents. Intrinsic lattice defects include vacancy and interstitial type point defects and clusters of such defects that are often present in nuclear particle irradiated alloys, but include also dislocations, stacking faults, and grain boundaries.

Extrinsic defects include all solute atoms, coherent precipitates, incoherent particles, anti-phase boundaries in ordered alloys, and gas-filled micro cavities in nuclear particle irradiated alloys.

On a fine scale, lattice defects interact with glide dislocations through their misfit strain fields, or incompatibility strain fields, through the different elastic constants between particles and their surroundings, or through the increases in free energy that result from the extension or modification of these defects when they are sheared by glide dislocations.

On a coarse scale, microstructure includes phase domains, large pores and cavities, all acting as continua. These heterogeneities interact with each other through their elastic moduli and plastic resistances.

3. KINEMATICS OF CRYSTAL PLASTICITY

Plastic deformation in crystals is primarily a consequence of the glide motion of dislocations, and secondarily results from twinning and other martensitic shear transformations. All such deformation processes are inherently inhomogeneous. Their essential kinematic details are discussed in considerably greater detail in Chapter 5. Here, we concentrate only on dislocation glide that for the passage of each dislocation translates by one lattice vector b, two parts of an elastically strained crystal lattice, across the slip plane. As Fig. 1 shows, the glide displacement of one dislocation by a distance δx in a crystal of thickness t is equivalent to the addition of a dipole of dislocation with a spacing δx to the former. This results, by very elementary considerations, in a plastic shear strain increment $\delta\gamma$ in a portion of the body having a volume V, and a thickness t, which is

$$\delta\gamma = \frac{b\,\delta x\,t}{V} \tag{1}$$

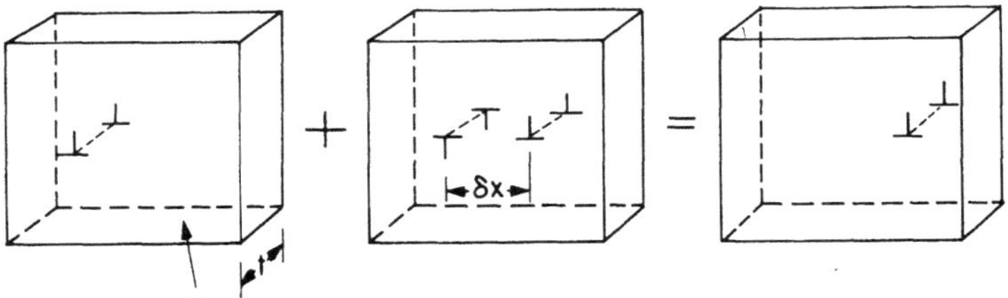

Fig. 1. The glide motion of a dislocation by distance δx is achieved by the addition of a dislocation dipole with spacing δx to it.

where b is the magnitude of the shortest lattice translation vector, the so-called Burgers vector, and the other symbols have their previously defined meaning. If the maximum distance δx that a dislocation can glide is Λ, which is fixed by physical considerations, then additional plastic strain results only from the motion of other dislocations of total length $t\,\delta n$ and the strain increment becomes

$$\delta\gamma = \frac{b\,\Lambda\,\delta(t\,n)}{V} \tag{2}$$

Clearly, the important dimension of the volume V is the mean-free-path length Λ of glide of dislocations. In an equiaxed volume of much smaller dimensions, the glide of dislocations will produce major perturbations and the plastic strain will increase in a jerky manner. Only for volume elements larger than $V \approx \Lambda^3$ will the straining of the solid, by repeated glide events of dislocations, approach a smooth behaviour that could be associated with that of a continuum.

Thus, when an explanation of plastic flow in volume elements much smaller than Λ^3 is required, e.g. the physical processes of dislocation interaction with slip plane obstacles, then the deformation must be analysed by dislocation mechanics. Above this scale, continuum mechanics of plastic deformation should be the analysis of choice. The critical length scale Λ itself is a property of crystal plasticity and indicates the range of influence of elementary glide events. It will be discussed somewhat further in Chapter 5.

4. MECHANISMS OF INELASTIC DEFORMATION

4.1. Inelastic Shear Producing Configurational Changes

In crystalline materials, inelastic strain can be produced in several different ways, but the most prominent of these of interest to us here is that which results from the glide motion of dislocations. As discussed earlier in Section 3, but now viewed in a more general sense, when a dislocation segment with Burgers displacement b, that defines the smallest lattice translation vector, sweeps out an area Δa, as shown in Fig. 2, this results in an average plastic shear strain increment of

$$\Delta\gamma = \frac{b}{h}\frac{\Delta a}{A} = \frac{b\Delta a}{V} \tag{3}$$

in a volume V†.

†In this chapter, where we are discussing inelastic behaviour exclusively, all strains γ or ε and strain rates $\dot{\gamma}$ or $\dot{\varepsilon}$ refer to inelastic strains, i.e. potentially plastic strains. Therefore, a superscript p will be dispensed with.

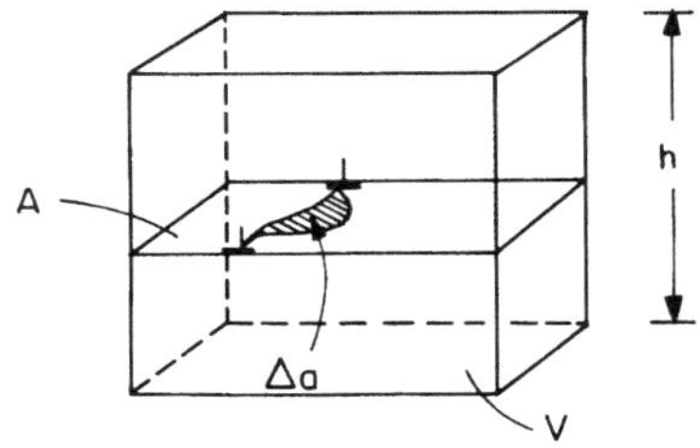

Fig. 2. A portion of a dislocation segment sweeping out an area Δa by glide producing a small increment of shear strain in the macro sample of volume V.

Another mechanism of strain production in crystals is that which involves the production or annihilation of point defects — primarily vacancies. As shown in Fig. 3, when a vacancy is produced by removing a surface, or interface, atom and adding it to the surface, or accommodating it in the atomically thin layer of disorder of an interface, this produces a normal strain increment $\Delta\varepsilon_{yy}$ that is:

$$\Delta\varepsilon_{yy} = \frac{b}{h}\left(\frac{b^2}{A}\right) = \frac{b^3}{V} = \frac{\Omega}{V} \tag{4}$$

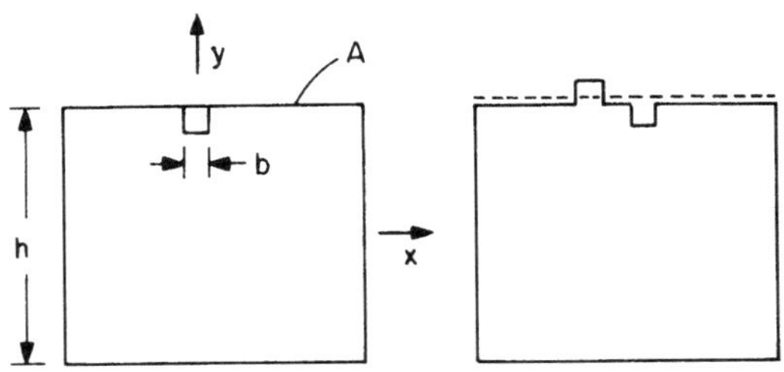

Fig. 3. The removal of a surface atom, and its addition to the top of the surface, producing a small increment of normal strain in the macro sample.

where b^3 is taken as equal to the atomic volume Ω. The annihilation event of a vacancy results in an equal strain increment of opposite sign. If such an annihilation occurred on the vertical faces, the resulting normal strain would be $\Delta\varepsilon_{xx} = (-\Omega/V)$, and the two events together will produce a shear strain increment $\Delta\gamma$ referred to axes at 45° with respect to x and y, which is, per pair of vacancies produced and annihilated:

$$\Delta\gamma = (\Delta\varepsilon_{yy} - \Delta\varepsilon_{xx}) = 2\frac{\Omega}{V} \tag{5}$$

Clearly, the random wandering of an equiaxed point defect in the lattice does not produce any strain.

In many crystal structures under an applied stress, transformations are possible, such as twinning or martensitic shears, in usually lenticular volume elements. Such shears may be accompanied with a dilatation as well. Thus, if the unconstrained transformation shear strain is $\Delta\gamma^T$, and the accompanying dilatation is $\Delta\varepsilon^T$, in a volume element of size Ω_f, as shown in Fig. 4, then the sample of total volume V would undergo an overall shear $\Delta\gamma$ and dilatation $\Delta\varepsilon$ that are, respectively, per transformation:

$$\Delta\gamma = (\Omega_f/V)\,\Delta\gamma^T \tag{6a}$$

$$\Delta\varepsilon = (\Omega_f/V)\,\Delta\varepsilon^T \tag{6b}$$

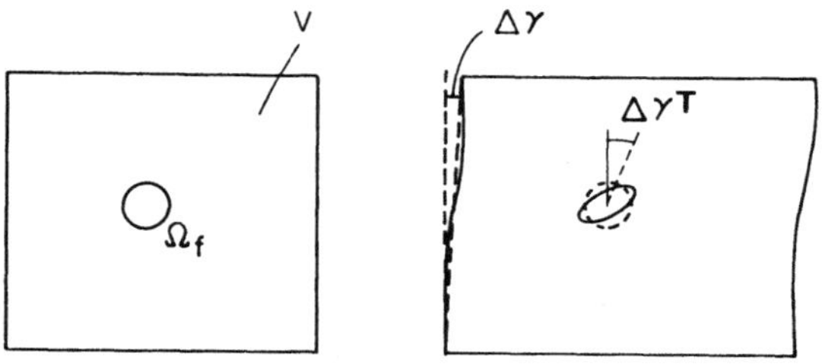

Fig. 4. A localized shear transformation with a transformation shear strain $\Delta\gamma^T$ occurring inside a small volume element Ω_f producing a small increment of shear strain in the macro sample.

4.2. General Classification of Resistance

The plastic resistance of a crystalline solid is directly linked to the resistance of strain production by the mechanisms discussed in Section 4.1. In the case of dislocation glide, this resistance is due to the slip plane obstacles that the dislocations encounter. These resistances may be of an *intrinsic* nature, governed, e.g., by the binding of the dislocation core to close-packed rows of atoms in the plane or by the need for glide dislocations to intersect forest dislocations threading through the slip plane. On the other hand, resistances may be of an *extrinsic* nature, being governed by solute atoms, or by second-phase particles acting as discrete obstacles.

At low temperatures where diffusion is negligibly slow, slip plane obstacles are generally stable and long lived. In this range, the higher energies of configurations assumed by glide dislocations in overcoming the slip plane obstacles can often be reduced by random thermal motion of the lattice. This leads to a general decrease in the plastic resistance with increasing temperature for any given state of obstacle structure and rate of enforced deformation. When the strain results from shear transformations of twinning or the formation of martensite shear lamellae, the resistance to deformation principally results from the nucleation of such transformations.

Above about half the absolute melting temperature, where diffusion is significantly rapid, the slip plane obstacles are no longer stable, but decay at characteristic times in diffusion-controlled recovery processes. This makes the resistance to dislocation glide more temperature and strain rate dependent in this high temperature range.

At still higher temperatures where point defect nucleation and mobility become quite easy, strain can be produced by the removal or addition of matter at interfaces, atom by atom, followed by point defect transport from sources to sinks.

These flow stress mechanisms, or sources of inelastic deformation resistance, have many subtleties which we cannot elaborate here. The interested reader will find them discussed elsewhere (Ashby and Frost, 1975; Kocks *et al.,* 1975; Argon, in press). Here, we will merely list the major sources of resistance that can exist singly or in superposition, and present information on their temperature and strain rate dependence.

4.3. The Inelastic Strain Rate

The basic form of the temperature and strain rate dependence of the deformation resistance results from the thermally-activated nature of the unit processes of strain production. These are all of similar kinematic nature and can be written in a common generic form in their principal coordinates of shear as

$$\dot{\gamma} = \Delta\gamma N(1/t_a) \tag{7}$$

Here, $\Delta\gamma$ represents the total shear strain increment derivable from a unit process in a sample of volume V, as defined by one of the kinematic relations of eqns (3), (5) or (6a); N is the total number of 'fertile' process regions (dislocation segments, effective number of atomic sites on internal incoherent interfaces, volume elements with requisite properties to initiate shear transformations) in the volume V; and t_a is the effective time of successful activation of a characteristic process event.

The important stress and temperature dependence of the strain resides in the reciprocal activation time, or more correctly labelled, the effective activation frequency. Thus, if the excess enthalpy of the strain producing activation configuration is ΔH, and the applied shear stress can produce an amount of work $\sigma\Omega_a$ during the activation to reduce the activation enthalpy, then the activation frequency for events in the direction of the applied stress becomes

$$1/t_a = \nu_a \exp\left(-\frac{\Delta H - \sigma\Omega_a}{kT}\right) \tag{8}$$

Here, ν_a is a characteristic 'normal mode' frequency of the strain-producing unit in the direction of the activation path and incorporates a term that measures the excess entropy of the activated state. The term ΔH incorporates the effects of the pressure component of the applied stress, and Ω_a is the so-called activation volume that is a product of the actual volume Ω_f of the strain-producing unit event and the average transformation shear strain that occurs in it. In nearly every case, eqn (8) provides the correct limiting form of the activation frequency at high stresses and finite temperatures. At very low stresses, however, where $\sigma\Omega_a \ll \Delta H$, reverse thermal fluctuations tending to undo the produced strain occur at frequencies comparable to those of the forward event, leading to a reverse rate that must also be considered. This modifies eqn (8) into the form:

$$1/t_a = \nu_a \exp\left(-\frac{\Delta H}{kT}\right) \sinh\left(\frac{\sigma\Omega_a}{kT}\right) \tag{8a}$$

In nearly all cases of interest, however, the form of eqn (8) governs in the plastic limit of high stress, and eqn (8a) becomes only of interest to give vanishing deformation rates at vanishing levels of stress.

It is instructive to rewrite eqns (7) and (8) to highlight the limiting form of behaviour:

$$\dot{\gamma} = \dot{\gamma}_0 \exp\left(-\frac{\Delta H}{kT}\left(1 - \frac{\sigma}{\hat{\tau}}\right)\right) \tag{9}$$

where

$$\dot{\gamma}_0 = \Delta\gamma N \nu_a \tag{9a}$$

$$\hat{\tau} = \Delta H/\Omega_a \tag{9b}$$

are, respectively, the pre-exponential factor of strain rate and the athermal plastic threshold stress where the strain-producing process can be activated without any assistance from thermal motion.

Equation (9) can be inverted to calculate the dependence of the plastic resistance on temperature and strain rate to obtain

$$\sigma = \hat{\tau}\left(1 - \frac{T}{T_0}\right) \tag{10}$$

where

$$T_0 = \Delta H/k \ln(\dot{\gamma}_0/\dot{\gamma}) \tag{10a}$$

is an extrapolated temperature where thermal energy is large enough to provide for the excess enthalpy of the activated state under a vanishing applied stress for the given imposed strain rate.

In many instances, due to the randomness of the distribution of obstacles and distributions in their strength, the stress dependence of the net activation free energy of eqn (9) is non-linear. This more general dependence can be written in a useful phenomenological form as

$$\dot{\gamma} = \dot{\gamma}_0 \exp\left(-\frac{\Delta H}{kT}\left(1 - \left(\frac{\sigma}{\hat{\tau}}\right)^p\right)^q\right) \tag{11}$$

where the ranges of the phenomenological exponents $0 < p \leqslant 1$ and $1 \leqslant q \leqslant 2$ are governed by some basic thermodynamic restrictions (Kocks *et al.*, 1975).

4.4. Evolution of Deformation Resistances

There is a peak deformation resistance $\hat{\tau}$ for any component flow stress mechanism, and these mechanisms superpose according to principles discussed below, for any given state of obstacle structure. Finite increments of plastic strain, and thermal recovery processes without and with diffusion will alter these resistances according to laws of evolution that can be written symbolically as

$$\dot{\hat{\tau}} = \dot{\hat{\tau}}(\hat{\tau}, \dot{\gamma}, T, p) \tag{12}$$

where the dependence on the applied shear stress acts through its effect on the strain rate, and the effect of pressure p is included to the extent it affects the static recovery process through its effect on diffusion.

4.5. Anisotropy and Three-Dimensional Effects

The treatment of the strain rate expression and the evolution law is intended to refer to individual slip systems. In real crystals, there can be families of slip systems of one or more classes with different obstacle structures, intrinsic and extrinsic resistances obeying different laws of evolution and always interacting with each other. Thus, the overall deformation resistance of an isolated grain to an imposed pure deformation rate will be governed by a multi-dimensional yield surface for the specific partition of the overall deformation rate among the appropriate slip rates for the systems lying on the multi-dimensional yield surface. The proper treatment of this subject is quite complex (Kocks and Chandra, 1982; Asaro, 1983) and beyond the scope of our treatment here. A somewhat expanded treatment of some of the kinematical aspects of the subject will be given in Chapter 5.

On a somewhat higher level of complexity is the problem of the deformation of assemblies of grains as polycrystalline aggregates, where compatibility conditions between deforming grains in internal mechanical equilibrium enforce a different partition of the applied deformation rate among the slip systems of the individual grains of different orientation. The proper treatment of this subject for continuing plastic flow requires careful consideration of evolution of lattice rotations into deformation textures and requires consideration of finite plastic strain increments (Kocks and Chandra, 1982). We will merely note these complexities here and develop only relations for quasi-isotropic plastic flow at constant state, by the various different mechanisms of strain production that dominate the early phases of deformation in different ranges of applied stress and temperature.

4.6. Flow Stress Mechanisms

(a) Low Temperature Plastic Flow

At temperatures below one-half of the absolute melting temperature T_m, where obstacle structures remain relatively stable, the overall plastic resistance could result from a number of contributions, such as the intrinsic resistances of the lattice, or the cutting of the forest dislocations, and the extrinsic resistances of precipitate particles.

For the case of discrete obstacles, including precipitate particles and even forest dislocations, the strain rate expression for a given state of obstacles can be approximated well by a choice of $p = 1$ and $q = 1$ in eqn (11) to give (Frost and Ashby, 1982):

$$\dot{\gamma}_d = \dot{\gamma}_0 \exp\left[-\frac{\Delta H}{kT}\left(1 - \frac{\sigma}{\hat{\tau}_d}\right)\right] \quad \text{(discrete obstacles)} \qquad (13)$$

For a lattice resistance contribution, the best fit requires a choice of $p = 3/4$ and $q = 4/3$, to give (Frost and Ashby, 1982):

$$\dot{\gamma}_l = \dot{\gamma}_0 \left[-\frac{\Delta H}{kT} \left(1 - \left(\frac{\sigma}{\hat{\tau}_l} \right)^{3/4} \right)^{4/3} \right] \quad \text{(lattice resistance)} \qquad (14)$$

When different resistances are present at low temperatures, superposing them requires care. Whenever the scales of the resistance mechanisms are very different, the resistances are directly additive. This is the case for a lattice resistance and discrete obstacles such as precipitates and forest dislocations, where the activation configuration of double kinks along dislocation lines to overcome the lattice resistance is very much smaller in length than the spacing of precipitate obstacles along a dislocation line. When several different types of discrete obstacles are present on the slip plane, the overall resistance is the square root of the sum of the squares of the individual resistance due to individual mechanisms, when they are present in isolation.

(b) Dislocation Creep at High Temperatures

Above about half the absolute melting temperature, effects of diffusion become significant in altering or removing slip plane obstacles that were produced as a result of earlier solid state reactions, or have been produced during earlier phases of plastic straining. Steady creep deformation becomes possible when a balance is reached between the rate of strain-induced obstacle production and the rate of obstacle removal by thermal recovery. This results in dislocation creep at a rate of (Frost and Ashby, 1982):

$$\dot{\gamma}_c = A_c \frac{D_{\mathrm{eff}}\,\mu b}{kT} \left(\frac{\sigma}{\mu} \right)^n \qquad (15)$$

where σ is the applied shear stress, D_{eff} is an effective diffusion constant combining the effects of volume diffusion and dislocation core diffusion in a weighted average, based on the diffusion cross-sections of both paths, between sources and sinks, A_c is a numerical constant determined from experiments, and n is a creep exponent, typically around 5, for sub-grain forming metals and alloys.

In solid solution alloys which do not form sub-grains, the exponent n is usually 3 and the effective diffusion constant is that of diffusion of the solute in the solvent lattice.

(c) Diffusional Flow

At high temperatures, close to the melting point, deformation by atom fluxes along grain interiors and grain boundaries, from those grain boundaries under compression to those grain boundaries under tension, occurs and results in a shear strain rate of (Frost and Ashby, 1982):

$$\dot{\gamma}_{df} = A_{df} \frac{\sigma\Omega.}{kTd^2} D_{eff} \tag{16}$$

where σ is the applied shear stress, d is the average grain size, D_{eff} is an effective diffusion constant combining the effect of volume diffusion

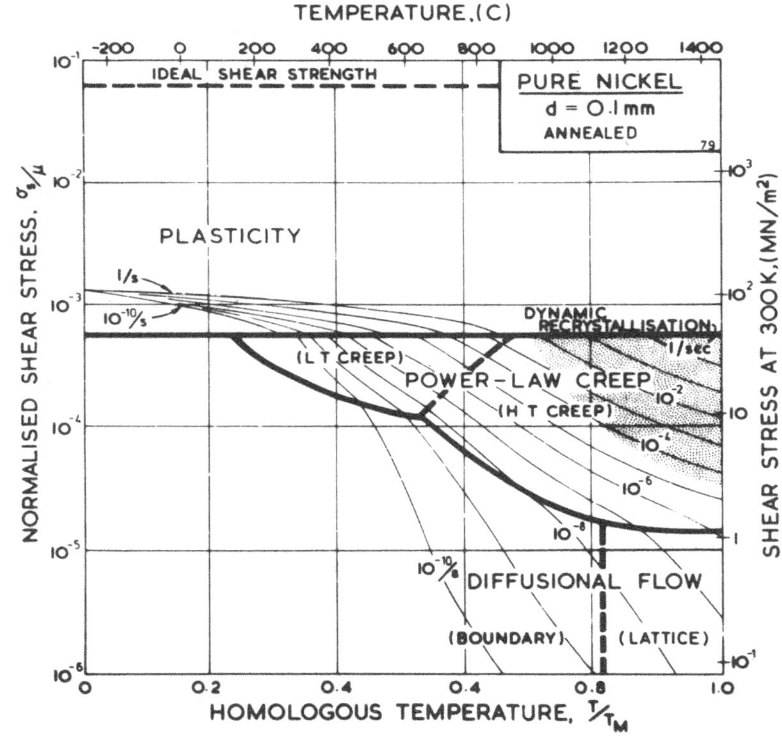

Fig. 5. A deformation-mechanism map for pure fcc nickel with a grain size of 100 μm. The bold curves delineate regions of dominance of low temperature plasticity, dislocation creep (of both low temperature and high temperature types), and diffusional flow. From Frost and Ashby, 1982; courtesy of Pergamon Press.

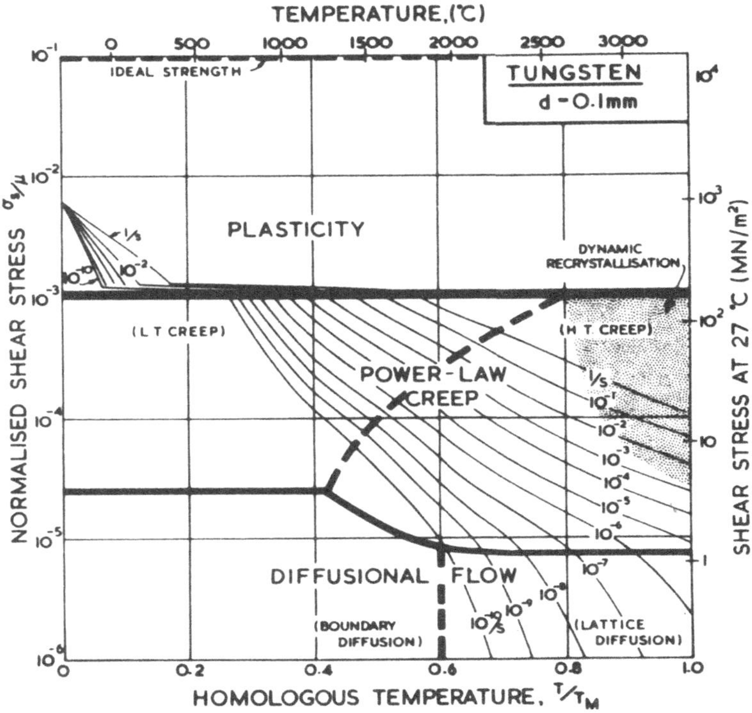

Fig. 6. A deformation-mechanism map for pure bcc tungsten with a grain size of 100 μm. The bold curves delineate the regions of dominance of the various mechanisms of inelastic deformation. Note the steep rise in resistance in the low temperature plasticity region. This is due to a lattice resistance. From Frost and Ashby, 1982; courtesy of Pergamon Press.

and grain boundary diffusion in a weighted average based on the diffusion cross-sections of these two diffusion paths, and A_{df} is a numerical constant typically of the order of 50 for equiaxed grains.

While these are the major mechanisms of deformation, there are interesting variations in different alloys that have been discussed by Frost and Ashby (1982) to which the reader is referred for additional detail.

(d) Deformation Mechanism Maps

Each of these mechanisms has a definite region of dominance in the stress and temperature plane, for a given state of obstacle structures, for a given strain rate. These regions of dominance are best presented in graphical form as deformation mechanism maps. Such maps have been assembled by Frost and Ashby (1982) for most of the important engineering metals and alloys, ceramics, and earth materials, and have been given together with an extensive collection of associated data on the crystallographic, elastic and diffusion constants for these materials. Two examples for fcc nickel and bcc tungsten, both for a grain size of 100 μm, are given in Figs 5 and 6.

REFERENCES

Argon, A. S. (in press). In *Encyclopedia of Materials Science and Engineering*, M. B. Bever (Ed.), Pergamon Press, Oxford.

Asaro, R. J. (1983). In *Advances in Applied Mechanics*, Vol. 23, Academic Press, New York, p. 1.

Ashby, M. F. and Frost, H. J. (1975). In *Constitutive Equations in Plasticity*, A. S. Argon (Ed.), MIT Press, Cambridge, Mass., p. 117.

Frost, H. J. and Ashby, M. F. (1982). *Deformation-Mechanism Maps*, Pergamon Press, Oxford.

Hart, E. W., Li, C. Y., Yamada, H. and Wire, G. L. (1975). In *Constitutive Equations in Plasticity*, A. S. Argon (Ed.), MIT Press, Cambridge, Mass., p. 149.

Kocks, U. F. and Chandra, H. (1982). *Acta Met.*, **30**, 695.

Kocks, U. F., Argon, A. S. and Ashby, M. F. (1975). *Progress in Materials Science*, Vol. 19, B. Chalmers *et al.* (Eds), Pergamon Press, Oxford.

ANNEXE TO CHAPTER 1

Physical Basis for Plasticity and Viscoplasticity

J. ZARKA
Laboratoire de Mécanique des Solides, Ecole Polytechnique, Palaiseau, France

Here, an abstract of the physical elements is given. It is, to a large extent, derived from Jaoul (1965).

1. CLASSICAL EXPERIMENTAL RESULTS

1.1. One-dimensional Experiments on Polycrystals

(a) Classical Plasticity
For almost any metal, at ordinary temperatures (~ 20 °C), and at rather small rates of loading, we obtain a mean curve $\sigma(\varepsilon)$ during a tensile experiment: this is the static curve. The variations of the strain rate $\dot{\varepsilon}$ or of the temperature T imply only very small deviations in the curves relative to the experimental dispersion (Fig. 1).

The whole static curve may be analysed with the help of several successive mathematical relations, such as for example:

$$\left.\begin{array}{lll} \sigma = E\varepsilon & \text{if } \varepsilon < \varepsilon_0 & \text{elastic domain} \\ \sigma = \sigma_0 + A_0(\varepsilon - \varepsilon_0)^{\alpha_0} & \text{if } \varepsilon_0 \leqslant \varepsilon < \varepsilon_1 & \\ \sigma = \sigma_1 + A_1(\varepsilon - \varepsilon_1)^{\alpha_1} & \text{if } \varepsilon_1 \leqslant \varepsilon < \varepsilon_2 & \\ \vdots & & \end{array}\right\} \quad (1.1)$$

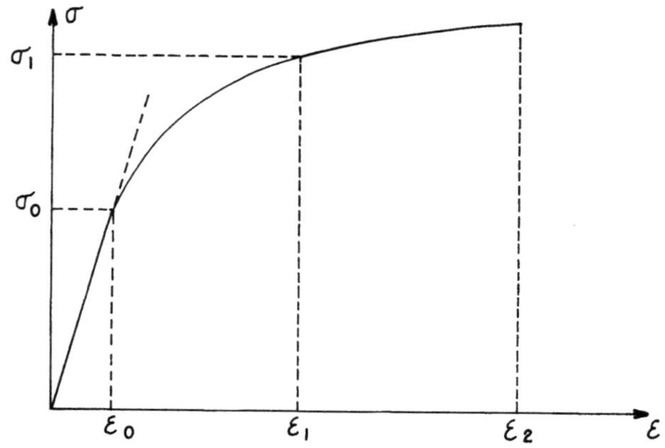

Fig. 1. Schematic tensile curve for a metallic polycrystal.

The constants are dependent on the metallurgical factors. It should be noted that the static curve may also present some irregularities due to instability phenomena.

(b) Viscoplasticity

(i) When we leave the domain of ordinary temperatures or classical strain rates (either higher than $10^{-1}\ s^{-1}$ or lower than $10^{-6}\ s^{-1}$) we can see very important variations in the different curves. Generally, the tensile curves keep their parabolic behaviour and become higher and higher for increasing strain rates or decreasing temperatures (Fig. 2). In some cases there may be regular oscillations. The experimental curves are very often analysed by mathematical relations such as

$$\frac{\partial \sigma}{\partial t} = f(\sigma, \varepsilon)\frac{\partial \varepsilon}{\partial t} + g(\sigma, \varepsilon) \tag{1.2}$$

The 'static' curve ($\dot{\varepsilon} \simeq 0$) is implicitly defined by $g(\sigma, \varepsilon) = 0$, while the 'dynamic' curve is obtained by the differential equation

$$\frac{d\sigma}{d\varepsilon} = f(\sigma, \varepsilon)$$

Relaxation experiments, when the total strain is maintained constant, also exhibit viscoelastic properties.

(ii) During creep experiments, we impose constant stress and temperature. According to the amplitude of the applied stress and temperature, the behaviour of the curves seems different:

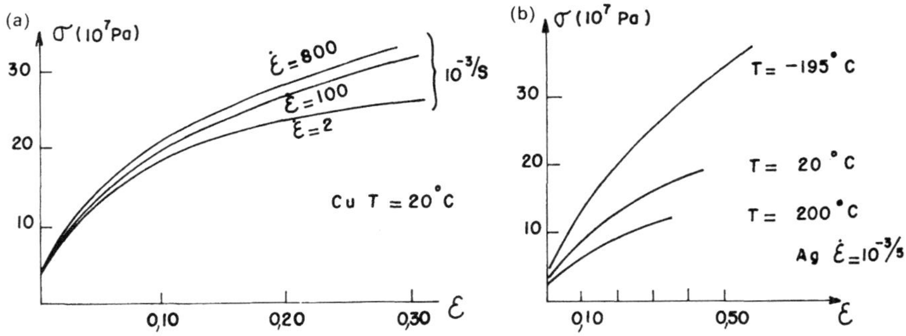

Fig. 2. (a) Influence of the strain rate; (b) influence of the temperature.

we have first a logarithmic creep which is very often analysed by a relation such as

$$\varepsilon = \varepsilon_a + \alpha\,(\log(1 + \gamma t))^{\beta} \tag{1.3}$$

For a higher stress, generally we separate

— transitory creep

$$\varepsilon = \varepsilon_m + \beta_m t^m$$

with m lying between 0·15 and 0·8.

— a stationary creep, at almost constant strain rate $\dot{\varepsilon}$ given by the Norton–Hoff law

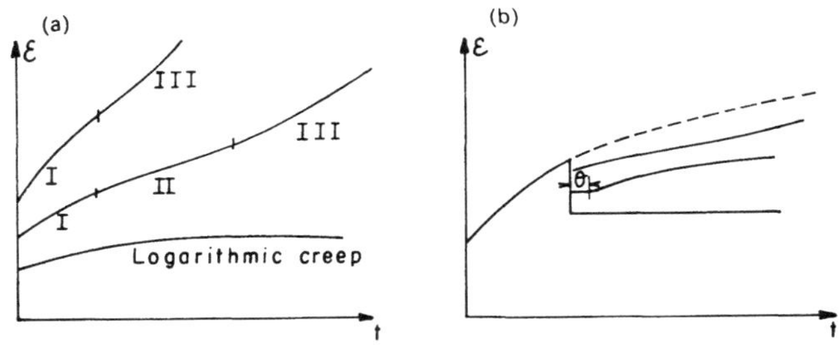

Fig. 3. (a) Classical stages during creep: I, transitory creep; II, stationary creep; III, tertiary creep. (b) Delay time on creep: θ, delay time.

During such creep experiments we can also see some transitory effects (delay-time in creep . . .). Several phenomenological constants were introduced and they were assumed to be dependent on metallurgical factors.

1.2. One-dimensional Experiments on the Single Crystal

In order to try to specify the nature of the constants for the polycrystal, experiments were performed on the single crystal. They were very difficult to perform and to interpret. Only a few data from tensile experiments are available. Generally the initial elastic limit and the work-hardening modulus are much lower than those of the polycrystal, and three stages are distinguished (Fig. 4).

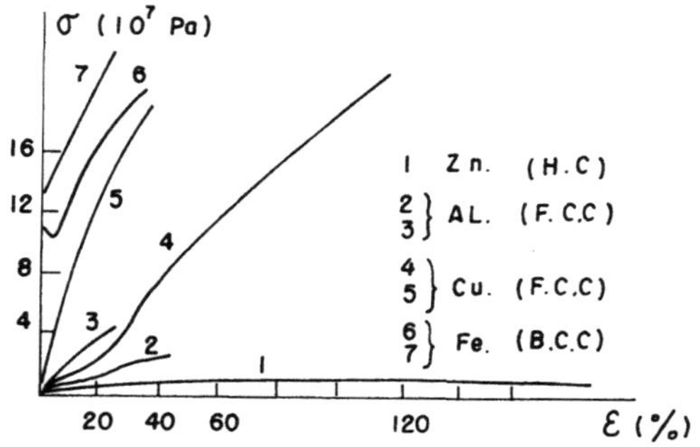

Fig. 4. Classical tensile curves on particular single crystals.

In consideration of the intrinsic anisotropy, the orientation of the crystal relative to the tensile axis has a fundamental influence. As for the polycrystal, the strain rates and the temperature are important.

1.3. Morphology of the Plastic Strain of Single Crystals and Polycrystals

(a) Macroscopic Observations on the Single Crystal

(i) Modes of deformation. After a plastic strain, we can easily see small steps on the surface of the crystal with the help of an optical microscope or even with the naked eye. They are called the slip lines.

Metals are crystalline solids: this means that they consist of atoms arranged in a pattern which is periodically repeated in three dimensions. This space is built with unit cells (or lattice). We can say also that we have a regular stacking of crystallographic planes which contain atoms.

We can see that generally, the slip lines are in crystallographic planes, the slip plane; the two parts of the crystal, which are on both sides of this plane, were submitted to a small relative displacement along a crystallographic direction, the slip direction. However, these two parts keep the same crystallographic structure and orientation. This mode is the slip process (Fig. 5). The set of the slip plane and the slip direction characterizes a slip system.

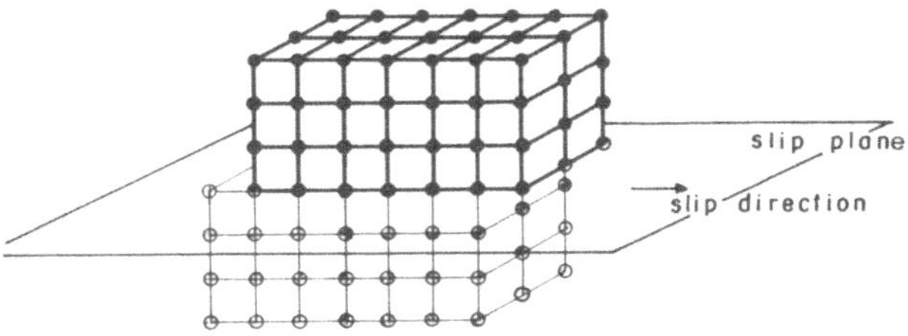

Fig. 5. Slip mode.

On polishing the surface of the strained crystal, permanent bands may become apparent. The band zone is the twinned zone. In this case, the twinned zone no longer has the same orientation; the atoms were submitted to a translation along a twinning-direction the amplitude of which is proportional to their distance from the twinning plane. This mode is the twinning process (Fig. 6).

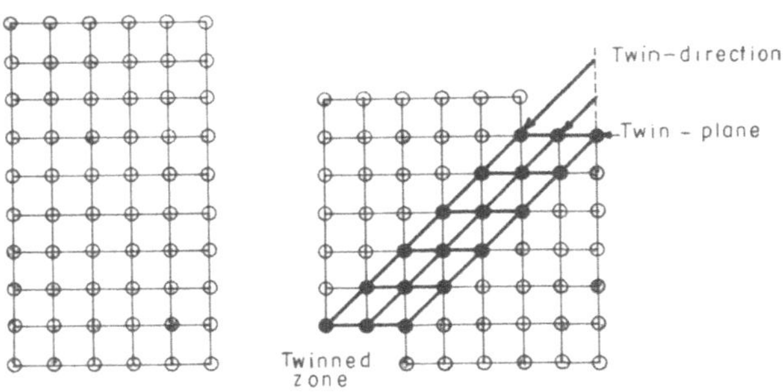

Fig. 6. Twin mode.

(ii) Principal slip systems and Schmid's law. Generally, the face-centred cubic metals (like Cu, Al, Ag, ...) all have the close packed {111} planes and the close packed <101> directions for their slip systems. For the body-centred cubic metal (Fe, ...) we find the {110} planes, but also {112} and {123} planes, ..., and the <111> directions. For the hexagonal

close packed (Zn, Cd, . . .) there are the (0001) basal plane and the $\langle 11\bar{2}0 \rangle$ directions; but also in some cases the pyramidal planes $\{10\bar{1}1\}$ and the prismatic ones $\{10\bar{1}0\}$.

The experimental data during tensile tests show that, as there are several slip systems for each metal, the slip system on which slip first occurs (the principal system) has a special position relative to the tensile axis. If we define as X and λ the angles between this axis and the normal to the slip plane and the slip direction, respectively, the principal system is such that the quantity $\cos X \cos \lambda$ is maximum. This means that if σ is the tensile stress, what is important for the slip process is the resolved shear stress

$$\tau = \sigma \cos X \cos \lambda \tag{1.6}$$

This is Schmid's law.

(b) Macroscopic Observations on Polycrystals

It may be seen that metals are usually polycrystalline, comprising an aggregate of crystals of different shape, orientation and physical state which are separated by grain boundaries. The modes of deformation inside the crystal stay the same but the grain boundaries introduce supplementary modes, chiefly at high temperatures and stresses.

In the general case we observe that the strain is different for each crystal of the polycrystal and, even, it is not homogeneous inside one crystal.

1.4. Complex Loading of the Polycrystal and the Single Crystal

The general case is one in which the structure is subjected to complex loading. Some experiments have been performed to study the elastic domain, work-hardening and the relations between the stress and strain tensors, chiefly in classical plasticity.

(a) Polycrystals

(i) Classical triaxial tests have been performed on cylindrical samples. They show that the hydrostatic pressure (first invariant of the stress tensor) does not influence the plastic behaviour of most metals.

(ii) In order to determine the elastic domain in stress space, thin tubular samples are often used and are subjected to tension (or compression) together with torsion.

It was pointed out that the border between the elastic and plastic domains is, at the beginning, nearly an ellipse and that it takes thereafter a very complicated shape: there are three effects: an expansion, a global translation towards the actual stress point and a local deformation near the stress point which is very pronounced (Fig. 7). It was pointed out also that the vector strain increment is almost normal to the elastic boundary when represented in stress space.

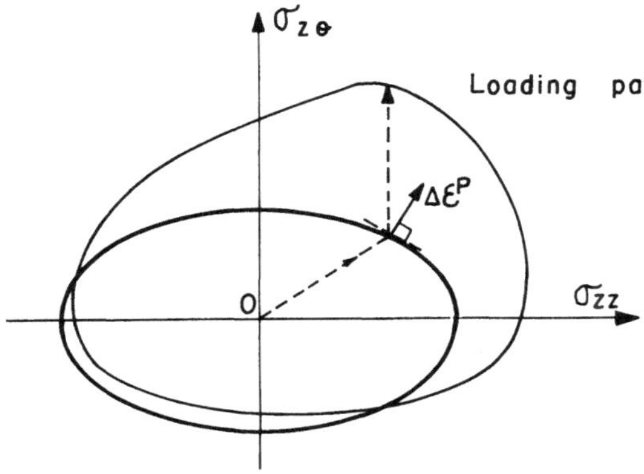

Fig. 7. Evolution of the elastic domain during a tension–torsion test (from H. D. Bui).

These results are essentially qualitative and very sensitive to the apparatus used to perform the experiment.

(b) Single Crystals

As for polycrystals it is found that the hydrostatic pressure has an unimportant effect. Tubular samples are not available but it is possible to cut small samples, out of a prestrained large sample, in different directions. With these few experiments it may be shown that for Al single crystals, when only the principal system is active during the strain, the work-hardening in this system may be higher than in the other systems (Fig. 8).

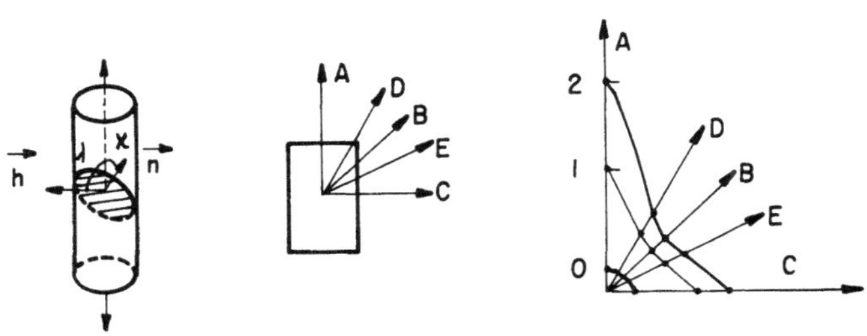

Fig. 8. Influence of the angle of the tensile axis on the elastic limit of the single crystal.

2. PHYSICAL BASIS FOR THE PLASTICITY OF SINGLE CRYSTALS

2.1. Some Results from the Physical Theory of Dislocations

(a) Introduction

This theory was born out of the contradiction between certain experimental results on the elastic yield limit of the crystal and the simple interpretation of the plastic strain of the relative glide between the two parts which are on either side of the slip plane (Fig. 9).

Fig. 9. Theoretical slip for a perfect single crystal.

Slip occurs by the movement of defects in the crystal lattice. The defect separates the slipped region from the unslipped region and it is termed a *dislocation.*

Dislocations are now routinely observed with the help of electron microscopes. Indeed, a crystal is almost never perfect; it always contains plenty of defects in its structure: linear defects as dislocation lines, 'point' defects as vacancies or interstitial atoms, surface defects as twins or particular types of dislocations (Figs 10 and 11). It is seen that the number of defects increases rapidly with plastic straining.

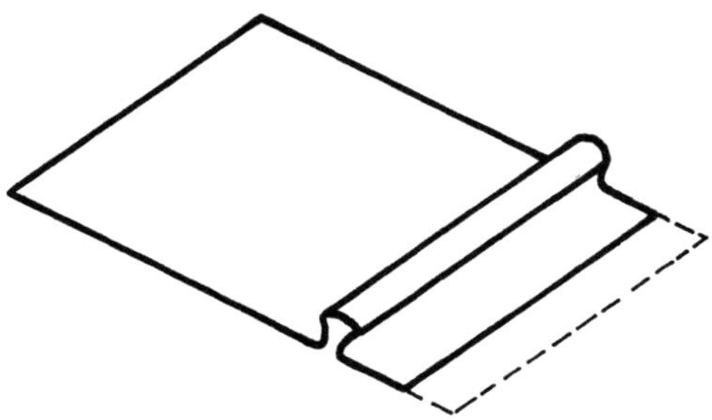

Fig. 10. Progressive slip due to the defect.

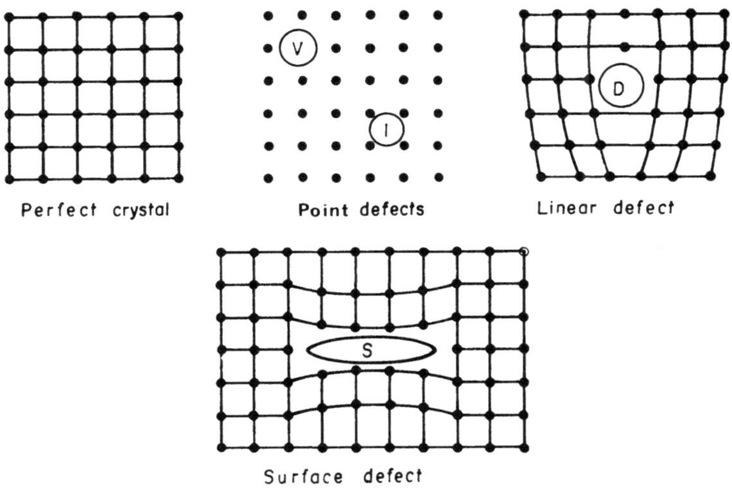

Fig. 11. Main defects in single crystals.

(b) What Exactly Does a Dislocation Represent?

(i) In an homogeneous elastic medium, let us consider a surface S_0 taken on a curve L (Fig. 12). We suppose that it is possible to separate the two lips of this surface, to suppress the discontinuity of matter (by adding or taking off a certain quantity of the same material) and to stick the matter together again along the lips. We do that in such a way that after the relaxation of the exterior forces, the stresses in the medium are continuous even when going through the lips. We then have a dislocation along the discontinuity line L. It may be shown that the only relative displacements that are possible between the two lips are translations and rotations (rigid motions) independent of the surface S_0.

(ii) In a crystal, generally, the dislocations that we meet are dislocations of translation; they are characterized by their vector of translation $\vec{b}$: the Burgers vector.

If the Burgers vector is an integer multiple of the period of the lattice, at a certain distance from the line L, the crystal structure is again found to be perfect and the surface S_0 does not matter; then the dislocation is said to be perfect. If, on the contrary, this vector is not an integer multiple, the surface S_0 does matter and the dislocation is said to be imperfect. Very often, perfect dislocations are split into several imperfect or partial dislocations.

(iii) It is obvious that a dislocation provokes internal stresses. In order to evaluate these stresses we, generally, treat the crystal like an

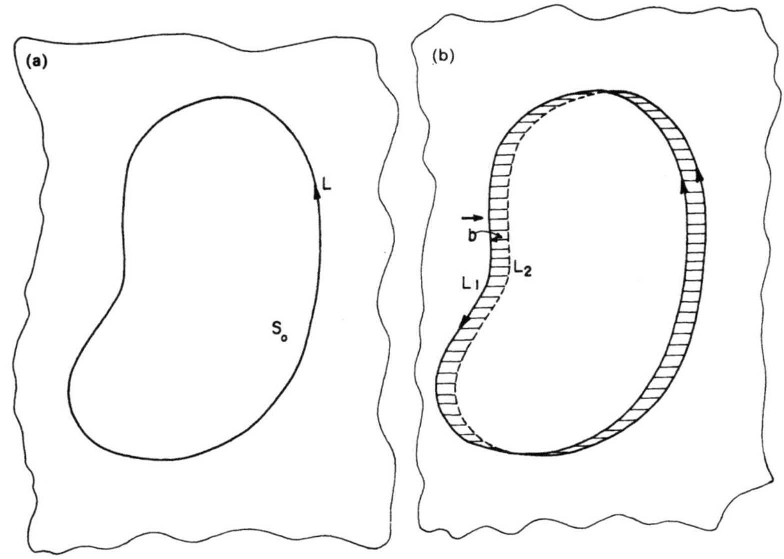

Fig. 12. (a) Burgers loop in the perfect crystal; (b) Burgers loop and vector in the dislocated crystal.

homogeneous elastic and isotropic medium (μ = shear modulus and ν = Poisson's ratio). For the analyst, this problem is simple. We consider a displacement field $\vec{u}$ (r) which is discontinuous through the surface S_0 ($\vec{b}$ is the discontinuity); we then obtain the stresses.

In the particular case of an infinite straight dislocation line in an infinite medium, we find for the stress tensor at any point (r, θ):

$$\left.\begin{aligned}
\sigma_{xx} &= -\frac{\mu b}{2\pi(1-\nu)}\cos\varepsilon\cdot\frac{\sin\theta}{r}(2+\cos 2\theta)\\
\sigma_{yy} &= \frac{\mu b}{2\pi(1-\nu)}\cos\varepsilon\cdot\frac{\sin\theta}{r}\cos 2\theta\\
\sigma_{zz} &= -\frac{2\nu\mu}{2\pi(1-\nu)}\cos\varepsilon\cdot\frac{\sin\theta}{r}\\
\sigma_{xy} &= \frac{\mu b}{2\pi(1-\nu)}\cos\varepsilon\cdot\frac{\cos\theta}{r}\cos 2\theta\\
\sigma_{xz} &= -\frac{\mu b\sin\varepsilon}{2\pi}\frac{\sin\theta}{r};\quad \sigma_{yz} = \frac{\mu b}{2\pi}\sin\varepsilon\frac{\cos\theta}{r}
\end{aligned}\right\} \tag{2.1}$$

when the line is along the z-axis, the y-axis is perpendicular to the plane defined by the line and the Burgers vector $\vec{b}$ and we put $(\overrightarrow{Ox}, \vec{b}) = \varepsilon$ (Fig. 13). ($\varepsilon = 0$, for an edge dislocation; $\varepsilon = \pi/2$, for a screw dislocation).

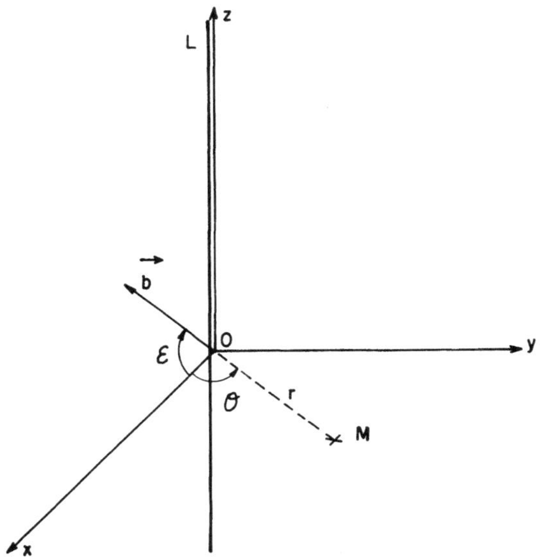

Fig. 13. Geometry for a straight dislocation.

(iv) We define the self-energy of a dislocation by the stored energy of the elastic strain due to it. It may be shown that we have for the self-energy per unit length

$$w_0 \simeq \frac{\mu b^2}{c} \tag{2.2}$$

where c lies between 1 and 5.

(c) How Can We Interpret Plastic Strain With the Help of the Dislocation?

(i) In order to move the dislocation line from L to $L + \Delta L$, the surface S_0 has to be extended by ΔS_0 and the two lips of ΔS_0 have to be displaced, one in respect to the other by a translation, the vector of which is the Burgers vector. This displacement creates a discontinuity of matter, the volume of which is ΔV.

When $\Delta V = 0$, we have the 'glide' of the dislocation on the slip plane. When $\Delta V \neq 0$, we have the 'climb' of the dislocation. ($\Delta V > 0$ means that there is a creation of vacancies; on the contrary, $\Delta V < 0$ means a creation of interstitial atoms.)

Hence the displacement of the dislocation on a whole plane section of the crystal provokes a relative displacement between the two parts situated on each side of this section. This displacement is equal to the Burgers vector. Obviously, this displacement of the two parts, which is obtained progressively with the help of the dislocation, occurs more easily than the homogeneous inblock-displacement. Nevertheless we have to make use of forces in order to move the dislocation (Fig. 14).

(ii) Now, suppose that there are N identical dislocations per unit volume and that each of them sweeps an area A in parallel planes (usually A is infinitely small in comparison with the section S of the crystal). An average relative displacement Γ, along the $\vec{h} = \vec{b}/b$ direction, between two displacement planes which are unit length distant may be defined by:

$$\Gamma \simeq \frac{b}{1} \cdot \frac{A}{S} \cdot N1S = bAN \quad \text{(see Fig. 15)} \qquad (2.3)$$

(d) How Do the External Forces Act on the Dislocation?

It is possible to define a conventional force per unit length of the dislocation. It is split into two parts:

- one is due to the dislocation itself; this is the 'line tension'.
- the other one is fundamental, and it is due to stresses which are independent of the dislocation (for example, applied stresses); it is equal to:

$$\vec{F} = (\vec{b}\,\sigma)\,\Lambda\vec{t} \qquad (2.4)$$

where $\vec{t}$ is the unit tangent vector to the line.

In any plane which contains the vector $\vec{t}$, the unit normal vector of which is $\vec{n}$, the component of this force is given by (Peach–Koelher formula):

$$F = \vec{b}\,\sigma\vec{n} = b^i n^j \sigma_{ij} = bh^i n^j \sigma_{ij} = b\tau \qquad (2.5)$$

with

$$\tau = h^i n^j \sigma_{ij}$$

(when $h^i n_i = 0$, τ is the resolved shear stress).

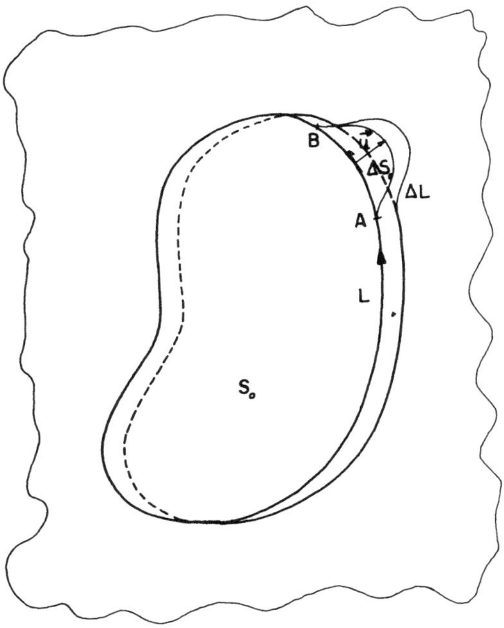

Fig. 14. Displacement of a dislocation.

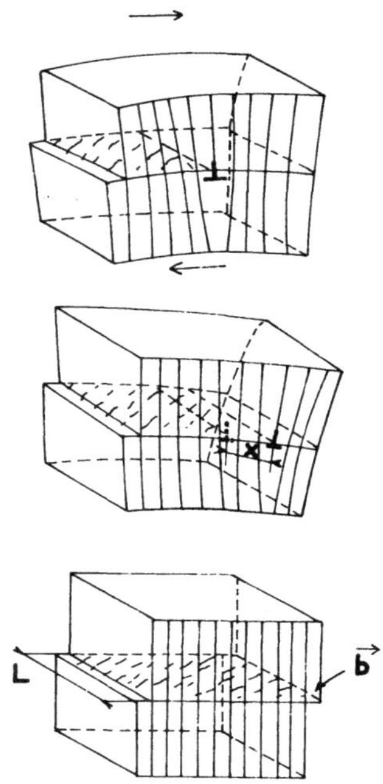

Fig. 15. Slip induced by the progressive glide of a dislocation.

This formula may be easily understood in computing the work done when a straight dislocation glides through the whole section of the crystal (Fig. 15): if τ is the shear stress in the glide direction, the work is equal to

$$w = \tau L x b$$

the conventional force is defined in order to preserve the equality

$$F \cdot L \cdot x = w$$

thus we get $F = \tau \cdot b$.

(e) When Can a Dislocation be Moved?

First of all, let us suppose that the dislocation is isolated in the crystal. To be able to displace the dislocation, the force due to the applied stresses has to overcome the actions of the atoms on the dislocation; this is the Peierls–Nabarro force $\mathscr{F}_{PN}$. It depends on the displacement plane and on the dislocation. Usually this force is negligible. In the real crystal, there are many defects. The applied force has to overcome not only the force $\mathscr{F}_{PN}$ but also the forces due to the other defects. By the mathematical theory of dislocations it is possible to evaluate the elastic interactions. The physical and particular inter-reactions may be also determined.

(f) How Can the Number of Dislocations Change During Plastic Straining?

(i) With the help of the applied stresses, a dislocation segment, which is fixed at its extremities, is bent in its glide plane (defined by the segment and the Burgers vector). If it passes beyond an unstable position, it can create a new loop of dislocation. This is a Frank–Read source (Fig. 16). So, the number of dislocations can increase very quickly.

Fig. 16. Various stages of a Frank–Read source.

(ii) A dislocation may meet another dislocation with opposite Burgers vector and may partly annihilate it (Fig. 17) (or reach the surface of the crystal).

A dislocation may also be set into a wall (or subgrain) with other dislocations; such a wall hardly provokes internal stresses and it is very stable.

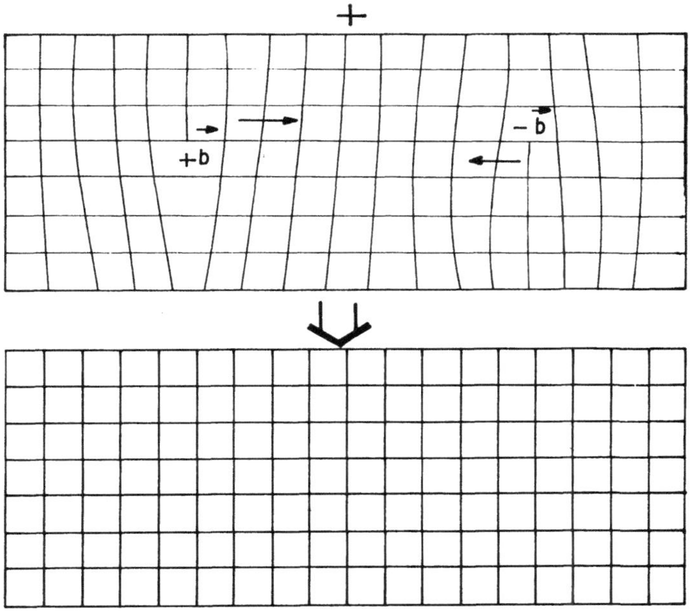

Fig. 17. Annihilation of two opposite straight dislocations.

(g) How Are the Dislocations Distributed?

Several observations with the help of electron microscopes have shown that the dislocation lines are randomly distributed and generally split into small straight segments which stretch out in some particular directions. These segments are arrested in front of other segments or intersect them, forming tangles. These dislocation interactions are named after the workers who first described them (Hirsch reaction, Lomer–Cottrell locks, etc.).

In fcc crystals, for example, such dislocations are often found on the {111} planes, their Burgers vectors are of type <101> and they are chiefly parallel to the <110> directions (Figs 18 and 19).

2.2. Interpretation of Plasticity of Single Crystals

We see at once how plasticity may be interpreted: when the applied stresses are sufficient, there are displacements and dislocations (thus plastic strain); some Frank–Read sources are active (giving rise to work-hardening).

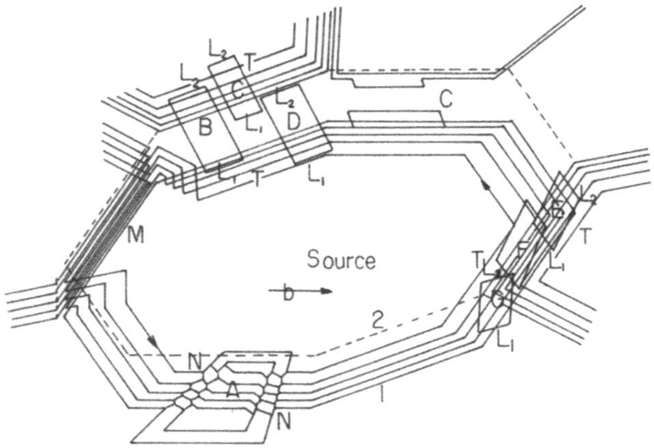

Fig. 18. Schematisation of the dislocation distribution (from J. W. Steeds).

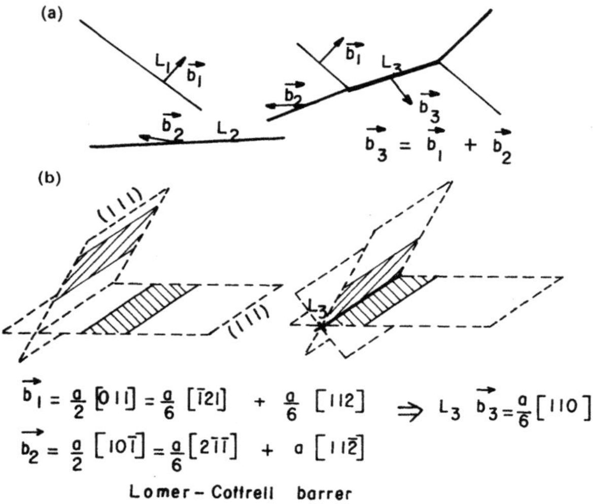

Fig. 19. (a) Hirsch's reaction. (b) Lomer–Cottrell's barrier.

(a) Taylor's Model

This author supposed a fictitious regular distribution of infinitely straight dislocations with density per unit surface equal to ρ (Fig. 20). The elastic limit τ_i is reached when a dislocation may be moved. His calculations gave:

$$\tau_i \simeq \frac{\mu b}{\beta} \rho^{1/2} \tag{2.6}$$

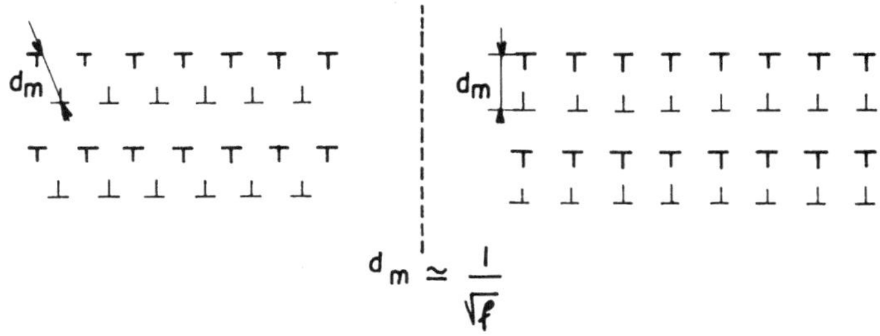

Fig. 20. Perfect distribution of an array of straight dislocations.

with β roughly equal to 4π. This is in agreement with the experimental results $\tau_i \simeq 10^{-3}$ to $10^{-4}\,\mu$, since $b \simeq 10^{-8}$ cm and $\rho^{1/2} \simeq 10^4, 10^5\ \text{cm}^{-1}$.

He assumed also that the glide Γ is proportional to the density ρ. Hence, he reached a parabolic work-hardening.

(b) Other Models

Seeger, Hirsch, Saada, . . . and many other research workers proposed more elaborate models, but they are all similar to that of Taylor:

The elastic limit is given by eqn (2.6) with β between 2 and 12; the relation between Γ and ρ is very often taken as

$$\Gamma = c\rho^{1/2} \tag{2.7}$$

where c is a constant.

They have a linear work-hardening.

The most important fact: is that the density of dislocation ρ is a parameter which allows us to specify the internal state of the crystal.

3. PHYSICAL BASIS FOR THE VISCOPLASTICITY OF SINGLE CRYSTALS

3.1. Thermal Activation

(a) Consider first the problem of the diffusion of a point defect (Fig. 21). Atoms in the crystal vibrate around their equilibrium position at a frequency of ν_0 which is linked to the temperature. If, in order to move one defect from A to B, we need the energy U, the frequency occurrence of this motion will be roughly equal to

$$\nu \simeq \nu_0 \exp\left\{-\frac{U}{kT}\right\} \tag{3.1}$$

where k is Boltzmann's constant.

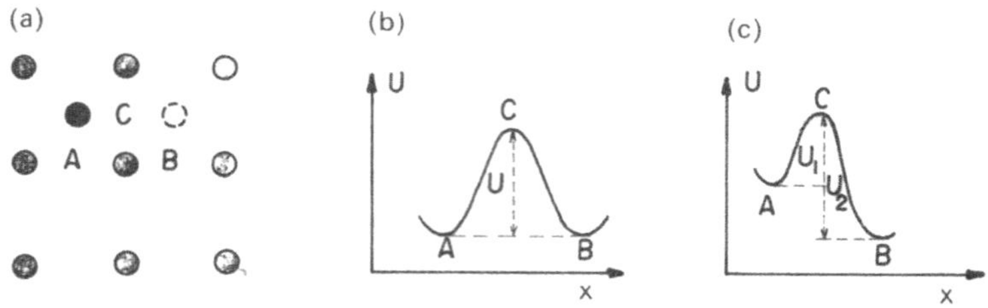

Fig. 21. (a) Diffusion of a point defect. (b) Thermal energy. (c) Variation of the thermal energy due to an applied stress.

We define a mean time interval as $t = 1/\nu$.

(i) When there is no applied force, the energy needed to go from B to A is equal to U. Thus, the frequency of movement in one sense equals that in the other sense and there is no net flow.

(ii) When we apply forces, movement in one sense becomes easier than the other. We have different energies $U_1 < U_2$, and thus we have a diffusion rate which is proportional to:

$$\nu = \nu_1 - \nu_2 = \nu_0 \left[\exp\left\{-\frac{U_1}{kT}\right\} - \exp\left\{-\frac{U_2}{kT}\right\} \right] \tag{3.2}$$

or to the mean time

$$t = \frac{1}{\nu} \tag{3.2'}$$

(iii) Often if it is assumed for an applied stress σ:

$$U_1 = U - \sigma v_a \qquad \text{and} \qquad U_2 = U + \sigma v_a \tag{3.3}$$

where v_a is the activation volume.

The formulae (3.2) are written thus:

$$\nu = \nu_0 \exp\left\{-\frac{U}{kT}\right\} 2 \sinh\left\{\frac{\sigma v_a}{kT}\right\} \tag{3.4}$$

or

$$t = \frac{1}{2\nu_0} \exp\left\{\frac{U}{kT}\right\} \frac{1}{\sinh\left\{\frac{\sigma v_a}{kT}\right\}}$$

where U is the activation energy; it is obvious that it is always positive and rather small. As time occurs in the expression we have viscosity effects.

(b) As the previous case, any process may be thermally activated. However, this activation may occur only for a very small volume. Indeed, it is not possible that the motions of a thousand particles in the same chain are activated together.

If the process needs such a global displacement, that implies that the thermal activation is unimportant; that means that the viscosity and the temperature do not take part in the process.

In summary, we can say that thermal activation implies:

— a rather small thermal energy of the process;
— a very small volume involved in the process;
— effects of temperature and time and hence *viscosity*.

3.2. Processes During Dislocation Motions Which May be Thermally Activated

(a) Glide of a Dislocation

(i) Glide of a dislocation is only possible when the applied force is higher than the actions of Peierls–Nabarro and the other defects since its usual length is about 10^3 to 10^4 times the interatomic distance. Again, we find the yield limit of the plasticity.

(ii) However, during its glide on a great area, the dislocation encounters other dislocations which pierce its slip plane. The intersection of two dislocations is a local fundamental process; there is the creation of two jogs (small segments of dislocation the length of which is roughly equal to the Burgers vector of the opposite dislocation). It was shown that this may be thermally activated (Fig. 22).

(iii) Some other processes which limit the glide speed of the dislocation may also intervene.

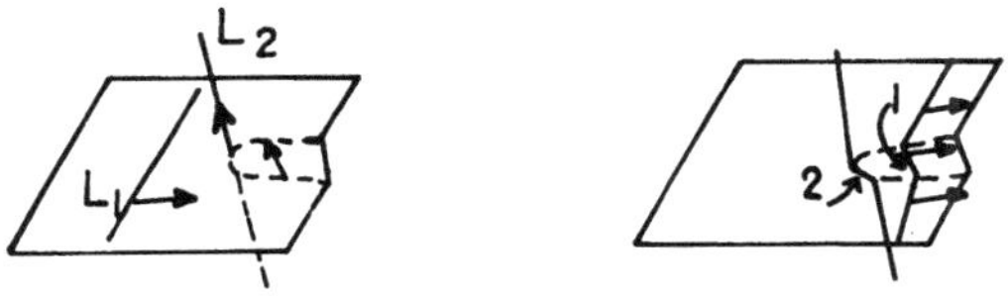

Fig. 22. Crossing of two dislocations. Formation of two jogs.

(b) Cross-slip of a Screw Dislocation

(i) A perfect screw dislocation has no defined slip plane. If it moves from its initial plane p to another one p', this motion is called cross-slip (Fig. 23).

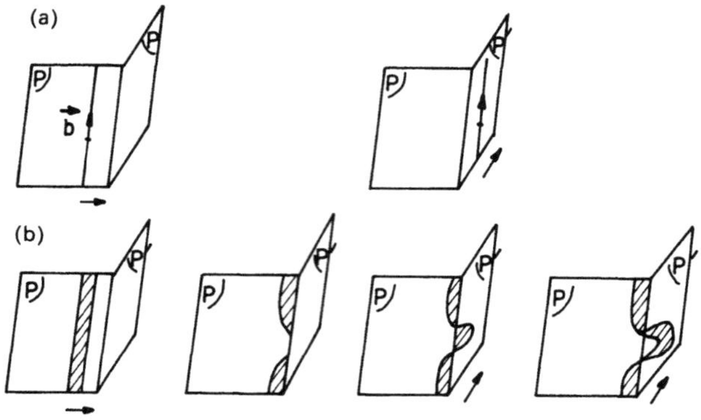

Fig. 23. (a) Cross-slip of a perfect screw dislocation. (b) Various stages for the cross-slip of a dissociated screw dislocation.

(ii) Since very often a perfect dislocation is split into imperfect dislocations, the set of these dislocations cannot easily leave the initial plane. We have to recombine these dislocations back to the perfect one. This process may be thermally activated.

(b) Climb of a Dislocation (Fig. 24)

The climb into a plane p' provokes a creation of point defects; it needs a high energy and, thus, occurs rather seldomly. However, progessive climb, in which a jog is created and propagated along the line, may be thermally activated.

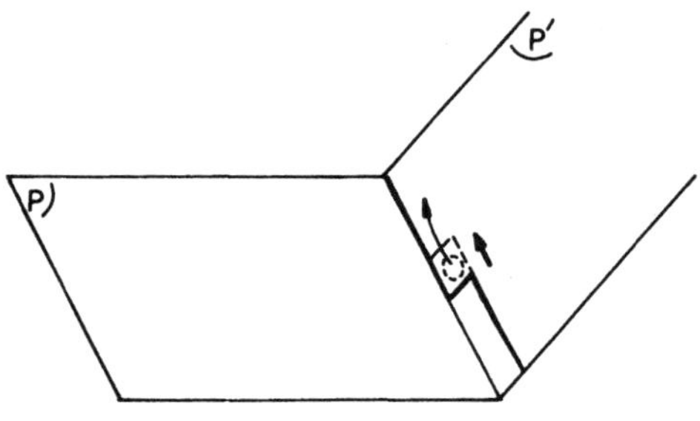

Fig. 24. Progressive climb of a dislocation.

(d) Influences of Impurities

Sometimes these impurities are randomly distributed (Fig. 25). The intersection of a dislocation with such an impurity may be thermally activated. In other cases, the impurities concentrate near the dislocation producing a 'Cottrell atmosphere'. At low temperatures the motion of the dislocation is only possible when the stress allows it to outstrip its atmosphere. At high temperature the dislocation may draw along with it the impurities, which diffuse.

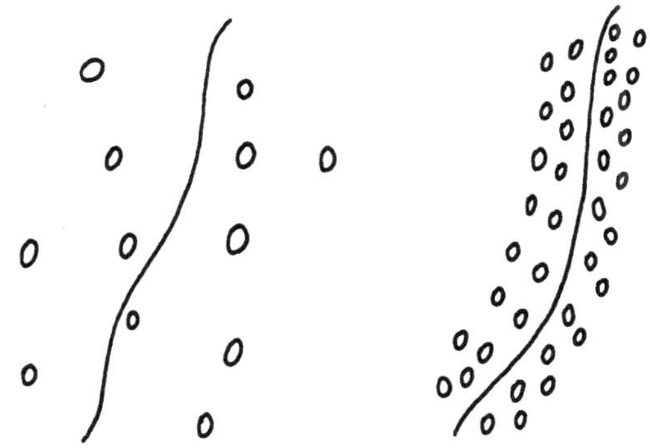

Fig. 25. Interaction of a dislocation with impurities.

3.3. Interpretation of Viscoplasticity of Single Crystals

With these few elements, several one-dimensional models were proposed in order to describe, at least qualitatively, viscoplasticity.

The slip rate is usually written as:

$$\dot{\Gamma} = f(\tau, T, \rho) \sinh \left\{ \frac{<\tau - \tau_c> v_a}{kT} \right\} \tag{3.5}$$

where f, τ_c and v_a have different forms according to the model. (This relation is similar to eqn (3.4).) $<x>$ means positive part of

$$x \begin{cases} = x \text{ if } x > 0 \\ = 0 \text{ if } x < 0 \end{cases}$$

The yield limit τ_c is introduced and $\tau - \tau_c$ represents the effective stress for the process.

4. PHYSICAL BASIS FOR POLYCRYSTALS

The polycrystal is an aggregate of crystals. Hence, it has not only properties which derive directly from those of the crystal but also some intrinsic properties of its own. Indeed, the plastic strain in one crystal is not compatible with the plastic strain of other surrounding crystals and this incompatibility is a source of internal stresses. There are interactions between the crystals in the same way that there are interactions between the defects inside one crystal.

Also, the surfaces of contact between crystals, the grain boundaries, may play a part during straining. For these reasons polycrystalline behaviour is not a simple 'mean' of the single crystal behaviour.

REFERENCE

Jaoul, B. (1965). *Etude de la Plasticité et Applications aux Métaux,* Dunod, Paris.
See also references in Chapter 10.

CHAPTER 2

Overall Stresses and Strains in Solids with Microstructure

S. NEMAT-NASSER
Department of Applied Mechanics and Engineering Sciences, University of California, San Diego, USA

ABSTRACT

Within the context of the usual small deformation theory, the notions of overall stresses and strains associated with representative samples of solids with microstructure, e.g. polycrystalline aggregates and composites, are reviewed. The overall response of a finite solid with microstructure, and that of an infinitely extended solid with periodic microstructure, is examined, considering both linearly elastic and nonlinearly creeping microconstituents. Then, estimates are given for the overall properties of solids containing randomly distributed microdefects; and, finally, the elastoplastic response of polycrystalline solids is discussed.

1. INTRODUCTION

In practice, it is often convenient to regard engineering materials as continua, even though, in reality, most materials are highly heterogeneous at the micro-level, consisting of a variety of constituents. Polycrystals, composites, rocks, soils, and plain and reinforced concrete are typical examples. To apply the continuum theory to the analysis of boundary value problems associated with materials of this kind, a 'homogenisation' of one kind or another must be introduced, whereby an equivalent locally homogeneous (but not necessarily isotropic) continuum is substituted for the locally[†] heterogeneous solid. Since the pioneering work of Voigt (1889) and Reuss (1929), considerable effort has been devoted to the formulation and estimate of the properties of such an equivalent locally homogeneous continuum for materials with a variety of microstructural characteristics. A recent book by Mura (1982) provides detailed accounts of many topics in this area associated with infinitesimal deformations, and gives a comprehensive list of references. Certain aspects of composite materials are reviewed by Hashin (1983) and Willis (1983). Questions pertaining to polycrystalline aggregates have been discussed by Bishop and Hill (1951), Budiansky and Wu (1962), Mandel (1966), Hill (1967), Hutchinson (1970), Bui *et al.* (1973), Zarka (1973), and, more recently, by Nemat-Nasser (1983) and Iwakuma and Nemat-Nasser (1984). In view of these, we shall emphasize in the present chapter only selected results on the overall properties of solids with certain microstructures, which have not been thoroughly dealt with in the above-mentioned sources. This will include the overall response of, for example, a finite cuboid containing a finite number of cavities of arbitrary geometry or inclusions. The material of the cuboid and its inclusions may be linearly elastic or follow a nonlinear creep law. The results are also applied to a solid with a periodic microstructure.

The organization of this chapter is as follows. In Section 2, we briefly review the notions of overall stress and strain associated with a representative sample of a solid with microstructure, including granular masses which support the overall applied loads through isolated intergranular contacts. In Section 3 we examine the overall response of a finite solid with microstructure, and an infinitely extended solid with periodic microstructure. In this section, attention is focused on linearly elastic and nonlinearly creeping constituents. Section 4 is devoted to the estimate of the overall properties of solids which contain randomly distributed microdefects. Finally, in Section 5 we give a brief review of the elastoplastic response of polycrystalline solids. Throughout this chapter, only infinitesimal deformations are considered.

2. AVERAGE QUANTITIES

The continuum theory of materials deals with objects which on a local level can be regarded as homogeneous, although globally they may be heterogeneous. In this approach, at a typical station (or a material point)

[†]The term 'locally' is emphasized here to bring out the fact that the equivalent continuum need not be globally homogeneous.

one considers an infinitesimal material element within which the strain and the stress are regarded as being homogeneous. As a prerequisite, therefore, the *continuum* itself must be locally homogeneous in a certain sense.

In reality, however, most materials are highly heterogeneous (locally as well as globally). Therefore, whether or not a continuum approach provides useful information depends on the values of two basic relative length scales. These are the dimension of the microconstituent of the aggregate relative to the dimension of the elementary continuum material neighbourhood, and the dimension of the elementary continuum material neighbourhood relative to the dimension of the overall continuum. For example, in a polycrystalline solid consisting of crystals of tens of microns, each crystal is viewed as a microconstituent. The elementary continuum material neighbourhood must then be of the order of fractions of a millimetre, and the overall solid, i.e. the structural component, may have dimensions of the order of centimetres or larger. On the other hand, an earth dam may be viewed as a continuum with aggregates as microconstituents and continuum elements of the order of metre size. In either case, the objective is to define an equivalent continuum material element which, in a certain sense, has the same average mechanical response as the actual heterogeneous material element; for a more detailed discussion and references, see Hashin (1983).

To this end, the concept of a 'representative volume element' (RVE) is introduced. In this approach, the actual dimensions of an RVE are of no concern. It is only its dimension relative to the dimension of its *essential* microstructure that is of importance. If the former is l and the latter is d, Hill (1956) suggests that an effective continuum may be produced for d/l less than 10^{-3}.

2.1. Average Stresses, Average Strains

Throughout this chapter we deal with the usual small deformation continuum theory; we use a fixed rectangular Cartesian coordinate system with coordinate axes x_i, $i = 1, 2, 3$, and denote second-order tensors by bold-typeface letters ($\boldsymbol{\varepsilon}$ for strain and $\boldsymbol{\sigma}$ for stress, for example), but for convenience and in order to avoid any confusion, mostly we give results in the component form (e.g. ε_{ij} and σ_{ij} for the components of the strain and stress tensors, respectively).

Consider an RVE in the shape of a cuboid, with faces parallel to the coordinate planes. When the cuboid is subjected on its boundaries to *uniform* displacements compatible with an overall *average* (constant and symmetric) strain field, ε_{ij}^0, neither the strain, ε_{ij}, nor the stress, σ_{ij}, fields will be uniform within the cuboid, if the material has a microstructure. Let the volume of the cuboid be V, and the region it occupies be denoted by D, being bounded by ∂D. If x_i^0 stands for the points on ∂D, the applied uniform surface displacements may be defined by

$$u_i^0 = \varepsilon_{ij}^0 x_j^0 \qquad \text{on } \partial D \tag{2.1}$$

where repeated indices are summed. With $\varepsilon_{ij}(\mathbf{x})$ denoting the *variable* strain field *within D,* it is clear that

$$\begin{aligned} \langle \varepsilon_{ij} \rangle_D \equiv \bar{\varepsilon}_{ij} &\equiv \frac{1}{V}\int_D \varepsilon_{ij}(\mathbf{x})\,\mathrm{d}\mathbf{x} = \frac{1}{V}\int_D \tfrac{1}{2}(u_{i,j} + u_{j,i})\,\mathrm{d}\mathbf{x} \\ &= \frac{1}{V}\int_{\partial D} \tfrac{1}{2}(u_i^0 n_j + u_j^0 n_i)\,\mathrm{d}\mathbf{x} = \varepsilon_{ij}^0 \end{aligned} \tag{2.2}$$

where n_i is the exterior unit normal of the considered face, and comma followed by a subscript denotes partial differentiation with respect to the corresponding coordinate. Note that eqn (2.2) is valid for D of any shape as long as the condition of uniform data on ∂D, i.e. eqn (2.1), is satisfied.

The unweighted volume averages of the stress and strain fields are useful overall dynamical and kinematical quantities which lead to a simple energy theorem (2.10). The overall average (symmetric) stress is defined by†

$$\langle \sigma_{ij} \rangle_D \equiv \bar{\sigma}_{ij} = \frac{1}{V}\int_D \sigma_{ij}(\mathbf{x})\,\mathrm{d}\mathbf{x} \tag{2.3}$$

With uniform displacement data (2.1), the tractions which must be supplied on the faces of the cuboid are, in general, not uniform, but their average value, $\bar{T}_i^{(k)}$, over the kth face of the cuboid is

$$\bar{T}_i^{(k)} = \bar{\sigma}_{ji} n_j^{(k)} \tag{2.4}$$

where $\mathbf{n}^{(k)}$ is the exterior unit normal to the considered face.

In a similar manner, when the cuboid is subjected to uniform surface tractions T_i^0, compatible with a uniform stress field, σ_{ij}^0,

$$T_i^0 = \sigma_{ji}^0 n_j \qquad \text{on} \quad \partial D \tag{2.5}$$

†Throughout this chapter, the stress is symmetric and the term strain refers to the symmetric part of the displacement gradient.

then the actual fluctuating stress field, $\sigma_{ij}(\mathbf{x})$, satisfies

$$\langle\sigma_{ij}\rangle_D \equiv \bar{\sigma}_{ij} \equiv \frac{1}{V}\int_D \sigma_{ij}(\mathbf{x})\,\mathrm{d}\mathbf{x} = \frac{1}{V}\int_{\partial D} x_i T_j^0\,\mathrm{d}\mathbf{x} = \sigma_{ij}^0 \tag{2.6}$$

Clearly, the associated strain field will be variable. Moreover, the surface displacements will not be uniform. The average strain is then defined by

$$\langle\varepsilon_{ij}\rangle_D \equiv \bar{\varepsilon}_{ij} \equiv \frac{1}{V}\int_D \varepsilon_{ij}(\mathbf{x})\,\mathrm{d}\mathbf{x} \tag{2.7}$$

and the average displacements of the kth face of the cuboid, $\bar{u}_i^{(k)}$, relative to its opposing face (taken to be stationary and coincident with a coordinate plane), relate to the average strain by

$$\bar{\varepsilon}_{ij} = \tfrac{1}{2}[\bar{u}_i^{(j)}/x_j^0 + \bar{u}_j^{(i)}/x_i^0] \qquad (i, j \text{ not summed}) \tag{2.8}$$

The validity of eqns (2.5) and (2.8) can be verified by considering special linear displacement and constant stress fields, respectively, and then applying the virtual work theorem (2.11).

Finally, from the identity

$$\langle\sigma_{ij}u_{i,j}\rangle - \langle\sigma_{ij}\rangle\langle u_{i,j}\rangle \equiv \frac{1}{V}\int_{\partial D}(u_i - \langle u_{i,k}\rangle x_k)(\sigma_{ij} - \langle\sigma_{ij}\rangle)\,n_j\,\mathrm{d}\mathbf{x} \tag{2.9}$$

it follows that

$$\langle\sigma_{ij}\varepsilon_{ij}\rangle = \langle\sigma_{ij}\rangle\langle\varepsilon_{ij}\rangle \tag{2.10}$$

for either boundary data (2.1) or (2.5); see Hill (1963, 1967) and Mandel (1980); above and in the sequel, the subscript D in $\langle\ldots\rangle_D$ is left out when it is felt that no confusion would result.

The relation (2.10) may also be deduced by the application of the virtual work theorem. Let u_i be a kinematically admissible displacement field with associated strain field ε_{ij}, and consider an *unrelated* statically admissible stress field σ_{ij} associated with tractions $T_i = \sigma_{ji}n_j$ on ∂D. According to the virtual work theorem, we have

$$\frac{1}{V}\int_{\partial D} T_i u_i\,\mathrm{d}\mathbf{x} = \frac{1}{V}\int_D \sigma_{ij}\varepsilon_{ij}\,\mathrm{d}\mathbf{x} \equiv \langle\sigma_{ij}\varepsilon_{ij}\rangle \tag{2.11}$$

Now, with $u_i = \varepsilon_{ij}^0 x_j$, the left-hand side of eqn (2.11) reduces to $\langle\sigma_{ij}\rangle\varepsilon_{ij}^0$, and with $T_i^0 = \sigma_{ji}^0 n_j$ it becomes $\sigma_{ij}^0\langle\varepsilon_{ij}\rangle$. In either case, we arrive at eqn (2.10).

2.2. Overall Stresses in Granular Masses

When the dimensions of typical particles in a particulate body are suitably small relative to the overall size of the assembly of the particulates (or the granular mass), it may often be convenient to treat the assembly as a continuum. The overall compressive forces produce contact forces at contacting points among the constituent particles. Whatever the nature of these contacts may be, a basic objective is to describe the overall average stress tensor in terms of the contact forces and some geometric characteristics of the granules. This problem has been satisfactorily solved by Christoffersen *et al.* (1981). A brief account is given below.

Consider an RVE subjected on its boundary to uniform tractions T_i^0, compatible with the overall average stresses $\bar{\sigma}_{ij}$ and overall equilibrium. Let the contact force at a typical contact α be denoted by f_i^α, and consider an infinitesimal virtual displacement, u_i, which produces a *virtual* separation Δ_i^α at this typical contact. With M denoting the total number of contacts in the RVE, the virtual work theorem yields

$$\int_{\partial D} T_i^0 u_i \,\mathrm{dx} = \sum_{\alpha=1}^{M} f_i^\alpha \Delta_i^\alpha \tag{2.12}$$

Since the virtual displacement field can be chosen arbitrarily, consider a linear field

$$u_i = \phi_{ij} x_j + c_i \tag{2.13}$$

where ϕ_{ij} is an arbitrary symmetric constant matrix and c_i is an arbitrary constant vector. Compatible with eqn (2.13), we have

$$\Delta_i^\alpha = \phi_{ij} l_j^\alpha \tag{2.14}$$

where l_j^α is the *branch vector* which connects the centroids of the two granules that are in contact at α. If these two granules are denoted by A and B, and if the contact force exerted by A on B at α is $\mathbf{f}^\alpha$, then the orientation of the branch $\mathbf{l}^\alpha$ is from the centroid of A to the centroid of B. In this manner, at each contact a pair of vectors $\mathbf{f}^\alpha$ and $\mathbf{l}^\alpha$ with a *unique* tensor product $f_i^\alpha l_j^\alpha$ (no sum on α) is defined.

If we now substitute from eqns (2.13) and (2.14) into eqn (2.12), note $T_i^0 = \bar{\sigma}_{ji} n_j$, and assume a symmetric stress tensor, we arrive at the following relation between the overall stress tensor and the contact forces and branch vectors:

$$\bar{\sigma}_{ij} = \tfrac{1}{2}\sum_{\alpha=1}^{N}(f_i^\alpha l_j^\alpha + f_j^\alpha l_i^\alpha) \equiv \frac{N}{2}\langle f_i l_j + f_j l_i\rangle \tag{2.15}$$

where $\langle \ldots \rangle \equiv \frac{1}{N}\sum_{\alpha=1}^{N}(\ldots)$, and $N = M/V$ is the number of contacts per unit volume of the RVE. It is shown by Mehrabadi *et al.* (1982) that this definition of the stress tensor for granular materials is completely equivalent to the definition of the stress tensor in terms of the average force transmitted per unit area of three mutually orthogonal imagined planes passing through a point in the RVE.

With minimal assumptions, the average stress in a granular mass consisting of *spherical* granules can be related to a second-order tensor which represents the 'fabric' of the granular mass.

To this end, let $\mathbf{n}^\alpha$ be the unit contact normal at α, a^α be the *effective* contact area (or representative area characterizing the size of the microscopic contact region), and observe that for *spherical* granules, $\mathbf{l}^\alpha = l^\alpha \mathbf{n}^\alpha$ (α not summed) with l^α denoting the branch length. Now we can express, without a loss in generality, the contact force $\mathbf{f}^\alpha$ in terms of a symmetric second-order tensor τ^α, as follows:

$$f_i^\alpha = a^\alpha \tau_{ij}^\alpha n_j^\alpha \qquad (\alpha \text{ not summed}) \tag{2.16}$$

Substitution into eqn (2.15) yields

$$\bar{\sigma}_{ij} = \tfrac{1}{2}\langle \tau_{ik} h_{kj} + \tau_{jk} h_{ki}\rangle, \qquad h_{ij}^\alpha = N a^\alpha l^\alpha n_i^\alpha n_j^\alpha \quad (\alpha \text{ not summed}) \tag{2.17}$$

We now assume that the symmetric tensor τ_{ij}^α represents an 'effective' *local* stress tensor in a locally homogeneous equivalent material that represents the corresponding contacting granules. This, in effect, is a substitution of the local microstructure by a *microcontinuum*. In this sense, it may then be reasonable to assume that τ_{ij}^α is a local stress field defined *statistically* independently of the contact normal $\mathbf{n}^\alpha$ or, in fact, of the *kinematical* quantity h_{ij}^α. With this assumption, we may regard h_{ij}^α and τ_{ij}^α in eqn (2.17) to be globally uncorrelated, and obtain

$$\bar{\sigma}_{ij} = \tfrac{1}{2}\,[\langle \tau_{ik}\rangle\langle h_{kj}\rangle + \langle \tau_{jk}\rangle\langle h_{ki}\rangle] \tag{2.18}$$

Since $\bar{\sigma}_{ij}$, $\langle \tau_{ij} \rangle$, and $\langle h_{ij} \rangle$ are symmetric, and since the symmetry of $\bar{\sigma}_{ij}$ implies

$$\langle \tau_{ik} \rangle H_{kj} = \langle \tau_{jk} \rangle H_{ki}, \qquad H_{ij} = \langle h_{ij} \rangle \tag{2.19}$$

it follows that the three tensors, $\bar{\boldsymbol{\sigma}}$, $\langle \boldsymbol{\tau} \rangle$ and $\mathbf{H}$, are coaxial and, therefore, eqn (2.18) becomes

$$\bar{\sigma}_{ij} = A_0 \delta_{ij} + A_1 H_{ij} + A_2 H_{ik} H_{kj} \tag{2.20}$$

where the coefficients, A_β, $\beta = 0, 1, 2$, are functions of the basic invariants of $\bar{\boldsymbol{\sigma}}$ and $\mathbf{H}$, δ_{ij} being the Kronecker delta.

The symmetric tensor H_{ij} is a measure of the microstructure or the fabric of the granular mass. The fabric itself is defined by the distribution of contacts, contact normals and branch lengths. It is interesting to note that, in eqn (2.20), $\mathbf{H}$ plays the same role as a suitable strain tensor does in finite elasticity.

3. ESTIMATE OF OVERALL PROPERTIES

In this section we first present some general results on estimating the overall properties of linearly elastic solids containing cavities or linearly elastic inclusions of *arbitrary shapes*. We then generalize some of the results for application to solids consisting of nonlinearly creeping materials.

3.1. A Basic Result

Consider a sample of a linearly elastic solid containing a set of inclusions, occupying the region $\boldsymbol{\Omega}$ with boundary $\partial\boldsymbol{\Omega}$ and total volume V_Ω. The solid is subjected to the self-equilibrating overall tractions T_i^0, compatible with the uniform overall stress $\bar{\sigma}_{ij}$; $T_i^0 = \bar{\sigma}_{ji} n_j$ on ∂D. With $f = V_\Omega / V$ denoting the volume fraction of the inclusions, we express the average stress and strain tensors as

$$\begin{aligned} \bar{\sigma}_{ij} &= (1-f)\langle \sigma_{ij} \rangle_{D-\Omega} + f \langle \sigma_{ij} \rangle_\Omega, \\ \bar{\varepsilon}_{ij} &= (1-f)\langle \varepsilon_{ij} \rangle_{D-\Omega} + f \langle \varepsilon_{ij} \rangle_\Omega \end{aligned} \tag{3.1}$$

By the application of the divergence theorem, we also have

$$\langle \varepsilon_{ij} \rangle_\Omega = \frac{1}{V_\Omega} \int_{\partial\Omega} \tfrac{1}{2}(u_i n_j + u_j n_i)\, \mathrm{d}\mathbf{x} \tag{3.2}$$

An interesting exact result pointed out by Horii and Nemat-Nasser

(1983) is that, in general and for arbitrarily shaped inclusions, one has

$$\frac{1}{V}\int_{\partial\Omega}\tfrac{1}{2}(u_i n_j + u_j n_i)\,\mathrm{dx} = H_{ijkl}\bar{\sigma}_{kl} \tag{3.3}$$

where H_{ijkl} is a constant tensor. When Ω is a single ellipsoid and D is unbounded, then VH_{ijkl} coincides with the tensor $V_\Omega Q_{ijkl}{}^{-1}$, introduced by Hill (1965), consistent with Eshelby's (1957) fundamental result; see also Budiansky (1965).

In many applications it is possible to calculate the left-hand side of eqn (3.3) directly and, in this manner, to obtain the tensor H_{ijkl} for arbitrarily shaped inclusions or cavities. Several examples are worked out by Horii and Nemat-Nasser (1983). As an illustration, let Ω be a set of cavities, and denote by D_{ijkl} the constant compliance tensor of the matrix material, i.e. that of $D - \Omega$. Then

$$\varepsilon_{ij} = D_{ijkl}\,\sigma_{kl} \qquad \text{in } D - \Omega \tag{3.4}$$

and substitution from this and eqn (3.3) into eqn (3.1) immediately yields

$$D^*_{ijkl} = D_{ijkl} + H_{ijkl} \tag{3.5}$$

where D^*_{ijkl} is the overall compliance of the solid, defined by

$$\bar{\varepsilon}_{ijkl} = D^*_{ijkl}\bar{\sigma}_{kl} \tag{3.6}$$

Even when the matrix is linearly elastic, the overall response of the solid will be nonlinear and load path-dependent if during the course of loading the solid cavitates by void nucleation or if the geometry of the cavities changes because of the void growth. The results presented above apply to this case, provided that one considers incremental loading. The overall compliance, in this case, is calculated for each incremental loading, and it will change with the stress history. Illustrative examples of elastic solids with microcracks are given by Horii and Nemat-Nasser (1983).

In some applications, the direct calculation of the left-hand side of eqn (3.3) may present major difficulties. An alternative formulation of the problem will now be presented.

3.2. Periodic Microstructure, or Finite Samples with Inhomogeneities

Let D be a rectangular parallelepiped with dimensions Λ_i, measured along the coordinate axes. For a solid with periodic microstructure, D

then constitutes a single cell. The formulation presented in the sequel applies to both a finite sample D or an unbounded solid consisting of unit cells identical to D.

Let there be M inhomogeneities, Ω_r, $r = 1, 2, \ldots, M$, in D. The geometry of Ω_r is arbitrary, but for simplicity, we assume that each Ω_r is bounded by a suitably regular piecewise continuous surface $\partial\Omega_r$.

The basic theory is presented for an elastic solid containing elastic inclusions (or cavities), and then it is generalized for application to inelastic cases such as nonlinear creep.

3.3. Elastic Solid with Elastic Inclusions

Denote by $\mathbf{C}$ the elastic modulus tensor of the 'matrix', and by $\mathbf{C}^r$ that of the rth inclusion, Ω_r, $r = 1, 2, \ldots, M$. Suppose the boundaries, ∂D, of D are subjected to uniform displacements, $\mathbf{u}^0$, compatible with overall strains $\boldsymbol{\varepsilon}^0$. It is convenient to denote by $\mathbf{u}$ and $\boldsymbol{\varepsilon}$ the perturbations in $\mathbf{u}^0$ and $\boldsymbol{\varepsilon}^0$ due to the presence of inhomogeneities, and write the *total* strain tensor as

$$\boldsymbol{\varepsilon}^{\mathrm{T}} = \boldsymbol{\varepsilon}^0 + \boldsymbol{\varepsilon} \tag{3.7}$$

where $\mathbf{u} = \mathbf{0}$ on ∂D. The stresses then are

$$\sigma_{ij}^T = \begin{cases} C_{ijkl}\,(\varepsilon_{kl}^0 + \varepsilon_{kl}) & \text{in } D - \Omega \\ C_{ijkl}^r\,(\varepsilon_{kl}^0 + \varepsilon_{kl}) & \text{in } \Omega_r \end{cases} \tag{3.8}$$

where Ω is the union of all Ω_r's.

The average overall stress is $\bar{\boldsymbol{\sigma}} = \langle \boldsymbol{\sigma}^{\mathrm{T}} \rangle_D$, and we also have $\bar{\boldsymbol{\varepsilon}} \equiv \boldsymbol{\varepsilon}^0 = \langle \boldsymbol{\varepsilon}^{\mathrm{T}} \rangle_D$ and $\langle \boldsymbol{\varepsilon} \rangle_D = \mathbf{0}$. The overall stress–strain relations are now written as

$$\bar{\sigma}_{ij} = C_{ijkl}^{*}\,\varepsilon_{kl}^0 \tag{3.9}$$

The objective is to calculate the overall modulus tensor $\mathbf{C}^*$.

In a similar manner, when uniform tractions are prescribed on ∂D in such a manner that they are compatible with overall stresses $\boldsymbol{\sigma}^0$, we write

$$\boldsymbol{\sigma}^{\mathrm{T}} = \boldsymbol{\sigma}^0 + \boldsymbol{\sigma} \tag{3.10}$$

The total strains then are

$$\varepsilon_{ij}^{\mathrm{T}} = \begin{cases} D_{ijkl}\,(\sigma_{kl}^0 + \sigma_{kl}) & \text{in } D - \Omega \\ D_{ijkl}^r\,(\sigma_{kl}^0 + \sigma_{kl}) & \text{in } \Omega_r \end{cases} \tag{3.11}$$

and we have $\bar{\varepsilon} = \langle \varepsilon^{\mathrm{T}} \rangle_D$, with $\sigma^0 \equiv \bar{\sigma} = \langle \sigma^{\mathrm{T}} \rangle_D$ and $\langle \sigma \rangle_D = \mathbf{0}$, where $\mathbf{D}$ and $\mathbf{D}^r$'s are the corresponding compliance tensors. Instead of eqn (3.9), we now have

$$\bar{\varepsilon}_{ij} = D^*_{ijkl} \sigma^0_{kl} \tag{3.12}$$

where $\mathbf{D}^*$ is the overall compliance tensor which, for consistency, must equal the inverse of $\mathbf{C}^*$.

Overall Moduli

As discussed by Nemat-Nasser and Taya (1981) and Nemat-Nasser *et al.* (1982), it is convenient to introduce Eshelby-type transformation strains, ε^{*r}, and rewrite eqn (3.8) as

$$\sigma^{\mathrm{T}}_{ij} = C_{ijkl} (\varepsilon^0_{kl} + \varepsilon_{kl} - \varepsilon^{*r}_{kl}), \qquad \varepsilon^{*r} = 0 \quad \text{in } D - \Omega_r \tag{3.13}$$

The transformation strains ε^{*r} are functions of $\mathbf{x}$ in each Ω_r. From eqns (3.8) and (3.13) we obtain the consistency conditions

$$C^r_{ijkl} (\varepsilon^0_{kl} + \varepsilon_{kl}) = C_{ijkl} (\varepsilon^0_{kl} + \varepsilon_{kl} - \varepsilon^{*r}_{kl}), \qquad r = 1, 2, \ldots, M \tag{3.14}$$

Furthermore, equilibrium requires

$$\sigma^{\mathrm{T}}_{ij,i} = C_{ijkl} (\varepsilon^0_{kl} + \varepsilon_{kl} - \varepsilon^{*r}_{kl})_{,i} = 0 \tag{3.15}$$

To obtain an expression for ε^{*r}, we expand all field quantities in the Fourier series; we use eqns (3.13)–(3.15) and obtain (Nemat-Nasser *et al.*, 1982):

$$\begin{aligned}
\varepsilon^0_{mn} &= A^r_{mnkl}\, \varepsilon^{*r}_{kl}(\mathbf{x}) - \varepsilon_{mn}(\mathbf{x}) \qquad \text{in } \Omega_r \quad (\text{no sum on } r) \\
\varepsilon_{jk}(\mathbf{x}) &= \frac{1}{V} \sum_{n_p=0}^{\pm\infty}{}' g_{jkmn}(\xi) \sum_{r=1}^{M} \int_{\Omega_r} \varepsilon^{*r}_{mn}(\mathbf{x}')\, e^{i\xi\cdot(\mathbf{x}-\mathbf{x}')}\, d\mathbf{x}' \\
A^r_{mnkl} &= (C_{mnpq} - C^r_{mnpq})^{-1} C_{pqkl} \\
g_{jkmn} &= [C_{pqjs}\xi_p\xi_s]^{-1} [C_{rqmn}\xi_r\xi_k] \\
\xi_j &= 2\pi n_{\underline{j}}/\Lambda_{\underline{j}} \quad \text{and} \quad i = \sqrt{-1}
\end{aligned} \tag{3.16}$$

here, a prime on $\sum$ indicates that $n = (n_p n_p)^{1/2} = 0$ is excluded in the summation, and repeated indices are summed except when underlined.

When the matrix (but not necessarily the inclusions) is isotropic, the tensor g_{ijkl} becomes

$$\mathbf{g} = \mathbf{g}^{(1)} - \frac{1}{1-\nu}\mathbf{g}^{(2)} + \frac{\nu}{1-\nu}\mathbf{g}^{(3)}$$

$$g^{(1)}_{ijkl} = \tfrac{1}{2}\{\bar{\xi}_j(\delta_{il}\bar{\xi}_k + \delta_{ik}\bar{\xi}_l) + \bar{\xi}_l(\delta_{jl}\bar{\xi}_k + \delta_{jk}\bar{\xi}_l)\} \tag{3.17}$$

$$g^{(2)}_{ijkl} = \bar{\xi}_i\bar{\xi}_j\bar{\xi}_k\bar{\xi}_l, \quad g^{(3)}_{ijkl} = \bar{\xi}_i\bar{\xi}_j\delta_{kl}, \quad \bar{\xi}_i = \xi_i/\xi, \quad \xi^2 = \xi_i\xi_i$$

where ν is the Poisson ratio of the matrix material; in this case, only ν enters the specification of $\mathbf{g}$ and $\boldsymbol{\varepsilon}(\mathbf{x})$ in eqn $(3.16)_2$.

We now average eqn (3.8), and using eqns (3.16) and (3.14), we obtain

$$\langle\sigma^{\mathrm{T}}_{ij}\rangle_r = C^r_{ijkl}\langle\varepsilon^0_{kl} + \varepsilon_{kl}\rangle_r = C_{ijmn}(A^r_{mnkl} - I_{mnkl})\bar{\varepsilon}^{*r}_{kl}$$

$$r = 1, 2, \ldots, M \quad (r \text{ not summed}) \tag{3.18}$$

$$\langle\sigma^{\mathrm{T}}_{ij}\rangle_{D-\Omega} = C_{ijkl}\langle\varepsilon^0_{kl} + \varepsilon_{kl}\rangle_{D-\Omega} \tag{3.19}$$

where $\langle\ldots\rangle_r$ denotes volume average over Ω_r. From these equations and eqn (3.9), it follows that

$$(C_{ijkl} - C^*_{ijkl})\,\varepsilon^0_{kl} = \sum_{r=1}^{M}(C_{ijkl} - C^r_{ijkl})f_r\langle\varepsilon^0_{kl} + \varepsilon_{rl}\rangle_r$$

$$= C_{ijkl}\sum_{r=1}^{M} f_r\bar{\varepsilon}^{*r}_{kl} \tag{3.20}$$

which defines the overall modulus tensor $\mathbf{C}^*$, where

$$f_r = V_r/V \tag{3.21}$$

is the volume fraction of the rth inclusion, and $\bar{\boldsymbol{\varepsilon}}^{*r}$ is the average value of $\boldsymbol{\varepsilon}^{*r}$ on Ω_r.

The integral equations $(3.16)_1$ can be solved for $\boldsymbol{\varepsilon}^{*r}$ in various ways (Nemat-Nasser and Taya, 1981; Nemat-Nasser *et al.*, 1982). The simplest method is to substitute for $\boldsymbol{\varepsilon}^{*r}$ in the right-hand side of $(3.16)_2$ its average over Ω_r and obtain, from eqn $(3.16)_1$,

$$\varepsilon^0_{jk} = A^r_{jkmn}\bar{\varepsilon}^{*r}_{mn} - \sum_{s=1}^{M}\left\{\sum_{n_p=0}^{\pm\infty}{}' g_{jkmn}(\xi)Q^r(\xi)Q^s(-\xi)\right\}f_s\bar{\varepsilon}^{*s}_{mn}$$

$$= \sum_{s=1}^{M} \left[\frac{1}{f_r} A^r_{jkmn} \delta^{rs} - S^{rs}_{jkmn} \right] f_s \bar{\varepsilon}^{*s}_{mn}, \qquad r = 1, 2, \ldots, M \qquad (3.22)$$

where

$$Q^r(\xi) = \langle e^{i\xi \cdot x} \rangle_r \quad \text{and} \quad S^{rs}_{jkmn} = \sum_{n_p = 0}^{\pm\infty}{}' g_{jkmn}(\xi) Q^r(\xi) Q^s(-\xi) \qquad (3.23)$$

The system of linear equations (3.22) can be solved for $f_s \bar{\varepsilon}^{*s}$. In this manner one obtains

$$\sum_{s=1}^{M} f_s \bar{\varepsilon}^{*s}_{mn} = \left\{ \sum_{s,r=1}^{M} \left[\frac{1}{f_r} A^r_{mnkl} \delta^{rs} - S^{rs}_{mnkl} \right]^{-1} \right\} \varepsilon^0_{kl} \qquad (3.24)$$

Direct substitution into eqn (3.20) then provides

$$C^*_{ijkl} = C_{ijkl} - C_{ijmn} \sum_{s,r=1}^{M} \left[\frac{1}{f_r} A^r_{mnkl} \delta^{rs} - S^{rs}_{mnkl} \right]^{-1} \qquad (3.25)$$

which defines the overall modulus tensor.

When the inclusions are large, the use of a constant transformation strain for each inclusion in the right-hand side of eqn $(3.16)_2$ may not provide the required accuracy.† In this case, the solution can be improved to any desired degree of accuracy by subdividing each inclusion into a finite number of subregions, and then regarding each subregion as an inclusion. Since $\boldsymbol{\varepsilon}^{*r}$ is a smooth function of $\mathbf{x}$, for inclusions with regular bounding surfaces, this piecewise constant approximation procedure can be quite effective.

Overall Compliances

Suppose instead of the overall strains, the overall stresses are prescribed to be $\boldsymbol{\sigma}^0$. Then the overall strains are defined by eqn (3.12), and from eqn (3.11) we have

$$\langle \varepsilon^{\mathrm{T}}_{ij} \rangle_r = D^r_{ijkl} \langle \sigma^0_{kl} + \sigma_{kl} \rangle_r, \qquad r = 1, 2, \ldots, M \quad (r \text{ not summed}) \qquad (3.26)$$

$$\langle \varepsilon^{\mathrm{T}}_{ij} \rangle_{D-\Omega} = D_{ijkl} \langle \sigma^0_{kl} + \sigma_{kl} \rangle_{D-\Omega} \qquad (3.27)$$

†Recently, Nunan and Keller (1984) have shown, for spherical cavities, that the error associated with this assumption does not exceed 4% for void volume fractions up to 30%.

Hence, it follows that

$$(D^*_{ijkl} - D_{ijkl})\sigma^0_{kl} = \sum_{r=1}^{M} f_r (D^r_{ijkl} - D_{ijkl}) \langle \sigma^0_{kl} + \sigma_{kl} \rangle_r \tag{3.28}$$

where eqn (3.12) is also used. The consistency condition (3.14), on the other hand, yields

$$\sum_{r=1}^{M} f_r (D^r_{ijkl} - D_{ijkl}) \langle \sigma^0_{kl} + \sigma_{kl} \rangle_r = \sum_{r=1}^{M} f_r \bar{\varepsilon}^{*r}_{ij} \tag{3.29}$$

which reduces eqn (3.28) to

$$(D^*_{ijkl} - D_{ijkl})\sigma^0_{kl} = \sum_{r=1}^{M} f_r \bar{\varepsilon}^{*r}_{ij} \tag{3.30}$$

When the overall stresses are given, ε^0_{jk} in eqns (3.22) and (3.24) must be replaced by $D^*_{jkmn}\sigma^0_{mn}$. Then eqns (3.24) and (3.30) give

$$D^*_{ijkl} = D_{ijmn}\left\{I_{mnkl} - \sum_{s,r=1}^{M}\left[\frac{1}{f_r}A^r_{mnkl}\delta^{rs} - S^{rs}_{mnkl}\right]^{-1}\right\}^{-1} \tag{3.31}$$

which defines the overall compliance.

From eqns (3.25) and (3.31) it is clear that $\mathbf{D}^*$ is the inverse of $\mathbf{C}^*$. Furthermore, the calculation of the overall moduli or the compliances requires only the values of the infinite series S^{rs}_{jkmn} defined by eqns (3.23), which depend on the geometry of the inclusions and on the Poisson ratio of the matrix when this matrix is isotropic; see eqns (3.17). A number of illustrative examples are worked out by Nemat-Nasser *et al.* (1982) and Iwakuma and Nemat-Nasser (1983).

3.4. Nonlinear Creep

In Section 3.3, the term $\boldsymbol{\varepsilon}$ is used to denote the *strain* tensor. In this section, $\boldsymbol{\varepsilon}$ will be used to denote the *strain rate* tensor. Hence, if v_i is the *velocity* field, we have

$$\varepsilon_{ij} = \tfrac{1}{2}(v_{i,j} + v_{j,i}) \tag{3.32}$$

Consider the general nonlinear creep relation

$$\boldsymbol{\varepsilon} = \mathbf{F}(\boldsymbol{\sigma}) \tag{3.33}$$

and for the *increment of the strain rate,* obtain

$$d\varepsilon = \frac{\partial F}{\partial \sigma_{ij}} d\sigma_{ij} \tag{3.34}$$

For example, if power law creep is assumed, we have

$$\varepsilon' = \eta J^n \sigma', \qquad J = (\tfrac{1}{2}\sigma'_{ij}\sigma'_{ij})^{1/2} \tag{3.35}$$

where η (a dimensional parameter) and n are constants, and prime denotes the deviatoric part. Then eqn (3.34) becomes

$$d\varepsilon'_{ij} = \eta J^n \left\{ \delta_{ik}\delta_{jl} + \frac{n\sigma'_{ij}\sigma'_{kl}}{2J^2} \right\} d\sigma'_{kl} \tag{3.36}$$

In general, therefore, we can write

$$d\varepsilon_{ij} = D_{ijkl}\, d\sigma_{kl} \quad \text{or} \quad d\sigma_{ij} = C_{ijkl}\, d\varepsilon_{kl} \tag{3.37}$$

where **D** and **C** are stress-dependent.

Consider now a composite consisting of a matrix whose incremental creep law is defined by eqn (3.37) with $\mathbf{D} = \mathbf{C}^{-1}$ given functions of the stress state. Let the sample D of this composite contain M inclusions with creep laws defined by parameters $\mathbf{D}^r = \mathbf{C}^{r-1}$, $r = 1, 2, \ldots, M$, which also may depend on the stress state. Since the constitutive parameters of the matrix, eqns (3.37), as well as those for the inclusions, depend on the stress, it is necessary to subdivide both the matrix and the inclusions into a finite number of subregions and then regard the stress spatially constant within each subregion. In this approach, each subregion is regarded as an inclusion. For creep of polycrystalline solids, the grain boundary sliding is believed to be governed by a linear creep law (Kê, 1947; Zener, 1948; McLean, 1957). In this case, the grain boundaries must be assigned a finite thickness in order for the method to apply. Then it is convenient to use the grain boundaries as the 'matrix'. In the general case, the periodicity or other factors suggest a natural choice for the matrix.

From the above considerations, and in view of the linear relations between the stress increment and the increment of the strain *rate,* we immediately conclude that all the equations in Section 3.3 also apply to the incremental calculation of nonlinear creep, provided we replace all ε's by $d\varepsilon$ and all σ's by $d\sigma$ (Nemat-Nasser and Iwakuma, 1983). The solution procedure is as follows.

Suppose we incrementally prescribe the overall strain rate, ε^0. At a given state, the stresses are known within all subregions. From constitutive relations similar to eqn (3.37), the moduli $\mathbf{C}^r = \mathbf{C}(\boldsymbol{\sigma}^r)$ can be calculated for $\boldsymbol{\Omega}_r$, assuming a constant $\boldsymbol{\sigma}^r$ within $\boldsymbol{\Omega}_r$. If the 'matrix' is linear, then $\mathbf{S}^{rs}$ depends on the geometry of the subregions and can be calculated once and for all. Equations (3.25) then give $\mathbf{C}^*$, from which we obtain $d\bar{\boldsymbol{\sigma}}$. The local stress increments are obtained from the incremental versions of eqns (3.18) and (3.19), the stresses are up-dated, and the entire procedure is repeated. A similar approach applies when the stress increments are prescribed.

4. SOLID WITH RANDOMLY DISTRIBUTED INHOMOGENEITIES

For simplicity in presentation, we consider an elastic solid with elastic inhomogeneities. The results apply directly to a nonlinearly creeping solid, if an incremental formulation in line with that discussed in Section 3.4 is employed.

Consider an isotropic linearly elastic matrix with elasticity tensor $\mathbf{C}$, and let it include M inhomogeneities $\boldsymbol{\Omega}_r$, $r = 1, 2, \ldots, M$, with the common elasticity tensor $\bar{\mathbf{C}}$. Let the overall applied strain be $\boldsymbol{\varepsilon}^0$. Equations (3.8), with C^r_{ijkl} replaced by $\bar{C}_{ijkl}$, define the total stress in the matrix and the inclusions. Averaging these equations and using eqn (3.9), one readily obtains

$$(C_{ijkl} - C^*_{ijkl})\varepsilon^0_{kl} = (C_{ijkl} - \bar{C}_{ijkl}) \sum_{r=1}^{M} f_r \langle \varepsilon^0_{kl} + \varepsilon_{kl} \rangle_r \tag{4.1}$$

This is an exact result, independent of the nature of the distribution or the shape of the inclusions. However, for it to be useful, the average strains, $\langle \varepsilon^0_{ij} + \varepsilon_{ij} \rangle_r = \varepsilon^0_{ij} + \langle \varepsilon_{ij} \rangle_r$, must be calculated. In the preceding section this was done either using the assumption of periodically distributed unit cells or considering a finite cuboid containing M inclusions. When the number of inclusions is very large and their distribution is not known *a priori*, one cannot easily employ the accurate method of the preceding section. In this case, the self-consistent method proposed by Mackenzie (1950), Hershey (1954), Kerner (1956), Kröner (1958), Budiansky (1965) and Hill (1965) may be used effectively. In this method, one inclusion, say $\boldsymbol{\Omega}_r$, with elasticity $\bar{\mathbf{C}}$, is embedded in a homogeneous, linearly elastic unbounded solid, with elasticity $\mathbf{C}^*$ (which is yet to be calculated), and the average strain in the inclusion is calculated when the solid is subjected to the overall uniform applied strains at infinity. Almost all calculations of this kind have been made with the assumption of an ellipsoidal inclusion. In this case it is easy to show that

$$\langle \varepsilon_{ij}^0 + \varepsilon_{ij} \rangle_r = A_{ijmn}^* (A_{mnkl}^* - S_{mnkl}^*)^{-1} \varepsilon_{kl}^0 \tag{4.2}$$

where S_{ijkl}^* is Eshelby's tensor, calculated for an inclusion embedded in the matrix of elasticity $\mathbf{C}^*$, and

$$A_{ijkl}^* = (C_{ijmn}^* - \bar{C}_{ijmn})^{-1} C_{mnkl}^* \tag{4.3}$$

Equations (4.1) now become

$$C_{ijkl} - C_{ijkl}^* = f(C_{ijmn} - \bar{C}_{ijmn}) A_{mnpq}^* (A_{pqkl}^* - S_{pqkl}^*)^{-1} \tag{4.4}$$

This yields the overall modulus tensor $\mathbf{C}^*$. It can easily be shown that equivalent results are obtained if one assumes that the overall stresses are prescribed.

5. POLYCRYSTALLINE AGGREGATES

The results presented in the preceding sections apply to crystalline solids consisting of aggregates of single crystals, limited, however, to small deformations only. (For a detailed account and illustrative examples of cases involving finite strains and rotations, see Hill (1972), Havner (1982), Nemat-Nasser (1983) and Iwakuma and Nemat-Nasser (1984).) In this section we briefly examine the overall elastic as well as inelastic properties of polycrystalline solids at infinitesimal strains, for which considerable information has been presented in the literature; see, e.g., Taylor (1938), Kröner (1958), Budiansky and Wu (1962), Mandel (1966), Hill (1967), Bui (1970), Hutchinson (1970) and Bui *et al.* (1973).

5.1. Overall Elasticity

If the single crystals are assumed to be isotropic, having a common elasticity tensor, then, trivially, the overall elastic response of the polycrystal will be isotropic, having the same elasticity tensor. When single crystal constituents are anisotropic, as is usually the case, then, whether or not the overall elastic response of the polycrystal will be isotropic depends on the distribution of the single crystals. For a random distribution, the overall response will be isotropic, whereas this response will be anisotropic if the distribution of crystal orientations is biased, for example, by metal processing or prior plastic deformation.

To obtain the overall elastic modulus tensor $\mathbf{C}^*$, and the compliance tensor $\mathbf{D}^*$, consider a typical crystal $\boldsymbol{\Omega}$ of elasticity $\mathbf{C}^{\Omega}$ and compliance $\mathbf{D}^{\Omega}$, and let $\mathbf{A}^{\Omega}$ and $\mathbf{B}^{\Omega}$ be the *concentration tensors* which yield the local strains and stresses in the crystal $\boldsymbol{\Omega}$, in terms of the overall uniform strains and stresses,

$$\varepsilon_{ij} = A^{\Omega}_{ijkl}\,\bar{\varepsilon}_{kl}, \qquad \sigma_{ij} = B^{\Omega}_{ijkl}\,\bar{\sigma}_{kl} \tag{5.1}$$

Then, averaging the strains and stresses over all crystal orientations and shapes, and using the overall stress and strain relations, i.e. eqn (3.6) and its inverse, we immediately obtain

$$C^*_{ijkl} = \langle C^{\Omega}_{ijmn}\,A^{\Omega}_{mnkl}\rangle, \qquad D^*_{ijkl} = \langle D^{\Omega}_{ijmn}\,B^{\Omega}_{mnkl}\rangle \tag{5.2}$$

where $\langle \ldots \rangle$ now stands for averaging over all crystal orientations and shapes. Consistency requires

$$\langle A^{\Omega}_{ijkl}\rangle = \langle B^{\Omega}_{ijkl}\rangle = I_{ijkl} \tag{5.3}$$

the identity tensor. Furthermore, the concentration tensors are related by

$$B^{\Omega}_{ijmn}\,C^*_{mnkl} = C^{\Omega}_{ijmn}\,A^{\Omega}_{mnkl}, \qquad A^{\Omega}_{ijmn}\,D^*_{mnkl} = D^{\Omega}_{ijmn}\,B^{\Omega}_{mnkl} \tag{5.4}$$

as can be verified by direct substitution (Hill, 1965, 1967; Walpole, 1969).

The above results are general, but their usefulness depends on whether or not one is able to calculate the concentration tensors, $\mathbf{A}^{\Omega}$ and $\mathbf{B}^{\Omega}$, which satisfy restriction (5.3). A commonly used approach is the self-consistent method based on the following proposal: to calculate the concentration tensors, embed a single crystal (invariably assumed to be ellipsoidal or even spherical) in a homogeneous, infinitely extended elastic solid with overall elasticity (compliance) $\mathbf{C}^*$ ($\mathbf{D}^*$) which is yet to be calculated, and obtain the strains (stresses) in the crystal in terms of the overall prescribed average strains (stresses). For an ellipsoidal crystal, Eshelby (1957) has shown that the corresponding strains (stresses) will be uniform. However, condition (5.3) is satisfied only if the axes of all considered ellipsoids are properly aligned (this only refers to the shape of the crystals and not to the orientation of the crystals' axes which characterize their anisotropy). In this case, the concentration tensors can be calculated directly and, indeed, one obtains simple expressions when the overall elastic response of the crystal is assumed to be isotropic. Detailed results for both isotropic and many important anisotropic cases are reported by Mura (1982). In the general case of an inclusion Ω (a crystal) of elasticity tensor $\mathbf{C}^{\Omega}$, embedded in a uniform solid of elasticity $\mathbf{C}^*$, we have

$$A^{\Omega}_{ijkl} = [I_{ijkl} + J^{\Omega}_{ijmn}(C^*_{mnkl} - C^{\Omega}_{mnkl})]^{-1} \tag{5.5}$$

where

$$J^{\Omega}_{ijmn} = \int_{\Omega} \mathcal{G}_{in,mj}(\mathbf{x}-\mathbf{x}')\,\mathrm{d}\mathbf{x}'$$

$$\mathcal{G}_{ji}(\mathbf{x}-\mathbf{x}') = (2\pi)^{-3}\int_{-\infty}^{\infty} \frac{N_{ij}(\xi)}{D(\xi)}\, \mathrm{e}^{i\xi\cdot(\mathbf{x}-\mathbf{x}')}\,\mathrm{d}\xi$$

$$N_{ij} = \text{cofactor}\,(K_{ij}), \qquad D = \det\,(K_{ij})$$

$$K_{ij} = C^*_{mijn}\xi_m\xi_n$$

$\mathbf{J}^{\Omega}$ is independent of $\mathbf{x}$ when Ω is an ellipsoid. The expression for $\mathbf{B}^{\Omega}$ is now obtained by substitution for $\mathbf{A}^{\Omega}$ into eqn (5.4).

When the axes of the ellipsoids considered in the self-consistent approach are not properly aligned, one may still use the self-consistent approach, but must modify the concentration tensors, so that they do satisfy restriction (5.3). For example, Walpole (1969) suggests using $A^{-\Omega}_{ijkl} = A^{\Omega}_{ijmn}\langle A^{\Omega}_{mnkl}\rangle^{-1}$, with a similar modification for $\mathbf{B}^{\Omega}$.

5.2. Overall Elastoplasticity

For an elastoplastic polycrystal, all the equations presented above continue to apply if they are properly interpreted in terms of instantaneous elastoplastic moduli of the crystal constituents and those of the associated overall polycrystal. For the polycrystal, the overall stress rate–strain rate relations may be expressed as

$$\dot{\bar{\sigma}}_{ij} = C^*_{ijkl}\dot{\bar{\varepsilon}}_{kl}, \qquad \dot{\bar{\varepsilon}}_{ij} = D^*_{ijkl}\dot{\bar{\sigma}}_{kl} \tag{5.7}$$

and for a typical single crystal we write

$$\dot{\sigma}_{ij} = C^{\Omega}_{ijkl}\dot{\varepsilon}_{kl}, \qquad \dot{\varepsilon}_{ij} = D^{\Omega}_{ijkl}\dot{\sigma}_{kl} \tag{5.8}$$

The individual crystal moduli will be the elastic ones if the crystal is instantaneously deforming elastically, and they will be the elastoplastic ones otherwise.

A simple rate-independent elastoplastic theory of crystals is obtained if one assumes that the plastic deformation of each crystal stems from slip over crystallographic planes, accompanied by lattice elastic distortion to produce a compatible deformation. Depending on the lattice structure, one identifies a finite number of slip systems and invokes simple rules which indicate whether or not a slip system is active. A commonly used rule is the Schmid law, according to which a slip system is inactive if the associated shear stress falls (or tends to fall) below a critical value. By assuming that this critical value changes with the accumulation of plastic strain due to slip on all or some of the slip systems, one can incorporate various workhardening rules in characterizing the elastoplastic deformation of single crystals; see, e.g., Hill (1966).

Let $\dot{\varepsilon}^{e}$ and $\dot{\varepsilon}^{p}$ denote the elastic and plastic strain rates in a typical crystal Ω,

$$\dot{\varepsilon}_{ij} = \dot{\varepsilon}_{ij}^{e} + \dot{\varepsilon}_{ij}^{p} \tag{5.9}$$

In accordance with the slip model of crystal plasticity, the plastic strain rate is given by

$$\dot{\varepsilon}_{ij}^{p} = \sum_{\alpha=1}^{M} \tfrac{1}{2}(s_i^{\alpha} n_j^{\alpha} + s_j^{\alpha} n_i^{\alpha})\,\dot{\gamma}^{\alpha} \equiv \sum_{\alpha=1}^{M} p_{ij}^{\alpha}\,\dot{\gamma}^{\alpha} \tag{5.10}$$

where s_i^{α} is a unit vector in the slip direction of the αth slip system on the slip plane with unit normal n_i^{α}, $\dot{\gamma}^{\alpha}$ is the corresponding slip rate, and the definition of p_{ij}^{α} is clear from the context. The rate of change of resistance to slip is governed by the hardening rule:

$$\dot{\tau}^{\alpha} = \sum_{\beta=1}^{M} h^{\alpha\beta}\,\dot{\gamma}^{\beta} \tag{5.11}$$

where $h^{\alpha\beta}$ is the hardening matrix and $\dot{\tau}^{\alpha}$ is the rate of change of the resolved shear stress $\tau^{\alpha} = \sigma_{ij} n_i^{\alpha} s_j^{\alpha}$ (no sum on α); $\dot{\gamma}^{\alpha}$ is nonzero if eqn (5.11) is satisfied and $\sigma_{ij} n_i^{\alpha} s_j^{\alpha} = \tau_y^{\alpha}$, where τ_y^{α} is the current yield stress in shear for the αth slip system.

The stress rate in crystal Ω is given in terms of the crystal elasticity tensor $\mathscr{C}^{\Omega}$ and the elastic strain rate ε^{e},

$$\dot{\sigma}_{ij} = \mathscr{C}_{ijkl}^{\Omega}\,\dot{\varepsilon}_{kl}^{e} \tag{5.12}$$

(To distinguish the instantaneous elastoplastic modulus tensor $\mathbf{C}^{\Omega}$ from the *elastic* modulus tensor of the crystal, the latter is denoted by $\mathscr{C}^{\Omega}$. For elastic loading and unloading, and in the context of small deformation theory, $\mathbf{C}^{\Omega} = \mathscr{C}^{\Omega}$.) From eqns (5.9), (5.10) and (5.12) we obtain

$$\dot{\sigma}_{ij} = \mathscr{C}^{\Omega}_{ijkl}\,\dot{\varepsilon}_{kl} - \mathscr{C}^{\Omega}_{ijkl}\sum_{\alpha=1}^{M} p^{\alpha}_{kl}\,\dot{\gamma}^{\alpha} \tag{5.13}$$

and substitution into eqn (5.11) yields

$$\sum_{\beta=1}^{M}\left[p^{\alpha}_{ij}\mathscr{C}^{\Omega}_{ijkl}\,p^{\beta}_{kl} + h^{\alpha\beta}\right]\dot{\gamma}^{\beta} = p^{\alpha}_{ij}\mathscr{C}^{\Omega}_{ijkl}\,\dot{\varepsilon}_{kl} \tag{5.14}$$

These are a system of M linear equations for the M slip rates $\dot{\gamma}^{\alpha}$, $\alpha = 1, 2, \ldots, M$. When these equations are linearly independent, a unique set of slip rates is obtained. But, in general, this may not be the case (Hill, 1966), depending on the structure of the hardening matrix $h^{\alpha\beta}$. When the system (5.14) does not yield a unique solution, other considerations must govern the choice of the active slip systems and hence the solution. Havner (1982), for example, suggests that this choice be made on the basis of the minimum rate of associated plastic work. At any rate, one is always able to select five active slip systems (assuming incompressible flow in a three-dimensional setting) which lead to a unique set of $\dot{\gamma}^{\alpha}$ in eqn (5.14). For this set, let

$$M^{\alpha\beta} = \left[p^{\alpha}_{ij}\mathscr{C}^{\Omega}_{ijkl}\,p^{\beta}_{kl} + h^{\alpha\beta}\right]^{-1} \tag{5.15}$$

and obtain

$$\dot{\gamma}^{\alpha} = \left\{\sum_{\beta=1}^{M} M^{\alpha\beta}\,p^{\beta}_{ij}\mathscr{C}^{\Omega}_{ijkl}\right\}\dot{\varepsilon}_{kl} \equiv Q^{\alpha}_{kl}\,\dot{\varepsilon}_{kl} \tag{5.16}$$

where the definition of Q^{α}_{kl} is clear from the context.

From eqns (5.16) and (5.13) we have

$$\dot{\sigma}_{ij} = C^{\Omega}_{ijkl}\,\dot{\varepsilon}_{kl} = \left\{\mathscr{C}^{\Omega}_{ijkl} - \mathscr{C}^{\Omega}_{ijmn}\sum_{\alpha=1}^{M} p^{\alpha}_{mn}\,Q^{\alpha}_{kl}\right\}\dot{\varepsilon}_{kl} \tag{5.17}$$

so that the instantaneous elastoplastic modulus tensor is

$$C^{\Omega}_{ijkl} = \mathscr{C}^{\Omega}_{ijkl} - \mathscr{C}^{\Omega}_{ijmn} \sum_{a=1}^{M} p^{a}_{mn} Q^{a}_{kl} \tag{5.18}$$

The overall instantaneous elastoplastic modulus tensor C^* is now given by the first part of eqn (5.2).

ACKNOWLEDGEMENT

This work has been supported by the US Army Research Office under contract DAAG29-82-K-0147 to Northwestern University, and by the National Science Foundation under a grant to the University of California, San Diego, NSF MSM 85-17056.

REFERENCES

Bishop, J. F. W. and Hill, R. (1951). (a) A theory of the plastic distortion of a polycrystalline aggregate under combined stresses. (b) A theoretical derivation of the plastic properties of a polycrystalline face-centered metal, *Phil. Mag.*, Series 7, **42**, 414 and 1298.

Budiansky, B. and Wu, T. T. (1962). Theoretical prediction of plastic strains *J. Mech. Phys. Solids*, **13**, 223.

Budiansky, B., and Wu, T. T. (1962). Theoretical prediction of plastic strains of polycrystals. In: *Proc. 4th US National Congress of Applied Mechanics*, p. 1175.

Bui, H. D. (1970). Évolution de la frontière du domaine élastique des métaux avec l'écrouissage plastique et comportement élastoplastique d'un agrégat de cristaux cubiques, *Mém. Artillerie Franç.: Sci. Tech. Armement*, **1**, 141.

Bui, H. D., Zaoui, A. and Zarka, J. (1973). Sur le comportement élastoplastique et viscoplastique des monocristaux et polycristaux métalliques de structure cubique á faces centrées. In: *Foundations of Plasticity*, A. Sawczuk (Ed.), Noordhoff, Leyden, p. 51.

Christoffersen, J., Mehrabadi, M. M. and Nemat-Nasser, S. (1981). A micro-mechanical description of granular material behaviour, *J. Appl. Mech.*, **48**, 339.

Eshelby, J. D. (1957). The determination of the elastic field of an ellipsoidal inclusion, and related problems, *Proc. Roy. Soc. Lond.*, **A241**, 376.

Hashin, Z. (1983). Analysis of composite materials — a survey, *J. Appl. Mech.*, **50**, 481.

Havner, K. S. (1982). Minimum plastic work selects the highest symmetry of deformation in axially-loaded, fcc crystals, *Mechanics of Materials*, **1**, 97.

Hershey, A. V. (1954). (a) The elasticity of an isotropic aggregate of anisotropic cube crystals. (b) The plasticity of an isotropic aggregate of anisotropic face-centered cubic crystals, *J. Appl. Mech.*, **21**, 236 and 241.

Hill, R. (1956). The mechanics of quasi-static plastic deformation in metals, In: *Surveys in Mechanics — The G. I. Taylor 70th Anniversary Volume*, G. K. Batchelor and R. M. Davies (Eds), Cambridge University Press, p. 7.

Hill, R. (1963). Elastic properties of reinforced solids: some theoretical principles, *J. Mech. Phys. Solids*, **11**, 357.
Hill, R. (1965). (a) Continuum micro-mechanics of elastoplastic polycrystals. (b) A self-consistent mechanics of composite materials, *J. Mech. Phys. Solids*, **13**, 89 and 213.
Hill, R. (1966). Generalized constitutive relations for incremental deformation of metal crystals by multislip, *J. Mech. Phys. Solids*, **14**, 95.
Hill, R. (1967). The essential structure of constitutive laws for metal composites and polycrystals, *J. Mech. Phys. Solids*, **15**, 79.
Hill, R. (1972). On constitutive macro-variables for heterogeneous solids at finite strain, *Proc. Roy. Soc. Lond.*, **A326**, 131.
Horii, H. and Nemat-Nasser, S. (1983). Overall moduli of solids with microcracks: load-induced anisotropy, *J. Mech. Phys. Solids*, **31**, 155.
Hutchinson, J. W. (1970). Elastic–plastic behaviour of polycrystalline metals and composites, *Proc. Roy. Soc. Lond.*, **A319**, 247.
Iwakuma, T. and Nemat-Nasser, S. (1983). Composites with periodic microstructure, *Computers and Structures*, **16**, 13.
Iwakuma, T. and Nemat-Nasser, S. (1984). Finite elastic–plastic deformation of polycrystalline metals, *Proc. Roy. Soc. Lond.*, **A394**, 87.
Kê, T.-S. (1947). Experimental evidence of the viscous behaviour of grain boundaries in metals, *Phys. Rev.*, **71**, 533.
Kerner, E. H. (1956). The elastic and thermo-elastic properties of composite media, *Proc. Phys. Soc.*, **B69**, 808.
Kröner, E. (1958). Berechnung der elastischen Konstanten des Vielkristalls aus den Konstanten des Einkristalls, *Z. Phys.*, **151**, 504.
Mackenzie, J. K. (1950). The elastic constants of a solid containing spherical holes, *Proc. Phys. Soc.*, **B63**, 2.
Mandel, J. (1966). Contribution théorique à l'étude de l'écrouissage et des lois de l'écoulement plastique. In: *Proc. 11th International Congress of Applied Mechanics*, H. Görtler (Ed.), Springer-Verlag, Berlin, p. 502.
Mandel, J. (1980). Généralization dans R9 de la régle du potential plastique pour un élément polycrystallin, *Compt. Rend. Acad. Sci. Paris*, **290**, 481.
McLean, D. (1957). *Grain Boundaries in Metals*, Oxford University Press, London.
Mehrabadi, M. M., Nemat-Nasser, S. and Oda, M. (1982). On statistical description of stress and fabric in granular materials, *Int. J. Numer. Anal. Methods Geomech.*, **6**, 95.
Mura, T. (1982). *Micromechanics of Defects in Solids*, Martinus Nijhoff, The Hague–Boston.
Nemat-Nasser, S. (1983). On finite plastic flow of crystalline solids and geomaterials, *J. Appl. Mech. (50th Anniversary Issue)*, **50**, 1114.
Nemat-Nasser, S. and Iwakuma, T. (1983). Micromechanically based constitutive relations for polycrystalline solids. In: *Nonlinear Constitutive Relations for High Temperature Applications*, NASA Conference Publication No. 2271, Scientific and Technical Information Office, NASA, p. 113.
Nemat-Nasser, S. and Taya, M. (1981). On effective moduli of an elastic body containing periodically distributed voids, *Quart. Appl. Math.*, **39**, 43.
Nemat-Nasser, S., Iwakuma, T. and Hejazi, M. (1982). On composites with periodic structure, *Mechanics of Materials*, **1**, 239.

Nunan, K. C. and Keller, J. B. (1984). Effective elasticity tensor of a periodic composite, *J. Mech. Phys. Solids,* **32**, 259.
Reuss, A. (1929). Berechnung der Fliessgrenze von Mischkristallen auf Grund der Plastizitätsbedingung für Einkristalle, *Z. angew. Math. Mech.,* **9**, 49.
Taylor, G. I. (1938). Plastic strain in metals, *J. Inst. Metals,* **62**, 307.
Voigt, W. (1889). Über die Beziehung zwischen den beiden Elastizitätskonstanten isotroper Körper, *Wied. Ann.,* **38**, 573.
Walpole, L. J. (1969). On the overall elastic moduli of composite materials, *J. Mech. Phys. Solids,* **17**, 235.
Willis, J. R. (1983). The overall elastic response of composite materials, *J. Appl. Mech.,* **50**, 1202.
Zarka, J. (1973). Étude du comportement des monocristaux métalliques. Application á la traction du monocristal c.f.c., *J. Mécanique,* **12**, 275.
Zener, C. (1948). *Fracturing of Metals,* ASM, Metals Park, Cleveland, OH.

CHAPTER 3

Basic Theorems of Elastoplasticity

NGUYEN QUOC SON
Laboratoire de Mécanique des Solides, Ecole Polytechnique, Palaiseau, France

ABSTRACT

A review of principal results concerning the quasi-static response of an elastic-plastic or elastic-viscoplastic structure undergoing infinitesimal transformation is given. It is concerned principally with the perfect plastic model and with a straightforward generalization denoted as the generalized standard model. Both of these models are based upon the classical assumptions of normality and convexity. Firstly, the rate problem is recalled and associated minimum principles are formulated. Secondly, a global description of the evolution is introduced and global results are derived such as uniqueness and existence theorems. Elastic and plastic shake-down theorems are then presented from an asymptotic analysis of the response of the structure under cyclic or periodic loading.

1. INTRODUCTION

Basic ideas of plasticity theory as a feasible modelling of the behaviour of ordinary materials were introduced very early on, almost at the same time as linear elasticity. But their development as a satisfactory theory only began with the fundamental works of Melan (1936), Prager (1937), Geiringer (1938), Mandel (1942), Hill (1950), Drucker (1964), Koiter (1960), etc. The mathematical theory of plasticity may be considered as having been completely formulated in the 1950s. At least for small transformations plasticity has become progressively more popular and more widely applied in the resolution of engineering problems, especially with the progress of computation facilities. The difficult problem of describing finite plastic deformation is also being overcome thanks to numerous interesting discussions due principally to Hill (1958), Hutchinson (1973), Lee (1969), Mandel (1971) and Rice (1971).

The objective of this chapter is to present, in the context of small transformations, some basic results of the mathematical theory of elastoplasticity. Our attention will be focussed on the most simple models involving two basic ideas of standard plasticity, namely normality and convexity. In this way, not only the classical model of perfect plasticity will be considered but also a simple generalization of this model in order to include certain effects of strain-hardening (Halphen and Nguyen, 1975).

2. CONSTITUTIVE EQUATIONS IN PERFECT PLASTICITY

Let us recall first the material behaviour as defined in perfect plasticity. Classical equations of perfect plasticity can be schematically written in the following form:

$$\left.\begin{array}{lll} \varepsilon = \varepsilon^{e} + \varepsilon^{p} & & \text{(strain decomposition in elastic and plastic parts)} \\ \sigma = E \cdot \varepsilon^{e} & & \text{(elastic law between stress and elastic strain)} \\ \sigma \in C & & \text{(convex of plastic domain)} \\ \dot{\varepsilon}^{p} = n_{c}(\sigma) & & \text{(normality law)} \end{array}\right\} \quad (2.1)$$

where E denotes elastic coefficients, C is a convex domain in the stress space and defines the plastically admissible states of stress, $n_c(\sigma)$ is a normal to the convex C at the point σ, and $(\dot{\ })$ is the forward time derivative. Normality law is also usually written in an equivalent variational form:

$$\dot{\varepsilon}^{p}.(\sigma - \sigma^{*}) \geqslant 0 \qquad \forall \sigma^{*} \in C \tag{2.2}$$

classically denoted as Hill's maximum principle.

Because normality relates to plastic strain rate, eqns (2.1) define an incremental behaviour. If a strain path $\varepsilon(t)$ is prescribed, the associated stress path $\sigma(t)$ and the plastic strain path $\varepsilon^{p}(t)$ can be defined incrementally by the correspondence $\dot{\varepsilon} \to \dot{\sigma}(\dot{\varepsilon}), \dot{\varepsilon}^{p}(\dot{\varepsilon})$ when the present state of stress and strain is assumed to be known. These incremental equations can be defined from eqns (2.1) in the following way. The development

$$\varepsilon_{t'} = \varepsilon + \tau\dot{\varepsilon} + \ldots, \qquad \varepsilon^{p}_{t'} = \varepsilon^{p} + \tau\dot{\varepsilon}^{p} + \ldots, \qquad \sigma_{t'} = \sigma + \tau\dot{\sigma} + \ldots$$

with $\tau = t' - t > 0$ gives, from the relations (2.2),

$$(\dot{\varepsilon}^{p} + \ldots).(\sigma + \tau\dot{\sigma} + \ldots - \sigma^{*}) \geqslant 0 \qquad \forall\sigma^{*} \in C$$

and thus, if we take $\sigma^{*} = \sigma$ and multiply by τ^{-1}, the following inequality results when $\tau \to 0$:

$$\dot{\varepsilon}^{p}.\dot{\sigma} \geqslant 0 \tag{2.3}$$

On the other hand, if $N_c(\sigma)$ denotes the normal cone at σ to convex C, it is clear that $\forall\beta \in N_c(\sigma)$, $\beta.(\sigma - \sigma^{*}) \geqslant 0$.

Thus, if we take $\sigma^{*} = \sigma_{t'}$, the following inequality results when $\tau \to 0$:

$$\beta.\dot{\sigma} \leqslant 0 \tag{2.4}$$

Since $\dot{\varepsilon}^{p} \in N_c(\sigma)$, one obtains from the inequalities (2.3) and (2.4) the well known equality

$$\dot{\sigma} \cdot \dot{\varepsilon}^{p} = 0 \tag{2.5}$$

and the variational inequality

$$(\mathrm{E}\dot{\varepsilon} - \mathrm{E}\dot{\varepsilon}^{p}) \cdot (\dot{\varepsilon}^{p} - \beta) \geqslant 0 \qquad \forall\beta \in N_c(\sigma) \tag{2.6}$$

The correspondence $\dot{\varepsilon} \to \dot{\varepsilon}^{p}$ can now be expressed from the relations (2.6) as a minimization problem.

$$\left.\begin{array}{l}\text{The associated plastic strain rate } \dot{\varepsilon}^{p}(\dot{\varepsilon}) \text{ minimizes the function} \\ I(\dot{\varepsilon}^{p}) = \frac{1}{2}\dot{\varepsilon}^{p} \cdot E \cdot \dot{\varepsilon}^{p} - \dot{\varepsilon} \cdot E \cdot \dot{\varepsilon}^{p} \qquad \text{in} \quad N_c(\sigma)\end{array}\right\} \tag{2.7}$$

which admits one unique solution from the convexity of C and the positivity of E.

For example, for multiple plastic potential, the plastic domain is described by m inequalities:

$$C = \{\sigma^{*} | f^{i}(\sigma^{*}) \leqslant 0, \qquad i = 1, m\}$$

the normal cone $N_c(\sigma)$ can be expressed as

$$N_c(\sigma) = \left\{n | n = \lambda^{i}\frac{\partial f^{i}}{\partial\sigma}, \quad \lambda^{i} \geqslant 0 \text{ if } f^{i} = 0, \quad \lambda^{i} = 0 \text{ if } f^{i} < 0\right\}$$

the associated plastic rate $\dot{\varepsilon}^p = \lambda^i(\partial f^i/\partial\sigma)$ is such that the function

$$\frac{1}{2}\lambda^i \frac{\partial f^i}{\partial\sigma} \cdot E \cdot \frac{\partial f^i}{\partial\sigma}\lambda^j - \dot{\varepsilon} \cdot E \cdot \frac{\partial f^i}{\partial\sigma}\lambda^i$$

attains its minimum in the admissible set of plastic multipliers λ^i, $\lambda^i \geqslant 0$ if $f^i = 0$ and $\lambda^i = 0$ if $f^i < 0$.

If $m = 1$, one obtains the classical expression of simple plastic potential:

$$\dot{\varepsilon}^p = \frac{\left\langle \frac{\partial f}{\partial\sigma} \cdot E \cdot \dot{\varepsilon} \right\rangle}{\frac{\partial f}{\partial\sigma} \cdot E \cdot \frac{\partial f}{\partial\sigma}} \cdot \frac{\partial f}{\partial\sigma} \quad \text{when } f(\sigma) = 0 \tag{2.8}$$

where $\langle a \rangle$ denotes the positive part of a. ($\langle a \rangle = 0$ if $a \leqslant 0$, $\langle a \rangle = a$ if $a > 0$.)

The relation $\dot{\sigma} = E(\dot{\varepsilon} - \dot{\varepsilon}^p(\dot{\varepsilon}))$ gives finally the correspondence $\dot{\varepsilon} \to \dot{\sigma}$. One obtains for simple plastic potential:

$$\left.\begin{aligned} &\dot{\sigma} = \frac{\partial\phi}{\partial\dot{\varepsilon}}(\dot{\varepsilon}) \quad \text{with} \quad \phi(\dot{\varepsilon}) = \frac{1}{2}\dot{\varepsilon} \cdot E \cdot \dot{\varepsilon} - \frac{1}{2}\theta \frac{\left\langle \frac{\partial f}{\partial\sigma} \cdot E \cdot \dot{\varepsilon} \right\rangle^2}{\frac{\partial f}{\partial\sigma} \cdot E \cdot \frac{\partial f}{\partial\sigma}} \\ &\theta = 1 \quad \text{if} \quad f(\sigma) = 0 \\ &\theta = 0 \quad \text{if} \quad f(\sigma) < 0 \end{aligned}\right\} \tag{2.9}$$

It is interesting to note that the existence of a rate potential $\phi(\dot{\varepsilon})$ such that $\dot{\sigma} = \partial\phi/\partial\dot{\varepsilon}$ is also available in the general case, although its expression is not straightforward. Formally, if the notations $\dot{e} = E \cdot \dot{\varepsilon}$, $\dot{e}^p = E \cdot \dot{\varepsilon}^p$ are introduced, one obtains $\dot{\sigma} = \text{proj}(\dot{e}, T_c(\sigma))$, $\dot{e}^p = \text{proj}(\dot{e}, N_c(\sigma))$ and $\phi(\dot{\varepsilon}) = \frac{1}{2}\|\dot{e}\|^2 - \frac{1}{2}\|\dot{e}^p\|^2$ where $T_c(\sigma)$ denotes the tangent cone at the actual state of stress and $\|\cdot\|$ the energy norm $\|\dot{e}\|^2 = \dot{e} \cdot E^{-1} \cdot \dot{e} = \dot{\varepsilon} \cdot E \cdot \dot{\varepsilon}$; the projection operator proj must be expressed in the energy sense (Nguyen, 1973).

The stress–strain rate relations are non-linear. For example, in the simplest case of simple plastic potential, eqns (2.9) can also be expressed explicitly as

$$\left.\begin{aligned} \dot{\sigma} = E_t \cdot \dot{\varepsilon} \quad \text{with} \quad & E_t = E \\ & \text{if } f < 0 \text{ or if } f = 0 \text{ and } \frac{\partial f}{\partial \sigma} \cdot E \cdot \dot{\varepsilon} < 0 \\ & E_t = E - \frac{\partial f}{\partial \sigma} \cdot E \otimes E \cdot \frac{\partial f}{\partial \sigma} \Big/ \frac{\partial f}{\partial \sigma} \cdot E \frac{\partial f}{\partial \sigma} \\ & \text{if } f = 0 \text{ and } \frac{\partial f}{\partial \sigma} \cdot E \cdot \dot{\varepsilon} > 0 \end{aligned}\right\} \quad (2.10)$$

The tangent modulus E_t depends on the present state of stress and on the direction of strain rate $\dot{\varepsilon}/\|\dot{\varepsilon}\|$.

The rate potential $\phi(\dot{\varepsilon})$ is a convex function of $\dot{\varepsilon}$. Indeed one obtains the following inequality:

$$\phi(\dot{\varepsilon}^*) - \phi(\dot{\varepsilon}) \geqslant \frac{\partial \phi}{\partial \dot{\varepsilon}}(\dot{\varepsilon}) \cdot (\dot{\varepsilon}^* - \dot{\varepsilon}) \quad (2.11)$$

which characterizes convex functions in the strain rate space.

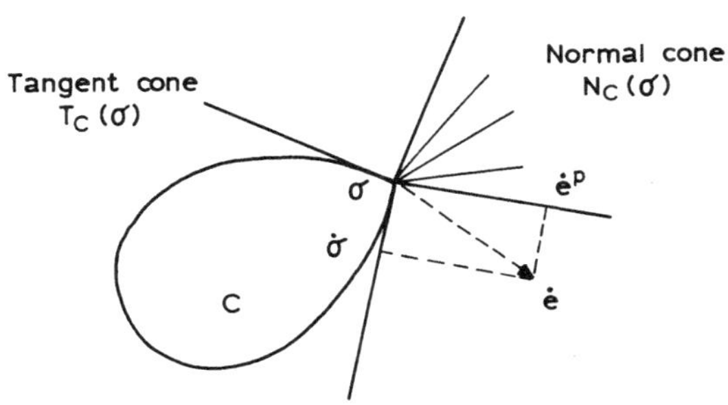

Fig. 1. Rate relations.

3. RATE PROBLEM AND ASSOCIATED MINIMUM PRINCIPLES

We are now principally concerned with the quasi-static response of an elastoplastic structure submitted to a given classical loading defined by a prescribed volume force $\mathbf{g}(t)$ in the volume $\boldsymbol{\Omega}$ of the structure, by a prescribed surface force $\mathbf{T}^d(t)$ on a portion S_T of the boundary $\partial\boldsymbol{\Omega}$ and by a prescribed displacement $\mathbf{u}^d(t)$ on the complementary part S_u,

$t \in [0, T]$ represents a kinematical parameter. The quasi-static response is such that the aforementioned constitutive relations and the equilibrium equations must be satisfied at each time t.

Since elastoplastic constitutive laws are incremental equations, if the associated rate $\dot{\mathbf{u}}, \dot{\sigma}$ can be obtained from the rate data $(\dot{\mathbf{g}}, \dot{\mathbf{T}}^{\mathrm{d}}, \dot{\mathbf{u}}^{\mathrm{d}})$ when the present state of the structure $\mathbf{u}, \sigma$ is assumed to be known, then, by a step-by-step time integration, one can *hope* to determine the global response $\mathbf{u}(t), \sigma(t)$ with respect to the loading path, from a given initial state $\mathbf{u}(0), \sigma(0)$. The derivation of $\dot{\mathbf{u}}, \dot{\sigma}$ as a function of given data $(\dot{\mathbf{g}}, \dot{\mathbf{T}}^{\mathrm{d}}, \dot{\mathbf{u}}^{\mathrm{d}})$ corresponds to the resolution of a boundary value problem specified as the rate problem.

3.1. Local Description of the Rate Problem

The rate problem can be defined in the following way. Find $\dot{\mathbf{u}}$ and $\dot{\sigma}$ such that the following equations must be verified:

$$\left.\begin{array}{lr}
\text{Kinematic conditions:} & \\
\dot{\varepsilon} = \tfrac{1}{2}(\nabla \dot{u} + \tilde{\nabla} \dot{u}) = \dot{\varepsilon}(\dot{u}) & \forall x \in \Omega \\
\dot{u} = \dot{u}^{\mathrm{d}} & \forall x \in S_u \\
\text{Static conditions:} & \\
\operatorname{div} \dot{\sigma} + \dot{g} = 0 & \forall x \in \Omega \\
\dot{\sigma} \cdot n = \dot{T}^{\mathrm{d}} & \forall x \in S_T \\
\text{Constitutive equations:} & \\
\dot{\sigma} = \dfrac{\partial \phi}{\partial \dot{\varepsilon}}(\dot{\varepsilon}) & \forall x \in \Omega
\end{array}\right\} \tag{3.1}$$

These local equations show that the rate problem is a non-linear problem of classical elasticity. The non-linearity is due to the form of constitutive equations:

$$\dot{\sigma} = \frac{\partial \phi}{\partial \dot{\varepsilon}}$$

The stress rate depends linearly on $\dot{\varepsilon}$, $\dot{\sigma} = E_t \cdot \dot{\varepsilon}$, but the tangent modulus E_t depends on the direction of strain rate, $E_t = E_t(\dot{\varepsilon}/\|\dot{\varepsilon}\|)$, since one must distinguish between loading and unloading conditions.

3.2. Variational Formulation of the Rate Problem

It is well known in elasticity that the existence of a convex elastic potential $\sigma = (\partial \phi / \partial \varepsilon) \cdot (\varepsilon)$ leads to the variational principles of total potential energy and complementary potential energy. Since the function $\phi(\dot{\varepsilon})$ is also convex, the following variational formulation of the rate problem can be associated in the same way.

Greenberg's Minimum Principle (Greenberg, 1949)
The actual velocity $\dot{\mathbf{u}}$ minimizes the functional

$$A(\dot{\mathbf{u}}) = \int_{\Omega} \phi(\dot{\varepsilon}(\dot{u}))\, \mathrm{d}\Omega - \int_{\Omega} \dot{g}\dot{u}\, \mathrm{d}\Omega - \int_{S_T} \dot{T}^{\mathrm{d}} \cdot \dot{u}\, \mathrm{d}S \tag{3.2}$$

among the set of kinematically admissible velocity fields $\dot{\mathbf{u}}^*$, i.e. $\dot{\mathbf{u}}^* = \dot{\mathbf{u}}^{\mathrm{d}}$ on S_u.

Hodge–Prager Minimum Principle (Prager and Hodge, 1951)
It is necessary to introduce the dual potential $\phi^*(\dot{\sigma})$, image of $\phi(\dot{\varepsilon})$ by Legendre–Fenchel transform:

$$\phi^*(\dot{\sigma}) = \sup_{\dot{\varepsilon}^*} \dot{\sigma}\dot{\varepsilon}^* - \phi(\dot{\varepsilon}) \tag{3.3}$$

One obtains easily

$$\phi^*(\dot{\sigma}) = \begin{cases} \frac{1}{2}\dot{\sigma} \cdot E^{-1} \cdot \dot{\sigma} & \text{if } f(\sigma) < 0 \\ \frac{1}{2}\dot{\sigma} \cdot E^{-1} \cdot \dot{\sigma} & \text{if } f(\sigma) = 0 \text{ and } \frac{\partial f}{\partial \sigma}\dot{\sigma} \leqslant 0 \\ +\infty & \text{if } f(\sigma) = 0 \text{ and } \frac{\partial f}{\partial \sigma}\dot{\sigma} > 0 \end{cases} \tag{3.4}$$

If only plastically admissible stress rates are considered, $(\partial f/\partial\sigma) \cdot \dot{\sigma} \leqslant 0$ if $f(\sigma) = 0$, the following minimum principle is derived. The actual stress rate $\dot{\boldsymbol{\sigma}}$ minimizes the functional

$$B(\dot{\sigma}) = \int_{\Omega} \frac{1}{2}\dot{\sigma} \cdot E^{-1} \cdot \dot{\sigma}\, \mathrm{d}\Omega - \int_{S_u} \dot{u}^{\mathrm{d}} \cdot \dot{\sigma} \cdot n\, \mathrm{d}S \tag{3.5}$$

among the set of plastically and statically admissible stress rates $\dot{\boldsymbol{\sigma}}^*$, i.e.

$$\left.\begin{aligned} &\frac{\partial f}{\partial \sigma} \cdot \dot{\sigma}^* \leqslant 0 \quad \text{if } f(\sigma) = 0 \\ &\operatorname{div} \dot{\sigma}^* + \dot{g} = 0 \qquad \forall x \in \Omega \\ &\dot{\sigma}^* \cdot n = \dot{T}^{\mathrm{d}} \qquad \forall x \in S_T \end{aligned}\right\}$$

This is a quadratic programming problem under linear equality and inequality constraints.

4. QUASI-STATIC EVOLUTION AND PRINCIPAL RESULTS

The study of rate problems is naturally useful in the sense that, since the elastoplastic deformation process is incremental, at first approximation, strain and stress rates define strain and stress increments. However, the full evolution of a structure under quasi-static transformation can be described in a global and more satisfactory formulation.

In this section, principal results concerning the quasi-static evolution are presented. The uniqueness theorem is obtained concerning the stress field when the given data are compatible with plasticity conditions. The response in displacement is more singular because of the possibility of localization of plastic strain; surfaces of displacement discontinuity are frequently observed. This is an essential feature of perfect plasticity. As will be shown later, the presence of a positive hardening effect regularizes completely the response and these singularities no longer exist.

4.1. Equations of Evolution for Stress and Displacement

For the sake of clarity, let us introduce the following notations:

$$\left.\begin{aligned} S(t) &= \{\sigma \mid \operatorname{div}\sigma + \mathbf{g}(t) = 0 \quad \text{in} \quad \Omega, \quad \sigma \cdot n = T^{\mathrm{d}}(t) \quad \text{on} \quad S_T\} \\ P &= \{\sigma \mid f(\sigma) \leqslant 0 \quad \text{in} \quad \Omega\} \\ U(t) &= \{\mathbf{u} \mid \mathbf{u} = \mathbf{u}^{\mathrm{d}} \quad \text{on} \quad S_u\} \end{aligned}\right\} \quad (4.1)$$

These sets represent respectively the set of statically admissible stresses, the set of plastically admissible stresses and the set of kinematically admissible displacements. Because of strain singularity, the notation $\int_\Omega \sigma \cdot \dot{\varepsilon}^{\mathrm{p}}\, \mathrm{d}\Omega$ may be ambiguous; this difficulty may be avoided by a global expression of the relations (2.2) as suggested by Suquet (1978).

The solution $(\sigma, \mathbf{u})$ must verify for almost all t the following conditions:

$$\left.\begin{aligned} &\text{(i)} \quad \sigma \in P \cap S(t), \quad \mathbf{u} \in U(t) \\ &\text{(ii)} \quad \int_\Omega \dot{\sigma} \cdot E^{-1} \cdot (\sigma^* - \sigma)\, \mathrm{d}\Omega - \int_{S_u} \dot{u}^{\mathrm{d}} \cdot (\sigma^* - \sigma) \cdot n\, \mathrm{d}s \\ &\qquad - \int_{S_T} \dot{u} \cdot (\sigma^* - \sigma) \cdot n\, \mathrm{d}s - \int_\Omega \dot{u} \cdot (\operatorname{div}\sigma^* - \operatorname{div}\sigma)\, \mathrm{d}\Omega \geqslant 0 \\ &\qquad\qquad \forall \sigma^* \in P \\ &\text{(iii)} \quad \sigma(0) = \sigma^0 \end{aligned}\right\} \quad (4.2)$$

The mathematical discussion of these has been given by Suquet (1978). The existence theorem is obtained if $U(t)$ includes surfaces of discontinuity, the functional framework is DB (Ω), the space of velocities $v \in L^2(\Omega)$ such that associated strain rates $\dot{\varepsilon} = \frac{1}{2}(\nabla v + \tilde{\nabla} v)$ are bounded measures on $\overline{\Omega}$.

4.2. Equations of Evolution for Stress

In fact, the velocity $\dot{\mathbf{u}}$ may be completely eliminated in the expressions (4.2). For clarity, the purely elastic response of the structure ($\sigma^E, \mathbf{u}^E$) is usually introduced; these fields represent the response to the same loading ($\mathbf{g}(t)$, $\mathbf{T}^d(t)$, $\mathbf{u}^d(t)$) of the same solid when its material behaviour is purely elastic, with the same elastic coefficients E. They are solutions of the following elastic problem:

$$\left.\begin{aligned} \mathbf{u}^E(t) &\in U(t) \\ \sigma^E(t) &\in S(t) \\ \sigma^E(t) &= \mathbf{E} \cdot \varepsilon^E(t) \end{aligned}\right\} \qquad (4.3)$$

In the relations (4.2), if σ^* is chosen such that $\sigma^* \in P \cap S(t)$, one obtains

$$\int_\Omega \dot{\varepsilon}^E \cdot (\sigma^* - \sigma)\, d\Omega = \int_\Omega \dot{u}^d \cdot (\sigma^* - \sigma) \cdot n\, ds - \int_{S_T} \dot{u} \cdot (\sigma^* - \sigma) \cdot n\, ds$$

$$- \int_{S_u} \dot{u}^d \cdot (\sigma^* - \sigma) \cdot n\, ds$$

and thus the following inequality is derived:

$$\left.\begin{aligned} &\int_\Omega (\dot{\sigma} - \dot{\sigma}^E) \cdot E^{-1} \cdot (\sigma^* - \sigma)\, d\Omega \geqslant 0 \\ &\forall t, \quad \forall \sigma^* \in P \cap S(t) \end{aligned}\right\} \qquad (4.4)$$

This variational inequality governs the evolution of the stress field $\sigma(t)$. It is a special case of multi-valued differential equations studied by Moreau (1973) and Brézis (1973) from the mathematical point of view.

Uniqueness of the stress response results immediately from the inequality (4.4). Indeed, if σ^1 and σ^2 denote two possible solutions, we can take in (4.4) respectively $\boldsymbol{\sigma} = \boldsymbol{\sigma}^1$, $\boldsymbol{\sigma}^* = \boldsymbol{\sigma}^2$ and $\boldsymbol{\sigma} = \boldsymbol{\sigma}^2$, $\boldsymbol{\sigma}^* = \boldsymbol{\sigma}^1$ to obtain

$$\int_\Omega (\dot{\sigma}^1 - \dot{\sigma}^2) \cdot E^{-1} \cdot (\sigma^1 - \sigma^2)\, \mathrm{d}\Omega < 0$$

and thus a contraction property holds for all t:

$$\| \boldsymbol{\sigma}^1(t) - \boldsymbol{\sigma}^2(t) \|^2 < \| \boldsymbol{\sigma}^1(0) - \boldsymbol{\sigma}^2(0) \|^2 \tag{4.5}$$

where $\| \boldsymbol{\sigma} \|^2 = \int_\Omega \boldsymbol{\sigma} \cdot E^{-1} \cdot \boldsymbol{\sigma}\, \mathrm{d}\Omega$ is the energy norm in the stress space. This inequality shows that $\boldsymbol{\sigma}^1(t) = \boldsymbol{\sigma}^2(t)$ since $\boldsymbol{\sigma}^1(0) = \boldsymbol{\sigma}^2(0) = \boldsymbol{\sigma}^0$.

4.3. Equations of Evolution of the Plastic Strain

Equations of evolution of the plastic strain can be given, at least formally, in the following way. Let $(\boldsymbol{\sigma}^{\mathrm{r}}, \mathbf{u}^{\mathrm{r}})$ be the response of the associated elastic structure under a given residual strain $\boldsymbol{\varepsilon}^{\mathrm{p}}$ and homogeneous boundary conditions over S_u, S_T and Ω:

$$\left.\begin{array}{lll} \varepsilon^{\mathrm{r}}_{ij} = \frac{1}{2}(u^{\mathrm{r}}_{i,j} + u^{\mathrm{r}}_{j,i}), & & \mathbf{u}^{\mathrm{r}} = 0 \ \text{ on } \ S_u \\ \operatorname{div} \boldsymbol{\sigma}^{\mathrm{r}} = 0 \ \text{ in } \ \Omega, & & \boldsymbol{\sigma}^{\mathrm{r}} \cdot \mathbf{n} = 0 \ \text{ on } \ S_T \\ \boldsymbol{\sigma}^{\mathrm{r}} = E(\boldsymbol{\varepsilon}^{\mathrm{r}} - \boldsymbol{\varepsilon}^{\mathrm{p}}) & & \end{array}\right\} \tag{4.6}$$

then, formally, one obtains

$$\boldsymbol{\sigma}^{\mathrm{r}} = -\mathbf{Z} \cdot \boldsymbol{\varepsilon}^{\mathrm{p}} \tag{4.7}$$

where $\mathbf{Z}$ is a linear operator, namely the projection in the energy sense of $\boldsymbol{\varepsilon}^{\mathrm{p}}$ on the set of self-consistent stresses. If one remarks that

$$\boldsymbol{\sigma}(t) = \boldsymbol{\sigma}^{\mathrm{E}}(t) + \boldsymbol{\sigma}^{\mathrm{r}}(\boldsymbol{\varepsilon}^{\mathrm{p}}(t)) = \boldsymbol{\sigma}^{\mathrm{E}}(t) - Z \cdot \boldsymbol{\varepsilon}^{\mathrm{p}}(t) \tag{4.8}$$

then, from the normality law, $\dot{\boldsymbol{\varepsilon}}^{\mathrm{p}} = n_{\mathrm{c}}(\boldsymbol{\sigma})$,

$$\dot{\boldsymbol{\varepsilon}}^{\mathrm{p}} \in N_{\mathrm{p}}(\boldsymbol{\sigma}^{\mathrm{E}}(t) - \mathbf{Z} \cdot \boldsymbol{\varepsilon}^{\mathrm{p}}(t)) \tag{4.9}$$

The evolution of the plastic strain field $\boldsymbol{\varepsilon}^{\mathrm{p}}$ is thus characterized by (4.9) and initial condition $\boldsymbol{\varepsilon}^{\mathrm{p}}(0) = \boldsymbol{\varepsilon}^{\mathrm{p}}_0$. In the same spirit, the reader may refer to Maier (1972) for a detailed expression of (4.9) using matrix notations in the case of discrete elastoplastic systems.

5. ILLUSTRATION

The problem of circular bending of a cylindrical beam is here presented as an illustration of the quasi-static evolution of elastoplastic structures (Koiter and Mandel's solutions) (Mandel, 1972).

A homogeneous cylindrical beam is submitted at the two extreme sections to a constant bending moment (Fig. 2). The material is assumed to be elastic-perfectly plastic and to obey the Tresca criterion. A closed form solution is to be constructed such that classical assumptions of beam theory are verified:

— plane sections remain plane and normal to the mean fibre;
— uniaxial stress.

Boundary conditions at the end sections are then

$$\left.\begin{aligned} u_x &= cxy \\ \sigma_{xy} &= \sigma_{xz} = 0 \end{aligned}\right\} \tag{5.1}$$

$c = c(t)$, $\dot{c} > 0$ is a given function and defines the loading path. It is assumed that $c(0) = 0$ and the initial state is the natural one.

The only non-zero component of stress is $\sigma_{xx} = \sigma$. Strain components are $\varepsilon_{xy} = \varepsilon_{yz} = \varepsilon_{zx} = 0$, ε_{xx}, ε_{yy}, ε_{zz} do not depend on x and $\varepsilon_{xx} = cy$ compatibility of strain reduces to one equation:

$$\varepsilon_{yy,zz} + \varepsilon_{zz,yy} = 0 \tag{5.2}$$

In the plastic zone defined by $|cy| \geqslant k/E$, the rate equations are

$$\begin{aligned} \dot{\varepsilon}_{xx} &= \frac{\dot{\sigma}}{E} + \lambda^1 + \lambda^2 \\ \dot{\varepsilon}_{yy} &= -\frac{\nu}{E}\dot{\sigma} - \lambda^2 \\ \dot{\varepsilon}_{zz} &= -\frac{\nu}{E}\dot{\sigma} - \lambda^1 \end{aligned} \tag{5.3}$$

(λ^1, λ^2 are plastic multipliers and have the sign of σ); they give, after integration,

$$\left.\begin{aligned} cy &= \frac{\nu}{E} + \left\langle |cy| - \frac{k}{E} \right\rangle \frac{y}{|y|} && \text{with} \quad \Lambda_i = \Lambda_i(y, z), \quad i = 1, 2 \\ \varepsilon_{yy} &= -\frac{\nu}{E}\sigma - \Lambda_2 && \Lambda_1 + \Lambda_2 = \left\langle |cy| - \frac{k}{E} \right\rangle \frac{y}{|y|} \\ \varepsilon_{zz} &= -\frac{\nu}{E}\sigma - \Lambda_1 && \end{aligned}\right\} \tag{5.4}$$

In the elastic zone defined by $|cy| < k/E$, eqn (5.2) is identically verified. In the plastic zone it implies that

$$\Lambda_{2,zz} + \Lambda_{1,yy} = 0$$

From eqns (5.4), an auxiliary function $L(y, z)$ can be introduced as

$$\Lambda_1(y, z) = L(y, z) + \nu \left\langle |cy| - \frac{k}{E} \right\rangle \frac{y}{|y|}$$

$$\Lambda_2(y, z) = L(y, z) + (1 - \nu) \left\langle |cy| - \frac{k}{E} \right\rangle \frac{y}{|y|}$$

It follows then that

$$L_{,zz} - L_{,yy} = 0 \quad \Rightarrow \quad L = \alpha'(y - z) + \beta'(y + z)$$

From these expressions of stress and strain ($y > 0$):

$$\left.\begin{aligned} \sigma &= k \\ \varepsilon_{yy} &= -k\frac{\nu}{E} + \alpha'(y - z) + \beta'(y + z) - (1 - \nu)\left(cy - \frac{k}{E}\right) \\ \varepsilon_{zz} &= -k\frac{\nu}{E} - \alpha'(y - z) - \beta'(y + z) - \gamma\left(cy - \frac{k}{E}\right) \end{aligned}\right\}$$

displacement components are ($y > 0$):

$$\left.\begin{aligned} u_x &= cxy \\ u_y &= -\frac{c}{2}\left[x^2 + \nu(y^2 - z^2) + (1 - 2\nu)\left(\frac{cy^2}{2} - \frac{ky}{2} + \frac{k^2}{cE}\right)\right. \\ &\quad + \alpha(y - z) + \beta(y + z) \\ u_z &= -\nu cyz + \alpha(y - z) - \beta(y + z) \end{aligned}\right\} \quad (5.5)$$

The continuity of displacement on the elastic–plastic boundary $y = k/cE$ implies

$$\alpha = 0 \quad \text{if} \quad y - z < \frac{k}{cE} + z\left(\frac{k}{cE}\right)$$

$$\beta = 0 \quad \text{if} \quad y + z < \frac{k}{cE} + z\left(\frac{k}{cE}\right)$$

This example shows clearly that, if $\boldsymbol{\sigma}$ is uniquely determined, $\mathbf{u}$ is not unique. In the region DEC (Fig. 2) the two functions α and β are arbitrary, the only restriction being that $\Lambda_1 \geqslant 0$, $\Lambda_2 \geqslant 0$, and thus

$$-\nu\left(cy - \frac{k}{E}\right) \leqslant \alpha'(y - z) + \beta'(y + z) \leqslant (1 - \nu)\left(cy - \frac{k}{E}\right) \qquad (5.6)$$

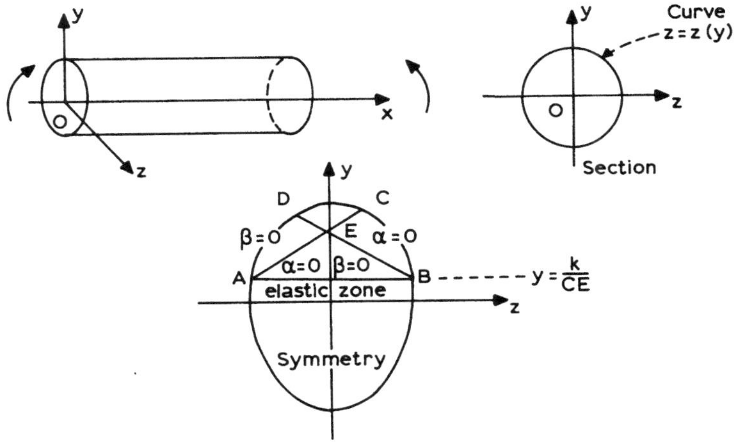

Fig. 2. Circular bending of an elastoplastic beam.

6. ASYMPTOTIC ANALYSIS: ELASTIC AND PLASTIC SHAKE-DOWN

In the previous example the present state of the beam was given in terms of the loading path, from a given initial condition. However, in many practical applications it may happen that both the loading and the initial state are not known with sufficient precision. The knowledge of the whole evolution in that case has only limited interest. If the asymptotic behaviour of the load is given, it may be interesting to know the asymptotic behaviour of the solution, whatever the initial condition of the structure.

Shake-down theorems give a satisfactory answer to this question and represent the basis of the analysis of the cyclic behaviour of structures under cyclic loads. Failure of elastoplastic solids by cyclic or progressive plastic strains, for example, may be estimated nowadays after numerous discussions such as the early work of Melan (1936) and the decisive contribution of Koiter (1960).

6.1. Elastic Shake-down

There is elastic shake-down under a given loading path if, for $t \to \infty$, the response of the structure becomes purely elastic. This definition implies that

$$\left.\begin{array}{ll} \text{(i)} & \lim\limits_{t\to\infty} \mathbf{u}^{\mathrm{r}}(t) \longrightarrow \mathbf{u}^{\mathrm{r}}_{\infty} \\ \text{(ii)} & \lim\limits_{t\to\infty} \boldsymbol{\sigma}^{\mathrm{r}}(t) \longrightarrow \boldsymbol{\sigma}^{\mathrm{r}}_{\infty} \\ \text{(iii)} & \lim\limits_{t\to\infty} \boldsymbol{\varepsilon}^{\mathrm{p}}(t) \longrightarrow \boldsymbol{\varepsilon}^{\mathrm{p}}_{\infty} \end{array}\right\} \tag{6.1}$$

i.e. the *existence* of limit state $(\mathbf{u}^{\mathrm{r}}_{\infty}, \boldsymbol{\sigma}^{\mathrm{r}}_{\infty}, \boldsymbol{\varepsilon}^{\mathrm{r}}_{\infty})$. (This limit may depend naturally on the initial conditions!) Such a condition is interesting since it rules out the possibility of failure due to cyclic plastic strain (oligocyclic fatigue, etc).

It is necessary to specify precisely for continuous systems the nature of the convergence in the limits (6.1). In perfect plasticity, as elastic energy $\frac{1}{2}\boldsymbol{\sigma} \cdot \boldsymbol{E}^{-1} \cdot \boldsymbol{\sigma}$ is well defined, the limit (6.1.ii) can be understood in the sense of energy norm. To avoid mathematical difficulty, (6.1.ii) will be accepted here as the definition of elastic shake-down. One obtains then the following theorem.

Melan's Theorem (1936)

Elastic shake-down happens, whatever the initial state, if there exists a residual state of stress $\boldsymbol{\sigma}^{\mathrm{r}}_{*}$ and a safety coefficient $m > 1$ such that $m(\boldsymbol{\sigma}^{\mathrm{r}}_{*} + \boldsymbol{\sigma}^{\mathrm{E}}(t))$ verifies the plasticity condition $\forall t > t_0$:

$$m(\boldsymbol{\sigma}^{\mathrm{r}}_{*} + \boldsymbol{\sigma}^{\mathrm{E}}(t)) \in P \tag{6.2}$$

The proof is obtained in two steps. It is shown first that the dissipative energy is bounded. Indeed, plastic dissipation is given by

$$D_{\mathrm{p}} = \int_S \dot{u} \cdot \sigma \cdot n \,\mathrm{d}s + \int_\Omega g\dot{u} - \int_\Omega \dot{\sigma} \cdot E^{-1} \cdot \sigma \,\mathrm{d}\Omega \tag{6.3}$$

This quantity is naturally positive as can be shown from the condition

(4.2.ii) with $\boldsymbol{\sigma}^* = 0 \in P$. If $\boldsymbol{\sigma}^*$ is taken as $m(\boldsymbol{\sigma}^{\mathrm{r}}_* + \boldsymbol{\sigma}^{\mathrm{E}}(t)) \in P$ in the same inequality, one obtains

$$\int_\Omega \dot{\sigma} E^{-1}[m(\sigma^{\mathrm{r}}_* + \sigma^{\mathrm{E}}(t)) - \sigma]\,\mathrm{d}\Omega - \int_{S_u} \dot{u}^{\mathrm{d}}[m(\sigma^{\mathrm{r}}_* + \sigma^{\mathrm{E}}) - \sigma]\cdot n\,\mathrm{d}s$$

$$- \int_{S_T} (m-1)\, T^{\mathrm{d}} \cdot \dot{u}\,\mathrm{d}s - \int_\Omega (m-1) g\dot{u}\,\mathrm{d}\Omega \geqslant 0$$

or

$$\frac{m-1}{m}\left[\int_{S_u} \dot{u}^{\mathrm{d}} \cdot \sigma \cdot m\,\mathrm{d}s + \int_{S_T} \dot{T}^{\mathrm{d}} \cdot \dot{u}\,\mathrm{d}s + \int_\Omega g\dot{u}\,\mathrm{d}\Omega - \int_\Omega \dot{\sigma} \cdot E^{-1} \cdot \sigma\,\mathrm{d}\Omega\right]$$

$$\leqslant \int_\Omega \dot{\sigma} \cdot E^{-1} \cdot (\sigma^{\mathrm{r}}_* + \sigma^{\mathrm{E}} - \sigma)\,\mathrm{d}\Omega - \int_{S_u} \dot{u}^{\mathrm{d}} \cdot (\sigma^{\mathrm{r}}_* + \sigma^{\mathrm{E}} - \sigma)\cdot m\,\mathrm{d}s$$

Since

$$\int_\Omega \dot{\sigma}^{\mathrm{E}} \cdot E^{-1} \cdot (\sigma^{\mathrm{r}}_* + \sigma^{\mathrm{E}} - \sigma)\,\mathrm{d}\Omega = \int_{S_u} \dot{u}^{\mathrm{d}} \cdot (\sigma^{\mathrm{r}}_* + \sigma^{\mathrm{E}} - \sigma)\cdot m\,\mathrm{d}s$$

by the definition $(\sigma^{\mathrm{E}}, u^{\mathrm{E}})$, it follows that

$$\frac{m-1}{m} D_{\mathrm{p}} \leqslant - \int_\Omega (\dot{\sigma}^{\mathrm{E}} - \dot{\sigma}) \cdot E^{-1} \cdot (\sigma^{\mathrm{r}}_* + \sigma^{\mathrm{E}} - \sigma)\,\mathrm{d}\Omega$$

or

$$\frac{m-1}{m} D_{\mathrm{p}} \leqslant - \frac{d}{\mathrm{d}t} \int_\Omega (\sigma^{\mathrm{r}}_* + \sigma^{\mathrm{E}} - \sigma) \cdot E^{-1} \cdot (\sigma^{\mathrm{r}}_* + \sigma^{\mathrm{E}} - \sigma)\,\mathrm{d}\Omega$$

and thus the dissipative energy $W_{\mathrm{p}} = \int_{t_0}^{t} D_{\mathrm{p}}\,\mathrm{d}t$ is bounded:

$$\frac{m-1}{m} W_{\mathrm{p}} \leqslant \| \boldsymbol{\sigma}^{\mathrm{r}}_* + \boldsymbol{\sigma}^{\mathrm{E}}(t_0) - \boldsymbol{\sigma}(t_0) \|^2 \qquad \forall t > t_0 \tag{6.4}$$

It is shown in the second step that the residual stress $\boldsymbol{\sigma}^{\mathrm{r}}(t)$ converges when $t \longrightarrow +\infty$. Let $t_0 < t_1 < t_2$ and $\boldsymbol{\sigma}^{\mathrm{r}}_1 = \boldsymbol{\sigma}^{\mathrm{r}}(t_1)$, $\boldsymbol{\sigma}^{\mathrm{r}}_2 = \boldsymbol{\sigma}^{\mathrm{r}}(t_2)$; the result holds if the following estimate is true (Nayroles, 1977):

$$\int_{t_1}^{t_2} D_p \, d\Omega \geqslant c \, \| \sigma_1^r - \sigma_2^r \|^2 \tag{6.5}$$

since the sequence $\sigma^r(t_i)$, $t_i \to \infty$ is a Cauchy one.

To establish the relation (6.5), let us consider first the expression

$$\| \sigma_1^r - \sigma_2^r \|^2 = \int_{t_1}^{t_2} \langle \dot{\sigma}^r | \sigma^r - \sigma_1^r \rangle \, dt = - \int_{t_1}^{t_2} \int_{\Omega} \dot{\varepsilon}^p (\sigma^r - \sigma_1^r) \, d\Omega \, dt$$

or

$$-\dot{\varepsilon}^p (\sigma^r - \sigma_1^r) = -\dot{\varepsilon}^p [\sigma - (\sigma^E + \sigma_*^r)] + \dot{\varepsilon}^p \sigma_1 - \dot{\varepsilon}^p (\sigma_1^E + \sigma_*^r)$$

The first term on the right-hand side is non-positive because of the normality law, the second term is bounded by $\sigma \dot{\varepsilon}^p$, the third is also bounded by $\sigma \dot{\varepsilon}^p$ *if the convex C is symmetric with respect to the origin O.* This assumption is verified for usual criteria (Mises, Tresca, etc.) and enables us to obtain the inequality (6.5) with $c = \frac{1}{2}$.

Melan's theorem is concerned principally with the static aspect. The complementary theorem dealing with the kinematic aspect is due to Koiter (1960). The reader may also refer to Débordes and Nayroles (1976) for a mathematical discussion of the subject.

6.2. Plastic Shake-down

When the load is periodic, there is plastic shake-down if, for $t \to \infty$, the response of the structure becomes periodic. This definition implies that

$$\left. \begin{aligned} \lim_{t \to \infty} \mathbf{u}^r(t) &= \mathbf{u}_P^r(t) \\ \lim_{t \to \infty} \sigma^r(t) &= \sigma_P^r(t) \\ \lim_{t \to \infty} \varepsilon^p(t) &= \varepsilon_P^p(t) \end{aligned} \right\} \tag{6.6}$$

Where $\mathbf{u}_P^r(t)$, $\sigma_P^r(t)$, $\varepsilon_P^p(t)$ are periodic functions. Such a condition is interesting since it gives an estimate of the final response, especially the amount of energy dissipated by plastic work per cycle.

Since the correspondence between stress and strain histories is rather complex in perfect plasticity, again it is necessary to distinguish stress response from strain response.

The stress response can be obtained from the following statement:

Moreau–Halphen Theorem (1978)

It is assumed that plastically admissible deviation stresses are bounded, i.e. if $f(\boldsymbol{\sigma}) \leqslant 0 \Rightarrow \|\boldsymbol{\sigma}'\| \leqslant M$ then:

(i) There exists at least one periodic *stress* response for allowable loads.

(ii) The convergence of stress field $\boldsymbol{\sigma}$ from any initial condition $\boldsymbol{\sigma}^0$ to one periodic response is practically ensured. (6.7)

In the first part of this proposition, allowable load means admissible load in the sense of limit analysis. In the second part the word practically expresses only the technical difficulty of establishing the convergence in the case of continuous systems. The mathematical discussion of this convergence is not always clear unless the system is discrete. The reader may refer to the work of Halphen (1978) for a more detailed discussion.

Concerning the strain response, it is important to note that a periodic stress response does not ensure necessarily a periodic strain one. Progressive plastic strain ratcheting may of course occur and thus theorem (6.7) does not give a definite answer to the plastic shake-down problem (6.6).

If the stress response converges to a periodic solution $\boldsymbol{\sigma}_{\mathrm{p}}(t) = \boldsymbol{\sigma}_{\mathrm{p}}^{\mathrm{r}}(t) + \boldsymbol{\sigma}^{\mathrm{E}}(t)$ and *if the strain rate $\dot{\boldsymbol{\varepsilon}}$ can be expressed in terms of the stress rate $\dot{\boldsymbol{\sigma}}$*, one obtains also periodic response of strain $\boldsymbol{\varepsilon}$ and displacement $\mathbf{u}$. Only under this additional condition does theorem (6.7) ensure plastic shake-down behaviour of the structure. The additional condition $\dot{\boldsymbol{\varepsilon}} = \dot{\boldsymbol{\varepsilon}}(\dot{\boldsymbol{\sigma}})$ means that rate equations (2.9) can be inverted, which is not true in perfect plasticity.

However, if the Bauschinger effect is taken into account, for *positive hardening material of modulus $h > 0$,* this inversion is straightforward (cf. for example Mandel, 1961). In this case a complete answer to the plastic shake-down problem can be obtained from theorem (6.7), in particular if the formalism of perfect plasticity can be generalized as will be shown in the following section.

7. GENERALIZED STANDARD MODELS

The perfect plastic model is basically characterized by normality and convexity assumptions. Within this framework the model can be generalized in a very simple way to include a large class of material hardening. Indeed, eqns (2.1) can be written as

$$\left.\begin{aligned} \sigma &= \frac{\partial W}{\partial \varepsilon}(\varepsilon, \varepsilon^{\mathrm{p}}), \qquad W = \tfrac{1}{2}(\varepsilon - \varepsilon^{\mathrm{p}}) \cdot E \cdot (\varepsilon - \varepsilon^{\mathrm{p}}) \\ \sigma' &= -\frac{\partial W}{\partial \varepsilon^{\mathrm{p}}}(\varepsilon, \varepsilon^{\mathrm{p}}) \\ \sigma' &\in C \\ \dot{\varepsilon}^{\mathrm{p}} &= n_{\mathrm{c}}(\sigma) \end{aligned}\right\} \tag{7.1}$$

In this way there exists an *energy function* $W(\varepsilon, \varepsilon^{\mathrm{p}})$ (which is the stored elastic energy in perfect plasticity) such that the associated force to ε^{p} must verify the plastic condition. The same idea can be generalized immediately by the introduction of an internal parameter $\alpha = (\varepsilon^{\mathrm{p}}, \ldots)$ and associated energy function $W(\varepsilon, \alpha)$ such that

$$\left.\begin{aligned} W &= W(\varepsilon, \alpha) && \text{(reversible stored energy)} \\ \sigma &= \frac{\partial W}{\partial \varepsilon} && \text{(generalized forces associated to } \varepsilon \text{ and } \alpha) \\ A &= -\frac{\partial W}{\partial \alpha} && \\ A &\in C && \text{(normality law and convex of plasticity)} \\ \dot{\alpha} &= n_{\mathrm{c}}(A) && \end{aligned}\right\} \tag{7.2}$$

For example, the Prager–Ziegler kinematic hardening model is defined by

$$W = \tfrac{1}{2}(\varepsilon - \varepsilon^{\mathrm{p}}) \cdot E \cdot (\varepsilon - \varepsilon^{\mathrm{p}}) + \tfrac{1}{2}\varepsilon^{\mathrm{p}} \cdot H \cdot \varepsilon^{\mathrm{p}}$$

and the plasticity criterion

$$C = \{A \,|\, f(A) \leqslant 0\}$$

By a proper choice of the function W and the state variable α, practically any experimental curve can be identified and thus this description covers most of the usual materials.

7.1. Material Behaviour

Rate equations related stress and strain rates follow simply from the relations (7.2), in the same spirit as before. If the plastic criterion can be expressed as $f(A) \leqslant 0$, then one obtains in the plastic range:

$$\dot{\sigma} = E\dot{\varepsilon} + F\dot{\alpha}, \quad \text{with } E = W_{,\varepsilon\varepsilon}, \quad F = W_{,\varepsilon\alpha}, \quad G = W_{,\alpha\alpha}$$

$$-\dot{A} = {}^{\mathrm{T}}F\dot{\varepsilon} + G\dot{\alpha}$$

$$\frac{\partial f}{\partial A}\dot{A} = 0 \quad \text{and} \quad \dot{\alpha} = \lambda\frac{\partial f}{\partial A}, \quad \lambda > 0$$

and thus

$$\dot{\varepsilon}^{\mathrm{p}} = -E^{-1} \cdot F \cdot \dot{\alpha} \tag{7.3}$$

represents plastic strain rate. Again the following equations are derived:

$$\left.\begin{aligned} \lambda &= \frac{1}{{}^{\mathrm{T}}\dfrac{\partial f}{\partial A} \cdot G \cdot \dfrac{\partial f}{\partial A}} \left\langle -{}^{\mathrm{T}}\frac{\partial f}{\partial A} \cdot {}^{\mathrm{T}}F \cdot \dot{\varepsilon} \right\rangle \\ \dot{\sigma} &= \frac{\partial \phi}{\partial \dot{\varepsilon}} \quad \text{with} \quad \phi = \frac{1}{2}\dot{\varepsilon} \cdot E \cdot \dot{\varepsilon} - \frac{1}{2}\frac{\left\langle -{}^{\mathrm{T}}\dfrac{\partial f}{\partial A} \cdot F \cdot \dot{\varepsilon} \right\rangle^2}{{}^{\mathrm{T}}\dfrac{\partial f}{\partial A} \cdot G \cdot \dfrac{\partial f}{\partial A}} \end{aligned}\right\} \tag{7.4}$$

If the matrix

$$\begin{bmatrix} W''_{,\varepsilon\varepsilon} & W''_{,\varepsilon\alpha}\dfrac{\partial f}{\partial A} \\ {}^{\mathrm{T}}\dfrac{\partial f}{\partial A}W''_{,\alpha\varepsilon} & {}^{\mathrm{T}}\dfrac{\partial f}{\partial A}W''_{,\alpha\alpha}\dfrac{\partial f}{\partial A} \end{bmatrix} \tag{7.5}$$

is positive-definite, $\dot{\varepsilon}$ and λ can be expressed in terms of $\dot{\sigma}$, and Drucker's inequality $\dot{\sigma}\dot{\varepsilon}^{\mathrm{p}} > 0$ holds. The positivity of the matrix (7.5) characterizes positive hardening materials in the classical sense. In particular, the matrix of the second derivative $[W'']$ must be positive-definite.

The dependence of W on α is not arbitrary since Illyushin's inequality must be verified. Let us prove that the convexity of W with respect to α ensures Illyushin's condition, i.e.

$$\int_0^1 \sigma\dot{\varepsilon}\,\mathrm{d}t \geqslant 0 \quad \text{for any arbitrary closed strain path } \varepsilon(t), \; \varepsilon(0) = \varepsilon(1) = \varepsilon_0 \tag{7.6}$$

Indeed, one obtains

$$\int_0^1 \sigma\dot{\varepsilon}\,\mathrm{d}t = \int_0^1 \dot{W}\,\mathrm{d}t - \int_0^1 W_{,\alpha}\,\dot{\alpha}\,\mathrm{d}t$$
$$= W(\varepsilon_0, \alpha_1) - W(\varepsilon_0, \alpha_0) + \int_0^1 A\dot{\alpha}\,\mathrm{d}t$$

Since

$$(A - A_0)\,\dot{\alpha} \geqslant 0 \quad \forall t, \quad \int_0^1 A\dot{\alpha}\,\mathrm{d}t \geqslant \int_0^1 A_0\dot{\alpha}\,\mathrm{d}t$$

and thus

$$\int_0^1 \sigma\dot{\varepsilon}\,\mathrm{d}t \geqslant W(\varepsilon_0, \alpha_1) - W(\varepsilon_0, \alpha_0) - \frac{\partial W}{\partial \alpha}(\varepsilon_0, \alpha_0)\cdot(\alpha_1 - \alpha_0) \geqslant 0$$

The last inequality results from the convexity of W with respect to α.

7.2. System Behaviour

If W is convex with respect to (ε, α) and quadratic, the analysis of quasi-static evolution of a given structure follows directly from perfect plasticity by the introduction of generalized stress and strain (Nguyen, 1973). Global results are obtained in the same way, the only difference being that there is uniqueness for stress and strain, because of the possibility of computing strain rate in terms of stress rate for positive hardening materials; this possibility follows from the matrix (7.5).

As an illustration, let us consider the case of linear kinematic hardening material following the Prager–Ziegler model with Von Mises criterion, i.e.

$$f(A) = \| A \| - k \leqslant 0, \quad A = -\frac{\partial W}{\partial \varepsilon^{\mathrm{p}}} = \sigma' - H\varepsilon^{\mathrm{p}}$$

and

$$\| A \|^2 = A_{ij}A_{ij}, \quad \mathrm{tr}(\varepsilon^{\mathrm{p}}) = \mathrm{tr}(H\cdot\varepsilon^{\mathrm{p}}) = 0$$

Once the expressions of generalized stresses and strains

$$\left.\begin{aligned} &\bar{\sigma} = \begin{bmatrix} \sigma \\ -H\varepsilon^{\mathrm{p}} \end{bmatrix}, \quad \bar{\varepsilon} = \begin{bmatrix} \varepsilon \\ 0 \end{bmatrix}, \quad \bar{\varepsilon}^{\mathrm{e}} = \begin{bmatrix} \varepsilon^{\mathrm{e}} \\ -\varepsilon^{\mathrm{p}} \end{bmatrix}, \quad \bar{\varepsilon}^{\mathrm{p}} = \begin{bmatrix} \varepsilon^{\mathrm{p}} \\ \varepsilon^{\mathrm{p}} \end{bmatrix} \\ &\bar{E} = \begin{bmatrix} E & 0 \\ 0 & H \end{bmatrix} \end{aligned}\right\} \quad (7.7)$$

are introduced, equations of perfect plasticity (2.1) follow immediately and thus the evolution of a structure is again described by evolution equations such as (4.2) or (4.4) or (4.9). Since the generalized elastic coefficient tensor $\bar{E}$ is positive-definite in the case of positive hardening, all global results concerning the quasi-static evolution hold; in particular, uniqueness, existence and shake-down theorems are always available. Since the plastic field $\boldsymbol{\varepsilon}^{\mathrm{p}}$ has now the same regularity as the stress field $\boldsymbol{\sigma}$, in the limits (6.6), all three of the conditions of elastic shake-down are equivalent. The plastic response $\boldsymbol{\varepsilon}^{\mathrm{p}}$ is unique. Its evolution in the same spirit as (4.9) is given by

$$\dot{\boldsymbol{\varepsilon}}^{\mathrm{p}} \in N_{\mathrm{p}}[\mathbf{D} \cdot \boldsymbol{\sigma}^{\mathrm{E}}(t) - \mathbf{D} \cdot (\mathbf{Z} + \mathbf{H}) \cdot \mathbf{D} \cdot \boldsymbol{\varepsilon}^{\mathrm{p}}(t)] \tag{7.8}$$

where $\mathbf{D}$ denotes the deviatoric operator, $\mathbf{D}$ and $\mathbf{Z} + \mathbf{H}$ are symmetric and $\mathbf{Z} + \mathbf{H}$ is positive-definite.

8. EXTENSION TO VISCOPLASTIC MATERIALS

Time-dependent behaviour such as that of viscoplastic materials can also be discussed in the same way if a convex and simple plastic potential is assumed. Let us consider the following viscoplastic materials:

$$\left.\begin{aligned} W &= W(\varepsilon, \alpha) && \text{(reversible stored energy)} \\ \sigma &= \frac{\partial W}{\partial \varepsilon}, \quad A = -\frac{\partial W}{\partial \alpha} && \text{(generalized forces)} \\ \dot{\alpha} &= \frac{\partial \Omega}{\partial A}(A) && \text{(viscoplastic potential)} \end{aligned}\right\} \tag{8.1}$$

It is assumed here that the viscoplastic potential Ω depends only on the generalized force A and $\Omega(A)$ is convex, $\Omega(0) = 0$.

For example, the Prager–Ziegler viscoplastic model with kinematic hardening is defined by

$$W(\varepsilon, \varepsilon^{\mathrm{p}}) = \tfrac{1}{2}(\varepsilon - \varepsilon^{\mathrm{p}}) \cdot E \cdot (\varepsilon - \varepsilon^{\mathrm{p}}) + \tfrac{1}{2}\varepsilon^{\mathrm{p}} \cdot H \cdot \varepsilon^{\mathrm{p}}$$

and

$$\Omega(A) = \frac{1}{\eta}\langle A_{ij}A_{ij} - k^2 \rangle$$

As has been pointed out by Prager (1937), for a vanishing viscoplastic coefficient η, the behaviour of this material tends to the aforementioned elastic plastic model (7.2). If $H = 0$, one obtains Perzyna's model of perfect viscoplasticity.

8.1. Material Behaviour

Rate equations follow from the relations (8.1) by derivation with respect to time:

$$\left.\begin{aligned}\dot{\sigma} &= \frac{\partial^2 W}{\partial\varepsilon\partial\varepsilon}\cdot\dot{\varepsilon} + \frac{\partial^2 W}{\partial\varepsilon\partial\alpha}\cdot\dot{\alpha} \\ \dot{\alpha} &= \frac{\partial\Omega}{\partial A}(A)\end{aligned}\right\} \tag{8.2}$$

If the present state of the material is assumed to be known in eqns (8.2), the rate $\dot{\alpha}$ is known and the correspondence $\dot{\varepsilon} \to \dot{\alpha}$ is the classical relation of linear elasticity with initial stress or residual strain.

As before, let us prove that the convexity of $W(\varepsilon, \alpha)$ with respect to α and the convexity of $\Omega(A)$ ensures Illyushin's inequality (7.6) *when the initial state is in equilibrium.*

Indeed, one obtains for an arbitrary closed strain path defined on the time interval [0, 1] and starting from equilibrium:

$$\int_0^1 \sigma\dot{\varepsilon}\, dt = W(\varepsilon_0, \alpha_1) - W(\varepsilon_0, \alpha_0) + \int_0^1 A\dot{\alpha}\, dt$$

If $\Omega(A)$ is convex, the inequality

$$\left(\frac{\partial\Omega}{\partial A} - \frac{\partial\Omega}{\partial A_0}\right)(A - A_0) \geqslant 0$$

gives

$$\left(\dot{\alpha} - \frac{\partial\Omega}{\partial A_0}\right)(A - A_0) \geqslant 0 \quad \text{or} \quad A\dot{\alpha} \geqslant A_0\dot{\alpha} + \frac{\partial\Omega}{\partial A_0}(A - A_0)$$

Since the initial state is in equilibrium, $\partial\Omega/\partial A_0 = 0$; thus

$$\int_0^1 \sigma\dot{\varepsilon}\, dt \geqslant W(\varepsilon_0, \alpha_1) - W(\varepsilon_0, \alpha_0) - \frac{\partial W}{\partial\alpha}(\varepsilon_0, \alpha_0)\cdot(\alpha_1, \alpha_0) \geqslant 0$$

as in the preceding paragraph.

8.2. System Behaviour

The rate response of a viscoplastic structure follows simply from eqns (8.2) which are linear equations of elasticity. It is obtained by the resolution of a linear elastic boundary value problem with initial strain and thus there is no such difficulty as in the case of elastic plastic material.

The global response can be formulated in different ways. In particular, the plastic response, in the case of the Prager–Ziegler model, is given by ordinary differential equations:

$$\left.\begin{aligned} \dot{\boldsymbol{\varepsilon}}^{\mathrm{p}} &= \frac{\partial \Omega}{\partial \mathbf{A}}(\mathbf{D} \cdot \boldsymbol{\sigma}^{\mathrm{E}}(t) - \mathbf{D} \cdot (\mathbf{Z} + \mathbf{H}) \cdot \mathbf{D} \cdot \boldsymbol{\varepsilon}^{\mathrm{p}}(t)) \\ \boldsymbol{\varepsilon}^{\mathrm{p}}(0) &= \boldsymbol{\varepsilon}_0^{\mathrm{p}} \end{aligned}\right\} \qquad (8.3)$$

Equations (8.3) do not present any mathematical difficulty and have been discussed by many authors (see, for example, Suquet, 1978). Existence and uniqueness theorems can be derived in the classical framework of functional analysis.

The asymptotic behaviour of a structure under cyclic loading can be again considered as in plasticity. For example, if the given load is periodic, $\boldsymbol{\sigma}^{\mathrm{E}}(t)$ is a periodic function of period T. As in elastoplasticity, the following results have been obtained by Halphen (1978):

— First, it is established that at least one periodic solution exists concerning the stress response.
— Secondly, from any initial condition $\boldsymbol{\varepsilon}_0^{\mathrm{p}}$, the stress response of the structure will become periodic.

If it is clear that the strain rate $\dot{\boldsymbol{\varepsilon}}$ will become periodic, progressive plastic ratcheting may again occur with the perfect viscoplastic model ($H = 0$). But if $H > 0$, elastic or plastic shake-down (6.6) is always obtained as can be shown directly from eqns (8.3).

REFERENCES

Brézis, H. (1973). *Opérateurs Maximaux Monotones et Semi-groupes Nonlinéaires dans les Espaces de Hilbert,* North-Holland, Amsterdam.
Débordes, O. and Nayroles, B. (1976). Sur la théorie et le calcul à l'adaptation des structures élastoplastiques. *J. Mécan.,* **15**, 1, 1–54.
Drucker, D. C. (1964). On the postulate of stability of material in the mechanics of continua, *J. Mécan.,* **3**, 235–50.
Geiringer, H. (1938). Fondements mathématiques de la théorie des corps plastiques isotropes, *Mém. Sci. Math.,* **86**.
Greenberg, G. (1949). Complementary minimum principles for an elastic plastic material, *Quart. Appl. Math.,* **7**, 85.

Halphen, B. (1978). Problèmes quasi-statiques en viscoplasticité. Thèse, Paris.
Halphen, B. and Nguyen, Q. S. (1975). Sur les matériaux standards généralisés, *J. Mécan.*, **14**, 1, 21–39.
Hill, R. (1950). *Mathematical Theory of Plasticity*, Oxford Ed.
Hill, R. (1958). A general theory of uniqueness and stability in elastic plastic solids, *J. Mech. Phys. Solids*, **6**, 236–49.
Hutchinson, J. (1973). Plastic buckling, *Adv. Appl. Mech.*, **14**, 67–144.
Koiter, W. T. (1960). General theory for elastic plastic solids. *Progress in Solid Mechanics*, Vol. 4, North-Holland, Amsterdam, p. 165.
Lee, E. H. (1969). Elastic-plastic deformation at finite strain, *J. Appl. Mech.*, **36**, 1–6.
Maier, G. (1972). A shake-down matrix theory allowing for work-hardening and second order geometric effect. *Foundations of Plasticity*, Varsovie.
Mandel, J. (1942). Equilibre par trasches planes des solides à la limite d'écoulement. Thèse, Paris.
Mandel, J. (1961). *Cours de Mécanique des Milieux Continus*, Gauthier-Villars, Paris.
Mandel, J. (1971). Plasticité et viscoplasticité. CISM Course.
Mandel, J. (1972). Note sur l'application du critère de Tresca au problème de la flexion circulaire d'un cylindre élastoplastique, *Arch. Mech. Stos.*, **24**, 863–72.
Melan, E. (1936). Theorie statisch unbestimmter Systeme aus ideal-plastichen Baustoff, *Sitz. Ber. AK. Wiss. Wien*, **IIa**, 145–95.
Moreau, J. J. (1973). On unilateral constraints, friction and plasticity, CIME, Bressarone.
Nayroles, B. (1977). Tendances récentes et perpectives à moyen terme en élastoplasticité asymptotique des constructions, 3ème Congrès Français de mécanique, Grenoble.
Nguyen, Q. S. (1973). Contribution à la théorie macroscopique de l'élastoplasticité avec écrouissage, Thèse, Paris.
Prager, W. (1937). Mécanique des solides isotropes au-delà du domaine élastique. *Mém. Sci. Math.*
Prager, W. and Hodge, P. G. (1951). *Theory of Perfectly Plastic Solids*, Dover, New York.
Rice, J. (1971). Inelastic constitutive relations for solids: an internal variable theory and its applications to metal plasticity, *J. Mech. Phys. Solids*, **19**, 433–55.
Suquet, P. (1978). Existence et régularité des solutions des équations de la plasticité parfaite, Thèse, Paris.
Zarka, J. and Casier, J. (1979). Elastic-plastic response of a structure to cyclic loading: practical rules, *Plasticity Today*, **6**.

CHAPTER 4

General Relationships between Micro and Macro Scales for the Non-linear Behaviour of Heterogeneous Media

C. STOLZ
Laboratoire de Mécanique des Solides, Ecole Polytechnique, Palaiseau, France

ABSTRACT

This chapter is mainly concerned with the determination of general relationships between microscopic and macroscopic mechanical properties for elastoplastic material with or without damage. The overall properties are determined in terms of the unknown properties of each constituent phase of the heterogeneous body. At first we must define a representative volume element (RVE) of the heterogeneous material for which the macroscopic mechanical fields are some spatial average of the microscopic one. The determination of local quantities is achieved by solving some particular boundary value problem on the RVE, from which macroscopic quantities are derived. The essential structure of micro–macro relationships is presented in the case of elastoplastic material. A generalisation is given to take the temperature field into account. We present also a model of damage and its averaging process. Finally we discuss the determination of macroscopic temperature through continuum thermodynamics.

INTRODUCTION

There are two main steps in the determination of the overall properties of a material in terms of the known properties of the constituent phases. First, the complex geometry of the n-phase body and the micromechanical behaviour of each of the phases being given, one has to solve a complicated boundary value problem. The determination of the local mechanical fields (stress σ, strain ε, plastic strain ε^p) is achieved by a localization process that requires some particular boundary conditions, from which macroscopic fields (stress Σ, strain $E, \ldots$) are derived. A comprehensive review of methods to solve some particular localization process is given by Willis (1981), Walpole (1969) and Hashin (1983).

The second step is concerned with the derivation of relationships between microscopic tensors and macroscopic tensors, which differ only by a length scale. Generally the macroscopic tensor is some spatial average of the microscopic one. Such a relationship is presented in continuum thermodynamics by Francfort *et al.* (1983) and Germain *et al.* (1983). This theoretical approach is summarized in the last section of this chapter.

This chapter is mainly concerned with the second basic problem, i.e. the determination of general relationships between microscopic and macroscopic fields for elastoplastic materials and for some model of damage. The essential structure of micro–macro relationships in elastoplastic materials is given in the previous works of Hill (1952, 1965) and Mandel (1964, 1971). After a general review and a second part concerned with the determination of the macroscopic elastic modulus of a heterogeneous material, we generalize the above-mentioned works and also our own previous results in plasticity to take the temperature field into account in Section 4.

In Section 5 we shall discuss the micromechanics of damage and its global average. Damage phenomena are due to microcracking in the body, for instance debonding between dissimilar materials, or at grain boundaries in polycrystals, and to the growth of holes or microcavities. Even if we assume small strain, the evolution of damage implies change of geometry, which is a non-linear and irreversible process.

In the last section we present a discussion about the determination of macroscopic temperature.

2. PRELIMINARY CONCEPTS

We consider a volume V, in which we distinguish two different scales, the micro one where the material properties vary from point to point like a highly heterogeneous medium, and the macro one where the properties are those of a homogeneous continuum.

To describe a heterogeneous medium it is essential to define a volume element which is small enough to allow us to distinguish the microscopic heterogeneities and sufficiently large to represent with accuracy the overall behaviour. This volume element is called a representative volume element (RVE); for example, the RVE of a periodic composite is simply the elementary cell (Suquet, 1982). For a randomly distributed composite of a polycrystalline aggregate it is necessary to have a minimum of information about the geometry of the constituent phases and to make assumptions like statistical homogeneity or ergodicity in order to define the RVE in a statistical sense (Beran, 1978; Fokin, 1979; Kröner, 1980; Willis, 1981; Hashin, 1983).

The problem is to characterize the overall behaviour only from a knowledge of the local constitutive equation of each constituent phase of the heterogeneous medium. On the microscopic level the mechanical properties vary from point to point with the local scale y while the average quantities vary smoothly with the macroscopic scale x.

With every microscopic quantity f (stress σ, mass per unit volume ρ, deformation ε, heat flux $q, \ldots$) we can associate its macroscopic value $\bar{f}$ by an averaging process on the RVE, extended eventually in the cases for which RVE contains holes or rigid inclusions:

$$\bar{f} = \langle f \rangle = \frac{1}{V}\int_V f(y)\,\mathrm{d}v_y \tag{2.1}$$

It is clear that eqn (2.1) defines a unique macro state quantity for each micro state.

Conversely, starting with a macro state, the definition of a suitable corresponding micro state requires complementary information or a localization process to determine the nearest possible micro state of the real one. For the choice of a representative volume element suitable boundary conditions must be prescribed and must satisfy some requirements (Francfort *et al.*, 1983; Germain *et al.*, 1983).

In particular, for study of the macroscopic behaviour these boundary conditions can be homogeneous; in this case one prescribes homogeneous stress conditions ($T_d = \Sigma.\vec{n}$ on ∂V) or homogeneous strain conditions ($U = E.y$ for $y \in \partial V$) (where Σ and E are two symmetric tensors).

For these boundary conditions, a local stress distribution σ and a local deformation ε are developed over the volume V.

In the case of homogeneous loading the stress σ verifies the equation of equilibrium and the boundary conditions

$$\left.\begin{aligned} \operatorname{div} \sigma &= 0 \quad \text{over } V \\ \sigma.\vec{n} &= T^{\mathrm{d}} \quad \text{on } \partial V \end{aligned}\right\} \tag{2.2}$$

In the other case the local deformation ε derives from a local displacement u which verifies

$$\left.\begin{aligned} \varepsilon_{ij} &= \frac{1}{2}\left(\frac{\partial u_i}{\partial x_j} + \frac{\partial u_j}{\partial x_i}\right) \quad \text{over } V \\ u &= U \quad \text{on } \partial V \end{aligned}\right\} \tag{2.3}$$

We require that the Hill-Mandel macrohomogeneity condition is fulfilled by all field σ^* with $\sigma^* = \langle\sigma^*\rangle$ and by all ε^* kinematically admissible so

$$\bar{\sigma}^* \bar{\varepsilon}^* = \langle\sigma^* \varepsilon^*\rangle \tag{2.4}$$

With these conditions it is obvious that, in the case of homogeneous loading,

$$\Sigma = \bar{\sigma} \tag{2.5}$$

and the microdisplacement u fluctuates around the homogeneous displacement U such that

$$\int_{\partial V} (U - u) \otimes \mathrm{n} \, \mathrm{d}a = 0 \tag{2.6}$$

and

$$E = \langle\varepsilon\rangle$$

Dually, if $u = U$ over ∂V it is clear that

$$E = \frac{1}{V}\int_{\partial V} \frac{1}{2}(u \otimes n + n \otimes u) \, \mathrm{d}a = \langle\varepsilon\rangle = \bar{\varepsilon} \tag{2.7}$$

and the microstress is such that

$$\int_{\partial V} (\bar{\sigma} n - \sigma n) \otimes y \, \mathrm{d}a = 0 \tag{2.8}$$

and the macrostress $\Sigma = \bar{\sigma} = \langle\sigma\rangle$.

Some other boundary conditions can be used; in particular, for periodic structures (Sanchez-Palencia, 1974; Suquet, 1982), the elementary cell V generates the geometry of the whole by periodicity, all the local fields are periodic functions and the Hill–Mandel macrohomogeneity condition is a necessary condition resulting from periodicity.

The construction of the macroconstitutive law from the microscopic behaviour can proceed as follows: starting from some macroscopic quantities, using a localization process to determine local fields, then the averaging process determines the missing macro quantities.

3. OVERALL PROPERTIES OF AN ELASTIC COMPOSITE MATERIAL

The main difficult problem is to determine effectively the overall property of a composite material. We shall restrict our discussion to some methods for the determination of the overall moduli for an elastic material.

The heterogeneous body is characterized by n phases with different elastic properties. The polycrystal aggregate is a particular case of a heterogeneous body, where the phases differ in the orientation of each crystal. Self-consistent theories which apply to polycrystal aggregates need knowledge of the properties of each grain (Hill, 1965; Berveiller and Zaoui, 1979). Kröner (1977, 1980), by a statistical approach, has compared the homogenization method of periodic media with the self-consistent scheme. The periodic model is a representative model of perfect order; one needs to solve a boundary value problem in a single cell, and the self-consistent method is more appropriate to model perfect disorder,

3.1. Bounds of the Elastic Modulus for a Heterogeneous Body

The medium is assumed to be elastic; each phase r has an elastic constitutive law with modulus Λ_r ($\varepsilon = \Lambda_r \sigma$, $\pi_r = \Lambda_r^{-1}$). We suppose that the characteristic function ϕ_r of each phase is known:

$$\left.\begin{aligned} \phi_r(y) &= 0 \quad y \text{ does not belong to phase } r \\ \phi_r(y) &= 1 \quad y \text{ belongs to phase } r \end{aligned}\right\} \tag{3.1}$$

We have the equality

$$\frac{1}{V}\sum_{r=1}^{n} \int_V \phi_r(y)\, \mathrm{d}v_y = 1 \tag{3.2}$$

and the concentration of the r phase is

$$c_r = \frac{1}{V}\int_V \phi_r(y)\,\mathrm{d}v_y \tag{3.3}$$

Several works (Beran and Molyneux, 1966; Hashin and Shtrikman, 1962; Hill, 1963) are related to the bounds of the moduli using minimum principles. For all displacements u^* kinematically admissible with given boundary conditions on ∂V the real field minimizes the energy:

$$U(u^*) = \frac{1}{V}\int \frac{1}{2}\varepsilon^* \,\pi\, \varepsilon^*\,\mathrm{d}v_y \tag{3.4}$$

then for a localization process, which describes a homogeneous deformation on ∂V ($u^* = \bar{\varepsilon}.y, y \in \partial V$), the following inequality holds:

$$\bar{U} \leqslant \frac{1}{2V}\bar{\varepsilon}.\sum_r \int_V \pi_r \phi_r(y)\,\mathrm{d}v_y\,.\,\bar{\varepsilon} \tag{3.5}$$

From the macrohomogeneity hypothesis we have

$$\bar{U} = \frac{1}{2}\,\overline{\varepsilon\,\pi\,\varepsilon} = \frac{1}{2}\overline{\sigma\,\varepsilon} = \frac{1}{2}\bar{\sigma}\,\bar{\varepsilon} \tag{3.6}$$

and, noting $\bar{\pi}$ the overall modulus ($\bar{\sigma} = \bar{\pi}\,\bar{\varepsilon}$), one can show that $\pi_v - \bar{\pi}$ is positive semi-definite, where the estimate π_v for the overall modulus is the Voigt average (Voigt, 1889):

$$\pi_v = \sum_{r=1}^{n} c_r \pi_r \tag{3.7}$$

Dually, we use the minimum of complementary energy for a localization process which prescribes homogeneous tensions on ∂V:

$$\bar{U} \leqslant \frac{1}{2V}\bar{\sigma}\sum_{r=1}^{n} \int_V \Lambda_r \phi_r(y)\,\mathrm{d}v\,\bar{\sigma} \tag{3.8}$$

Then, with the macrohomogeneity hypothesis, noting $\bar{\Lambda}$ the overall modulus ($\bar{\sigma} = \bar{\Lambda}\bar{\sigma}$), the inequality shows that $\Lambda_v - \bar{\Lambda}$ is positive semi-definite and Λ_v^{-1} is an estimate of $\bar{\pi}$ in the Reuss sense:

$$\pi_R = \Lambda_v^{-1} = \left(\sum_{r=1} c_r \Lambda_r\right)^{-1} \tag{3.9}$$

The two estimates π_v and π_R are lower and upper bounds. The best estimates are such that the localization process approaches the real solution. In the case of periodic media the answer is clear; for such a material Ω, with many elementary cells, if homogeneous tension or homogeneous deformation are applied on $\partial\Omega$, far from the boundary, the field solution of the boundary value problem is a periodic field; the averages of stress or strain on V are equal to the prescribed value. The real localization process is the periodic localization process, so the overall moduli obtained with homogeneous prescribed macrostress or macrostrain are equal.

For other composites, it is necessary to have complementary information. The calculus of overall properties is related to the determination of fields which satisfy equilibrium equations; perfect bonding across phase interfaces and some simplifications must be made to derive explicit results.

In the self-consistent scheme, for example, we consider that the inclusions are spherical or ellipsoidal, and when the medium is statistically uniform the equivalent body is assumed to be isotropic. To estimate the overall moduli, we can consider that each inclusion is embedded in a homogeneous matrix with the overall moduli Λ as elastic properties, and subject to a homogeneous field at infinity. The interaction problem between a matrix and a spherical or ellipsoidal inclusion is well known (Eshelby, 1957, 1961). The microstress in the inclusion is homogeneous and an explicit result can be given in the formal equation

$$\sigma(r) = \Sigma + J_r : (E - \varepsilon(r)) \tag{3.10}$$

The interaction is described by J_r which depends on the overall moduli Λ, on the inclusion geometry and on orientation (Berveiller and Zaoui, 1979).

If we take the statistical average and the local behaviour into account,

$$\left.\begin{aligned} \sigma_r &= \Lambda_r \varepsilon_r \\ \Sigma &= \langle \sigma_r \rangle \quad \text{and } E = \langle \varepsilon_r \rangle \\ \Sigma &= \Lambda E \end{aligned}\right\} \tag{3.11}$$

we have the equation

$$\Lambda = \langle \Lambda_r (\Lambda_r + J_r)^{-1} (\Lambda + J_r) \rangle \tag{3.12}$$

with unknown Λ. Usually this equation is complex to solve; the determination of J_r is not the easiest thing to do. Each phase seems to see the other phases through the homogeneous material; we have not given a good description of the interactions between all the inclusions. Some developments due to Berveiller and Zaoui (1979), Fokin (1979) and Andrieux (1983) can be used to solve this problem with greater accuracy.

The first estimates (Voigt and Reuss estimates) are established using the minimum energy principle, and the bounds are obtained for extremal values associated with homogeneous boundary conditions. A successful method is to determine a suitable displacement field to have a better estimate. For this reason it is convenient to introduce a homogeneous comparison material with modulus Λ_0 and to set the local modulus Λ equal to Λ_0 with some perturbation $\delta\Lambda$:

$$\Lambda = \Lambda_0 + \delta\Lambda \tag{3.13}$$

If ε is the real deformation over V, the stress polarization

$$p = \sigma - \Lambda_0\varepsilon = \delta\Lambda.\varepsilon \tag{3.14}$$

verifies

$$\mathrm{div}(p + \Lambda_0\varepsilon) = 0 \qquad \text{over } V \tag{3.15}$$

and if we know the Green's function for the comparison material we have

$$u(x) = u_0(x) + \int_V G(x,y)\,\mathrm{div}\,p(y)\,\mathrm{d}v_y \tag{3.16}$$

The deformation ε is the sum of the homogeneous deformation $E = \varepsilon_0$ and the last term of eqn (3.16), due to heterogeneity. In this integral formulation we can substitute the variational expression. Such methods are presented by Willis (1981) and Hashin and Shtrikman (1962*ab*, 1963). These authors have obtained bounds for overall moduli by different approximations of p and different averaging processes.

3.2. The Case of a Macrohomogeneous Body

In the last section we saw that the estimate of the elastic modulus is not the same when one prescribes macrohomogeneous strain or stress over V. But when the body can be considered as macrohomogeneous in the sense of Hill (1952, 1965) and Mandel (1964) the difference between the two estimates is negligible.

So when one prescribes homogeneous boundary conditions in stress ($T^d = \Sigma \cdot \vec{n}$ on ∂V, with Σ a symmetric tensor) the local stress is obtained as the solution of an elastic linear problem:

$$\left.\begin{array}{l}\text{Equation of equilibrium:}\\ \operatorname{div}\sigma = 0 \quad \text{over} \quad V\\ \sigma\cdot\vec{n} = T^{\mathrm{d}} \quad \text{on} \quad \partial V\\ \text{Elastic constitutive law:}\\ \varepsilon = \Lambda^0(y):\sigma \quad (\forall y \in V)\\ \text{Existence of a local displacement } u \text{ such that}\\ \varepsilon_{ij} = \frac{1}{2}\left(\frac{\partial u^j}{\partial x^i}+\frac{\partial u^i}{\partial x^j}\right)\end{array}\right\} \tag{3.17}$$

and we can prove the existence of a localization tensor A_{ijpq} such that

$$\sigma_{\mathrm{ij}} = A_{ijpq}\,\Sigma_{pq} \tag{3.18}$$

The tensor A is the elastic localization tensor introduced by Mandel (1964) in the case of a localization process in homogeneous macro-stresses.

As $\langle\sigma\rangle = \Sigma$ for all prescribed Σ, we have

$$\langle A_{ijpq}\rangle = \tfrac{1}{2}(\delta_{ip}\,\delta_{jq} + \delta_{iq}\,\delta_{jp}) \tag{3.19}$$

At fixed subscripts (p, q) A_{ijpq} satisfies the equations of equilibrium

$$A_{ijpq,j} = 0 \tag{3.20}$$

with the boundary conditions

$$A_{ijpq}n_j = \tfrac{1}{2}(\delta_{ip}\,\delta_{jq} + \delta_{iq}\,\delta_{jp})\cdot n_j \tag{3.21}$$

and the deformation

$$\varepsilon_{ij} = \Lambda^0_{ijkl}\,A_{klpq} \tag{3.22}$$

satisfies the condition of compatibility.

With the help of a macrohomogeneity condition, for all prescribed Σ we have

$$\Sigma\,E = \langle\sigma\,\varepsilon\rangle \tag{3.23}$$

The macroscopic modulus Λ ($E = \Lambda\,\Sigma$) is determined as a function of the local one ($\varepsilon = \Lambda^0\,\sigma$) by replacing σ in eqn (3.23) by $A\,\Sigma$:

$$\Lambda_{pqrs} = \langle A_{ijpq}\,\Lambda^0_{ijkl}\,A_{klrs}\rangle = \overline{{}^{\mathrm{t}}A\,\Lambda^0\,A} \tag{3.24}$$

In the following we consider only the case of a macrohomogeneous body.

4. STUDY OF A THERMOELASTOPLASTIC MATERIAL

4.1. The Macroscopic Constitutive Law

In this example we consider a thermoelastoplastic material and we assume that there is no effect of viscosity. The absolute temperature T is given and the total deformation ε is the sum of an elastic part ε^{e} and an irreversible part ε^{p}:

$$\varepsilon = \varepsilon^{e} + \varepsilon^{p} \tag{4.1}$$

The irreversible part has a standard evolution; the rate $\dot{\varepsilon}^{p}$ is normal to a yield surface delimited by a convex function ϕ of the stress σ:

$$\dot{\varepsilon}^{p} = \frac{\partial \phi}{\partial \sigma} \tag{4.2}$$

The reversible part ε^{e} is related to σ and to the increase of the temperature from its initial value T_0:

$$\varepsilon^{e} = \Lambda^{0} \sigma + \alpha (T - T_0) \tag{4.3}$$

4.2. Global Domain of Elasticity

At the local level there exists an elastic domain $\mathscr{C}$ defined by $\mathscr{C} = \{\sigma / \phi(\sigma) \leqslant 0\}$. At the macro level there exists an elastic domain $\mathscr{E}$ defined as follows.

Let two macrostresses Σ and Σ^* belong to $\mathscr{E}$, at each point of V; the corresponding microstresses σ and σ^* are elements of $\mathscr{C}$ and the difference $\sigma - \sigma^*$ is deduced from $\Sigma - \Sigma^*$ by means of a localization process with purely elastic behaviour.

So it is obvious that

$$\sigma - \sigma^* = A(\Sigma - \Sigma^*) \tag{4.4}$$

where A is the localization tensor introduced in the last section.

4.3. On the Decomposition of the Macroscopic Deformation

Let Σ be the real macrostress and σ the corresponding micro one. At the initial temperature T_0, the local response in purely elastic behaviour is

$$\sigma_{\mathrm{E}} = A\Sigma \tag{4.5}$$

The stress field $r = \sigma - \sigma_{\mathrm{E}}$ is a self-equilibrated stress and is the sum of two self-equilibrated stresses:

$$r = r_{\mathrm{p}} + r_{\theta} \tag{4.6}$$

such that the two microdeformations

$$\left.\begin{aligned} \varepsilon_1 &= \Lambda^0 r_\theta + \alpha(T - T_0) \\ \varepsilon_2 &= \Lambda^0 r_p + \varepsilon^p \end{aligned}\right\} \tag{4.7}$$

are kinematically admissible.

By definition the macroscopic elastic deformation due to the difference in absolute temperature $T - T_0$ and to the given macrostress Σ is the deformation deduced from a purely elastic unloading with return of the temperature to its initial value (Mandel, 1971; Halphen, 1977).

The local deformation defined by

$$\varepsilon_E = \Lambda^0 \sigma_E + \varepsilon_1 \tag{4.8}$$

is related to the macro one E_E by the lemma of macrohomogeneity and

$$E_E = \langle \varepsilon_E \rangle \tag{4.9}$$

and for all stresses σ' statically admissible with $\Sigma' = \langle \sigma' \rangle$ we have

$$\Sigma' E_E = \langle \sigma' \varepsilon_E \rangle \tag{4.10}$$

For the particular choice of $\sigma' = A\Sigma'$, and taking into account that r_θ is self-equilibrated, we can split E_E into two terms:

$$E_E = \overline{{}^tA\Lambda^0 A}\,\Sigma' + \overline{{}^tA\alpha}(T - T_0) \tag{4.11}$$

We recognize the macroscopic modulus $\Lambda = \overline{{}^tA\Lambda^0 A}$ and the result obtained is in agreement with the definition of the macroscopic elastic deformation.

In the same way, the global deformation $E = \langle \varepsilon \rangle$ must satisfy eqn (4.10), and for the difference $E_p = E - E_E$ we have the definition of the irreversible macroscopic deformation:

$$E_p = \overline{{}^tA\varepsilon_p} \tag{4.12}$$

The irreversible part of the macrodeformation is not the mean of the micro one because the field of plastic deformation is not kinematically admissible (Mandel, 1964, 1971; Hill, 1965).

The relations (4.11) and (4.12) describe the form of the stress–strain relation, and moreover these equations are true for quasi-static evolutions. If there exists a viscoplastic potential ϕ for the rate of the plastic deformation, such that

$$\dot{\varepsilon}^p = \frac{\partial \phi}{\partial \sigma} \tag{4.13}$$

it is obvious that at the macro level the mean potential (Mandel, 1971; Rice, 1971)

$$\Phi(\Sigma, r) = \overline{\phi(\sigma)} \tag{4.14}$$

is the macroviscoplastic potential associated with $\dot{E}_{\mathrm{p}}$: $(\sigma = A\Sigma + r)$

$$\left.\begin{aligned} \frac{\partial \Phi}{\partial \Sigma} &= \overline{\frac{\partial \phi}{\partial \sigma}\frac{\partial \sigma}{\partial \Sigma}} = \overline{{}^{\mathrm{t}}A\frac{\partial \phi}{\partial \sigma}} \\ \frac{\partial \Phi}{\partial r} &= \dot{\varepsilon}^{\mathrm{p}} = \frac{\partial \phi}{\partial \sigma} \end{aligned}\right\} \tag{4.15}$$

and we obtain

$$\dot{E}^{\mathrm{p}} = \overline{{}^{\mathrm{t}}A\dot{\varepsilon}^{\mathrm{p}}} \tag{4.16}$$

4.4. Analysis of the Dissipation

At the micro level, the second law of thermodynamics gives the rate of dissipation:

$$d = \rho T\dot{s} + \operatorname{div} q - q \cdot \xi \geqslant 0 \tag{4.17}$$

Using the conservation of energy,

$$\rho\dot{e} = \sigma\dot{\varepsilon} - \operatorname{div} q \tag{4.18}$$

and knowing the relation between the thermodynamical potentials e and $w(\varepsilon^{\mathrm{e}}, \Gamma)$, $e = w - Ts$, we can deduce

$$d = \sigma\dot{\varepsilon}^{\mathrm{p}} - q\xi \geqslant 0 \tag{4.19}$$

We have assumed that q is defined by $q = -\partial\omega/\partial\xi$ where ω is a convex function of ξ, therefore the term $-q \cdot \xi$ is positive, and when we admit the decomposition of the dissipation rate in two parts, the term due to the irreversible mechanism is also positive, $\sigma\dot{\varepsilon}^{\mathrm{p}} \geqslant 0$. By integration over V the total plastic dissipation rate $\mathcal{D}_{\mathrm{p}} = \overline{\sigma\dot{\varepsilon}^{\mathrm{p}}}$ can be given in terms of macroscopic variables:

$$\mathcal{D}_{\mathrm{p}} = \Sigma\dot{E}^{\mathrm{p}} - \frac{\mathrm{d}}{\mathrm{d}t} W_{\mathrm{b}} \geqslant 0 \tag{4.20}$$

where $W_{\mathrm{b}} = \frac{1}{2}\overline{r\Lambda_0 r}$ is the energy of the self-equilibrated stresses. The last result is obtained by application of eqn (4.10); for the plastic deformation rate and the real stresses $\sigma = A\Sigma + r$.

The physical meaning of this result is that perfect plastic behaviour at the micro level presents hardening behaviour on the macro scale.

5. THE CASE OF A SOLID HAVING LOCALLY DAMAGED ZONES

5.1. Description of the Model of Damage

It is observed that the reduction of material rigidity is generally due to the evolution of defects such as cavities, cracks, etc. Under external loading, these zones cannot support tensile stresses. It is proposed to characterize the damaged material only by the property that the stress tensor vanishes in the damaged zone. Therefore the stress vector must vanish on the boundary ∂z_i of any damaged zone z_i ($\sigma \cdot \vec{n} \equiv 0$ on ∂z_i). If the volume V contains many damaged parts with total volume $z(t)$, a function of the time t, the volume of the sound material is $\Omega(t) = V - z(t)$; we note $\partial\Omega_i$, the common boundary between Ω and z. The mechanical properties of the sound material are defined by the thermodynamical potential.

The definition of the average must be changed because f is only defined in Ω:

$$\bar{f} = \langle f \rangle_\Omega = \frac{1}{V}\int_\Omega f \, \mathrm{d}v \tag{5.1}$$

We suppose that the mass per unit volume is constant:

$$\forall x \in \Omega \qquad \rho(x) = \rho_0 \tag{5.2}$$

Our study is concerned with the behaviour of material introduced by Bui and co-workers (Bui and Ehrlacher, 1980; Bui *et al.*, 1981) to study fracture, pictured in Fig. 1.

To determine the evolution of damaged zones, it is necessary to introduce a rupture criterion. Rupture occurs when a threshold value (in stress, deformation or energy) is reached; when the strain is increasing the stress vanishes.

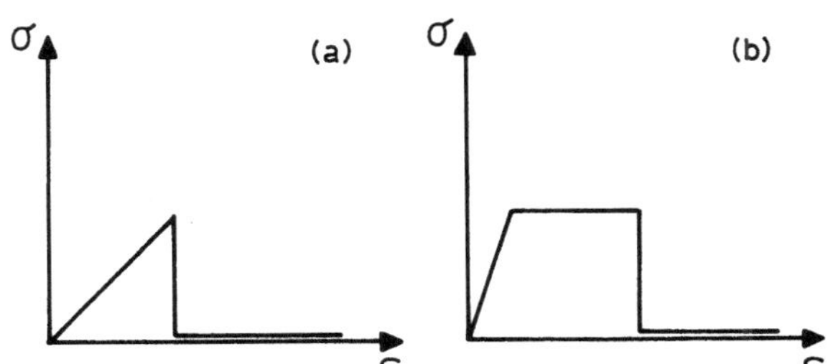

Fig. 1. (a) Elastic-brittle or (b) elastoplastic-brittle behaviour.

To define an accurate localization process, the boundary condition ($\sigma \cdot n = 0$ on $\partial\Omega_i$) must be taken into account. All the stresses σ must verify the equation of equilibrium and the new boundary conditions $\sigma n = 0$ at each point of $\partial\Omega_i(t)$. For the sake of simplicity, we assume that the heat flux is zero on $\partial\Omega_i$, and the damaged zones are considered to have no heat conductivity.

The localization process gives a new fourth rank tensor A of elastic concentration. This tensor has the properties (3.20)–(3.22) over Ω and verifies the boundary conditions $A_{ijpq} \cdot n_j = 0$ at each point of $\partial\Omega_i$.

One can define the general relations between the micro and macro deformations and stresses as follows:

$$\left.\begin{aligned} E &= \overline{{}^{t}A\varepsilon} = \bar{\varepsilon} - \frac{1}{2V}\int_{\partial\Omega_i} (u \otimes n)_s \, ds \\ \dot{E} &= \overline{{}^{t}A\dot{\varepsilon}} = \bar{\dot{\varepsilon}} - \frac{1}{2V}\int_{\partial\Omega_i} (v \otimes n)_s \, ds \\ \Sigma &= \bar{\sigma} \end{aligned}\right\} \tag{5.3}$$

The local compliance Λ^0 ($\varepsilon = \Lambda^0\sigma$) is related to the macro compliance ($E = \Lambda\Sigma$) by the equation

$$\Lambda = \overline{{}^{t}A\Lambda_0 A} \tag{5.4}$$

because the field $\Lambda_0 A$ is kinematically admissible and A is statically admissible with $A \cdot \vec{n} = \vec{0}$ on $\partial\Omega_i$.

However, the tensor A depends on the actual volume $\Omega(t)$ of the sound material and its rate $\dot{A}$ is related to the propagation of the damaged zones (Bui *et al.*, 1982).

5.2. Determination of the Evolution of the Damaged Zone for a Thermoelastic Material

Consider a thermoelastic material with free energy $w(\varepsilon, T)$ which depends only on the deformation ε and the absolute temperature T. At each point of V we define the energy density e and the entropy s per unit mass. For simplicity we assume that damaged zones have no heat conductivity and that they are inert ($e = 0$, $s = 0$, $v = 0$).

The first law of thermodynamics is the conservation of energy, at the macro level:

$$\dot{\bar{e}} + \dot{\bar{K}} = P_{\mathrm{cal}} + P_{\mathrm{ext}} \tag{5.5}$$

The variation of internal energy and kinetic energy $\overline{\mathrm{K}} = \frac{1}{2}\overline{\rho u^2}$ is equal to the heat flux and the power of external loading. We consider only quasi-static evolution so that the contribution of kinetic energy is negligible. The heat flux determines P_{cal}:

$$P_{\mathrm{cal}} = -\int_{\partial V} q \cdot \vec{n}\, \mathrm{d}a \tag{5.6}$$

and by means of the equilibrium equation we have, for the power of external loading,

$$P_{\mathrm{ext}} = \overline{\sigma \dot{\varepsilon}} - \int_{\partial \Omega_i} \sigma n v\, \mathrm{d}a \tag{5.7}$$

The localization process depends on the evolution of damaged zones and all the quantities are changed with the variation of Ω. If f is a regular mechanical quantity and defined on Ω, we have the equation

$$\dot{\bar{f}} = \bar{\dot{f}} - \frac{1}{V}\int_{\partial \Omega_i} fc\, \mathrm{d}a \tag{5.8}$$

where c is the normal velocity of the boundary $\partial\Omega_i$. So we have the local equation corresponding to (5.5):

$$\left.\begin{aligned} \rho_0 \dot{e} &= \sigma \dot{\varepsilon} - \operatorname{div} q && \text{over } \Omega \\ q\vec{n} &= -\rho_0 e c && \text{on } \partial\Omega_i \text{ with } \phi \geqslant 0 \end{aligned}\right\} \tag{5.9}$$

The production of entropy is necessarily positive:

$$\dot{\bar{s}} + \int_{\partial \Omega_i} \frac{q}{T} \cdot n\, \mathrm{d}a \geqslant 0 \tag{5.10}$$

By eqn (5.8) we can write the inequality (5.10) in another form:

$$\overline{\rho_0 \dot{s} + \frac{1}{V}\operatorname{div} q} - \overline{q \cdot \frac{\nabla T}{T^2}} - \frac{1}{V}\int_{\partial \Omega_i} \left(\frac{q}{T} n + \rho_0 s c\right) da \geqslant 0 \tag{5.11}$$

For an elastic material we have

$$\rho_0 \dot{s} + \frac{1}{T}\operatorname{div} q = 0 \tag{5.12}$$

With the conductivity law $q = -\partial\omega/\partial\xi$ where $\xi = \nabla T/T$ and ω convex, function of ξ, we have

$$-\frac{1}{T}\overline{q \cdot \nabla \xi} \geqslant 0 \tag{5.13}$$

Therefore the fracture dissipation rate is equal to the last term of (5.11):

$$\mathscr{D}_{\mathrm{f}} = \frac{1}{V}\int_{\partial\Omega_i} (qn + \rho_0 sc)\,\mathrm{d}a \geqslant 0 \tag{5.14}$$

Using eqns (5.9),

$$\mathscr{D}_{\mathrm{f}} = \frac{1}{V}\int_{\partial\Omega_i} \rho_0 \frac{W}{T} c\,\mathrm{d}a \geqslant 0 \tag{5.15}$$

Per unit of area the thermodynamical force associated with c has the dimensions of entropy. We assume a brittle criterion (Nguyen, 1978; Suquet, 1982; Stolz, 1984):

$$\left.\begin{aligned} \mathscr{S} &= \frac{W}{T} < \mathscr{S}_c \quad & c &= 0 \\ \mathscr{S} &= \mathscr{S}_c & c &\geqslant 0 \end{aligned}\right\} \tag{5.16}$$

The two criteria $\mathscr{S} = \mathscr{S}_c$ and $\sigma\vec{n} = 0$ must be satisfied along the moving part of the boundary $\partial\Omega_i$, and we must introduce the transport condition at a geometrical point of $\partial\Omega_i$ by the convected derivative $\mathscr{D}_{\mathrm{f}}$:

$$x \in \partial\Omega_i \qquad \mathscr{D}_{\mathrm{f}} = \lim_{\Delta t \to 0} \frac{f(x + c\Delta t, t + \Delta t) - f(x, t)}{\Delta t} \tag{5.17}$$

On $\partial\Omega_i$ the two conditions are written as follows:

$$\left.\begin{aligned} \mathscr{D}(\sigma\vec{n}) &= (\dot{\sigma} + c\vec{n} \cdot \nabla\sigma) \cdot \vec{n} + \sigma\mathscr{D}\vec{n} = 0 \\ T\mathscr{D}(\mathscr{S}) &= \dot{W} - W\frac{\dot{T}}{T} - Tc\vec{n}\,\nabla\mathscr{S} \end{aligned}\right\} \tag{5.18}$$

It is easy to show that, in the plane case (with normal $\vec{k}$),

$$\mathscr{D}\vec{n} = \frac{\mathrm{d}c}{\mathrm{d}s} \cdot \vec{\tau} \tag{5.19}$$

$\vec{\tau}$ is the tangent vector to $\partial\Omega_i$ with s the curvilinear coordinate.

Taking into account the equilibrium equation

$$\vec{\tau} \cdot \nabla\sigma \cdot \vec{\tau} + \vec{n} \cdot \nabla\sigma \cdot \vec{n} = 0 \tag{5.20}$$

and a particular form of $\sigma = \sigma\vec{\tau}\otimes\vec{\tau} + \sigma_z\vec{k}\otimes\vec{k}$, the equations (5.18) become

$$\left.\begin{aligned} \mathscr{D}(\sigma n) &= \dot{\sigma}n + \frac{\mathrm{d}}{\mathrm{d}s}(c\sigma\cdot\vec{\tau}) = 0 \\ T\cdot\mathscr{D}\mathscr{S} &= \sigma\frac{\dot{\varepsilon}}{\rho} - e\frac{\dot{T}}{T} - Tc\cdot\nabla\mathscr{S} = 0 \end{aligned}\right\} \quad (5.21)$$

Knowing at each time t the thermomechanical fields and the geometry of $\Omega(t)$, we search for the rate quantities and the normal velocity c on $\partial\Omega_i$. The rate equations are

$$\left.\begin{aligned} &\operatorname{div}\dot{\sigma} = 0 \qquad \rho T\dot{s} + \operatorname{div} q = 0 \\ &\dot{\varepsilon} = \tfrac{1}{2}(\Delta v + \Delta^{\mathrm{T}} v) \qquad \xi = \Delta\log T \qquad x\in\Omega \\ &\dot{\sigma} = \Lambda_0\dot{\varepsilon} + A\dot{T} \qquad -\dot{s} = A^{\mathrm{T}}\dot{\varepsilon} + Z\dot{T} \\ &q = -\frac{\partial\omega}{\partial\xi} \\ &(\Lambda_0 = \rho W''_{,\varepsilon\varepsilon};\quad A = \rho W''_{,T\varepsilon};\quad Z = \rho W''_{,TT}) \end{aligned}\right\} \quad (5.22)$$

with the boundary conditions

$$\left.\begin{aligned} &q\cdot n = Q^{\mathrm{d}} \text{ on } \partial V_q \qquad \dot{T} = \dot{T}^{\mathrm{d}} \text{ on } \partial V_T \\ &v = v^{\mathrm{d}} \qquad \text{on } \partial V_u \qquad \dot{\sigma}n = \mathscr{F}^{\mathrm{d}} \text{ on } \partial V_\sigma \end{aligned}\right\} \quad (5.23)$$

$(\partial V = \partial V_q \cup \partial V_T = \partial V_u \cup \partial V_\sigma,\ \partial V_q \cap \partial V_T = \phi,\ \partial V_\sigma \cap \partial V_u = \phi)$ and on $\partial\Omega_i$ we have

$$\left.\begin{aligned} &c = 0;\quad \dot{\sigma}\vec{n} = 0;\quad q\cdot n = 0 \\ &c \neq 0;\quad \mathscr{D}(\sigma n) = 0;\quad \mathscr{D}(\mathscr{S}) = 0;\quad q\cdot n = -\rho_0 e c \end{aligned}\right\} \quad (5.24)$$

The solution of this problem is a stationary point of the functional (Stolz, 1984):

$$\left.\begin{aligned} \mathscr{J}(v, c, \dot{T}) = &\int_\Omega \left[\frac{1}{2}\dot{\varepsilon}\Lambda_0\dot{\varepsilon} + \dot{T}A\dot{\varepsilon} + \frac{1}{2}\dot{T}z\dot{T} + \frac{\mathrm{d}}{\mathrm{d}t}\omega\right]\mathrm{d}v \\ &- \int_{\partial\Omega_i}\rho\left\langle\sigma\frac{\dot{\varepsilon}}{\rho} - \frac{\dot{T}}{T}e\right\rangle c\,\mathrm{d}a + \int_{\partial\Omega_i}\frac{c^2}{2}\rho T\vec{n}\cdot\nabla(\mathscr{S})\,\mathrm{d}a \\ &+ \int_{\partial V_\sigma}\dot{\mathscr{F}}^{\mathrm{d}} v\,\mathrm{d}a - \int_{\partial\Omega} Q^{\mathrm{d}}\frac{\dot{T}}{T}\,\mathrm{d}a \end{aligned}\right\} \quad (5.25)$$

On the set $V = \{v, \dot{T}/v = v^{\mathrm{d}} \text{ on } \partial V_u, \dot{T} = T^{\mathrm{d}} \text{ on } \partial V_T\}$.

Once the velocity c is determined, we are able to know the variation of all macro quantities by application of eqn (5.8). In the following part we suppose that c is determined, and our point of interest is the macroscopic behaviour of a RVE V, and we reduce our consideration on elastoplastic material, with free energy of the form $w(\varepsilon - \varepsilon^{\mathrm{p}})$. The influence of temperature could be treated as in the last example (Section 4).

5.3. Evolution of Macroscopic Quantities

Here we determine the evolution of the macroscopic modulus and we decompose the rate of deformation.

By comparison of the two expressions for E and $\dot{E}$ in eqns (5.3), the application of eqn (5.8) shows that

$$\overline{{}^{\mathrm{t}}\dot{A}\varepsilon} - \frac{1}{V}\int_{\partial\Omega_i} {}^{\mathrm{t}}A\varepsilon c \, \mathrm{d}a = 0 \tag{5.26}$$

for all kinematically admissible ε. One particular field ε is $\Lambda^0_{ijpq}A_{pqrs}$, so it is obvious to obtain (Bui *et al.*, 1982)

$$\dot{\Lambda} = \overline{{}^{\mathrm{t}}\dot{A}\Lambda_0 A} = \frac{1}{V}\int_{\partial\Omega_i} {}^{\mathrm{t}}A\Lambda_0 A c \, \mathrm{d}a \tag{5.27}$$

The elastic modulus varies with the propagation of the damaged zones; this variation must appear in the reversible part of the rate of the elastic macroscopic deformation.

When the damaged zone does not propagate in Ω the local stress field is $\sigma = A\Sigma' + r$ where r is a self-equilibrated stress on the actual geometry. For an increasing $\dot{\Sigma}$ of the macrostress from the value Σ such that the damaged zone has a normal velocity c of propagation, the local stresses are $\dot{\sigma}$ with a corresponding rate of deformation

$$\dot{\varepsilon} = \Lambda_0\dot{\sigma} + \dot{\varepsilon}^{\mathrm{p}} \tag{5.28}$$

in the case of an elastoplastic material. Let $\dot{\sigma}_{\mathrm{E}}$ be the fictitious stresses obtained with purely elastic behaviour with the same velocity of propagation c:

$$\dot{\sigma}_{\mathrm{E}} = A\dot{\Sigma} + \dot{A}\Sigma \tag{5.29}$$

where $\dot{A}$ is related to the variation of the localization tensor A due to the change in the geometry of Ω induced by the damaged zone propagation.

If the stress field r is self-equilibrated on $\Omega(t)$ we have, from eqn (5.8),

$$\bar{\dot{r}} = \frac{1}{V}\int_{\partial\Omega_i} rc \, \mathrm{d}a \tag{5.30}$$

which shows that $\dot{r}$ is generally not self-equilibrated; we have inside Ω a redistribution of residual stresses. The rate $\dot{\varepsilon}$ and $\Lambda^0\dot{\sigma}_E$ are two kinematically admissible fields of deformation, then the macroscopic deformations E and E_E are given by the relations (5.3). For all σ' statically admissible with $\Sigma' = \langle\sigma'\rangle$ in the localization process we have

$$\left.\begin{aligned} \overline{\sigma'\dot{\varepsilon}} &= \Sigma'\dot{E} \\ \overline{\sigma'\dot{\varepsilon}} &= \Sigma'\dot{E}_E \end{aligned}\right\} \tag{5.31}$$

For the particular choice of $\sigma' = A\Sigma'$ we have

$$\dot{E}_E = \overline{{}^tA\Lambda_0\dot{\sigma}_E} = \Lambda\dot{\Sigma} + \dot{\Lambda}\Sigma \tag{5.32}$$

which defines exactly the variations of the reversible part E_E of the macrodeformation ($E_E = \Lambda\Sigma$).

Moreover, by subtraction in eqns (5.31) (with the same choice $\sigma' = A\Sigma'$) one obtains

$$\dot{E}_{IR} = \overline{{}^tA\dot{\varepsilon}^p} + \overline{{}^tA\Lambda_0\dot{r}} \tag{5.33}$$

The complementary term is due to the redistribution of residual stresses inside Ω, when the geometry changes with the propagation of the damaged zones.

When no damage occurs eqn (5.30) shows that $\dot{r}$ is self-equilibrated and the classical relation (4.20) is recovered. However, the existence of a viscoplastic potential at the microlevel does not ensure the existence of a macro-viscoplastic potential E_{IR}. Only the first term can be deduced from the average potential $\Phi(\Sigma, \rho) = \overline{\phi(\sigma)}$.

5.4. Study of the Dissipation Rate

The global energy $\overline{W}$ is the sum of a reversible part and an embedded energy W_b in the residual stresses r. For a quadratic form $w(\varepsilon - \varepsilon^p)$, the dissipation $\mathscr{D}$ can be decomposed into three terms, the first due to heat conduction, the second due to plasticity $\overline{\sigma\dot{\varepsilon}^p}$ and the third due to fracture $\mathscr{D}_f$.

So the mechanical part of $\mathscr{D}$ is given by

$$\mathscr{D} = \overline{\sigma\dot{\varepsilon}^{\mathrm{p}}} + \frac{1}{V}\int_{\partial\Omega_i} wc\,\mathrm{d}s \geqslant 0 \tag{5.34}$$

With the hypothesis of macrohomogeneity applied to actual stresses, we can deduce

$$\Sigma\dot{E}_{\mathrm{IR}} = \overline{\sigma\dot{\varepsilon}^{\mathrm{p}}} + \overline{\sigma\Lambda_0\dot{r}} \tag{5.35}$$

with

$$\overline{W} = \frac{1}{2}\overline{\sigma\Lambda_0\sigma} = \frac{1}{2}\Sigma E_{\mathrm{E}} + \frac{1}{2}\overline{r\Lambda_0 r}$$

and the equation

$$\mathscr{D}_{\mathrm{f}} = \frac{1}{V}\int wc\,\mathrm{d}a = \overline{\sigma\dot{\varepsilon}} - \overline{\sigma\dot{\varepsilon}^{\mathrm{p}}} - \frac{\mathrm{d}}{\mathrm{d}t}\left(\frac{1}{2}\Sigma E_{\mathrm{E}} + w_{\mathrm{b}}\right) \tag{5.36}$$

Using the macrohomogeneity hypothesis $\overline{\sigma\dot{\varepsilon}} = \Sigma E$

$$\mathscr{D}_{\mathrm{f}} = \frac{1}{2}(\Sigma\dot{E}_{\mathrm{E}} - \dot{\Sigma}E_{\mathrm{E}}) + \overline{\sigma\Lambda_0\dot{r}} - \frac{\mathrm{d}}{\mathrm{d}t}w_{\mathrm{b}} \geqslant 0 \tag{5.37}$$

where $w_{\mathrm{b}} = \frac{1}{2}\overline{r\Lambda_0 r}$. If the damaged zones do not propagate, $\mathscr{D}_{\mathrm{f}} = 0$ and the macro compliance Λ is constant, we recover $\mathrm{d}w_{\mathrm{b}}/\mathrm{d}t = \overline{r\Lambda_0\dot{r}}$, i.e. the variation of the stored energy is only due to residual stresses, as in classical plasticity (Mandel, 1971; Bui *et al.*, 1982).

If there is propagation of damaged zones, the first term $\frac{1}{2}(\Sigma\dot{E}_{\mathrm{E}} - \dot{\Sigma}E_{\mathrm{E}})$ does not vanish. It is analogous to Irwin's energy release rate for crack propagation (Bui, 1978) and it contributes to the dissipation rate.

6. THERMODYNAMICAL CONSIDERATIONS

6.1. Relationships between Micro and Macro Scales

In the preceding sections we were concerned by means of averaging processes and applications to define what are macroscopic mechanical quantities. We have said nothing about the macroscopic temperature or we have implicitly admitted that the temperature is given or corresponds to the temperature of the equilibrium state.

We have restricted our exposition to macrohomogeneous bodies; can we then give general relations for other materials?

To be specific, one will be concerned with materials that admit thermodynamical potentials, $e(\varepsilon, \alpha, s)$ internal energy or $w(\varepsilon, \alpha, \theta)$ free energy, where ε is the deformation, α the internal variables, θ the absolute temperature and s the entropy. Then the local equation of state may be written e.g.

$$\sigma = \rho \frac{\partial w}{\partial \varepsilon}; \quad s = -\frac{\partial w}{\partial \theta}; \quad A = -\rho \frac{\partial w}{\partial \alpha} \tag{6.1}$$

where σ is the elastic stress (this supposes no viscosity), and A are the thermodynamical forces associated with the variation of the internal variables α.

To determine the field of temperature, we must add to the constitutive equations (6.1) the heat conduction law. Noting $\xi = \nabla(\log \theta)$, we assume the following form for the generalized Fourier relation between heat flux and temperature:

$$q = -\frac{\partial \omega}{\partial \xi} \tag{6.2}$$

when $\omega(\xi)$ is a convex no-negative function of ξ.

As in the first section, it is clear that with every microscopic quantity f we can associate its macroscopic value $\bar{f}$ by an averaging process on the RVE V extended eventually in the cases for which V contains holes, damaged zones, or rigid inclusions:

$$\bar{f} = \langle f \rangle = \frac{1}{V}\int_V f \, \mathrm{d}v \tag{6.3}$$

and we have remarked that some constraints are necessary to determine a micro state corresponding to a given macro one. For the choice of a representative volume element suitable boundary conditions must be prescribed and must satisfy special requirements (Francfort *et al.*, 1983; Germain *et al.*, 1983):

R1 — Assume that the macro state $(\bar{\sigma}, \bar{\varepsilon}, \bar{\xi}, \bar{w})$ is given; the microscopic quantities f must then satisfy:

- the local constitutive law;
- the boundary conditions on ∂V;
- the relation of averaging;

— the equation of equilibrium in stress and flux $\operatorname{div}\sigma = 0$ and $\operatorname{div} q = 0$ over V.

R2 — The Hill–Mandel macrohomogeneity condition must be fulfilled by all fields σ^*, q^* with $\bar{\sigma}^* = \langle\sigma^*\rangle$ and $\bar{q}^* = \langle q^*\rangle$ and by all fields ε^* and ξ^* kinematically and thermally admissible:

$$\left.\begin{aligned} \bar{\sigma}^*\bar{\varepsilon}^* &= \langle\sigma^*\,\varepsilon^*\rangle \\ \bar{q}^*\bar{\xi}^* &= \langle q^*\,\xi^*\rangle \end{aligned}\right\} \tag{6.4}$$

We have given particular localization processes for periodic structures and for homogeneous boundary conditions.

R3 — The process of localization must ensure the existence and uniqueness of the microscopic response.

The construction of the macro constitutive law from the microscopic behaviour can proceed as follows:

— starting from some macro quantities;
— using a localization process to determine local fields then the averaging process to determine the missing macro quantities.

6.2. Study of Systems in Equilibrium

In this case the homogeneous deformation $E = \bar{\varepsilon}$, the set of internal variables α and the temperature T of the macroscopic description are prescribed. Near the equilibrium we assume that the temperature is quasi-uniform $\theta = T$ (but the presence of heterogeneity implies that the gradient ξ is not uniform, $\xi \neq \bar{\xi}$).

We wish to find the local quantities $\sigma, \varepsilon, q, \xi$ as functions of the macro state E, T, α, $\Xi = \bar{\xi}$; they are solutions of the boundary value problem $(\mathscr{P})$:

$$\left.\left.\begin{aligned} E &= \langle\varepsilon\rangle \quad \sigma = \rho\frac{\partial w}{\partial\varepsilon}(\varepsilon,\alpha,T) \quad \operatorname{div}_y\sigma = 0 \\ \Xi &= \langle\xi\rangle \quad q = -\frac{\partial\omega}{\partial\xi} \quad \xi = \nabla\log\theta \quad \operatorname{div}_y q = 0 \end{aligned}\right\}\mathscr{P}\right\} \tag{6.5}$$

Boundary conditions of localization

For a given localization process this solution is unique (R3). Let

$$\bar{\rho}\,\bar{w}\,(E;\alpha,T) = \langle\rho w(\varepsilon,\alpha,T)\rangle \tag{6.6}$$

where ε is the solution of the problem $(\mathscr{P})$. Using the hypothesis of macrohomogeneity of Hill (R2), we obtain

$$\bar{\rho}\,\frac{\partial \bar{w}}{\partial E} = \left\langle \rho\,\frac{\partial w}{\partial \varepsilon}\frac{\partial \varepsilon}{\partial E} \right\rangle = \left\langle \sigma\,\frac{\partial \varepsilon}{\partial E} \right\rangle = \langle \sigma \rangle\,\frac{\partial \varepsilon}{\partial E}$$

and noting that $\varepsilon = E + \eta$ with $\langle \eta \rangle = 0$ for all E, so $\langle \partial\varepsilon/\partial E \rangle = 1$. The equation (6.6) gives the global behaviour

$$\bar{\rho}\,\frac{\partial \bar{w}}{\partial E} = \langle \sigma \rangle = \bar{\sigma} \tag{6.7}$$

and in the same way it can be proved that

$$\begin{aligned} \bar{s} &= -\partial \bar{w}/\partial T \\ T &= \partial \bar{e}/\partial \bar{s} \end{aligned} \tag{6.8}$$

The global description has the same form as the local one, and the equation (6.7) gives the constitutive stress–strain relation. Similarly it is possible to give the generalized Fourier relation between $\bar{q}$ and $\bar{\xi}$.

If we consider that near a state of equilibrium the two principles of thermodynamics are valid at the microlevel

$$\left.\begin{aligned} \rho\dot{e} &= \sigma\dot{\varepsilon} - \operatorname{div} q \\ \rho\theta\dot{s} &+ \operatorname{div} q - q\cdot\xi \geqslant 0 \end{aligned}\right\} \tag{6.9}$$

we can write the same relations at the macrolevel, if all the equations (6.6)–(6.9) are valid at each time, which means that the thermostatic description and the Gibbs equation are true for the equivalent homogeneous medium if they are valid for each of the constituent phases (Francfort *et al.*, 1983; Germain *et al.*, 1983).

6.3. Non-equilibrium Evolution

When the system is far from an equilibrium state, in particular when the local absolute temperature θ is not uniform on V, the equilibrium equation and the property of macrohomogeneity are not valid. Some complementary terms due to inertia phenomena must be taken into account. However, we can try to define the macroscopic temperature by the concept of a local accompanying equilibrium state.

At each time, the volume element V must be isolated and one has to study the dynamical evolution toward an equilibrium state and find the micro quantities which characterize this equilibrium. When the volume element V is isolated, quantities such as the strain $E = \bar{\varepsilon}$ and energy $\bar{e}$, together with some internal parameter α, must be kept constant. Inside the RVE various physical dissipative mechanisms have taken place and each of them is described by the internal variable α. The dimensions of the RVE and the inertia effect define a characteristic time τ, and we must distinguish two kinds of internal parameters α and α'. The first family α can be considered as constant; the second α', having a relaxation time shorter than τ, must be considered a function of time.

Under the assumption that in this transition towards equilibrium the two principles of thermodynamics are valid, the equilibrium state is the solution of the problem (6.5) where

$$E = \langle \varepsilon_t \rangle, \alpha = \alpha_t \quad \text{for} \quad \alpha \not\in \{\alpha'\} \text{ and } e = \langle e_t \rangle$$

are prescribed, and we obtain

$$\left. \begin{aligned} \Sigma^e &= \frac{\partial e}{\partial E}(E, \alpha, \alpha'^e, T^e) \\ S^e &= \frac{\partial e}{\partial T}(E, \alpha, \alpha'^e, T^e) \end{aligned} \right\} \tag{6.10}$$

where T^e and α'^e are the values of the macroscopic temperature and the values of internal parameters at the equilibrium state, respectively.

If the actual state is far from the equilibrium, it is obvious that $\Sigma^e \neq \langle \sigma_t \rangle$. Examples are given by Francfort (1982).

The major difficulty in this study is to define correctly the absolute temperature for a macroscopic body which has microheterogeneity. In the first part of this chapter we have studied only the quasi-static evolution of the RVE, for which the absolute temperature is defined as in Section 6.2.

7. CONCLUSION

We have presented some aspects of the general relations between micro and macro scales in heterogeneous media, and we have tried to exhibit some of the most significant lines of research in the mechanical behaviour of composite materials. This domain is still in a state of expansion, not only in solid mechanics but also in fluid mechanics and in mixtures theory. In the next few years progress may be expected in many directions: better mathematical consistency, deepening of the physical interpretation and significance.

ACKNOWLEDGEMENTS

The author wishes to thank H. D. Bui, K. Dang Van and Q. S. Nguyen for discussion and for guidance.

REFERENCES

Andrieux, S. (1983). Doct.-Ing. thesis, ENPC, Paris.
Beran, M. J. (1978). *Statistical Continuum Theories*, Wiley-Interscience, New York.
Beran, M. J. and Molyneux, J. (1966). Use of classical variational principles to determine bounds for the effective bulk modulus in heterogeneous media, *Q. Appl. Math.*, **24**, 107–18.
Berveiller, M. and Zaoui, A. (1979). *Mech. Phys. Solids*, **26**, 325.
Bui, H. D. (1978). *Mécanique de la Rupture Fragile*, Masson, Paris.
Bui, H. D. and Ehrlacher, A. (1980). Propagation dynamique d'une zone endommagée dans un solide élastique-fragile en mode III et en régime permanent, *Compt. Rend. Acad. Sci.* (B), **290**, 273–6.
Bui, H. D., Dang Van, K. and Stolz, C. (1981). Relations entre grandeurs microscopiques et macroscopiques pour un solide élastique-fragile ayant des zones endommagées, *Compt. Rend. Acad. Sci.* (II), **292**, 863–6.
Bui, H. D., Dang Van, K. and Stolz, C. (1982). Relations entre grandeurs microscopiques et macroscopiques pour un solide anélastique ayant des zones endommagées, *Compt. Rend. Acad. Sci.* (II), **294**, 1155–8.
Eshelby, J. D. (1957). The determination of the elastic field of an ellipsoidal inclusion and related problem, *Proc. Roy. Soc.* (London), **A241**, 376–96.
Eshelby, J. D. (1961). Elastic inclusion and inhomogeneities. In *Progress in Solids Mechanics*, I. N. Sneddon and R. Hill (Eds), North-Holland, Amsterdam, Vol. II, pp 87–140.
Fokin, A. G. (1979). Statistical properties of inhomogeneous solid media: central moment functions of material characteristics, *J. Appl. Math. Mech.* (Engl. trans.), **42**, 570–9.
Francfort, G. (1982*a*). Thèse Ph.D.
Francfort, G. (1982*b*). Homogénéisation et oscillations rapides en thermo-élasticité linéaire, *Compt. Rend. Acad. Sci.* (I), **295**, 367–70.
Francfort, G., Nguyen, Q. S. and Suquet, P. (1983). Thermodynamique et lois de comportement thermo-mécanique homogénéisées, *Compt. Rend. Acad. Sci.* (II), **296**, 1007–10.

Germain, P., Nguyen, Q. S. and Suquet, P. (1983). Continuum thermodynamics, *Appl. Mech.*, **50**, 1010–20.
Halphen, B. (1977). Sur la thermodynamique du polycristal métallique, Colloque Annuel 1977 du Groupe Français de Rhéologie (Paris).
Hashin, Z. (1983). Analysis of composite materials, *J. Appl. Mech.*, **50**, 481-505.
Hashin, Z. and Shtrikman, S. (1962*a*). On some variational principles an anisotropic and non-homogeneous elasticity, *J. Mech. Phys. Solids*, **10**, 335–42.
Hashin, Z. and Shtrikman, S. (1962*b*). A variational approach to the theory of the elastic behaviour of polycrystals, *J. Mech. Phys. Solids*, **10**, 343-55.
Hashin, Z. and Shtrikman, S. (1963). A variational approach to the theory of the elastic behaviour of multiphase materials, *J. Mech. Phys. Solids*, **11**, 127–40.
Hill, R. (1952). The elastic behaviour of a crystalline aggregate, *Proc. Phys. Soc.* (London), **A65**, 349-54.
Hill, R. (1963*a*). Elastic properties of reinforced solids: some theoretical principles, *J. Mech. Phys. Solids*, **11**, 357-72.
Hill, R. (1963*b*). New derivations of some elastic extremum principles. In *Progress in Applied Mechanics*, Prager Anniversary Volume, Macmillan, New York, pp 99-106.
Hill, R. (1965*a*). Continuum micromechanism of elastoplastic polycrystals, *J. Mech. Phys. Solids*, **13**, 89-101.
Hill, R. (1965*b*). A self consistent mechanics of composite materials, *J. Mech. Phys. Solids*, **13**, 213-22.
Kröner, E. (1977). Bounds of effective elastic moduli of disordered materials, *J. Mech. Phys. Solids*, **25**, 137-55.
Kröner, E. (1980). Linear properties of random media: the systematic theory, C. R. 15ème Coll. du Groupe Français de Rhéologie.
Mandel, J. (1964). Proc. 11th Int. Congr. Appl. Mech., Munich, p. 502.
Mandel, J. (1971). *Plasticité Classique et Viscoplasticité*, Courses and Lectures No 17, CISM, Udine/Springer-Verlag.
Nguyen, Q. S. (1978). Normal dissipativity and energy criteria in fracture, IUTAM symposium, Evanston, Illinois.
Rice, J. R. (1971). Inelastic constitutive relations for solids: an internal variable theory and its application to metal plasticity, *J. Mech. Phys. Solids*, **9**, 433-55.
Sanchez-Palencia, E. (1974). Comportement local et macroscopique d'un type de milieux physiques hétérogènes, *Int. J. Eng. Sci.*, **12**, 331-51.
Stolz, C. (1984). Analyse du comportement et formulation du problème en vitesse pour un solide thermoélastique fragile avec zone endommagée, *Compt. Rend. Acad. Sci.* (II), **299**, 491-3.
Suquet, P. (1982). Plasticité et homogénéisation, Thèse Université Pierre et Marie Curie, Paris.
Voigt, W. (1889). Ueber die Bezichung zwischen den beiden Elastizitäts-constanten isotrope Körper, *Ann. Phys.* (Leipzig), **38**, 573-87.
Walpole, L. J. (1969). On the overall elastic moduli of composite materials, *J. Mech. Phys. Solids*, **17**, 235-51.
Willis, J. R. (1981). Overall properties of composites. In *Advances in Applied Mechanics*, Vol 21, C. S. Yih (Ed), Academic Press, New York.

BIBLIOGRAPHY

Bensoussan, A., Lions, J. L. and Papanicolaou, G. (1978). *Asymptotic Analysis for Periodic Structures,* North-Holland, Amsterdam.

Beran, M. J. and McCoy, J. J. (1970). Mean field variations in a statistical sample of heterogeneous linearly elastic solids, *Int. J. Solids Struct.,* **6**, 1035–54.

Budiansky, B. (1965). On the elastic moduli of some heterogeneous materials, *J. Mech. Phys. Solids,* **13**, 223–7.

Budiansky, B. (1970). Thermal and thermoelastic properties of isotropic composites, *J. Compos. Mater.,* **4**, 286–95.

Budiansky, B. and O'Connell, R. J. (1976). Elastic moduli of cracked solids, *Int. J. Solids Struct.,* **12**, 81–97.

Christensen, R. M. (1979). *Mechanics of Composite Materials,* Wiley-Interscience, New York.

Hashin, Z. (1962). The elastic moduli of heterogeneous materials, *ASME, J. Appl. Mech.,* **29**, 143–50.

Hashin, Z. (1964). Theory of mechanical behaviour of heterogeneous media, *Appl. Mech. Rev.,* **17**, 1–9.

Herakovich, C. T. (1981). On the relationship between engineering properties and delamination of composite materials, *J. Mech. Phys. Solids,* **11**, 127–40.

Hershy, A. V. (1954). The elasticity of an isotropic aggregate of anisotropic cubic crystals, *J. Appl. Mech.,* **21**, 236–40.

Hill, R. (1964). Theory of mechanical properties of fibre strengthened materials: elastic behaviour, *J. Mech. Phys. Solids,* **12**, 199–212.

Kestin, J. and Bataille, G. (1980). *Thermodynamics of Solids, Continuum Models of Discrete System,* University of Waterloo Press, Freudenstadt, FRG.

Laws, H. (1973). On the thermostatics of composite materials, *J. Mech. Phys. Solids,* **21**, 9–17.

Leguillon, D. and Sanchez Palencia, E. (1982). On the behaviour of a cracked elastic body with or without friction, *J. Mécan.,* **1** (2).

Reuss, A. (1929). Calculation of the flow limit of crystals on the basis of the plasticity of monocrystals, *Z. Angew. Math. Mech.,* **9**, 49–58.

Walpole, L. J. (1966*a*). On bounds for the overall elastic moduli of inhomogeneous systems. I, *J. Mech. Phys. Solids,* **14**, 151–62.

Walpole, L. J. (1966*b*). On bounds for the overall elastic moduli of inhomogeneous systems. II, *J. Mech. Phys. Solids,* **14**, 289–301.

PART II

PHYSICAL MODELLING

CHAPTER 5

Modelling Plastic Response of Solids

A. S. ARGON
Department of Mechanical Engineering, Massachusetts Institute of Technology, Cambridge, USA

ABSTRACT

The elementary principles of modelling plastic response of crystalline solids, based on mechanistic details, are presented. A short discussion on the details of the kinematics of strain production and the possible attendant lattice rotations in crystal plasticity is followed by a discussion of the modelling of separate mechanisms governing deformation resistance in pure crystalline substances. A short presentation is also given of a continuum treatment of the plastic resistance of microcomposites with a quasi-random dispersal of separate phases.

INTRODUCTION

As discussed in Chapter 1, the plastic response of real crystalline solids, with microstructure, requires consideration at two different levels.

The plastic resistance of a metal crystal, governed by the long-range motion of dislocations through slip plane obstacles, requires understanding of the details of such interactions and can be dealt with only through the proper application of dislocation theory. In most instances of deformation, slip systems of different type intersect, requiring the dislocations on them to interpenetrate through each other. In many alloys, in addition, dislocations have to overcome precipitate particles in the slip plane. The overcoming of such slip plane obstacles, which govern the glide resistance, has been extensively modelled. In polycrystalline aggregates, where grains of all conceivable orientations must deform compatibly, multiplicities of slip systems must be active in each grain. When such constraints are considered and their effects averaged over all orientations, the plastic resistance of the polycrystalline aggregate is obtained. The effect of heterogeneities of a smaller scale than grains, such as precipitates interacting strongly with individual dislocations, must usually be taken care of by dislocation theory, as just mentioned.

There are, however, heterogeneities on a coarser scale, such as large phase domains with different plastic resistance that are found widely in structural alloys, in directionally solidified super alloys, and cavities in solids undergoing ductile fracture. In these cases, the heterogeneities act as local continua with plastic resistances governed by their internal details involving kinematics and kinetics of dislocation motion. They interact with each other, however, only through the interfacial traction and displacement border conditions. For such problems, the modelling approach is that of micromechanics.

As discussed in Chapter 1, the demarcation in scale between problems requiring the use of dislocation theory for the solution and those requiring micromechanics is rarely clear and well delineated. Since the procedures of micromechanical modelling are well established in applied mechanics and can directly use rapidly developing non-linear computer codes, this technique has been used quite widely even on a scale where its applicability is questionable. Nevertheless, the results of micromechanical modelling serve as an asymptotic limit of behaviour, and are therefore always useful. A characteristic feature of the results of micromechanical analysis is that they obey a rule of mixtures, i.e. the properties of the heterogeneous assembly are a linear function of the volume concentration c of the distinct constituent components. In the realm of dislocation theory, a much stronger concentration dependence in the form of c^n exists, with n being in the range of 0·5–0·66. In this chapter some examples of dislocation mechanics modelling and micromechanical modelling will be considered.

2. KINEMATICS OF CRYSTAL PLASTICITY

2.1. Transformations and Translations

As already discussed in Chapter 1, plastic deformation in crystals is inherently of an inhomogeneous nature. In the most general sense, it occurs by local shear transformations that may or may not be accompanied by a limited amount of transient or residual dilatation. Emission of dislocation loops from sources is a limiting form of such shear transformations, albiet of a very important and pervasive kind. Thus, the representation of plastic deformation as a smoothly varying field of distortions is only possible on the average over volume elements that must contain enough of such elemental transformations to remove the large fluctuations that would otherwise be present, and assure the required smoothness. When conditions are present for plastic flow to become inhomogeneous on a macroscale, such as in the formation of large zones of necking, shear faults, or 'localization'† of the otherwise homogeneous deformation, then a homogeneous representation of the deformation is not even possible on the scale of the entire part or structure. The conditions and forms of such localization on the macroscale will not be examined in detail.

Here, we separate the geometrical and kinematical aspects of plastic deformation by shear transformations on the microscale from their energetic aspects that describe the interactions of the transformations with the applied stress tensor and with each other. While such separation is instructive in clarifying the kinematics, it is not realistic in understanding the overall nature of the development and internal distribution of the transformations.

2.2. Average Strains and Rotations due to Transformations and Translations

Consider, as shown in Fig. 1(a), a shear transformation in a small volume element v of a large sample of volume V. If the transformation has occurred unconstrained by the surroundings, let it produce a homogeneous transformation strain $\boldsymbol{\varepsilon}^{\mathrm{T}}$. When constrained by the elastic surroundings, however, the transformation will produce only a residual uniform strain of $\boldsymbol{\varepsilon}^{\mathrm{C}}$ inside the region of volume v when all distant stresses are relaxed. If a total volume fraction ${}^{s}c$ of a similarly

†In comparison with elastic deformation, plastic deformation is non-local on the microscale but can be represented as quasihomogeneous on a coarser scale than the range of slip processes, etc. Under certain conditions, however, even on the macroscale, the deformation can become unstable to concentrate into narrow zones, to become inhomogeneous, and to localize itself over the background of the homogeneously represented field. Thus, the terms local vs. non-local, and localization are used descriptively, to represent two opposite tendencies but also at opposite sides of the scale of the phenomenon.

characterizable set of transformations occurs in the sample, then allowing for any diffuse interactions of the elemental transformations with each other in the representations of the constrained strain ${}^s\boldsymbol{\varepsilon}^{\mathrm{C}}$ in any one of the sets of transformations, the overall strain $\boldsymbol{\varepsilon}$ in the sample will be given by the volume average of the unconstrained strains, $\boldsymbol{\varepsilon}^{\mathrm{T}}$, as the surrounding material flexes to make up the difference, i.e.

$$\boldsymbol{\varepsilon} = \sum_s {}^s c\, {}^s\boldsymbol{\varepsilon}^{\mathrm{T}} \tag{1}$$

where the summation is over all the different sets of transformations which could, for example, be several systems of deformation twins or

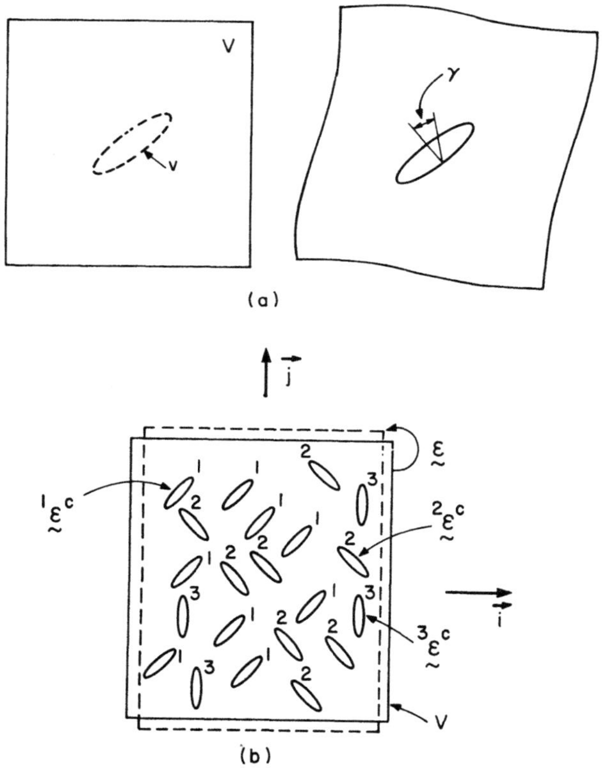

Fig. 1. Inelastic strain production: (a) shear transformation producing a constrained shear strain ε^{C} in a volume v of a large sample of volume V; (b) average strain ε produced in a volume V as a result of three sets of transformations producing constrained strains ${}^s\varepsilon^{\mathrm{C}}$ (s = 1, 2, 3) (from Argon, in press; courtesy of Pergamon Press, Oxford).

deformation-induced martensite lamellae, etc., as depicted in Fig. 1(b). We note that this incorporates all components of a local transformation in the form of shears and dilatation, and does not imply that in the relaxed form, under no applied stresses, the untransformed parts of the volume V are stress free.

The interactive computation of the constrained transformation strains inside the elastic body is a coupled procedure and is usually handled by the application of the well-known self-consistent method of composite property averaging.

When the plastic deformation is produced by a uniform shear $\delta\gamma$ in ellipsoidal regions with principal planes normal to a unit vector $\vec{\mathbf{n}}$ and a shear direction parallel to a unit vecord $\vec{\mathbf{d}}$ then the average plastic strain increment in the sample will still be the volume average of all plastic shears and is given as

$$\delta\varepsilon_{ij}^{\mathrm{p}} = \frac{h\delta\gamma}{V}\int_{A}(1/2)\,(n_j d_i + n_i d_j)\,\mathrm{d}A \qquad (2)$$
$$= c\delta\gamma(1/2)\,(n_j d_i + n_i d_j)$$

where h is the average thickness of the ellipsoids, d_i and n_j are the components of the unit vectors $\vec{\mathbf{d}}$ and $\vec{\mathbf{n}}$ in the directions of the macro-axes i and j, the area integral is over the principal plane areas of the entire set of sheared ellipsoids. Clearly, if the shear is produced by a set of s separate transformations having distinct plane normals ${}^s\vec{\mathbf{n}}$, shear directions ${}^s\vec{\mathbf{d}}$, and shear increments ${}^s\delta\gamma$, then the overall increment of plastic strain in the macrosample will be

$$\delta\varepsilon_{\mathrm{ij}}^{\mathrm{P}} = \sum_{s} {}^s c\,{}^s\delta\gamma\,{}^s\alpha_{ij} \qquad (3a)$$

where

$${}^s\alpha_{ij} = (1/2)\,({}^s n_j{}^s d_i + {}^s n_i{}^s d_j) \qquad (3b)$$

is a strain resolution factor.

Such simple shears are accompanied by local *average* rotations that might be definable for the region as†

$$\delta\omega_{ij} = \delta\gamma(1/2)\,(n_j d_i - n_i d_j)c \qquad (4)$$

†The subscript symbols on the rotation increment are to be interpreted as being in the positive sense about an axis $\vec{\mathbf{k}}$; that is, in the right-hand screw sense as axis $\mathbf{i}$ is wrapped into axis $\vec{\mathbf{j}}$, and other right-handed triad permutations.

Thus the average rotation in the entire body could be taken also as a volume average over all increments of rotation provided by the set of m separate transformations each occupying a volume fraction ${}^sc$,

$$\delta\omega_{ij} = \sum_s {}^sc{}^s\delta\gamma{}^s\alpha_{ij} \tag{5a}$$

$${}^s\alpha'_{ij} = (1/2)({}^sn_j{}^sd_i - {}^sn_i{}^sd_j) \tag{5b}$$

Finally, when the shears are due to individual translations associated with the expansion of dislocation loops or with the sweeping of dislocation segments over portions of slip planes having unit normal vectors $\vec{\mathbf{n}}$, with the dislocations having Burgers displacements b, parallel to the direction $\vec{\mathbf{d}}$, then the local shears degenerate into discontinuous translations and the plastic strain increments due to a separate set of slip systems transform, as a special case of eqns (3a) and (3b), into

$$\delta\varepsilon^{\mathrm{p}}_{ij} = \sum_s \frac{{}^sb{}^s\alpha_{\mathrm{ij}}{}^s\delta a}{V} \tag{6}$$

Here ${}^s\alpha_{ij}$ is of the same form as given in eqn (3b), and the swept-out area ${}^s\delta a$ is for the entire length of dislocation line of the specific set s under consideration. The average rotations that are produced by such translations can also be given by an extension of eqns (5a) and (5b), but as we will discuss below, have little utility in that form.

Often, single slip systems are of interest in mechanistic considerations for which eqn (6) simplifies to

$$\delta\varepsilon^{\mathrm{p}}_{ij} = \frac{b\delta a}{V}\alpha_{ij} \tag{7}$$

where δa is again the entire swept area increment by mobile segments of dislocations in the slip planes oriented in relation to the external area by α_{ij} as depicted in Fig. 2. When this motion can be well characterized as an average displacement δx of a total segment length Λ of dislocation, then

$$\delta\varepsilon^{\mathrm{p}}_{ij} = \frac{b\Lambda\delta x}{V}\alpha_{ij} = b\rho_m\delta x\alpha_{ij} \tag{8}$$

where in the final form $\rho_m(= \Lambda/V)$ has been introduced as the mobile dislocation (segment) density, i.e. mobile line length per unit volume.

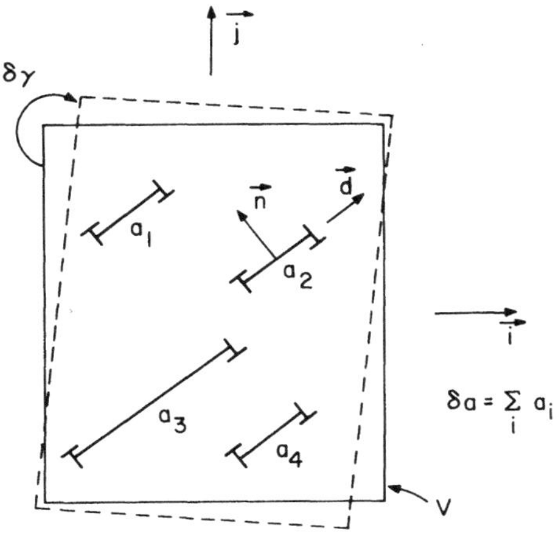

Fig. 2. A shear strain increment $\delta\gamma$ in a volume V resulting from a collection of slip processes over a total area δa with unit normal vectors parallel to $\vec{\mathbf{n}}$ and in the direction $\vec{\mathbf{d}}$ lying in the planes (from Argon, in press; courtesy of Pergamon Press, Oxford).

Finally, whenever time enters into the motion of dislocations to establish a kinematic basis of the process, eqn (8) readily transforms into a rate expression as

$$\dot{\varepsilon}_{ij}^{p} = b\rho_m \dot{x} \alpha_{ij} \tag{9}$$

We note that the transformations of the rate-independent kinematical relations that have been developed here leading to eqn (9) can all be similarly transformed into rate expressions by dividing all incremental quantities with an increment of time. This is often an operationally useful procedure, introducing time as a running (book-keeping) parameter for chronological ordering of modelling increments. Furthermore, while all the kinematical relations discussed above have been motivated on the basis of shears and translations of planes parallel to each other, they are also applicable without change to dilatations. For example, if the areas in eqn (6) or eqn (7) were to be separated (or brought together) by a magnitude b parallel to the unit normal vector $\vec{\mathbf{n}}$, then this is reflected directly in the definition of the resolution factors α_{ij}, where we note that in this case of thickening or thinning of the area, $\vec{\mathbf{n}} \,//\, \vec{\mathbf{d}}$. In this last sense, the developments find ready application in transformation-induced plasticity.

2.3. Lattice Rotations

Lattice rotations that result from slip produced by dislocation motion between crystallographic lamellae have kinematical subtleties arising from what is taken as the reference set of axes to characterize these rotations. Here, we follow a procedure presented by Kocks and Chandra (1982) that has been found useful in the accounting of texture development due to slip.

In crystal plasticity by slip along parallel crystallographic planes, the lattice rotates with respect to material lines or planes inscribed on the deforming solid as shown in Fig. 3. Here, slip translations have occurred alternatively on one or the other of two orthogonal slip systems, where in addition, an initial material line parallel to the 1-axis has been constrained to remain horizontal. The set of slip translations in Fig. 3(b) have produced no lattice rotation in relation to the initially horizontal material line, while the set in Fig. 3(c) has produced a large lattice rotation equal to the shear strain. Alternatively, the plane with unit normal initially parallel to the 1-axis has undergone a large rotation relative to the lattice in Fig. 3(b), and no rotation in the case of Fig. 3(c).

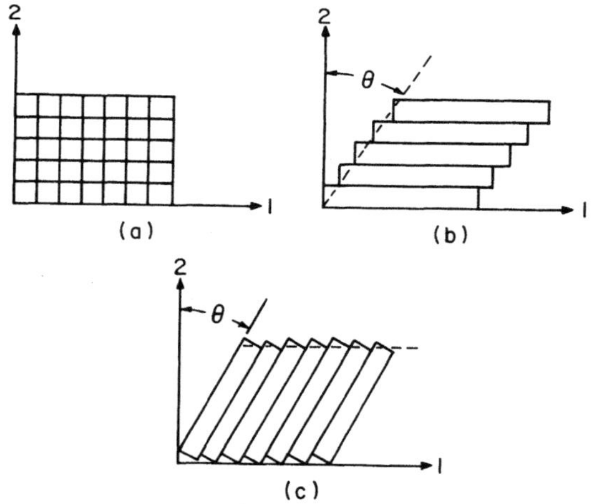

Fig. 3. Different lattice rotations resulting from the same average shear increment, θ, depending upon which slip systems are activated (from Argon, in press; courtesy of Pergamon Press, Oxford).

The most convenient way to describe the lattice rotations due to slip, in relation to material lines and planes inscribed on the body, is by considering the lattice itself as the frame of reference and determining the rotation of material lines and planes through this reference frame. Since crystallographic slip is a process of simple shear, the two separate parts of the strain resolution factors ${}^{s}\alpha_{ij}$ of eqn (3b) for any given slip system s acquire separate and distinct meanings. Therefore, we rewrite eqns (3a) and (3b) for the pure tensor shear strain increment as

$$\delta\varepsilon_{ij}^{p} = (1/2)\,(n_j d_i + n_i d_j)\delta\gamma = \frac{1}{2}(m_{ji} + m_{ij})\delta\gamma \tag{10}$$

for any single slip system, where we have introduced a single symbolic notation m_{ij} for the products of the direction cosines for the two unit vectors and have considered the magnitude of the shear strain $\delta\gamma$ as the basic scale for the tensor strain increment. We note that the m_{ij} are merely the deformation gradients per unit shear strain increment in the crystal frame of reference. We use the m_{ij} to define a distortion tensor $\mathrm{d}\beta_{ij}$ as

$$\mathrm{d}\beta_{ji} = m_{ji}\delta\gamma = \frac{\partial u_i}{\partial X_j} \tag{11a}$$

$$\mathrm{d}\beta_{ij} = m_{ij}\delta\gamma = \frac{\partial u_j}{\partial X_i} \tag{11b}$$

where u_i and X_i are the displacements and the coordinates of points in the basic cartesian lattice frame of reference. The increments of rotation of material *lines* $\mathrm{d}\Omega_{ij}^{l}$, and *plane* normals $\mathrm{d}\Omega_{ij}^{p}$ resulting from simple shears produced by slip translations are then related directly to the elements of the distortion tensor by the following simple relations based on examination of the results of diagrams, such as those in Fig. 3,

$$\mathrm{d}\Omega_{ij}^{l} = \mathrm{d}\beta_{ij} \tag{12a}$$

$$\mathrm{d}\Omega_{ij}^{p} = -\mathrm{d}\beta_{ji} \tag{12b}$$

We interpret these as the rotation of a material line initially parallel to the i-axis of the lattice towards the direction of the j-axis of the lattice, and the rotation of the unit normal of the material plane initially parallel to the i-axis of the lattice towards the direction of the j-axis of the lattice, respectively. Figure 4 gives an example that clarifies these rotations. Consider a rectangular block of material with its edges initially coinciding with the three principal lattice axes, x, y and z. Suppose the block suffers a simple shear strain $d\gamma$ by slip on the x–y planes (with unit normal parallel to the z-axis) parallel to the negative y-direction, as shown in the figure, producing the dotted external outline, while the principal lattice axes remain parallel to their initial directions. The components of the distortion tensor for this deformation gradient are

$$d\beta_{zy} = m_{zy} d\gamma = \frac{\partial u_y}{\partial X_z} = n_z d_y d\gamma = -d\gamma \tag{13a}$$

$$d\beta_{yz} = d\beta_{xy} = d\beta_{yx} = d\beta_{xz} = d\beta_{zx} = 0 \tag{13b–f}$$

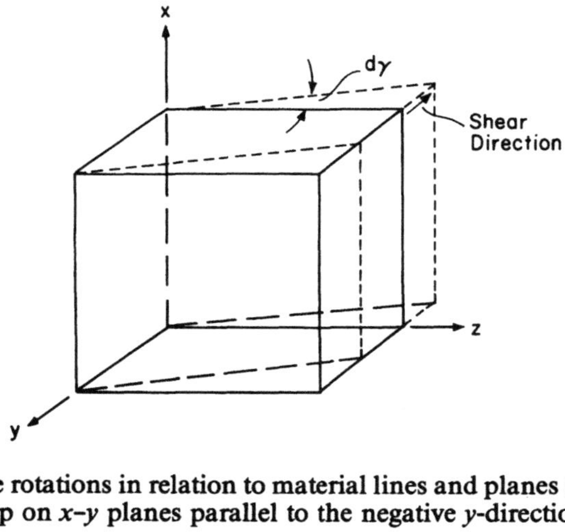

Fig. 4. Lattice rotations in relation to material lines and planes resulting from slip on x–y planes parallel to the negative y-direction.

Equations (12a) and (12b) indicate the existence of the following rotations of the three principal material lines initially parallel to the x-, y- and z-axes; and the three principal material planes initially normal to the unit vectors parallel to the x-, y-, and z-directions, all in reference to the lattice:

$$d\Omega^l_{zy} = d\beta_{zy} = -d\gamma \tag{14a}$$

$$d\Omega^l_{yz} = d\Omega^l_{xy} = d\Omega^l_{yx} = d\Omega^l_{xz} = d\Omega^l_{zx} = 0 \tag{14b–f}$$

and

$$d\Omega^{p}_{yz} = -d\beta_{zy} = d\gamma \tag{15a}$$

$$d\Omega^{p}_{zy} = d\Omega^{p}_{yx} = d\Omega^{p}_{xy} = d\Omega^{p}_{zx} = d\Omega^{p}_{xz} = 0 \tag{15b–f}$$

Thus, we are informed that only the material line initially parallel to the z-axis rotates toward the negative y-axis 'through' the lattice, and the material plane with unit normal vector initially parallel to the y-axis rotates toward the z-axis 'through' the lattice. All other principal material lines and planes remain fixed relative to the lattice.

We have discussed only the basic kinematical essence of lattice rotations arising from slip. The efficient operational implementation of lattice rotations for the purpose of modelling texture evolution requires some additional considerations that have been discussed by Kocks and Chandra (1982).

3. THE FLOW STRESS

3.1. The Lattice Resistance

Directionally bonded solids including body-centred cubic (bcc) solids and some hexagonal metals offer an intrinsic lattice resistance to the motion of dislocations at low temperatures. This arises because the structure of dislocation cores has significantly lower energy along specific lattice rows than between such rows. Thus, the dislocation cores are 'bound' to specific lattice rows, requiring a definite stress to move them from row to row. The problem was analysed first by Peierls (1940) in a generic way, and in more recent times by many special computer-simulation studies using specific inter-atomic pair potentials to represent the non-linear interactions of atoms at the large separations that occur inside dislocation cores.

Atomistic models of dislocation cores have been developed primarily for bcc metals and ionic solids. These have been reviewed comprehensively by Puls (1981). Several studies have also been performed recently on hexagonal close packed (hcp) metals (Bacon and Martin, 1981).

In all these modelling studies, the basic goal has been to obtain information on the structure of the dislocation core, and the stress required to displace the dislocation in the absence of thermal motion. The procedure for this involves replacing a certain portion of the singular region near the centre of the dislocation in the linear anisotropic solution of the strain field with the appropriate lattice of atoms interacting with each other by an atomic pair potential. The 'welding' condition of the discrete central region to the outer anisotropic linear elastic continuum field is an important feature affecting the accuracy of the solution. In the currently most widely used scheme, called Flex II, a feature is built into the relaxation process for the atoms of the central region to affect the outer continua field by the application of a series of Green's function solutions. The final solution is obtained by placing the atoms of the inner region at positions calculated from the anisotropic linear continuum solution and relaxing the excess energy of the atoms according to the requirements of the pair potentials. The relaxations are also applied to the border region between the inner discrete material and the outer continuum by the Green's function approach already mentioned. When fully relaxed, the positions of atoms in the central region give the structure of the dislocation core. To obtain the stress needed to move the dislocation, uniform shear is imposed on the outer region and its effect along the border region is used to seek the new structure of the core in equilibrium with the imposed shear. The procedure is carried out step-wise under gradually increasing shears until the centre of the dislocation or the variously disoriented and distorted core structure is observed to displace in a self-similar manner. This is taken as a measure of the so-called Peierls resistance to motion of the dislocation at 0 K.

A great deal of understanding has been derived from these models for the structure of dislocation cores, and whether edge, screw or special mixed dislocations feel the most resistance to their motion. This understanding has served to explain many of the slip anisotropies known in bcc metals that result in the selection of different slip systems in tension and in compression for a given uniaxial orientation of straining. Quantitative agreement between model predictions and experimental results for the lattice resistance in bcc metals, back-extrapolated to 0 K has, however, been quite poor. In most cases, the computations that have used empirical pair potentials have given Peierls resistances almost an order of magnitude larger than the best experimental measurements (Puls, 1981). Even in the most discriminating use of a study on bcc potassium, using the best pair potentials, the theoretical answer was found to be no closer than a factor of 3 above the experimental measurement performed in the same laboratory (Basinski and Duesbery, 1981). It is the opinion of the investigators in this field (Puls, 1981), that the principal cause of the discrepancy is that the empirical pair potentials that are used to model atomic interactions are inadequate. Even more sophisticated pseudo-potentials tailored to match many physical and structural bulk properties, as well as some lattice defects, and incorporating even non-central force contributions to the total binding potential, are still not good enough to model cohesive and ideal shear properties of atoms in the core region of the dislocation. Thus, in this area of modelling, further meaningful progress is likely to come only when the entire core region atoms of a dislocation can be treated by quantum mechanics or at least when suitable pair potentials are developed for specific materials from first principles (Puls, 1981).

The modelling of dislocation cores and the associated lattice shear resistances are meant to be applicable to conditions at 0 K, where thermal motion is completely absent. The values computed range as follows: (a) for bcc transition metals, $\hat{\tau}_l = 2\text{–}7 \times 10^{-2}\mu$; (b) for bcc alkali metals, $\hat{\tau}_l = 2\text{–}7 \times 10^{-3}\mu$; and (c) for alkali halides, NaCl, $\hat{\tau}_l = 3{\cdot}5 \times 10^{-3}\mu$; KCl, $\hat{\tau}_l = 2 \times 10^{-3}\mu$, and for MgO, $\hat{\tau}_l = 1{\cdot}5 \times 10^{-3}\mu$ (Puls, 1981), where μ in each case is the appropriate shear modulus. Of these results, all but those for alkali halides are likely to be substantially too high because of 'rigid' imposed strain border conditions that were used in the simulation. The Flex II border condition, which was described above, was used only in the case of alkali halides.

At finite temperatures, the entire dislocation line need not overcome the lattice resistance everywhere at the same time. Thermal fluctuations can produce a bulge that locally puts a short dislocation segment into the next lattice valley, as shown in Fig. 5. This 'nucleus' can then sweep into the next valley. The net frequency of such events is governed by an

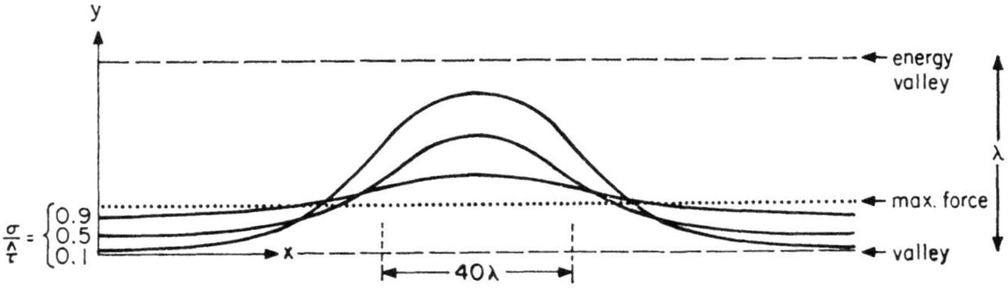

Fig. 5. Shapes of bulges in a dislocation line, representing critical activation configurations over the lattice resistance potential, for three different levels of stress — 0·1, 0·5 and 0·9 — of the athermal resistance at 0 K (from Kocks *et al.*, 1975; courtesy of Pergamon Press).

Arrhenius expression in which the activation free energy is reduced with increasing stress, and eventually vanishes when the stress reaches the level of the lattice resistance at 0 K (adjusted for reduction in elastic constants above 0 K). This thermally assisted overcoming of the lattice resistance results in a monotonic and, generally, sharp reduction of the overall plastic resistance of the solid that is characteristic of covalent solids and bcc metals. Models for the thermally activated overcoming of the lattice resistance have been reviewed by Kocks *et al.* (1975). As we have already discussed in Chapter 1, the resulting temperature dependence of the yield stress is often quite well characterized by a phenomenological form of

$$\frac{\sigma}{\hat{\tau}} = \left(1 - \left(\frac{T}{T_0}\right)^{3/4}\right)^{4/3} \tag{16a}$$

where

$$T_0 = \frac{\Delta H}{k} \frac{1}{\ln(\dot{\gamma}_0/\dot{\gamma})} \tag{16b}$$

In eqns (16a) and (16b) T_0 is the temperature where the lattice resistance effectively vanishes due to the intensity of thermal motion, and where $\hat{\tau}$ is the lattice resistance in the absence of thermal motion. For the latter, a good measure is the product of the lattice resistance at 0 K with the ratio of the shear modulus at the test temperature to that at 0 K.

3.2. The Particle Resistance

When additional constituents are present in an alloy in the form of second-phase particles, an extrinsic glide resistance appears on the slip plane as the motion of dislocations are impeded by discrete obstacles. This type of discrete obstacle resistance is also present as an intrinsic process when forest dislocations penetrating through a slip plane have to be cut by the glide dislocations. When the lattice resistance that was discussed in Section 3.1 vanishes above T_0, the discrete obstacle resistance is the only remaining contribution to the plastic resistance. This resistance is often dominant, and together with solid solution resistance, it is the principal means of obtaining hardened alloys.

When a glide dislocation encounters discrete obstacles in its plane, it is locally held back at these obstacles, as the remainder of the unsupported dislocation bows out between these pinning points, as shown in Fig. 6. As already mentioned above, the obstacles can be precipitate particles of different types, offering a resistance against penetration by a dislocation by various mechanisms, or they can be

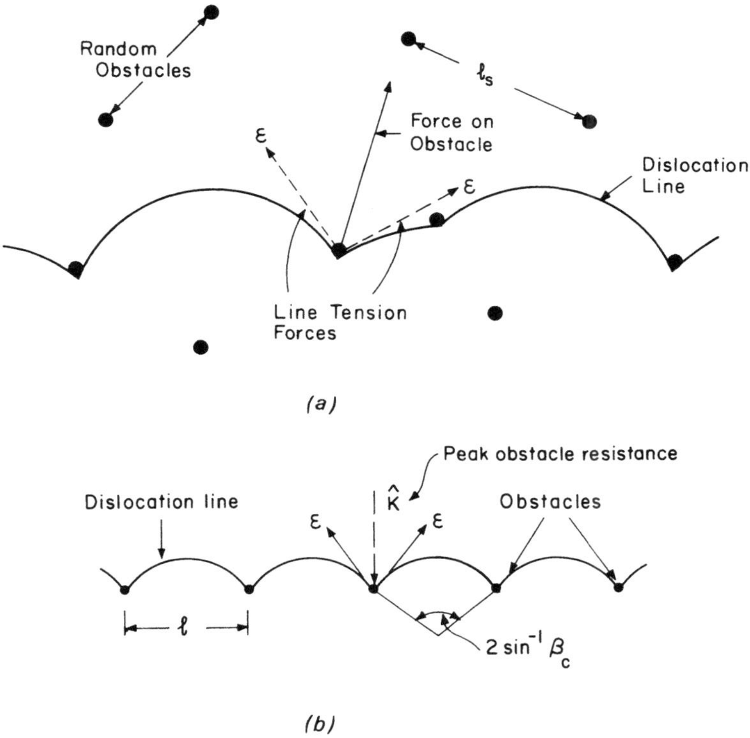

Fig. 6. Schematic, cuspy shape of a dislocation line: (a) pressed against a set of irregularly placed point obstacles; (b) pressed against equally spaced point obstacles strung up on a line.

forest dislocations of at least two different types that either repel or attract the glide dislocation. In all cases, the detail of the interaction between obstacles and the glide dislocation, as well as the overall level of the resistance is governed by the specific obstacle resistance to penetration by the glide dislocation.

This resistance is expressed as a characteristic curve of resistive force vs. distance of penetration. The most important single property of the resistive force–distance curve is the maximum resistive force $\hat{K}$, which is usually measured in units of twice the line tension of a dislocation $2\mathscr{E}$, and is abbreviated as β_c. In most statistical computer models, the differences between the energies of edge and screw dislocations are ignored and the line tension is taken to be equal to the average line energy of the edge and screw dislocations. In this so-called constant line tension approximation, the shapes of dislocation segments bowed out between obstacles are arcs of circles, and the overall force exerted on a dislocation segment by the applied resolved shear stress τ_a is equilibrated by the components of the line tension forces applied at the end points of the segment, opposite to the direction of principal bow-out indicated as the y-directions in Fig. 6(a). By continuity of force at every point along the line, these same line tension force components are also exerted on the obstacle. The bow-out condition of the largest curvature in the free dislocation segments is reached when the line tension components in the principal direction of motion exerted on a typical obstacle from both sides reach the peak obstacle strength $\hat{K}$. The maximum force that can be exerted on obstacles by such bowing out of segments around obstacles is $2\mathscr{E}$, and is the reason for normalization of peak obstacle resistance by $2\mathscr{E}$. Clearly, obstacles with peak strength $\beta_c > 1$ cannot be penetrated by dislocations. Such obstacles are termed impenetrable, but are nevertheless still overcome when the dislocation that wraps itself around the obstacle pinches off around it and continues to move ahead.

If all obstacles were strung up along linear rows with equal spacing between them, then the obstacles could be overcome by a cuspy configuration of the dislocation in which each bowed-out segment subtends an angle of $2\sin^{-1}\beta_c$, as shown in Fig. 6(b). In reality, quite often, second-phase particles are nearly randomly distributed in space (Argon and Im, 1975) or on crystallographic planes, and forest dislocations can be idealized to also have randomly placed points of penetration through slip planes, at least when this point density in the plane is small so that mutual interactions between these forest dislocations are weak. In this case, the glide plane resistance to the motion of dislocations is a statistical percolation problem that has been discussed quite exhaustively, starting with Kocks (1966) and Foreman and Makin (1966). We will review here the results of some elegant computer modelling of this problem by Morris and coworkers (Hanson and Morris, 1975*a*, *b*, 1978; Altintas, 1978) which have been independently verified by even more elegant analytical approaches by Labusch (1977).

In this percolation problem, point obstacles of characteristic normalized peak strengths β_c are placed randomly on a plane and a dislocation line is introduced on one edge of the plane, as shown in Fig. 7, and is advanced under increasing stress, if necessary, until the entire plane can be swept by the dislocation. The maximum resistance of the plane occurring at a certain configuration where the largest linear density of pinning points is touched is taken as the terminal glide plane resistance at 0 K, if the analysis treats the points with temperature- and time-independent strength. These computer results are somewhat sensitive to the extent of the point field considered. By the use of periodic boundary conditions and exploration of the size dependence of the answer, it has been found that the peak glide resistance of the plane becomes independent of size when the roughly square field contains about 3–4 × 10^5 points with relatively low strength ($\beta_c \approx 0{\cdot}01$) (Altintas, 1978). Far smaller numbers are adequate for points of high strength.

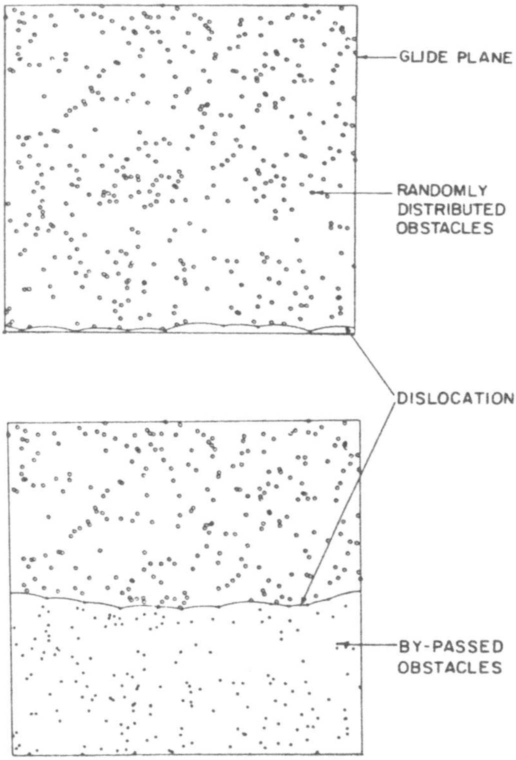

Fig. 7. Two successive positions in the passage of a dislocation line across a random field of point obstacles in a computer simulation (from Altintas, 1978; courtesy of University of California, Berkeley).

The computer simulations have shown that the glide plane resistance $\hat{\tau}_p$ for obstacle points with an average spacing l_s in the plane, and peak strength β_c is given by

$$\hat{\tau}_p = \alpha(\beta_c)^{3/2}(2\mathscr{E}/l_s b) \tag{17}$$

where $\alpha = 0{\cdot}887$ is the principal result of the computer simulation, while the remainder of the terms gives the reference glide resistance of the same obstacle points when they are strung up along a line with constant spacing l as idealized in Fig. 6(b), where $l = l_s(1/\beta_c)^{1/3}$ is the so-called Friedel sampling length. An instructive result of the analysis is the frequency distribution of the segment lengths (distances between pinning points) in the initial configuration. An example of this is shown in Fig. 8, obtained from a simulation for $\beta_c = 0{\cdot}1$ (moderately weak obstacles) for which the mean segment length $\bar{l} = 1{\cdot}5\,l_s$ along the critical configuration of the dislocation line (Hanson and Morris,

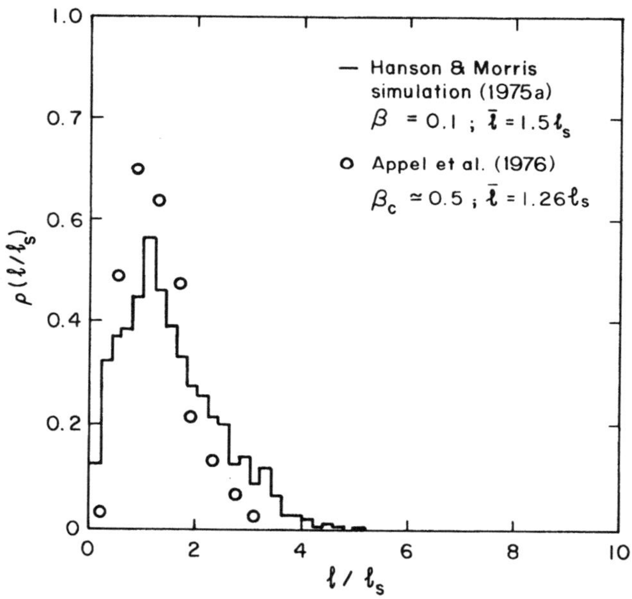

Fig. 8. Distribution of segment lengths along dislocation lines at various configurations traversing through a field of random point obstacles with normalized peak strength $\beta_c = 0{\cdot}1$, as calculated by Hanson and Morris (1975*a*). The points are from experiments of Appel *et al.* (1976) on MgO crystals.

1975*a*). This is compared with the experimental distribution obtained by Appel *et al.* (1976) in MgO crystals from micrographs similar to that shown in Fig. 9, for which the peak obstacle strength was much higher ($\beta_c \simeq 0{\cdot}5$), giving a mean segment length $\bar{l} = 1{\cdot}26 l_s$. The similarity in the shapes of the segment distributions is satisfying.

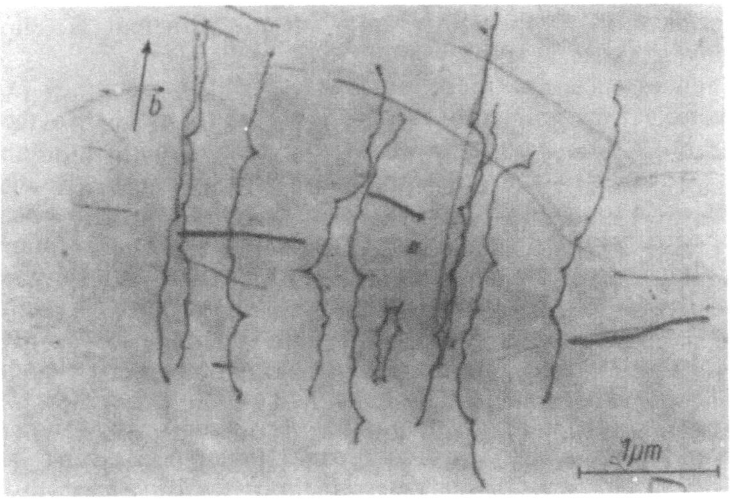

Fig. 9. Transmission electron micrograph of dislocations in an MgO crystal pressed against point obstacles of approximate peak strength of $\beta_c \simeq 0{\cdot}5$ (from Appel *et al.*, 1976; courtesy of *Phys. Stat. Sol.*).

Such computer simulations with their large volumes of associated graphical outputs have been very instructive in understanding the nature of the motion of dislocations through slip plane obstacles. Among the many useful results is the slip plane glide resistance for two or more different types of obstacles of different peak strengths. For example, for the case of a set of weak and a set of strong obstacles with peak strengths β_w and β_s making up fractions c_w and c_s of the total point density l_s^{-2}, the overall slip plane glide resistance $\hat{\tau}_p$ at 0 K is given by (Altintas, 1978):

$$\hat{\tau}_p \simeq \alpha(2\mathscr{E}/l_s b)(c_w\beta_w^3 + c_s\beta_s^3)^{1/2} \tag{18}$$

where $\alpha \simeq 0{\cdot}887$ as in the case of eqn (17) above. This combination law for two types of obstacles is very useful in the case of forest dislocations where those forest dislocations that attract glide dislocations by reacting with them at the point of contact act as if they were impenetrable obstacles with $\beta_s = 1{\cdot}0$ and constitute a set of strong obstacles. Alternatively, those that repel glide dislocations and can be intersected readily with the production of two simple jogs for which β_w is generally of the order of 0·1 act as a set of weak obstacles. The ratio c_s/c_w of the densities of these obstacles in the plane depends somewhat on the orientation of the slip plane with the principal direction of deformation and what other slip planes are also active.

When obstacles are penetrable with $\beta_c < 1{\cdot}0$, then they can be overcome at finite temperatures $T > 0$ K with the partial help of thermal fluctuations. Since, during a thermal fluctuation, the dislocation line segment pinned in front of the obstacle is made to traverse through the obstacle at a constant cusp angle, the actual resistive force–penetration distance curves of obstacles need to be known, to determine the activation free energy that governs the rate of activation through the forced obstacles. Since the interaction of the obstacle with the dislocation may have a number of different component mechanisms, force–distance curves are usually composite sums of elementary resistive force–distance curves. The determination of such force–distance curves from first principles has been discussed by a number of investigators and an account of some examples has been given by Kocks *et al.* (1975). The experimental determination of resistive force–distance curves is, in principle, possible when additional metallographic information is available for the density of obstacles. An example of the determination of force–distance curves for penetrable forest dislocations in pure Cu, where the combination law given by eqn (18) has been used in conjunction with the overall density of forest dislocations obtained by an etch pit technique, has been discussed by Argon and East (1980).

Thermally-activated penetration of randomly positioned point obstacles in the plane has also been extensively simulated by a stochastic release of dislocations from pinning points subject to cutting forces by the cusped dislocations. In such simulations, the pinning points are treated as being subject to thermal release, according to an Arrhenius expression for the release rate that is a specific function of temperature and the force applied to the obstacle. The overall velocity of the dislocation is obtained by dividing the distance moved by a quasi-straight dislocation over many mean obstacle spacings by the total integrated time (Altintas, 1978).

4. PLASTIC RESISTANCE OF MICROCOMPOSITES

4.1. Plastic Resistance of Coarse Heterogeneous Continua

As discussed in Section 1, when the scale of the heterogeneities is much larger than the influence range of individual dislocations, i.e. grain size or slip band length, the solid appears to deform as a continuum in the stable range in which there are no constitutive instabilities. If such a solid is heterogeneous, with prescribed plastic resistances of the individual component phases, its overall plastic resistance can be modelled by micromechanics where interactions between the compatibly deforming heterogeneities are prescribed by the traction and displacement border conditions between the different phases. A large number of such problems have been modelled to give the behaviour of plastic media containing more or less deformable heterogeneities of all types and shapes. In such studies, when the heterogeneities have large aspect ratios, occupy a large volume fraction or are aligned, the specific features of the heterogeneities become important and must be dealt with by special numerical procedures utilizing, for example, finite element approaches. On the other hand, when the heterogeneities are relatively equiaxed, take up a small volume fraction, or are randomly mixed with the majority phase in the form of topologically interpenetrating domains, the overall deformation resistance can be determined by the self-consistent method developed by Hershey (1954), Kröner (1958), and more generally by Hill (1965*a, b*). Here, we will discuss only two such limiting problems of interest.

The first is the plastic resistance of a two-component solid, in which the stress–strain laws of the individual components are given from information derived from studies of the individual components in isolation, and the assumption is made that the compatible deformation of the two components in contact does not alter these stress–strain laws. A limiting case of this problem is when a topologically isolated, included phase is rigid.

The second problem is the case where the topologically isolated, included phase is a collection of voids.

4.2. A Two-Component Heterogeneous Plastic Medium

We consider the incremental plastic response of a heterogeneous alloy of a coarse microstructure with a random and relatively equiaxed mixture of phases having power-law hardening (or power-law creep) behaviour. One phase of a volume fraction c_i has a higher initial yield strength but low hardening rate (strong stress dependence of creep rate), while the second with a volume fraction $c_o = 1 - c_i$ has a low initial yield stress but a higher hardening rate (weak stress dependence of creep rate). We consider the initially harder (more creep resistant at small strain rates) phase with lower hardening rate (strong stress dependence of creep rate) as the included phase, and the second phase the majority phase, i.e. $c_i < c_o$. The equivalent strain increment (strain rate), equivalent stress relation for these phases can thus be given as:

$$d\varepsilon_i = (\text{or } \dot{\varepsilon}_i) = F_i(\sigma_i)^{m_i} \tag{19a}$$

and

$$d\varepsilon_o = (\text{or } \dot{\varepsilon}_o) = F_o(\sigma_o)^{m_o} \tag{19b}$$

where

$$F_i = d\varepsilon_{io}/(\sigma_{io})^{m_i} = (\text{or } \dot{\varepsilon}_{io}/(\sigma_{io})^{m_i}) \tag{20a}$$

$$F_o = d\varepsilon_{oo}/(\sigma_{oo})^{m_o} = (\text{or } \dot{\varepsilon}_{oo}/(\sigma_{oo})^{m_o}) \tag{20b}$$

are convenient decompositions of the coefficients in the deviatoric isotropic incremental, or rate relations of eqns (19a) and (19b). As stated above, we take

$$\sigma_{oo} < \sigma_{io}, \quad m_o < m_i \quad \text{and} \quad c_i < c_o$$

The overall incremental plastic (or creep rate) response of the heterogeneous plastic (or creeping) continuum has been determined by Chen and Argon (1979) by means of the self-consistent method in an incrementally linear manner as a problem of creeping continua. Their general result for different levels of F_i and F_o and different composition with different c_i is complex and a function of the level of equivalent stress. A series of such solutions have been discussed by these authors in which the two phases were treated as quasispherical domains.

We discuss here only the asymptotic results for the composite behaviour at very low stresses, $\sigma \ll \sigma_c$ and at very high stress, $\sigma \gg \sigma_c$, where σ_c is the deviatoric stress for which both components have the same incremental plastic response (or the same equivalent creep rate), and the heterogeneous solid deforms homogeneously.

Thus, in the low stress limit ($\sigma_e \ll \sigma_c$), the equivalent strain increments $d\varepsilon_e$ (or the equivalent creep rate $\dot{\varepsilon}_e$) can be given by the following analytical expressions for the composite alloy:

$$d\varepsilon_e = (\text{or } \dot{\varepsilon}_e) = F_o\left(1 - \frac{5}{2}c_i\right)^{m_o} c_o\left(\frac{\sigma_e}{C_o}\right)^{m_o} \text{ for } (c_i < 0{\cdot}4) \quad (21a)$$

$$d\varepsilon_e = (\text{or } \dot{\varepsilon}_e) = F_i\left(1 - \frac{5}{2}c_o\right)^{-1} c_i\left(\frac{\sigma_e}{C_i}\right)^{m_i} \text{ for } (c_i > 0{\cdot}6) \quad (21b)$$

Equations (21a) and (21b) give the behaviour of the composite alloy where the less deformable component (subscript i) is the topologically isolated minority phase and the topologically interconnected majority phase, respectively. At a volume fraction of $c_i = 0{\cdot}4$, the random touching probability of the minority phase becomes very high and a phase inversion occurs. Above $c_i > 0{\cdot}6$, the more resistant phase (at low stress) becomes the topologically continuous one. The region of compositions between $0{\cdot}4 < c_i < 0{\cdot}6$ is inaccessible to the one-step self-consistent method based on randomness of phase distribution and requires specific models for specific phase distributions.

In the high stress limit ($\sigma_e \gg \sigma_c$), the equivalent strain increments $d\varepsilon_e$ (or the equivalent creep rates $\dot{\varepsilon}_e$) for the composite alloy show a complementary behaviour to that of the low stress limit discussed above, and are

$$d\varepsilon_e = (\text{or } \dot{\varepsilon}_e) = F_o\left(1 - \frac{5}{2}c_i\right)^{-1} c_o\left(\frac{\sigma_e}{C_o}\right)^{m_o} \text{ for } (c_i < 0{\cdot}4) \quad (22a)$$

$$d\varepsilon_e = (\text{or } \dot{\varepsilon}_e) = F_i\left(1 - \frac{5}{2}c_o\right)^{m_i} c_i\left(\frac{\sigma_e}{C_i}\right)^{m_i} \text{ for } (c_i > 0{\cdot}6) \quad (22b)$$

From the asymptotic results of eqns (21a) and (21b) and eqns (22a) and (22b) we note that the topologically continuous phase dominates the behaviour of the heterogeneous composite at both low and high stresses. In eqns (21a) and (22a), both the coefficients of the strain increment expression, as well as the stress exponent, are those of the topologically continuous majority phase designated as (o). This behaviour is merely modulated by the presence of the included minority phase (i). At low stress levels, the included phase is essentially unrelaxed and acts as a filler to lower the gradient of the plastic potential of the composite (or raise the non-linear viscosity of the creeping solid); that is, the equivalent strain increment of the composite is lower than that of the pure matrix phase for the same level of equivalent stress. This is shown in Fig. 10 as the lower branch, as a function of the volume fraction c_i of the less deformable phase for a stress exponent of $m_o = 5$ of the majority phase. At high stress levels, where the phase with the higher stress exponent now becomes the more easily deformable phase, the included phase increases the gradient of the plastic potential to increase the equivalent strain increments in the composite above those

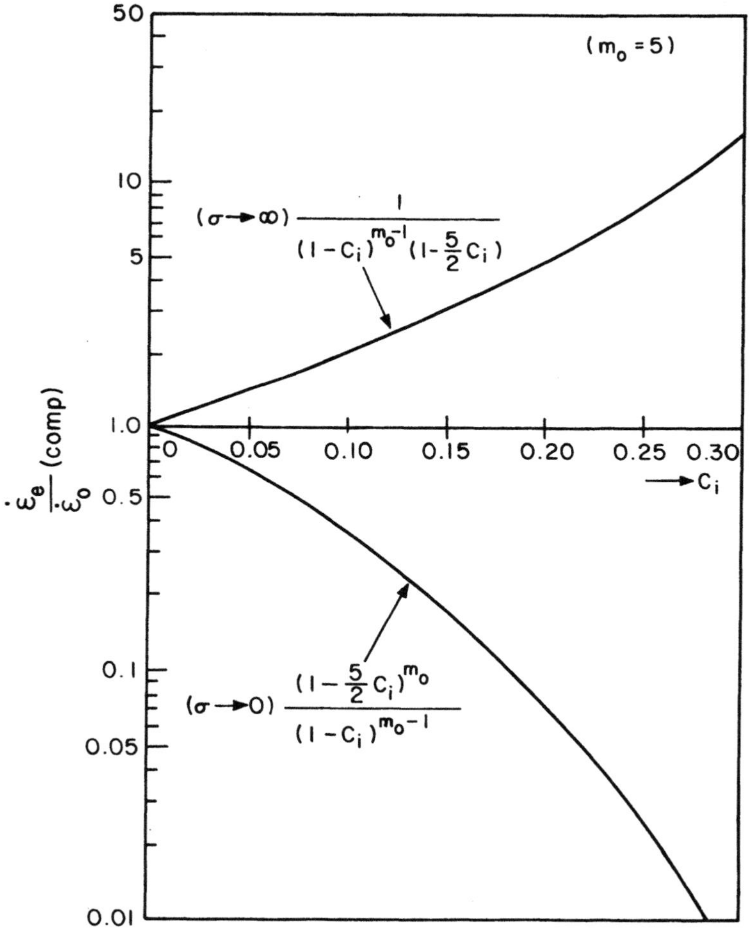

Fig. 10. Dominance of the behaviour of the majority phase at both low stress ($\sigma \ll \sigma_o$) and high stress ($\sigma \gg \sigma_o$) in a composite for $c_i < 0{\cdot}4$.

in the pure matrix phase for the same level of equivalent stress. Now, the included phase is relaxed and increases the equivalent stress on the surrounding majority phase. This behaviour is shown by the upper branch curve in Fig. 10 as a function c_i.

If the included phase were rigid, the composite would abruptly lock up at a volume fraction $c_i = 0{\cdot}4$, and would cease to deform above this volume fraction. This abrupt behaviour is a characteristic of the 'one step' self-consistent model used by Chen and Argon. A more relaxed limiting solution is obtainable by considering the effect of incremental additions of volume fraction of rigid phase on current composite properties. This has been done by Duva (1984), who found that the attenuation of the strain increments (strain rates) with increasing volume fraction of rigid inclusion is gradual and approaches the locking condition only asymptotically, as the volume fraction of filler phase approaches unity. The real behaviour of a solid clearly is between the abrupt locking behaviour at $c_i = 0{\cdot}4$ and the somewhat unrealistic very gradual asymptotic approach to vanishing strain increments (or strain rates).

4.3. Plastic Resistance of Randomly Voided Solids

The overall plastic resistance of a solid with a given volume concentration of voids and its response to different states of stress is a problem of long-range interest in the understanding of ductile fracture and creep fracture. The subject has received some notable attention in the past from McClintock (1968) and Rice and Tracey (1969) for the computation of the strain to fracture by quasi-homogeneous plastic cavity expansion from given levels of initial porosity, under states of stress with different triaxiality. In addition to these pioneering approaches, many other useful developments have been presented: for mutually interactive expansion of cavities along a plane to determine traction displacement laws, for ductile separation across the plane (Nagpal *et al.*, 1972); for generalized yield surfaces of porous plastic solids under combined shear and negative pressure (Gurson, 1977); for quasi-homogeneous cavitation of creeping solids (Hancock, 1976); and for intergranular cavitation by combined creep deformation and diffusional transport (Needleman and Rice, 1980).

Recently, more general solutions of the expansion of single ellipsoidal cavities in plastically deforming or creeping solids of infinite extent, under very general stress states, have been studied by Budiansky *et al.* (1981). These have lent themselves as so-called 'kernel problems' for the statement of general constitutive laws for the deformation of randomly and dilutely voided solids (Duva and Hutchinson, 1984), with important applications in the development of shear localizations by ductile cavitation, and in creep fracture (Argon *et al.*, 1985). We will consider these recent generalizations only briefly.

As has been discussed by Duva and Hutchinson (1984), the inelastic rigid behaviour of a cavitating solid is best described by a scalar potential in the framework of a rate-dependent non-linearly viscous behaviour from which specific application to incremental plastic behaviour is readily possible.

Consider the non-linear viscous (steady state creep) response of the solid to be represented by a relation between equivalent strain rate $\dot{\varepsilon}_e$ and equivalent stress σ_e in the form of a power-law as

$$\dot{\varepsilon}_e = \dot{\varepsilon}_0(\sigma_e/s_0)^m \tag{23}$$

where $\dot{\varepsilon}_0$ is the coefficient that can be temperature dependent, s_0 is a reference scalar deformation resistance that can in principle be subject to special evolutionary changes (Argon *et al.*, 1985) and m is the non-linearity exponent. As these authors show, the creep response of the basic continuum can be obtained from a potential function ϕ defined as (Hutchinson, 1983):

$$\phi = \frac{\dot{\varepsilon}_0 s}{(m+1)}\left(\frac{\sigma_e}{s_0}\right)^{m+1} \tag{24}$$

leading by differentiations to individual strain rates of

$$\dot{\varepsilon}_{ij} = \frac{\partial\phi}{\partial\sigma_{ij}} = \frac{3}{2}\dot{\varepsilon}_0\left(\frac{\sigma_e}{s_0}\right)^{m-1}\frac{s_{ij}}{s_0} \tag{25}$$

where s_{ij} are the deviator stress elements and the equivalent stress is given by the usual expression

$$\sigma_e = \left(\frac{3}{2}s_{ij}s_{ij}\right)^{1/2} \tag{26}$$

When a volume fraction ρ of randomly placed voids is present, the creep behaviour of the solid is perturbed, resulting in an additional potential function modifying eqn (24) for the uncavitated solid, giving for the cavitating solid:

$$\phi = \frac{\dot{\varepsilon}_0 s}{(m+1)}\left(\frac{\sigma_\varepsilon}{s_0}\right)^{m+1}[1 + \rho f(X,m)] \tag{27}$$

In this expression, $f(X,m)$ incorporates the principal results of the solution of problems involving ellipsoidal cavity growth in infinite media. In their general consideration, Duva and Hutchinson (1984) obtained several limiting forms for the function $f(X, m)$ in which $X = \sigma_m/\sigma_c$ is the triaxiality ratio in the distant field. The components of the tensor creep rates can be obtained again by differentiation operations given by eqn (25). The specific solutions for $f(X, m)$ that have been presented by Duva and Hutchinson will not be considered here. Instead, we will discuss only a specific application of this approach to the problem of acceleration of creep by formation and enlargement of separated grain boundary facets which act as very oblate spheroids normal to the direction of maximum principal tension.

For the problem of enlargement of cracked grain boundary facets by power-law creep flow, the potential function becomes

$$\phi = \frac{\dot{\varepsilon}_0 s_0}{(m+1)} \left(\frac{\sigma_e}{s_0}\right)^{m+1} \left[1 + \rho \left(X + \frac{2}{3}\right)^2\right] \tag{28}$$

where

$$\rho = \frac{4}{\pi\sqrt{14}} \frac{1+m}{\left(1 + \frac{3}{m}\right)^{1/2}} \frac{\alpha}{(1-\alpha^p)^{m/p}} \tag{29}$$

and α is the fraction of fractured grain boundary facets within a layer of one grain diameter, projected on a plane normal to the local maximum principal tensile stress (Argon *et al.*, 1985), and p (usually about 1·3) is a fitting exponent representing the interaction of neighbouring cavities.

Argon *et al* (1985) have adapted this formula to the problem of creep fracture and have stated creep constitutive relations that have incorporated specific evolution laws for the fraction of cracked grain boundary facets during creep. As a test case, the uniaxial tensile creep rate is of interest; this can be found from the potential function of eqn (28) by differentiation and is

$$\dot{\varepsilon}_{11} = \dot{\varepsilon}_0 \left(\frac{\sigma}{s_0}\right)^m (1 + \rho(\alpha,m)) \tag{30}$$

Figure 11 shows the resulting constant stress tensile creep curve computed for a special case of Type-304 stainless steel from the information on grain boundary cavitation rates obtained in earlier experiments by Argon *et al.* (1980). The computed curve compares well with the experimental creep curves obtained for the same alloy by Swindeman (1983).

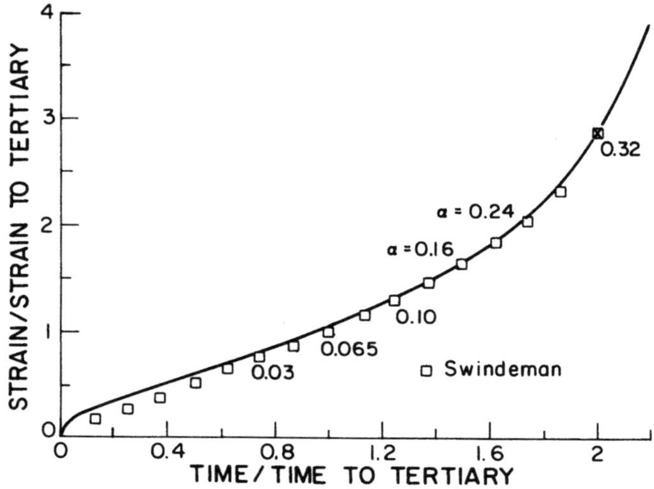

Fig. 11. Computed constant stress tensile creep curve for Type-304 stainless steel compared with data of Swindeman (1983) (from Argon *et al.*, 1985; courtesy of Cambridge University Press).

REFERENCES

Altintas, S. (1978). Plastic deformation of crystals: analytical and computer simulation studies of dislocation glide, PhD Thesis, University of California, Berkeley.

Appel, F., Bethge, H. and Messerschmidt, U. (1976). *Phys. Stat. Sol.,* **38**, 103.

Argon, A. S. (in press). In *Encyclopedia of Materials Science and Engineering*, M. B. Bever (Ed.), Pergamon Press, Oxford.

Argon, A. S. and East, G. H. (1980). In *Strength of Metals and Alloys*, Vol. 1, P. Haasen *et al.* (Eds), Pergamon Press, Oxford, p. 9.

Argon, A. S. and Im, J. (1975). *Met. Trans.,* **6A**, 839.

Argon, A. S., Chen, I. W. and Lau, C. W. (1980). In *Creep-Fatigue-Environment Interactions*, R. M. Pelloux and N. S. Stoloff (Eds), AIME, New York, p. 46.

Argon, A. S., Lau, C. W., Ozmat, B. and Parks, D. M. (1985). In *Fundamentals of Deformation and Fracture*, K. J. Miller (Ed.), Cambridge University Press, p. 243.

Bacon, D. J. and Martin, J. W. (1981). In *Dislocation Modeling of Physical Systems,* M. F. Ashby *et al.* (Eds), Pergamon Press, Oxford, p. 269

Basinski, Z. S. and Duesbery, M. S. (1981). In *Dislocation Modeling of Physical Systems*, M. F. Ashby *et al.* (Eds), Pergamon Press, Oxford, p. 273.
Budiansky, B., Hutchinson, J. W. and Slutsky, S. (1981). In *Mechanics of Solids*, H. G. Hopkins and M. J. Sewell (Eds), Pergamon Press, Oxford, p. 13.
Chen, I. W. and Argon, A. S. (1979). *Acta Met.*, **27**, 785.
Duva, J. M. (1984). *J. Eng. Mat. Tech.*, **106**, 317.
Duva, J. M. and Hutchinson, J. W. (1984). *Mech. Mat.*, **3**, 41.
Foreman, A. J. E. and Makin, M. J. (1966). *Phil. Mag.*, **14**, 911.
Gurson, A. L. (1977). *J. Eng. Mat. Tech.*, **99**, 2.
Hancock, J. W. (1976). *Met. Sci.*, **10**, 319.
Hanson, K. and Morris, J. W., Jr (1975*a*). *J. Appl. Phys.*, **46**, 983.
Hanson, K. and Morris, J. W., Jr (1975*b*). *J. Appl. Phys.*, **46**, 2378.
Hanson, K. and Morris, J. W., Jr (1978). *J. Appl. Phys.*, **49**, 3266.
Hershey, A. V. (1954). *J. Appl. Mech.*, **21**, 236.
Hill, R. (1965*a*). *J. Mech. Phys. Solids*, **13**, 89.
Hill, R. (1965*b*). *J. Mech. Phys. Solids*, **13**, 213.
Hutchinson, J. W. (1983). *Acta Met.*, **31**, 1079.
Kocks, U. F. (1966). *Phil. Mag.*, **13**, 541.
Kocks, U. F. and Chandra, H. (1982). *Acta Met.*, **30**, 695.
Kocks, U. F., Argon, A. S. and Ashby, M. F. (1975). In *Progress in Materials Science*, Vol. 19, B. Chalmers *et al.* (Eds), Pergamon Press, Oxford.
Kröner. E, (1958). *Z. Physik*, **151**, 504.
Labusch, R. (1977). *J. Appl. Phys.*, **48**, 4550.
McClintock, F. A. (1968). *J. Appl. Mech.*, **35**, 363.
Nagpal, V., McClintock, F. A., Berg, C. A. and Subdhi, M. (1972). In *Foundations of Plasticity*, A. Sawczuk (Ed.), Noordhoff, Groningen, p. 365.
Needleman, A. and Rice, J. R. (1980). *Acta Met.*, **28**, 1315.
Peierls, R. E. (1940). *Proc. Phys. Soc.*, **52**, 34.
Puls, M. P. (1981). In *Dislocation Modeling of Physical Systems*, M. F. Ashby *et al.* (Eds), Pergamon Press, Oxford, p. 249.
Rice, J. R. and Tracey, D. M. (1969). *J. Mech. Phys. Solids*, **17**, 201.
Swindeman, R. (1983). Results quoted in Argon *et al.* (1985); private communication.

CHAPTER 6

Physical Processes During Creep

J. GITTUS
Safety and Reliability Directorate, UKAEA, Warrington, UK

ABSTRACT

In polycrystals subjected to low stresses at high temperatures the individual grains move relative to one another rather like grains of flowing sand. This is the phenomenon known as superplasticity and it is a condition of high ductility and low strength. If the stress is increased, dislocations appear in increasing numbers and their movements give rise to permanent deformation of each grain. Now the deformation is more akin to the shear of a pack of playing cards. The phenomenon is termed 'dislocation creep' and differs from superplasticity, being more dependent upon stress and temperature.

In this chapter we review superplasticity and dislocation creep. During creep the crystal dislocations are arranged within each grain in the form of three-dimensional networks. Two types of network have been observed: cells which have the same kind of space-filling geometry as the grains themselves and a general three-dimensional network. In the former case during creep, dislocations glide across the cells causing strain; they glide and climb within the thin arrays that comprise the cell boundaries. It is shown how to calculate the steady-state creep rate from a knowledge of the dimensions of the cells, the dislocation density, the applied stress, the prevailing temperature and various parameters descriptive of the metal itself such as its self-diffusion coefficient. An important development is the calculation of cell diameter using reversible thermodynamics. The creep equation arrived at is one which does not contain any arbitrary constants: in cases where it has been compared with creep data and with microscopical observations of dislocation structure it has proved to yield accurate predictions.

Turning to superplasticity: this is exhibited when the stress is reduced or the temperature raised. A simple view might be that the material becomes superplastic when the dislocation cells become identical in size to the grains. However, an overriding consideration is free energy. Calculations show that superplasticity is a condition of even lower free energy than dislocation creep in the steady state. This is partly due to the fact that dislocations are virtually absent from within the grains during superplastic flow and accordingly the energy associated with these crystal defects is minimized. The condition for superplasticity to predominate is when the free energy for deformation by superplastic flow is lower than that for deformation by dislocation creep. A complete theory of superplastic flow in terms of the same parameters as those invoked for dislocation creep is developed. Here again, reasonable agreement between theoretical prediction and experimental results is shown to prevail.

1. INTRODUCTION

In this chapter two specific models for the deformation of polycrystalline materials are presented. The first describes dislocation creep and the second superplastic flow. The problem in the case of dislocation creep is to explain the fact that the deformation rate depends on a high power of the externally applied stress and is constant for long periods of time. The problem in the case of superplastic flow is to account for the low strength and very high ductility.

Often a material which is superplastic at low stresses deforms by dislocation creep at higher stresses and the transition occurs abruptly; this is another feature that successful theories must seek to explain. There are microscopic differences too: the grains flow like sand in an egg-timer during superplastic flow, changing neighbours but not their shape. During dislocation creep, on the other hand, every grain suffers the same deformation as the polycrystal in which it resides.

Most, if not all, polycrystals appear to be capable of deforming by either process; the determining factors are the magnitude of the applied stress, the temperature and the grain size. High temperatures, low stresses and a fine grain size favour superplastic flow although there is often a threshold below which superplastic flow does not occur, and this too needs to be explained.

The objectives of this chapter are limited to the development of microscopic models and their use to predict macroscopic behaviour for highly stylized polycrystals containing grain boundaries and crystal dislocations. Despite the simplifications introduced, the models are of sufficient accuracy to be of real and practical utility in a number of cases.

2. PHYSICAL PROCESSES DURING CREEP

As explained, the creep process analysed in this chapter is termed dislocation creep. It involves the glide and climb of crystal dislocations, processes which can give rise to strain if they occur in a co-operative sense. The simplest examples are the glide of edge or screw dislocations across a crystal, which produces a simple shear of magnitude b, and the climb of an edge dislocation through the same crystal, which again produces a deformation of magnitude b. The dislocation movements could in principle produce self-cancelling strains, and it is the applied *stress* which causes co-operative movements of dislocations leading to macroscopic strain.

Another creep mechanism is known as diffusion creep. It is believed to involve diffusion of lattice vacancies between adjacent grain boundaries, leading to a change in the shape of each grain. Again the changes are co-operative, so the polycrystal exhibits a macroscopic strain, equal to the microscopic strain exhibited by each individual crystal.

Superplastic flow is itself a form of creep. There are various views on its mechanism and in this chapter it is shown that these earlier ideas can be reconciled by a theory which invokes sliding at or near the grain boundaries by the climb and glide of dislocations residing in or near the boundaries. No doubt this process occurs during dislocation creep but at low stresses it contributes almost all of the strain, dislocations being largely absent from within the grains themselves.

3. MODEL OF THE DISLOCATION CREEP PROCESS

Consider a polycrystal which contains crystal dislocations. Let the macroscopic strain equal the strain in each grain. The strain will be calculated from a simple model which considers the glide and climb of the dislocations within that grain. An hypothesis is first made about the form of the dislocation assembly; it is later justified on thermodynamic grounds.

The physical hypothesis is that the dislocations are arranged, within the grain, in the form of three-dimensional 'cells' having the same kind of space-filling geometry as the grains themselves. Dislocations glide across the cells, causing strain, and they glide and climb within the thin arrays that comprise the cell boundaries.

In fact, of course, experiments *show* that characteristic networks of dislocations tend to form in creep. The networks outline cells within which there are relatively few free dislocations. As the networks do not exert significant long-range stresses, most of the free dislocations within the cells are unaffected by the stresses produced by dislocations in the cell walls (Friedel, 1964). They are still, however, subject to the internal stress produced by line tension.

3.1. Mathematical Modelling

The stress acting to cause glide of the mobile (within cell) dislocations is the difference between the applied stress σ and the internal stress $C_1\mu b/x'$ where

$$1 \geqslant C_1 \geqslant 1/(2\pi) \tag{3.1}$$

Here x' is the mean distance between adjacent dislocations and is therefore a sub-multiple of the cell diameter L.

Holt (1970) has reviewed data on the cell structure and has advanced a theory relating the cell diameter to the mean distance between dislocations. Both the data and his theory indicate that L is of the order of the cell diameter and that the latter is proportional to the mean spacing ($\rho^{-1/2}$) of the dislocations, imagining them to be uniformly spread throughout the volume of the material, i.e.

$$L = K\rho^{-1/2} \tag{3.2}$$

Here ρ is the dislocation density and K is a constant. For copper, deformed at temperatures between 25 and 700 °C under shear stresses of from 10·69 MN m^{-2} (at the highest temperature) to 78·97 MN^{-2} (at the lowest temperature), Staker and Holt (1972) later found that K had a mean value of 16. Other workers have reported values of this order of magnitude for other materials, including, for example, iron and aluminium (Gittus, 1975). We shall therefore make $x' = L$, justifying this on the grounds that this is the tendency indicated by metallography. The value of C_1 depends on the network geometry. In the one extreme case where the spacing of dislocations in the cell walls equals the cell diameter (i.e. the Frank network), the internal stress is predominantly that exerted on each dislocation by its neighbours. C_1 is then theoretically close to the lower bound set by the restriction (3.1). At the other extreme, where each cell contains a single dislocation which does not penetrate a cell wall but leaves lines of dislocations on the side walls as it advances, the line tension effect is the more important and in that case C_1 is near to the upper bound, i.e. close to unity. In this work, then, we take C_1 equal to unity.

The balance between the internal and external stresses can be written

$$F = \sigma - C_1 \mu b / x' \tag{3.3}$$

which for the case of cells each of which contains one mobile dislocation, and for which separation r_b of dislocations in the cell wall is much smaller than the cell diameter, becomes by the above arguments

$$F = \sigma - \mu b / L \tag{3.4}$$

where F is the net stress causing the mobile dislocation to glide. Here we have made use of the well known theoretical finding that the force acting on a crystal dislocation is σb when the macroscopic stress, globally applied to the specimen, is σ.

For the value of the friction stress F we consider the non-conservative motion of vacancy-absorbing and vacancy-emitting jogs in the screw components of glide dislocations which exert a retarding effect on glide. We review models which invoke this mechanism in Section 4.8 and in the case where the local vacancy concentrations are predominantly controlled by the rates at which vacancies can diffuse to or from the jogs we then obtain for the jog velocity

$$v = \frac{b \sinh(\alpha F)}{\alpha A} \tag{3.5}$$

which at low stresses reduces to

$$\alpha F \rightarrow 0: \quad V = \frac{Fb}{A} \tag{3.6}$$

In these equations, following Barrett and Nix (1965),

$$\alpha = \frac{b^2 l}{kT} \tag{3.7}$$

$$A = \frac{b^2}{20\alpha D_v} \tag{3.8}$$

where, as before, b is the Burgers vector, k is Boltzmann's constant, T is the absolute temperature, D_v is the volume self-diffusion coefficient, and l is a characteristic length.

The relationship between the smeared dislocation density ρ and the spacing of the cell boundary dislocations (which make up the majority of those present, if there is only a single mobile dislocation per cell) is given, to a reasonable approximation, by the expression

$$Lr_b = 1/\rho \tag{3.9}$$

whilst the expression suggested by Holt (1970), and shown by him and later workers to give a good representation of data, is

$$L = K/\sqrt{\rho} \tag{3.10}$$

Eliminating ρ between eqns (3.9) and (3.10) gives

$$L = K^2 r_b \tag{3.11}$$

and substituting eqn (3.11) in eqn (3.4) gives

$$F = \sigma - \mu b/(K^2 r_b) \tag{3.12}$$

Our scheme is to calculate the mobile dislocation density (i.e. that of the individual mobile dislocations, one of which we assumed above to be located inside each of the cells) and the velocity v of these mobile dislocations. The mobile dislocations are on average one cell diameter apart. So the mobile dislocation density ρ_m is of order

$$\rho_m = 1/L^2 \tag{3.13}$$

The steady-state creep rate $\dot{\varepsilon}$ is then given by

$$\dot{\varepsilon} = bv\rho_m = bv/L^2 \tag{3.14}$$

Not only is glide resisted by the internal stress, it is also resisted by the friction of the screw jogs. The velocity of a dislocation subjected to a net stress F is given by eqn (3.5). Using eqns (3.14) and (3.5) to define a value for L, and substituting this into eqn (3.4), we find

$$\sigma - F = \mu\sqrt{\left(\frac{\alpha A\dot{\varepsilon}}{\sinh(\alpha F)}\right)} \tag{3.15}$$

Consider now the recovery process which occurs in the cell boundary dislocation network. In many cases this network is three-dimensional and the cell boundaries have a small but finite thickness. We confine attention to such cases. As in the earlier sections, and following Friedel (1964), we consider the case in which the dislocations climb to shorten their total length and in this way reduce the energy of the network. The energy E for a dislocation in a cell boundary is of order $\mu b^2/2$ and the force causing climb is then $\mu b^2/2r_b$. The velocity of climb under this force is then given by the equation

$$\left(\frac{dr_b}{dt}\right)_{\text{recovery}} = \frac{K_1}{r_b} \tag{3.16}$$

with

$$K_1 = \frac{D_v \mu b^3 c_j}{2kT} \tag{3.17}$$

Let us write the creep equation in the form

$$\dot{\varepsilon} = \dot{\rho} bL \tag{3.18}$$

In this equation mobile dislocations are entering the boundary network at a frequency $\dot{\rho}$ after gliding across the cell diameter L. So

$$d\varepsilon = d\rho \cdot bL \tag{3.19}$$

Eliminating L between eqns (3.9) and (3.10) we obtain

$$\rho = \frac{1}{K^2 r_b^2} \tag{3.20}$$

so that

$$d\rho = -\frac{2r_b^{-3}}{K^2}\, dr_b \tag{3.21}$$

Using eqn (3.11) to substitute for L in eqn (3.19), and eqn (3.21) to substitute for $d\rho$, we obtain

$$\left(\frac{d\varepsilon}{dr_b}\right)_{\text{straining}} = -\frac{2b}{r_b^2} \tag{3.22}$$

At an equilibrium condition, with which we are concerned, straining must generate a length of dislocation exactly equal to the length which has been eliminated, during the straining process, by the simultaneous climb controlled coarsening of the boundary network:

$$\mathrm{d}r_\mathrm{b} = \left(\frac{\mathrm{d}r_\mathrm{b}}{\mathrm{d}t}\right)_{\text{recovery}} \mathrm{d}t + \left(\frac{\mathrm{d}r_\mathrm{b}}{\mathrm{d}\varepsilon}\right)_{\text{straining}} \mathrm{d}\varepsilon \tag{3.23}$$

but $\mathrm{d}r_\mathrm{b}/\mathrm{d}t = 0$ when $\dot{\varepsilon}$ = constant, so

$$\dot{\varepsilon} = \frac{(\mathrm{d}r_\mathrm{b}/\mathrm{d}t)_{\text{recovery}}}{(\mathrm{d}r_\mathrm{b}/\mathrm{d}\varepsilon)_{\text{straining}}} \tag{3.24}$$

Using eqns (3.16) and (3.22) in (3.24) we obtain

$$\dot{\varepsilon} = \frac{2K_1 b}{r_\mathrm{b}^3} \tag{3.25}$$

Using eqn (3.12) to define r_b in eqn (3.25), we find

$$\dot{\varepsilon} = 2K_1 b \left[\frac{K^2(\sigma - F)}{\mu b}\right]^3 \tag{3.26}$$

Rearranging eqn (3.26) and using eqn (3.15) to substitute for $(\sigma - F)$,

$$\left[\frac{\dot{\varepsilon}}{2K_1 b}\right]^{1/3} \frac{b}{K^2} = \left[\frac{\alpha A \dot{\varepsilon}}{\sinh(\alpha F)}\right]^{1/2} \tag{3.27}$$

i.e.

$$F = \frac{1}{\alpha}\sinh^{-1}\left[\alpha A \dot{\varepsilon}^{1/3} (2K_1 b)^{2/3} K^4/b^2\right] \tag{3.28}$$

Let

$$B = c_\mathrm{j} \frac{D_\mathrm{v} b}{\mu^2 kT} \tag{3.29}$$

and then from eqn (3.17)

$$2K_1 = \mu^3 b^2 B \tag{3.30}$$

From eqns (3.28) and (3.30)

$$F = \frac{1}{\alpha}\sinh^{-1}\left[\left(\frac{\dot{\varepsilon}}{B}\right)^{1/3} \mu^2 \alpha A B K^4\right] \tag{3.31}$$

From eqns (3.26) and (3.30)

$$\sigma - F = \left(\frac{\dot{\varepsilon}}{B}\right)^{1/3} \frac{1}{K^2} \quad (3.32)$$

From eqns (3.31) and (3.32), finally,

$$\sigma = \frac{1}{K^2}\left(\frac{\dot{\varepsilon}}{B}\right)^{1/3} + \frac{1}{\alpha}\sinh^{-1}\left[\left(\frac{\dot{\varepsilon}}{B}\right)^{1/3} \mu^2 \alpha ABK^4\right] \quad (3.33)$$

which is the required result, the equation for dislocation creep.

Thermodynamic Calculation of K

The approach to the calculation of K is a thermodynamic one; what is done is to calculate the value of K that minimizes the free energy of the creeping specimen. The free energy of the specimen comprises the elastic energy stored by Hookean deformation plus the energy of the dislocations. Details of this calculation appear elsewhere (Gittus, 1978) showing that, to a good approximation,

$$K = 1{\cdot}58 c_j^{-0{\cdot}4} \quad (3.34)$$

3.2. Examples Illustrating the Validity of the Model of Dislocation Creep

The validity of this model of dislocation creep has been extensively tested. In particular, one may make reference to the work of Povolo and co-workers (Povolo *et al.*, 1984; Povolo and Marzocca, 1984*a*, *b*; Povolo and Capitani, 1984) who have studied its predictions for experiments which they have conducted on creep and stress relaxation in Zircaloy-4. I am indebted to Professor Povolo for supplying details of this work, some of them in advance of actual publication elsewhere.

Creep of Stress Relieved Zircaloy-4

Figure 1 (Povolo, 1981) shows how the data of Povolo and Marzocca (1984*a*) conform to eqn (3.33) which they write in the following form:

$$a\sigma = (\dot{\varepsilon}/\dot{\varepsilon}^*)^{1/3} + \sinh^{-1}[(\dot{\varepsilon}/\dot{\varepsilon}^*)^{1/3}\beta] \quad (3.35)$$

where

$$\beta = \mu^2 ABK^6; \quad \dot{\varepsilon}^* = B(kT)^3K^6/b^6l^3; \quad a = b^2l/kT;$$

$$B = c_j D_v b/\mu^2 kT; \quad A = b^2/20\alpha D_v$$

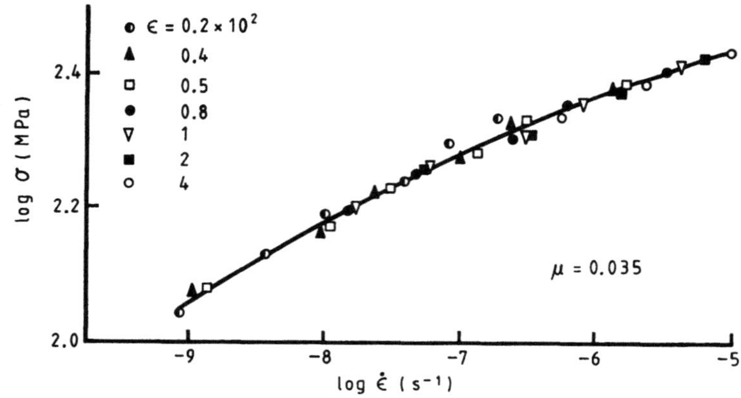

Fig. 1. Master log σ versus $\dot{\varepsilon}$ curve.

This equation is plotted in Fig. 2 as $\log(\alpha\sigma)$ versus $\log(\dot{\varepsilon}/\dot{\varepsilon}^*)$ for different values of β.

Equation (3.35) was deduced by Gittus under steady state conditions but the master curve of Fig. 1 includes also data taken in the transient creep regime. Povolo shows that the Gittus equation may be valid also during transient creep.

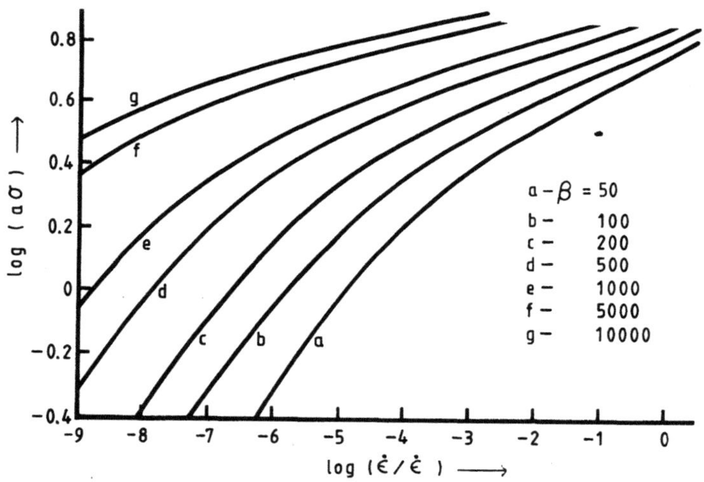

Fig. 2. Normalized plot of eqn (3.35).

It is assumed in the Gittus model that l depends only on the thermal jog concentration, which is clearly an oversimplification. As shown by Fig. 3, however, l changes only by one Burgers vector from low to high strains.

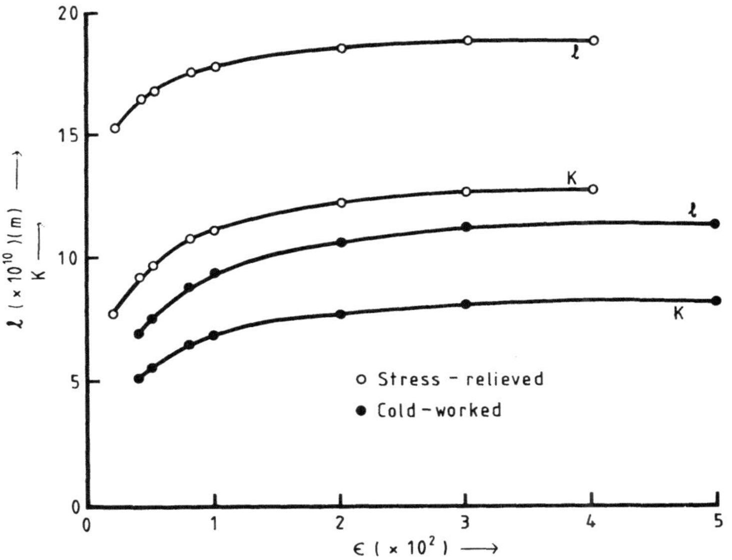

Fig. 3. Average spacing between neighbouring jogs, l, and ratio of cell diameter to dislocation spacing, K, as a function of the creep strain.

On differentiating eqn (3.35) it is easy to show that the parameter m is given by

$$m = 3\alpha\sigma(\dot{\varepsilon}/\dot{\varepsilon}^*)^{-1/3}\{1 + \beta/[1 + \beta^2(\dot{\varepsilon}/\dot{\varepsilon}^*)^{2/3}]^{1/2}\}^{-1} \qquad (3.36)$$

where

$$m = (\partial \ln \dot{\varepsilon}/\partial \ln \sigma)_T \qquad (3.37)$$

This equation shows that m depends on the applied stress, on the strain rate and on the parameters α, $\dot{\varepsilon}^*$ and β or, which is equivalent, on K and l. Since α, $\dot{\varepsilon}^*$ and β are known as a function of strain, eqn (3.36) can be used to calculate the predicted m values which can be compared

TABLE 1
Comparison between the Measured Values (m_{exp}) for the Parameter m and the Values (m_{th}) Predicted by eqn (3.36)

σ (MPa)	$\dot{\varepsilon}$ (s^{-1})	$\varepsilon \times 10^{-2}$	$\alpha \times 10^{-3}$ (MPa^{-1})	$\dot{\varepsilon}^* \times 10^{-4}$ (s^{-1})	β	m_{exp}	m_{th}
120	$8\cdot33 \times 10^{-10}$	0·70	19·45	2·23	228·4	6·8	7·1
147	$3\cdot33 \times 10^{-9}$	1·74	20·59	4·29	623·7	8·7	8·9
169·4	$8\cdot06 \times 10^{-9}$	2·80	21·05	5·32	807·2	10·3	10·4
188·9	$1\cdot31 \times 10^{-8}$	6·56	21·12	5·60	855·4	11·0	11·6

with the measured ones. Such a comparison is made in Table 1, where the strain rate, the stress and the strain level, before the small change in stress, and the parameters α, $\dot{\varepsilon}^*$ and β are indicated, m_{th} are the values predicted by eqn (3.36) and m_{exp} are those obtained by making a small change in the applied stress ($\Delta\sigma = 0\cdot1\sigma$). The agreement between the measured values and those predicted by the theoretical model is very good, confirming further that eqn (3.35) describes the creep behaviour of this alloy, at 673 K. In addition, preliminary observations of the microstructure by electron microscopy and studies of the evolution of the texture during creep confirm this conclusion (Povolo *et al.*, unpublished work).

Values of K from Thermodynamic Arguments

Povolo *et al.* (1983) have compared the present author's evaluation of the ratio of cell diameter to dislocation spacing with their creep and stress relaxation measurements on Zircaloy-4. They confirm the theoretical predictions.

The experimental value for K obtained from the creep experiments was found to increase with the creep strain, reaching a saturation value of the order of 13 at high strains. Since the data at high strains are obtained at large stresses and at large secondary-creep strain rates, where the steady state regime is obtained more rapidly, it is clear that K measured at large creep strains is more representative of equilibrium. The measured value lies in the range predicted by the Gittus theory.

In studies of stress relaxation in bending, Povolo and Capitani (1984) further explore the present author's theory and find good agreement between observation and prediction.

4. MODEL OF THE SUPERPLASTIC FLOW PROCESS

4.1. Theory

At low stresses and small grain sizes, Ashby and Verrall (1973) noted that the ability of the boundary to act as a source and sink for the diffusing point defects might become limiting. They consider the case where grain boundary dislocations are the sources and sinks of point defects and are also responsible for sliding. The creep rate then depends on the density of dislocations in the boundary. They suggested, by analogy with the spacing between dislocations in the body of a crystal, that the spacing l between boundary dislocations would be

$$l = \mu b/\beta\sigma \tag{4.1}$$

where β is a constant of order unity, μ is the shear modulus, σ is the applied stress, and b is the value of the Burgers vector of a boundary dislocation.

Interface reaction control coupled with a sink density which [see eqn (4.1)] increases with stress leads to a dependence of the strain rate on the square of the stress. This kind of stress dependence was in fact a characteristic of the earliest model of superplastic flow, put forward by Ball and Hutchison (1969). Their model, as they remarked, is virtually the same as one suggested earlier by Friedel (1964) to explain the fact that the activation energy for the creep of fine-grained zinc is that for grain boundary diffusion. The model may be regarded as a time-dependent version of that proposed by Hall (1951) and by Petch (1953) to explain the dependence of yield stress upon grain size d:

$$\sigma_e = \frac{2(\sigma - \sigma_0)^2 d}{\mu b} \tag{4.2}$$

Here σ_e is the compressive stress at the head of a dislocation pile-up; σ_0 is the dislocation friction stress. Equation (4.2) (with σ_e a materials parameter) has been widely verified by experiments in which the value of σ at which plastic yielding occurs has been correlated with the grain size.

Ball and Hutchison, following Friedel, suggested that in superplastic flow dislocations from the head of the pile-up would climb into the boundary against which they were piled up; strain is so produced. Mohamed *et al.* (1975) have examined the Ball and Hutchison model, particularly in relation to superplasticity in the Zn–Al eutectic, and have pointed to two drawbacks. Firstly, the model assumes that adjacent grains slide in groups and this is not supported by direct observations in the scanning electron microscope. Secondly, the model predicts the presence of many intergranular dislocations although transmission electron microscopy and internal marker experiments have been interpreted to indicate that there is very little within-grain dislocation activity during superplastic flow.

Neither of the two theories is entirely satisfactory, even for the case of superplastic flow in a single-phase alloy. In the case of a two-phase alloy there are additional difficulties due to the nature of the sliding and climb processes at heterophase boundaries. We consider this matter in Section 4.2.

4.2. Sliding, Growth and Dissolution at Heterophase Boundaries

Ishida and McLean (1967) and McLean (1971) have reviewed and extended the theory of the ways in which dislocations in grain boundaries can glide and climb; the climb component leads to a vacancy emission while the glide component causes grain boundary sliding. The boundaries with which these authors were concerned are those between like grains. The possibilities are, however, more limited at heterophase boundaries. It is, of course, established that particles of a second phase can precipitate and dissolve. In addition, processes such as 'ripening' are observed in which material from one particle of a second phase diffuses to another through the intervening matrix. Again, the shape of a particle can change, presumably by the diffusion of matter from one region to another. These mechanisms are capable then of producing the changes in particle shape (and possibly size) which have to occur in order to preserve continuity at the contacting faces of particles and matrix during superplastic flow. Aaronson *et al.* (1970) have summarized the mechanisms of these diffusional growth processes for precipitate crystals. The orientation relationship of the precipitate lattice with respect to the heterophase boundary varies with the orientation of the boundary. Dislocation structures appear at one or two orientations and disordered structures at all the others.

The dislocation boundaries can only migrate by the movement along their length of ledges. The narrow faces of these ledges often appear to be disordered, so there is no structural barrier to precipitation of additional material on to them. Similarly, there is no structural barrier to the growth (migration) of the disordered boundaries. In the ordered boundaries, on the other hand, addition of a new layer of atoms will generally alter the packing relationship between the two contacting lattices, introducing a measure of disorder. The energy needed to accomplish this is usually high. It is for this reason that precipitates are often in the form of long plates or rods; the long dimension is a dislocation boundary which can only thicken by the slow process of ledge growth, whereas elongation of the precipitate occurs preferentially by the addition of atoms to regions of the boundary that are already so disordered that the new atoms cannot raise their energy significantly.

Sliding at dislocation heterophase boundaries is impeded by the same type of energy barrier. Thus the passage of a glissile dislocation leaves in its wake a disordered structure of high energy. The energy increase could be so high as to prohibit the glide and with it any sliding at the heterophase boundary (HPB). That is to say, the threshold stress for HPB sliding could be very high (Ashby, 1972). Dislocation boundaries in single-phase materials can slide more easily. They do this, in most cases, by a combination of translation and migration (Ishida and McLean, 1967; McLean, 1971). Atoms are transferred across the interface during sliding to restore the boundary structure, which would otherwise be altered by the sliding process. In this way one of the grains grows at the expense of the other. If atoms are transferred across the interface between two different phases, the grain that grows does so by acquiring the wrong kind of atoms — those that constituted the other phase. Equilibrium can be restored by the diffusional redistribution of the components, but this introduces a large viscosity term since the diffusion path is of the order of the grain size.

There is at least one possible mechanism of sliding available on dislocation HPBs. It involves the glide of a superdislocation in the boundary plane. The first of a pair of boundary dislocations creates disorder but the second dislocation, when it passes along the same path, restores order once more. Another example comprises the climb in an HPB of a faulted dislocation loop or of a superdislocation whose Burgers vector is perpendicular to the plane of the boundary. Atoms diffuse from the trailing edge to the leading edge to produce this climb motion. In the wake of the climbing dislocation order is restored to the boundary structure. Passage of such a dislocation by pure climb cannot, however, cause boundary sliding; for this to occur there must be a component of the Burgers vector oriented in the plane of the boundary. Mixed climb and glide can then occur, the glide component giving rise to sliding and the climb component involving a diffusion distance equal only to the width of the dislocation.

A specific example of a known HPB with defined dislocations occurs in the work of Nakagawa and Weatherley (1972), who have actually observed pairs of dislocations in the interface between two crystal lattices. The lattices were those of Ni_3Al and Ni_3Nb in a directionally solidified eutectic alloy. The dislocation pairs had separations of 300–400 Å. They were epitaxial dislocations with $b = a/2(110)_{Ni_3Al}$ and the orientation relationship between the phases at their common boundary was $(110)_{Ni_3Al} \| (100)_{Ni_3Nb}$, $(113)_{Ni_3Al} \| (031)_{Ni_3Nb}$. An HPB between two pure substances (α and β) may be regarded as a plane of ordered solid solution of one phase in the other. The disordered region between the leading and the trailing components of the superdislocation is therefore an antiphase boundary (APB). Gleiter (1977) in fact shows that, even for grain boundaries of random orientation, the width of a boundary dislocation is theoretically always much larger than that of a dislocation of similar Burgers vector in the perfect lattice. The partial dislocations are only separated by a stacking fault, instead of an APB. Their movement does not disorder the structure. This may occur in superplastic alloys such as Ag–Cu where both phases are face-centred cubic.

The equilibrium width of the disordered region can be calculated from a knowledge of the elastic constants and a fault energy Γ. Then the force with which the fault tends to contract, acting as a surface, equals Γ per unit of length, which is equivalent to a resolved stress of magnitude Γ/b pulling the partial dislocations together. On the other hand, the partials being of like orientation and on the same slip plane tend to repel one another. For reasonably large distances of separation w (the width of the disordered zone) this repulsion is, in the first approximation, due to the resolved shear stress set up by the elastic stress field of each of the two partial dislocations at the position of the other. Now outside the core, the elastic displacements, strains, stresses, and energy of the grain boundary dislocation are identical with the well known results for a dislocation of the same Burgers vector in an isotropic elastic continuum. Provided then that the partials are sufficiently separated, that their cores do not overlap, the elastic part of their stress fields and the interaction energy between them can be calculated by treating them as elastic dislocations in a linear isotropic continuum. Of course allowance must be made for the fact that the shear moduli of the two phases differ. We shall assume that an adequate approximation is provided by the use of an effective modulus μ defined as a function of the two moduli μ_α and μ_β:

$$\mu = f(\mu_\alpha, \mu_\beta) \tag{4.3}$$

At equilibrium a dislocation HPB will have misfit dislocations spaced out along its length. They are necessary because their presence compensates for the differences between the lattice spacings of the two phases where they contact along their boundary. If the lattice parameters were identical then the resultant coherent boundary would contain no dislocations and would have the structure of a fully ordered solid solution. That is to say, the long-range order parameter S for a coherent boundary is unity. More generally, for the case where the lattice parameters differ and superdislocations occur at a separation l, we can write

$$S = 1 - 2w/l \tag{4.4}$$

so that, when the dislocations are so close together that the partials touch, $S = 0$, corresponding to the complete absence of long-range order (a disordered boundary). The width of the superdislocation is then given, to the first approximation, by an equation essentially similar to that for the width of the superdislocations in an ordered solid solution (see Haasen, 1965):

$$w = \frac{\mu b^2}{2\pi \Gamma S^2} \tag{4.5}$$

On substituting for S from eqn (4.4),

$$w = \left\{1 - \left[1 - \frac{8\mu b^2}{2\pi\Gamma l}\right]^{1/2}\right\}\frac{l}{4} \quad \text{for} \quad w \ll l \tag{4.6}$$

As an example, consider the case where the energy of the disordered boundary in the fault Γ is 0·10 J m^{-2} (100 erg cm^{-2}) and let $b = 2 \times 10^{-10}$ m (2 Å) and $\mu = 5 \times 10^4$ MPa. Let $S = 0{\cdot}75$. Then

$$w = 5 \times 10^{-9}\ \text{m (50 Å)} \tag{4.7}$$

The disordered region is identical in structure with the narrow face of a growth ledge, and we may in fact regard such ledges as superdislocations in which the fault and glide planes are no longer coparallel. Such 'staggered' superdislocations are certainly observed in grain interiors.

4.3. The Model of Superplastic Flow

In Section 4.2 we have argued that particles of differing phases can slide relative to one another by the movement of interfacial superdislocations.

We have also remarked that a particle can change its shape or size by the diffusion of matter to and from disordered regions of its boundary. We are going to use these principles in a model of the superplastic flow process.

Consider the sliding process which occurs on a dislocation boundary. Suppose that superdislocations can be generated by the operation of sources, e.g. by the expansion of intrinsic dislocation loops under the applied stress. In addition, there is the likelihood that matrix dislocations will enter the boundary. Thus the boundary dislocations have Burgers vectors that are equal to the sum or difference of those of matrix dislocations. So a matrix dislocation can enter a boundary provided that it splits into two components. One of these could then, for example, climb into a disordered region of the boundary while the other glides in a dislocation region. Once in the boundary the dislocations are effectively trapped and can only emerge into the matrix if they reassociate to form matrix dislocations of appropriate Burgers vector. The dislocation boundary is therefore potentially a very effective pile-up of length d. From eqn (4.2) we can calculate the stress at the head of such boundary pile-ups. μ is now defined by eqn (4.3); b, the Burgers vector of a dislocation boundary superdislocation, equals the sum or difference of Burgers vectors for the contacting phases. The movement of dislocations in this dynamic pile-up proceeds at a rate determined by the climb of the dislocations away from the head of the pile-up. They cannot generally climb out of the boundary; therefore they must climb in it. The most likely situation is one in which the dislocations pile up at a triple junction where a dislocation boundary, on which they are gliding, meets two disordered boundaries into which the dislocations then climb. For this climb to occur, matter must diffuse to or from the dislocations (diffusion coefficient = D_B). Let the diffusion distance be λd. The velocity of climb is then given by the following equation due to Friedel (1964):

$$V = \frac{D_B}{\lambda d}\left(\frac{\sigma_e b^3}{kT}\right) \tag{4.8}$$

The dislocations will have to move a distance approximately equal to the diffusion distance in order to encounter others climbing away from neighbouring pile-ups. As Friedel pointed out, many of these encounters will be between boundary dislocations whose Burgers vectors add up to zero and so the dislocations will annihilate one another. The time taken for a dislocation to climb this distance λ at velocity V is $t = \lambda d/V$, i.e.

$$t = \frac{\lambda^2 d^2 kT}{D_B \sigma_e b^3} \tag{4.9}$$

At intervals of t, therefore, two of the dislocations climbing away from each pile-up are annihilated (one as it climbs up one of the boundaries that meet at the triple junction, and the other as it climbs up another). Two new dislocations will therefore need to be generated in the pile-up in order to replace those that climb out of it in pursuit of the two that have been annihilated. In this way equilibrium is maintained. The strain contributed is that which would be produced if the dislocation boundary slid a distance $2b$ at intervals of t. The strain rate is now calculated from the equation

$$\dot{\varepsilon} = \bar{u}/d \tag{4.10}$$

Here $\bar{u}$ is the mean rate of sliding translation at the crystal boundaries.

So to produce a strain of ε, boundary sliding of εd must occur. A sliding displacement of magnitude $2b$ will therefore produce a strain of $2b/d$.

The strain rate is therefore given by

$$\dot{\varepsilon} = 4\left(\frac{d}{\lambda}\right)^2 \frac{D_B \mu b}{kT}\left(\frac{\sigma - \sigma_0}{\mu}\right)^2\left(\frac{b}{d}\right)^2 \quad \text{for} \quad \sigma > \sigma_0 \tag{4.11}$$

4.4. Boundary Layer Shear at Heterophase Boundaries

In some circumstances sliding occurs by the shear of a thin boundary layer. Recapitulating what is known about this phenomenon, Eberhardt (1979) has studied boundary sliding in two phased bicrystals of brass. One phase (α) is hard and non-deformable; the other phase (β) is ductile at the temperature of his sliding experiments. Below a certain stress, sliding occurred at the α–β (heterophase) boundary. Neither phase exhibited any deformation and the sliding rate was proportional to the first power of the shear stress. Sliding, he showed, is simply controlled by the rates of diffusion at these low stresses.

At higher stresses the sliding rate varied with the square of the shear stress (the class of behaviour which we examined in arriving at eqn (4.11), in fact). Metallography showed that sliding was occurring by the movement of matrix dislocations in a thin layer of β phase adjacent to the α–β boundary. Sliding at the boundary played a negligible role at these stresses.

The dislocations, in this instance, were piled up not in the boundary itself but in the boundary layer. In these circumstances there is no need to invoke the presence of HPB superdislocations; sliding does not occur at the boundary and accordingly the equilibrium of the boundary structure is not perturbed by the movement of interfacial dislocations. The equation for superplastic flow does not depend on where the dislocation pile-up is situated; if it is in a thin boundary layer this alters b and μ, but eqn (4.11) remains valid.

4.5. Structural Effects

The dislocation spacing on dislocation boundaries theoretically falls with increasing stress. Hence [see eqn (4.4)] the long-range order parameter S will fall as the stress increases. Now a completely disordered boundary ($S = 0$) has no preferred relation to the lattices of the contacting phases and tends to adopt a spherical shape to minimize its area (and thus to minimize the surface energy). At the other extreme a dislocation boundary ($S \rightarrow 1$) follows preferred crystallographic directions and is essentially planar. When a material is stressed, we expect the gradual build-up of HPB superdislocations towards the equilibrium density (and the associated fall in S) to make more and more of the boundary adopt a curved shape. This is observed (Gittus, 1975), and in the earlier literature led to the idea that the formation of 'bulbous' structures was a prerequisite of superplasticity.

4.6. Magnitude of the Threshold Stress

The threshold stress due to the presence of ledges on the dislocation boundary is one source of an estimate of the value of σ_0. To traverse a ledged boundary the dislocations will have to bend. Moreover, there will be a change in fault energy of magnitude Γwh per intersection, where h is the height of a ledge, when the narrow face of a ledge is a disordered boundary. The applied stress will have to do sufficient work to overcome this energy in order to detach a dislocation from a ledge. Garmong and Rhodes (1975) saw evidence of these interactions between HPB dislocations and ledges in their transmission electron microscope studies of Ni_3Al–Ni_3Nb composites. The interaction was manifested both as a local alteration of the line vector of the dislocation in certain circumstances and as a change in the response of the dislocation image to image contrast experiments. They remark that the presence of ledges will modify mechanical properties, such as creep, in several ways. Thus

Hirth (1972) has discussed the generation of matrix dislocations at interfacial defects. There will be a simultaneous production of boundary dislocations in order to conserve the Burgers vector. Garmong (1974) had observed this precise mechanism in operation in a Ni–W alloy of eutectic composition. For such a mechanism to operate, the boundary dislocations produced must be glissile in the plane of the boundary, otherwise the ledge source would not operate. Boundary ledges, by pinning the glissile boundary dislocations, could then stop the source from operating until the external stress was raised to a level at which the pinning forces could be overcome. The pinning of glissile boundary dislocations would in this manner reduce the ability of the boundary to creep, as Garmong and Rhodes pointed out. The ledges could be viewed as 'keying' the boundaries, but the boundary locking effect would arise not from physical interlocking but from the inhibition of boundary dislocation glide. They noted that in many respects the pinning of boundary dislocations by ledges, which they had observed in their work, resembles the pinning produced by boundary precipitates, illustrated by Ashby (1972, Fig. 17). Garmong and Rhodes pointed out that, in cases where the narrow edges of ledges are themselves dislocation boundaries, misfit dislocations lying in the narrow face could attract misfit dislocations lying on the contacting broad face on the ledge. This is clearly a supplementary source of ledge dislocation pinning. Aaronson *et al.* (1970) support the idea that a dislocation structure can develop at the narrow faces of ledges, and they point to the slow growth of certain ledges during recrystallization as evidence of the occasional presence of this kind of structure instead of the expected disordered boundary.

A suggested model of the ledge dislocation network is a two-dimensional dislocation network. In proposing this we are in effect following the suggestion of Garmong and Rhodes. In order to detach itself from two ledges, a boundary dislocation may be forced to adopt a curvature C as high as $2/L$, where L is the width of the broad face of a ledge (i.e. the distance between adjacent narrow faces). If the boundary defect is regarded as a flexible line whose energy per unit length is E, a potential

$$\frac{E\Omega C}{b} \tag{4.12}$$

or

$$\sim \frac{2E\Omega}{bL} \tag{4.13}$$

is required to do this (Ashby, 1972; Bardeen and Herring, 1952; Thomson and Balluffi, 1962). Here Ω is the atomic volume. This then gives as an estimate of the threshold stress due to the interaction of glissile HPB superdislocations and ledges

$$\sigma_0 \leqslant \frac{2E}{bL} \tag{4.14}$$

Precipitates lying in the boundary are another source of a threshold stress. A similar equation to (4.14) then applies but with L equal to the spacing between the precipitate particles. Solute drag may also tend to restrict the movement of the boundary superdislocations. However, this will reduce the rate of glide at all stresses rather than introduce a threshold stress.

4.7. Comparison with Experimental Results

Ball and Hutchison (1969) found that an equation algebraically identical with (4.11) fitted their data for the superplastic flow in an Al–Zn alloy of eutectoid composition. The factor which in our equation is $4(d/\lambda)^2$ had a value of approximately 100 for their alloy, so $\lambda = 0{\cdot}2d$, which compares with the value of $0{\cdot}3d$ calculated by Friedel. They did not find any evidence of a threshold stress.

Quite generally our theoretical equation does appear to provide a reasonable fit to data with λ/d in the right order of magnitude (Gittus, 1977*a,b*, 1978). Geckinli and Barrett (1976) (for example), in their experimental studies of the Pb–Sn eutectic, concluded that at room temperature the threshold stress was 1·3 MPa. From eqn (4.14) the value of L is then of the order of 4×10^{-7} m (4000 Å); this is of the order of magnitude observed on dislocation boundaries. That the ledges do impede the movement of boundary dislocations is established by the electron microscopy reported by Garmong and Rhodes (1975). These workers calculated that in their Ni_3Al–Ni_3Nb composites of eutectic composition the pinning stress acting to pin the HPB dislocations to the ledges would, theoretically, be $\mu/70$. The threshold stress for dislocation movement is expected to be about h/L times this value, since the applied stress acts on the entire unpinned length of dislocation (minimum length L) while the pinning stress acts only on the length of dislocation near to the ledge (height h). So

$$\sigma_0 \sim \frac{\mu h}{70L} \tag{4.15}$$

If this equation, derived for a different material, nevertheless holds good for a Pb–Sn alloy of eutectic composition, the observed value of the threshold stress in that material would imply that the ledges on the dislocation boundaries are about 100 times as long as they are high. This again is in reasonable accord with experiment.

For the case where the applied stress is much above the threshold stress, eqn (4.11) predicts that the strain rate will depend on the square of the stress. In the literature on superplastic flow the reciprocal of the exponent of stress (m) is often quoted. In the case of a material conforming to eqn (4.11), then, $m = 0{\cdot}5$. In the event, values in the range 0·3–0·7, with a mean near to 0·5, are commonly reported, as we shall see later. It is particularly to be noted that a value of $m = 1{\cdot}0$, predicted and observed in the Herring–Nabarro and Coble types of diffusion creep, is rarely if ever found in superplastic materials. The characteristic of these models of diffusion creep, of course, is the assumption that the boundaries are perfect sources and sinks for point defects.

4.8. The Diffusion Coefficient

Baudelet with co-workers (Herriot *et al*, 1976; Suery and Baudelet, 1975) found, in their studies of the two-phase Cu–P alloys, that the activation energy for superplastic flow, calculated with no attempt to correct for change in structure, was close to that for grain boundary diffusion in copper. When, however, allowance was made for changes in structure, the activation energy for flow rose to a value quite close to that for *volume* self-diffusion in copper. On the face of it this would seem to run contrary to our theory. However, Perinet (1975) and Bondy *et al.* (1971) have found that a high value of the activation energy is characteristic of self-diffusion at heterophase (as opposed to grain) boundaries. Thus Perinet found that the activation energy for diffusion in the boundary between copper and silver was 272·1 kJ mol^{-1} (65 kcal mol^{-1}), which is more than twice the grain boundary activation energy for either of the pure elements by itself. Again, a similar finding was reported for the diffusion of iron and silver in the interface between these two metals. These findings clearly accord, on the present theory, with the high value of activation energy for superplastic flow at constant structure found by Baudelet and his collaborators. Indeed, this may be a general feature of superplastic flow in two-phase materials. It may have passed unnoticed in some investigations because care has not always been taken to compensate for the effect of changing structure when calculating the activation energy for superplastic flow. When Baudelet omitted this precaution he obtained a spuriously low activation energy which is, coincidentally it now appears, similar to that for grain boundary diffusion (as opposed to phase boundary diffusion) in one of the pure components of which his Cu–P alloy was composed.

The theory predicts that the strain rate will depend on the square of the particle size. We shall see that this is borne out in the case of the data for the superplastic flow of a number of materials.

5. ORIGIN OF THE HIGH DUCTILITY IN THE SUPERPLASTIC STATE

Above we showed how, during dislocation creep, the dislocations frequently arrange themselves in cells. The cell size rises if the stress causing creep is reduced and eventually the cell size becomes equal to the grain size. Significantly, when this structure is attained it appears that superplastic flow then often becomes the main deformation mechanism, and in the present section we offer an explanation of the associated high ductility. The free energy of a specimen undergoing dislocation creep has been calculated and proves to be generally much higher than that of a specimen deforming superplastically at the same rate. Only at high strain rates does the situation reverse. Accordingly the superplastic condition, once established, persists, and neck propagation is strongly resisted until a late stage, by which time large elongations will have occurred.

Figure 4 (a) and (b) have been prepared to illustrate the comparison that we are going to undertake. Consider Fig. 4(b); at low strain rates the material is superplastic and the stress needed to deform it rises with increasing strain rate. At high strain rates it deforms predominantly by dislocation creep, and the strain dependence of applied stress is now characteristic of a material containing dislocation cell structure. At the point x the stress needed to deform the material is the same for both deformation mechanisms. The strain rate at which this equality exists is here termed $\dot{\varepsilon}_\sigma$; in effect the point x is a point on a deformation map where it would mark the boundary between the stress field where dislocation creep is held to dominate and the region of lower stresses where superplasticity is said to cause most of the strain. Figure 4(a) is similar except that here we plot free energy instead of stress. Then we can show that

$$\frac{\dot{\varepsilon}_F}{\dot{\varepsilon}_\sigma} = \left[1 + \left(\frac{K}{K'}\right)^2\right]^3 \tag{5.1}$$

where

$$K' = 1 + K/2 \tag{5.2}$$

We can more easily discuss the implications of eqn (5.1) if we substitute values for K and K'. For Zircaloy at 900 K (for example) theoretical values are

$$K = 20{\cdot}25 \tag{5.3}$$

and

$$K' = 11{\cdot}12 \tag{5.4}$$

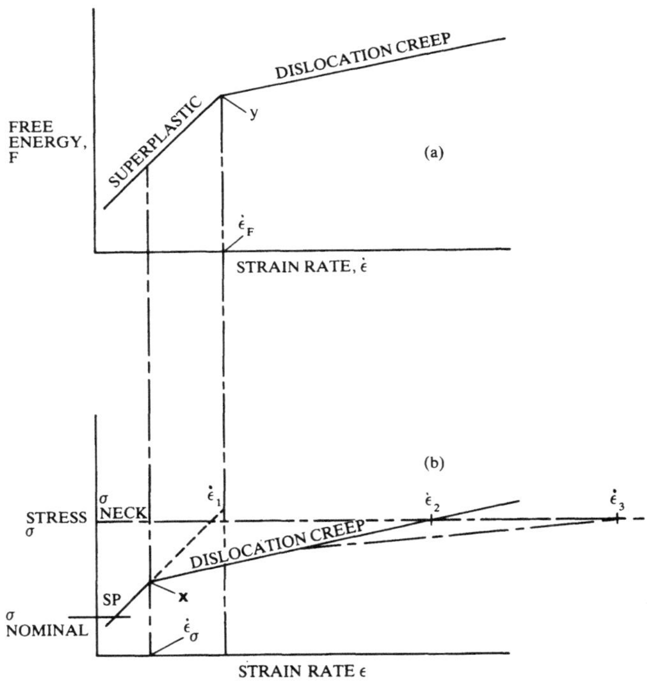

Fig. 4. Relation between strain rate and (a) free energy or (b) stress. Theoretically, $\dot{\varepsilon}_F \gg \dot{\varepsilon}_\sigma$ and both are virtually independent of temperature.

Using these values in eqn (5.1), then,

$$\frac{\dot{\varepsilon}_F}{\dot{\varepsilon}_\sigma} = 80{\cdot}24 \tag{5.5}$$

More generally, for large K, eqns (5.1) and (5.2) give

$$\lim_{K\to\infty} \frac{\dot{\varepsilon}_F}{\dot{\varepsilon}_\sigma} = 125 \tag{5.6}$$

What this means is that the transition to dislocation creep occurs at much higher strain rates than we would have expected. Accordingly a neck (thin region of the specimen), which locally increases the stress, will not necessarily cause dislocation creep with its (relatively) higher creep rate. Propagation of the neck is in this way resisted.

To clarify this point, consider again Fig. 4. Here we have marked two stresses. $\sigma_{nominal}$ is the nominal stress (force divided by uniform cross-section area) causing superplastic flow in a specimen. The specimen contains a neck which locally raises the stress to σ_{neck} and should, according to the bold lines relating σ with $\dot{\varepsilon}$, produce dislocation creep at the high rate $\dot{\varepsilon}_2$. Rapid straining of the necked region would then cause the neck to become more pronounced and eventually lead to fracture. However, σ_{neck}, although a high stress, is not high enough to cause a transition to dislocation creep when we consider the free energies involved (Fig. 4(a)), so the creep rate at the neck is only $\dot{\varepsilon}$ ($\ll \dot{\varepsilon}_2$) in Fig. 4(b). The neck propagates much more slowly and the ductility remains high as a consequence.

6. EXAMPLES OF SUPERPLASTIC MATERIALS

Having given theories for the mechanisms of deformation and the origin of ductility in superplastic alloys, we now summarize the data that have been accumulated during experimental studies of superplasticity.

6.1. Al–Zn–Zr and Al–Zn–Mg–Zr

Zinc is one of the few alloying additions soluble in aluminium in significant amounts at or near ambient temperature. It was therefore selected by Grimes *et al.* (1976) as a substitute for copper.

The Al–Zn–Zr alloys had good superplastic properties which increased with increasing zinc content. At 3%, 6% and 10% zinc (each with 0·5% zirconium), elongations at fracture of 360%, 580% and 620%, respectively, were obtained. The Al–Zn–Zr alloys are relatively weak, and it was for this reason that magnesium was added. Al–Zn–Mg alloys are known to be relatively strong. Good superplastic elongations were obtained, an Al–5%Zn–1%Mg–0·5%Zr alloy having an optimum elongation of 750%, and an Al–8%Zn–1%Mg–0·5%Zr alloy having more than 1100% elongation. In the equation $\sigma = K\dot{\varepsilon}^m$, where σ is the stress, $\dot{\varepsilon}$ is the strain rate, and K and m are constants, the Al–8%Zn–1%Mg–0·5%Zr alloy had an m-value similar to that of the Al–6%Cu–0·5%Zr alloy (both values are 0·5 at 535 °C). At a strain rate of 3×10^{-3} s^{-1} the flow stress was about 7 MPa.

6.2. MgO-Fe: Magnesia plus Ferric Ions

Knoch (1975) has studied the high temperature creep of single-phase and of two-phase magnesia with high iron contents. The deformation behaviour of the two-phase material depended on the volume fraction and the distribution of the second phase. In two-phase material of small grain size (10 μm) in which grain and phase boundaries were identical, the m-value lay between 0·3 and 0·5. This two-phase structure exhibited superplastic flow. In coarse-grained two-phase material, on the other hand, the grains deformed in the same way as in single-phase material and superplasticity did not then occur.

6.3. Sn-Pb: Tin-Lead Eutectic

Superplastic behaviour in the Pb–Sn eutectic (Pb–62%Sn) has been reported by many workers (Johnson, 1970). Ahmed and Langdon (1977) found an exceptional ductility of 4850%. The test had to be discontinued at this strain and the specimen had not failed. The material had a grain size of 6·9 μm and was pulled in tension at 413 K with an initial strain rate of $1{\cdot}33 \times 10^{-4}$ s^{-1}.

Vastava and Langdon (1979) investigated intercrystalline and interphase boundary sliding in the Pb–Sn eutectic. They found that the contribution of sliding to strain is 50% or 60% in region II. Sliding contributed about 20% of the strain in regions I and III. The largest sliding offsets occurred at the Sn–Sn intercrystalline boundaries; smaller offsets were recorded at the tin HPBs and there was little or no sliding at the Pb–Pb intercrystalline boundaries.

6.4. Ti-Al: Titanium–Aluminium Alloys

Superplasticity is exhibited by fine-structured two-phase Ti–Al alloys. One of these is an alloy of Ti–5%Al–2·5%Sn. Lee and Backofen (1967) have studied this material. When heated rapidly into the two-phase region just below the β phase boundary (about 0·65 Tm) the interparticle spacing varies between 6 and 24 μm depending on composition. Values of m of about 0·8 are obtained associated with ductilities of more than 1000%. At strain rates above 0·01 s^{-1}, m falls below 0·3 and there is no superplastic elongation. Both the flow stress and the strain rate sensitivity depend markedly upon temperature because temperature governs both the size and proportion of second phase in the two-phase structure. The m-value and the elongation have been correlated for temperatures between 25 and 1000 °C (Lee, 1969), and in all the alloys the maximum strain rate sensitivity occurs at temperatures just below the phase boundary of the β solid solution. The activation energy for the Ti–5%Al–2·5%Sn alloy is about 50–65 kcal mol^{-1} (209–272 kJ mol^{-1}).

6.5. Zn–Al: Zinc–Aluminium Alloys

Ball and Hutchison (1969) studied superplasticity in the Al–Zn eutectoid alloy. On the basis of their results they proposed a theory of superplastic flow which contains many of the elements of modern theories. Figure 5 shows the dependence of stress upon strain rate for specimens of 2·5 μm grain size deformed at 150 and 250 °C. An approximately linear relationship between log(stress) and the reciprocal of temperature was found, the slope being independent of grain size. The relationship between stress and grain size was linear at fixed temperature and strain rate. The stress depended upon the square root of the strain rate up to a limiting value of strain rate.

Mohamed *et al.* (1977) have studied factors that influence ductility in the superplastic Zn–Al eutectoid in some detail. Figure 6 summarizes some of their data relating elongation and stress with strain rate; here the marked peak in ductility in region II is shown very clearly and they have obtained elongations of up to nearly 3000%. Metallographic examination of fracture specimens shows extensive cavitation after deformation in the low stress region I and significant cavitation after deformation to 2600% in the superplastic region II.

Naziri and Pearce (1974) have rendered the Zn–0·4%Al alloy superplastic. A fine grain size of 0·6 μm was produced by rolling a cast slab to 90% reduction at room temperature. At this temperature m was about 0·45 at an initial strain rate of $6{\cdot}67 \times 10^{-4}\,s^{-1}$ and elongations exceeding 500% were obtained.

Samuelsson *et al.* (1976) have studied the superplastic Zn–40%Al alloy and the dislocation structures in it. The material was hot extruded into bars and solution treated. It exhibits superplasticity in the

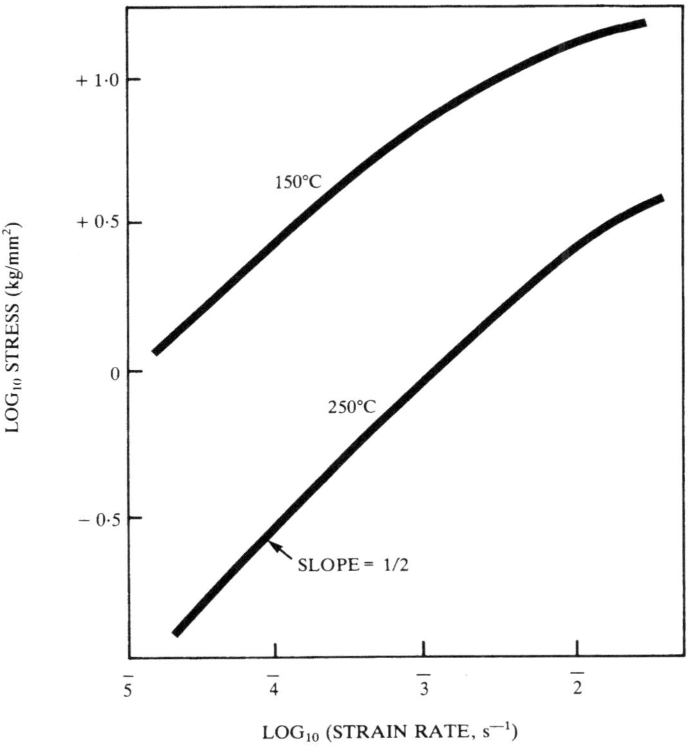

Fig. 5. Specimens of Al–Zn eutectoid of grain size 2·5 μm exhibit a quadratic dependence of strain rate upon stress.

temperature range 520–550 K with a maximum strain rate sensitivity of 0·48. Information on the role of grain boundary sliding derives from studies made by Melton *et al.* (1974), who studied textures in the superplastic Zn–40%Al alloy. They showed that slip occurs in both phases during superplastic flow even at maximum strain rate sensitivity. At all the strain rates that they investigated the zinc-rich phase deformed readily by slip, but large amounts of slip in the aluminium-rich phase only occurred at high non-superplastic strain rates. Grain boundary sliding reduced the overall level of texture after deformation at all strain rates. They concluded that superplastic flow in this alloy involves grain boundary sliding together with slip in both phases.

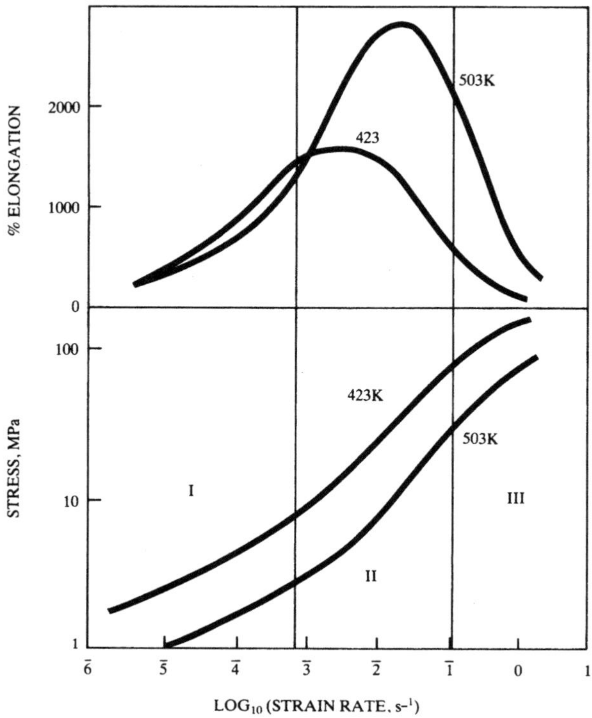

Fig. 6. Elongation, stress and strain rate for Zn–Al eutectoid of 2·5 μm grain size.

6.6. Zr–Sn: Zircaloy-4

A typical composition of Zircaloy-4 is given in Table 2. The main alloying element is tin (about 1·5%) with smaller amounts of iron and chromium. Bocek *et al.* (1976) have studied the superplastic behaviour of this material. Most of the investigations were performed in air between 750 and 1050 °C at cross-head speeds ranging over three orders of magnitude. Superplastic behaviour was observed between 860 and 1000 °C for a strain rate of the order of 10^{-4} s^{-1}. The strain rate sensitivity *m* depended upon strain. No effect of anisotropy upon the yield stress was observed above 550 °C, but in the superplastic region *m* revealed strong dependence upon sample orientation. At higher temperatures the total elongation increased with temperature up to 850 °C. The

TABLE 2
Composition of Zircaloy-4

Element		Sn	Fe	Cr	Fe + Cr	Zr
Wt.%		1·43	0·21	0·11	0·32	balance
Main impurities						
Element	C	HF	N	O	Pb	Si
ppm	170	46	30	980	< 50	56

apparent activation energy for superplastic flow agreed well with the energy of grain boundary diffusion in α-zirconium . In Fig. 7 the stress dependence of strain rate is shown for Zircaloy-4 at various temperatures. In the high-temperature tests the usual sigmoidal dependence emerges. Bocek *et al.* also found convincing evidence that in the ($\alpha + \beta$) phase region, α grain switching is the dominating deformation mode during superplastic flow at lower strains. Thus scanning micrographs were taken of the outer surface of internally pressurized Zircaloy-4 cladding tubes. The tubes were internally pressurized and distended by superplastic creep. To reveal the behaviour of individual grains during this ballooning or distension process, particular surface areas were provided with scratch markings. Besides simple grain translation, the grains rotated and grains from the interior appeared on the surface.

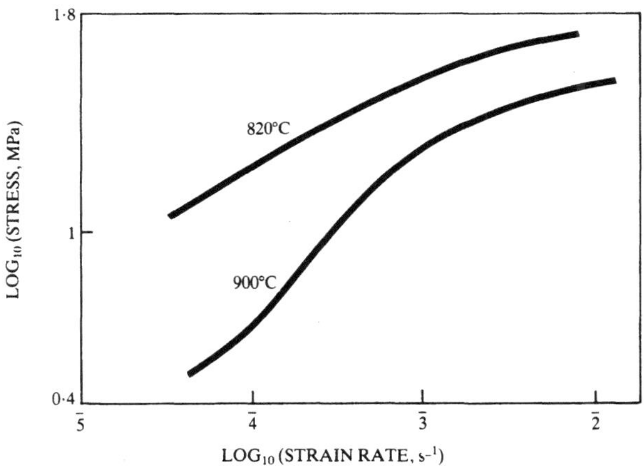

Fig. 7. Stress versus strain rate for Zircaloy-4.

7. DISCUSSION AND CONCLUSIONS

In this chapter we have given attention to theories of dislocation creep and of superplastic flow. The theories have been developed and comparison has been made with the results of experimental studies. The agreement between theory and experiment is particularly striking. Thus for both theories there are no arbitrary constants. Instead, all the parameters of the theories are derived from basic considerations and are therefore available for experimental tests.

In the case of the theory of dislocation creep there had been a deficiency of experimental information. The theory had, during its development, been compared with the data that were available. These data referred to the relationship between creep rate on the one hand and stress and temperature on the other. The theory is couched in terms which include the dislocation spacing and the mean diameter of dislocation cells. Some data had been available which permitted the theoretical predictions of these parameters to be compared with data, and the comparison had been encouraging. However, one can always level the criticism at this type of approach that the theory had been developed after the experiments were done. A more confident situation has now been attained, however, since additional data have been derived and can be compared with the theoretical predictions. Now that this has been done it appears that there are increasing grounds for believing that the theory is sound, not only in generality but also in detail. In particular, the new thermodynamic arguments which lead to identification of unique values for the ratio of dislocation cell diameter to smeared dislocation spacing are confirmed by data both from stress relaxation and from creep experiments.

Turning to the theory of superplastic flow, it is shown that this bears a systematic relationship to dislocation creep in the sense that as the stress is lowered there comes a point where the superplastic flow process should be preferred on thermodynamic grounds. At this stress the free energy of the system is lower if the superplastic process supervenes. Now the individual grains of the polycrystal, instead of maintaining their station, move relative to one another. The change in shape of the specimen is no longer reflected in similar changes in the shape of the individual grains. Instead, the grains tend to preserve their shape and size, flowing rather like grains of sand in an hour-glass. The theory considers how the movements of grain boundary dislocations can account for the dependence of superplastic flow upon stress, temperature, grain size, diffusion coefficient, etc. Here, again, excellent agreement between theory and experiment is observed. Superplastic behaviour has been observed in a wide range of alloys, and it is shown that the basic characteristics of these experimental observations are in good accord with the theoretical deductions.

REFERENCES

Aaronson, H. I., Laird, C. and Kinsman, K. R. (1970). *Phase Transformations,* American Society for Metals, Metals Park, Ohio, pp 313–96.

Ahmed, M. M. I. and Langdon, T. G. (1977). Exceptional ductility in the superplastic Pb–62%Sn eutectic, *Metall. Trans.,* **8A**, 1832–4.

Ashby, M. F. (1972). Boundary defects and atomistic aspects of boundary sliding and diffusional creep, *Surf. Sci.,* **31**, 498–537.

Ashby, M. F. and Verrall, R. A. (1973). Diffusion accommodated flow and superplasticity, *Acta Metall.,* **21**, 149–63.

Ball, A. and Hutchison, M. M. (1969). Superplasticity in the aluminium–zinc eutectic, *Metal Sci. J.,* **3**, 1–7.

Bardeen, J. and Herring, C. (1952). Diffusion in alloys and the Kirkendall effect, in *Imperfections in Nearly Perfect Crystals,* symposium held at Pocono Manor, October 1950, W. Shockley (Ed.), John Wiley, New York, pp 261–88.

Barrett, C. R. and Nix, W. D. (1965). A model for steady state creep based on the motion of jogged screw dislocations, *Acta Metall.,* **13**, 1247–58.

Bocek, M., Faisst, G. and Peterson, C. (1976). Examination of the plastic properties of Zircaloy-4 at elevated temperatures in air atmosphere, *J. Nucl. Mater.,* **62**, 26–36.

Bondy, A., Regnier, P. and Levy, V. (1971). Diffusion at interfaces, *Scr. Metall.,* **5**, 345–50.

Eberhardt, A. (1979). Ph.D. thesis, University of Metz, France.

Friedel, J. (1964). *Dislocations,* Pergamon Press, Oxford.

Garmong, G. (1974). Science Centre, Rockwell International Corporation (unpublished work).

Garmong, G. and Rhodes, C. G. (1975). Interfacial ledge structures in Ni_3Al/Ni_3Nb eutectic composites, *Metall. Trans. A.,* **6A**, 2209–16.

Geckinli, A. E. and Barrett, C. R. (1976). Superplastic deformation of the lead–tin eutectic, *J. Mater. Sci.,* **11**, 510–21.

Gittus, J. H. (1975). *Creep Viscoelasticity and Creep Fracture in Solids,* Applied Science Publishers, London.

Gittus, J. H. (1977*a*). Theory of superplastic flow in two-phase materials: roles of interphase boundary dislocation ledges and diffusion, *J. Engng Mater. Technol., Trans. ASME,* **99**, 244–51.

Gittus, J. H. (1977*b*). Theoretical value of the ratio *K* of cell diameter to dislocation spacing for a material (gold) undergoing dislocation creep, *Phil. Mag.,* **35**, 293–300.

Gittus, J. H. (1978). High temperature deformation of two phase structures, *Phil. Trans. Roy. Soc.,* **A288**, 121–46.

Gleiter, H. (1977). Private communication.
Grimes, R., Stowell, M. J. and Watts, B. M. (1976). Superplastic aluminium based alloys, *Metall. Technol.,* **3**, 154–60.
Haasen, P. (1965). *Physical Metallurgy,* North-Holland, Amsterdam, p. 867.
Hall, E. O. (1951). The deformation and ageing of mild steel, II: Characteristics of the Lüders deformation; III: Discussion of results, *Proc. Phys. Soc.,* **B64**, 742–7, 747–53.
Herriot, G., Baudelet, B. and Jonas, J. J. (1976). Superplastic behaviour of two-phase copper–phosphorus alloys, *Acta Metall.,* **24**, 687–94.
Hirth, J. P. (1972). The influence of grain boundaries on mechanical properties, *Metall. Trans.,* **3**, 3047–67.
Holt, D. L. (1970). Dislocation cell formation in metals, *J. Appl. Phys.,* **41**, 3197–201.
Ishida, Y. and McLean, D. (1967). (Solid solution strengthening) Effect of nitrogen and manganese on recovery rate and friction stress during creep in iron, *J. Iron Steel Inst.,* **205**, 88–93.
Johnson, R. H. (1970). Superplasticity, *Metall. Rev.,* **15**, 115–34.
Knoch, H. (1975). *High Temperature Creep of Single Phase and of Two Phase MgO with High Iron Contents,* ERAMS-I/TR26, University of Erlangen-Nürnberg, West Germany.
Lee, D. (1969). The nature of superplastic deformation in the Mg–Al eutectic, *Acta Metall.,* **17**, 1057–69.
Lee, D. and Backofen, W. A. (1967). Superplasticity in some titanium and zirconium alloys, *Trans. Metall. Soc. AIME,* **239**, 1034–40.
McLean, D. (1971). Grain boundary dislocations: their effect on vacancies and sliding, *Phil. Mag.,* **23**, 467–72.
Melton, K. N., Edington, J. W., Kallend, J. S. and Cutler, C. P. (1974). Textures in superplastic Zn–40 wt%Al, *Acta Metall.,* **22**, 165–70.
Mohamed, F. A., Shei, S. A. and Langdon, T. G. (1975). The activation energies associated with superplastic flow, *Acta Metall.,* **23**, 1443–50.
Mohamed, F. A., Ahmed, M. M. I. and Langdon, T. G. (1977). Factors influencing ductility in the superplastic Zn–22%Al eutectoid, *Metall. Trans.,* **8A**, 933–8.
Nakagawa, Y. G. and Weatherley, G. C. (1972). The nature of interface between two ordered lattices: the Ni_3Al–Ni_3Nb eutectic, *Mater. Sci. Engng,* **10**, 223–8.
Naziri, H. and Pearce, R. (1974). Superplasticity in a Zn–0·4%Al alloy, *Acta Metall.,* **22**, 1321–30.
Perinet, F. (1975). Contribution to the study of the interfacial diffusion, CEA-R-4657, Saclay, Gif-sur-Yvette, France.
Petch, N. J. (1953). The cleavage strength of polycrystals, *J. Iron Steel Inst.,* **174**, 23–28.
Povolo, F. (1981). On the analysis of stress relaxation data. *J. Nucl. Mater.,* **96**, 178.
Povolo, F. and Capitani, J. C. (1984). Influence of temperatures and stress-relieving treatment on the stress relaxation in bending of Zircaloy-4 near 673 K, *J. Mater. Sci.,* **19**, 2969.
Povolo, F. and Marzocca, A. J. (1984*a*). Creep of stress-relieved Zircaloy-4 at 673 K, *J. Nucl. Mater.,* **118**, 224.
Povolo, F. and Marzocca, A. J. (1984*b*). Apparent activation energy for creep controlled by jog-drag and cell-formation, *J. Nucl. Mater.,* **119**, 78.

Povolo, F., Marzocca, A. J. and Capitani, J. C. (1984). Ratio of cell diameter to dislocation spacing for creep of Zircaloy-4, *Phil. Mag.*, **48**, 759.
Povolo, F., Marzocca, A. J. and Hermida, D., to be published.
Samuelsson, L. C. A., Melton, K. N. and Edington, J. W. (1976). Dislocation structures in a superplastic Zn–40 wt% Al alloy, *Acta Metall.*, **24**, 1017–26.
Staker, M. R. and Holt, D. L. (1972). The dislocation cell size and dislocation density in copper deformed at temperatures between 25 and 700 °C, *Acta Metall.*, **20**, 569–79.
Suery, M. and Baudelet, B. (1975). Theoretical and experimental constitutive equations of superplastic behaviour: discussion, *J. Mater. Sci.*, **10**, 1022–8.
Thomson, R. M. and Balluffi, R. W. (1962). Kinetic theory of dislocation climb, *J. Appl. Phys.*, **33**, 803–25.
Vastava, R. B. and Langdon, T. G. (1979). An investigation of intercrystalline and interphase boundary sliding in the superplastic Pb–62% Sn eutectic, *Acta Metall.*, **27**, 251–7.

BIBLIOGRAPHY

Alden, T. H. (1967). The origin of superplasticity in the tin–5% bismuth alloy, *Acta Metall.*, **15**, 469–80.
Backofen, W. A., Azzarto, F. J., Murty, G. S. and Zehr, S. W. (1968). Superplasticity in ductility, paper presented at a seminar, Metals Park, Ohio, October 1967, American Society for Metals, Metals Park, Ohio, pp 279–310.
Bell, L. G. and Duncan, R. G. (1975). *Hydride Orientation in Zr–2·5%Nb,* AECL 5110, Atomic Energy of Canada Ltd.
Breen, J. E. and Weertman, J. (1955). Creep of polycrystalline tin, *Trans. Am. Inst. Min. Metall. Engrs,* **203**, 1230–4.
Chadwick, R. (1934). *J. Inst. Metals,* **54**, 131.
Chapman, J. A. and Wilson, D. V. (1962–3). The room temperature ductility of fine grain magnesium, *J. Inst. Metals,* **91**, 39–40.
Clark, M. A. and Alden, T. H. (1973). Deformation enhanced grain growth in a superplastic Sn–1%Bi alloy, *Acta Metall.*, **21**, 1195–1206.
Cline, H. E. (1967). The mechanical properties of the CoAl–Co eutectic, *Trans. Metall. Soc. AIME,* **239**, 710–4.
Cutler, C. P. and Edington, J. W. (1971). Textures in the superplastically deformed tin-lead eutectic alloy, *Metal Sci. J.*, **5**, 201–5.
Cutler, C. P., Edington, J. W., Kallend, J. S. and Melton, K. N. (1974). Quantitative texture studies of the superplastically deformed aluminium-copper eutectic alloy, *Acta Met.*, **22**, 665–71.
Dunlop, G. L. and Taplin, D. M. R. (1972*a*). The tensile properties of superplastic aluminium-bronze, *J. Mater. Sci.*, **7**, 84–92.
Dunlop, G. L. and Taplin, D. M. R. (1972*b*). A metallographic study of superplasticity in a micrograin aluminium-bronze, *J. Mater. Sci.*, **7**, 316–24.
Edington, J. W. (1976). Physical metallurgy of superplasticity, *Metall. Technol.*, **3**, 138–51.
Floreeen, S. (1967). Superplasticity in pure nickel, *Scr. Metall.*, **1**, 19–23.

Garfinkle, M., Witzke, W. R. and Klopp, W. D. (1967). Metall. Soc. AIME, *Abstract Bulletin,* **3**, 77.
Gifkins, R. C. (1958–9). Recrystallisation of lead during creep, *J. Inst. Metals,* **87**, 255–61.
Gifkins, R. C. (1977). The effect of grain size and stress upon grain boundary sliding, *Metall. Trans.,* **8A**, 1507–16.
Greenwood, G. W. and Cole, J. (1948). Influence of various factors on the creep of lead, *Metallurgia,* **37**, 285–9.
Greenwood, G. W. and Johnson, R. H. (1965). The deformation of metals under small stresses during phase transformations, *Proc. Roy. Soc.* (Lond.), **A283**, 403–22.
Hayden, H. W., Gibson, R. C., Merrick, H. F. and Brophy, J. H. (1967). Superplasticity in the Ni–Fe–Cr system, *Trans. Am. Soc. Metals,* **60**, 3–14.
Holt, D. L. and Backofen, W. A. (1966). Superplasticity in the aluminium–copper eutectic alloy, *Trans. Am. Soc. Metals,* **59**, 755–68.
Johnson, R. H. and Sykes, E. C. (1966). The enhancement of ductility in alpha-uranium, *Nature,* **209**, 192–3.
Kot, R. and Weiss, V. (1967). Transformation plasticity of steels and titanium in torsion, *Trans. Am. Soc. Metals,* **60**, 566.
Lee, D. (1973). Strengthening and superplasticity in the Ag–Cu and Mg–Al eutectics, *J. Soc. Mater. Sci. Japan,* **22**, 209–12.
Marder, A. R. (1969). Effect of carbon content, test temperature and strain rate on the strain sensitivity of iron–carbon alloy, *Trans. Metall. Soc. AIME,* **245**, 1337–45.
Moore, H. F., Betty, B. B. and Dollins, C. W. (1938). *Univ. Ill. Bull.,* **35**, 102.
Morrison, W. B. (1968). Superplasticity of low alloy steels, *Trans. Am. Soc. Metals,* **61**, 423–34.
Nuttall, K. (1976). Superplasticity in the Zr–2·5%Nb alloy, *Scr. Metall.,* **10**, 835–40.
Nuttall, K. and McCooeye, D. P. (1974). Phase transformation plasticity in zirconium–hydrogen alloy, *Proc. 2nd Symp. Mech. Behav. Mater.,* Society Material Science Japan, Kyoto, Japan, pp 129–40.
Oshida, Y. and Weiss, W. (1967). Research Inst. Rep. MET-E, Syracuse University, USA, pp 1504–67 FR.
Povolo, F., Marzocca, A. J. and Rubiolo, G. H., accepted for *Res. Mech.*
Sagat, S. and Taplin, D. M. R. (1976). The stability of plastic flow in strain-rate-sensitive materials, *Metal Sci.,* **10**, 94–100.
Schadler, H. W. (1968). The stress–strain rate behaviour of a manganese steel in the temperature range of the ferrite–austenite transformation, *Trans. Metall. Soc. AIME,* **242**, 1281–7.
Smith, C. I. and Ridley, N. (1974). Design of a superplastic alloy steel, *Metals Technol.,* **1**, 191–8.
Smith, C. I., Norgate, B. and Ridley, N. (1976). Superplastic deformation and cavitation in a microduplex stainless steel, *Metal Sci.,* **10**, 182–8.
Smith, C. S. (1948). Grains, phases and interfaces: an interpretation of microstructure, *Trans. Am. Inst. Min. Engrs,* **175**, 15–51.
Suery, M. (1979). Métallurgie et rhéologie des alliages superplastiques biphases, Ph.D. thesis, University of Metz, France.

Suery, M., Baudelet, B., Labulle, B. and Petipas, C. (1974). X-ray diffraction study of behaviour of 60/40 brass phases during superplastic deformation, *Scr. Metall.*, **8**, 703–6.

Takeuchi, S. and Argon, A. S. (1976). *J. Mater. Sci.*, **11**, 1542.

Yoder, G. R. and Weiss, V. (1972). Superplasticity in eutectoid steel, *Metall. Trans.*, **3**, 675–81.

CHAPTER 7

Quasi-Physical Modelling of the Plastic Behaviour of Polycrystals

A. ZAOUI
Laboratoire des Propriétés Mécaniques et Thermodynamiques des Matériaux, CNRS, Université Paris-Nord, Villetaneuse, France

ABSTRACT

The transition from single crystal to polycrystal rate-independent plasticity is investigated according to two distinct approaches. The first, which is based on experimental observations of the influence of grain boundaries and junctions on the plastic response of grains of a polycrystal, emphasizes the inhomogeneous character of the intragranular plastic strain field; consequently, the average plastic behaviour of a grain completely embedded within a polycrystalline aggregate does not reduce to that of an isolated single crystal. Though this conclusion can hardly lead, at the moment, to a quantitative estimate of such a difference, several attempts at quasi-physical modelling of the plastic behaviour of polycrystals are reported. The general principles of the corresponding procedure are first analysed, with special attention to the 'localization' problem, which is concerned with the relationship between local and overall mechanical variables. Several models are then discussed, which have a space phase distribution independence as a common feature. Typical illustrations of the predictive ability of these models are reported, with special emphasis on the self-consistent method. This method is finally shown to provide a possible guide for the introduction of space distribution effects, through the definition of multi-site self-consistent schemes.

1. INTRODUCTION

This chapter is concerned with rate-independent plasticity of metallic polycrystals. While such behaviour is simple enough to focus attention on restricted problems, it is complicated enough to point out the main difficulties that must be overcome when one tries to introduce some physical phenomena and structural information into the usual approach of continuum mechanics, in the field of polycrystal elasto-plasticity. The complexity of plastic behaviour is well known to originate in the fact that the internal state of the material is continuously changing during plastic flow, reflecting the corresponding evolution of the microstructure (dislocation distribution, lattice orientations, etc.). According to a phenomenological approach, this situation may be described by the introduction of a sufficient number of 'hidden' variables and the development of thermodynamical treatments and laws of evolution which are derived from an identification process with standard experiments. Powerful methods have been developed in this field, but they have to deal with each material as a new one and then need lengthy and difficult experimental investigations.

Part of these efforts could be saved if a deductive approach were available, which would permit the integration of minimal but significant structural information into an adequate mechanical treatment, so that changes in given physical data concerning the material or the initial state could be dealt with according to the same procedure without performing a whole new mechanical investigation. Such an attempt has to answer the critical questions: what are the good, significant physical data (parameters as well as mechanisms), and how can they be introduced into a consistent mechanical treatment?

The answer to the first question seems to depend on the observation technique. Optical (or even scanning electron) microscopy leads to the conclusion that the important feature is the aggregation of grains (or phases) with various shapes and sizes and that the most significant mechanism is the crystallographic plastic glide which occurs within the grains. A classical X-ray diffraction technique will focus attention on the lattice orientation distribution and stress the importance of crystallographic textures. If more local X-ray diffraction techniques (such as micro-Laue or Kossel) are available, they will emphasize the spatial distribution of grain orientations and the correlations between position, shape, size . . . and orientation . . . . If a thin foil is examined in a transmission electron microscope it is clear that the important features are the cell structure and the dislocation distribution. A refined analysis would point out the prominent part of the core structure of the dislocations, and so on. As a matter of fact, several scales may be defined and for each scale there are distinct significant parameters and determinative mechanisms, so that a truly deductive treatment would ideally incorporate a succession of scale transitions from the atomic (or sub-atomic) level to the macroscopic level which is the scale of continuum engineering mechanics. Obviously this is not a realistic objective for the current century, and we shall have to choose, in what follows, the starting scale of which the elements and properties will be described without too much detail using phenomenological (on this scale) characteristics and parameters.

Then the second question arises when the transition from this basic scale towards the macroscopic one is dealt with. Two kinds of treatment will be adopted successively in the following. According to the first (the more physical) we shall try to perform an analytical reconstruction of the polycrystal, starting from the single crystal and successively examining the influence of the grain boundaries, the grain junctions, etc. in bi-, tri- and multi-crystals. Although this procedure will appear to provide a good understanding of the problems and phenomena involved, it will lead to such a complicated picture that the transition to the macroscopic level will look out of range (Section 2). Consequently a more global and less physical approach (some 'quasi-physical modelling') will be preferred, omitting some aspects of the physical complexity and using a more or less elaborate description of interacting homogeneous elements (Section 3). This mixed physical and mechanical treatment will reveal a useful tool for the prediction of several aspects of polycrystalline plasticity which will be illustrated with a few examples, especially concerning the plastic behaviour of two-phase metals and the development of crystallographic texture and plastic anisotropy of initially isotropic polycrystals. Final comments (Section 4) will stress some still open questions and possible future advances in this field.

2. PHYSICAL ANALYSIS OF THE TRANSITION FROM SINGLE CRYSTAL TO POLYCRYSTAL

A logical — even if somewhat ingenuous — way to the modelling of polycrystal plasticity would be to start from the single crystal situation and to make it gradually more and more complicated by considering a bicrystal with a grain boundary, a tricrystal with a grain junction, and more and more complex aggregates with interacting grain boundaries and junctions, up to the usual polycrystal. We shall briefly describe some of the difficulties besetting this way which at present looks far from reaching the issue of the polycrystal.

2.1. Single Crystal Behaviour

The first step would seem to be the easiest, since the plasticity of single crystals has been the subject of intensive study by solid state physicists. If each stage of the tensile stress–strain curve of ordinary crystals is well understood in the light of the dislocation theory, it is hard to say that such an understanding leads to well established and verified constitutive equations which could be used for arbitrary strain or stress paths, in the continuous and tensorial form needed for further mechanical treatments. Evidently it is not yet time to consider only dislocation distributions and to derive the macroscopic behaviour from a comprehensive analysis of their interactions and evolution. Consequently mixed approaches are needed which simultaneously deal with microstructural variables such as dislocation density ρ, stacking fault energy etc. and more macroscopical elements such as slip systems, resolved shear strain γ, shear stress τ, etc. In this respect the Schmid law appears to be an essential ingredient in any mechanical description of single crystal plasticity, regardless of further refinements which would retain the property that the plastic activity of an easy glide system (g) is determined by the value of the resolved shear stress on it, τ^g, compared with some critical one, τ_c^g. If this (important) step is taken, several difficulties still remain regarding the hardening law and the flow rule.

The simplest form of hardening law (Hill, 1966; Mandel, 1965) surely consists in a direct relation between the critical resolved shear stress (CRSS) rate on any system (g), say $\dot{\tau}_c^g$, and the resolved shear strain rates $\dot{\gamma}^h$ in the form

$$\dot{\tau}_c^g = \sum_h h^{gh} \dot{\gamma}^h \tag{2.1}$$

where the hardening matrix h^{gh} correlates the slip activity on any system (h) with the hardening behaviour on any other (g). Such a relation is very convenient in that shear strain elements are directly taken as hardening parameters, which makes the derivation of the flow rule straightforward since the plastic strain rate $\dot{\varepsilon}^p$ is simply (within the assumption of small strain)

$$\dot{\varepsilon}^p = \sum_g (\vec{m}^g \otimes \vec{n}^g)^S \dot{\gamma}^g = \sum_g \mathbf{R}^g \dot{\gamma}^g \tag{2.2}$$

where the superscript S indicates the symmetrization operation, $\otimes$ indicates a tensorial product, and $\vec{m}^g$ and $\vec{n}^g$ are respectively the slip direction and the slip normal of system (g). In this respect the hardening matrix h^{gh} may depend on the resolved shear strains and stresses (Havner and Shalaby, 1977; Weng, 1979; Peirce *et al.*, 1982). Unfortunately this matrix is not a constant. Two kinds of test, at least, easily show this fact.

Latent hardening tests. Secondary specimens are cut from a primary one which has previously been pulled so as to make only the primary slip system active (i.e. within stage I, for crystals exhibiting such a stage); measurement of the CRSS τ_c^s on the secondary specimens, when stretched, clearly shows that it does not vary linearly with the primary resolved shear strain γ^p just before the cut. Latent hardening ratios, defined as

$$L^{sp} = \tau_c^s / \tau_c^p \tag{2.3}$$

where τ_c^p is the primary CRSS before the cut, have an increasing variation with γ^p and then a decreasing one, towards stationary values (always higher than unity), which is at variance with the monotonic variation expected from eqn (2.1) (Fig. 1). This behaviour is common to all the secondary slip systems that can be tested, depending on the orientation of the secondary specimens, regardless of the large discrepancies that exist between the absolute values of the L^{sp}, which reflects the anisotropy of the interactions between the slip systems when one of them has already been activated (see e.g. Jackson and Bazinski, 1967; Kocks, 1964; Nakada and Keh, 1966); this anisotropy, which is present even in materials known as isotropic (such as aluminium), increases as the stacking fault energy decreases. In cubic crystals it can be closely correlated with the anisotropic interactions between dislocation families on different slip systems, according to whether they can form sessile or glissile junctions or no junction at all (Franciosi *et al.*, 1980).

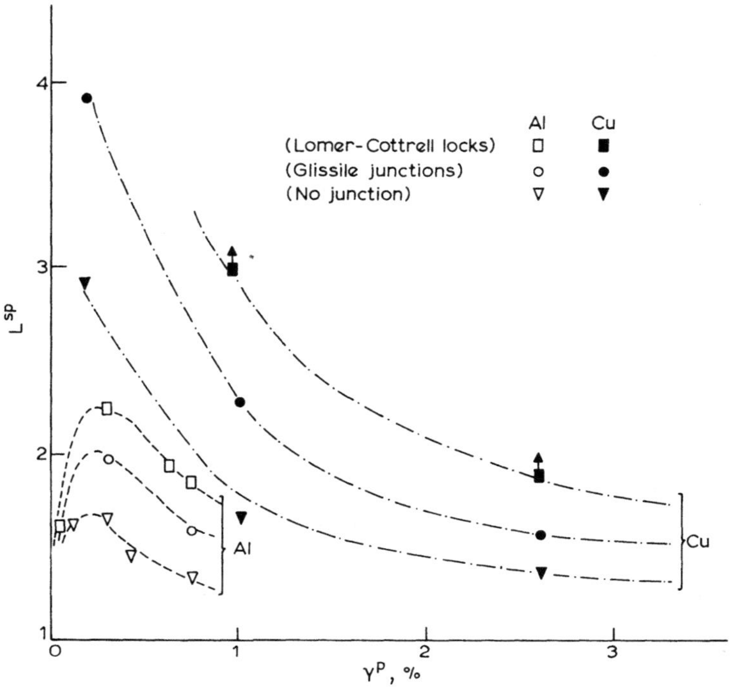

Fig. 1. Variation of the latent hardening ratios of various secondary slip systems (aluminium and copper crystals) with the primary rate of shearing.

Multislip tensile tests. In these the tensile axis lies along a symmetrical axis of the crystallographic structure. Without reporting the corresponding experimental results with all the details, it must be stressed that from such tests, when two (or more) critical systems suffer equal applied shear stresses, the corresponding sub-matrix h^{gh} — to the extent that it can be defined, in a narrow strain range — now looks roughly isotropic, whether the systems are or are not active simultaneously and whether their interactions, in contrast with what was discussed above, are strong or weak (Franciosi and Zaoui, 1982*a, b*). Consequently the hardening matrix, if it exists, clearly depends on the shear stress distribution on the slip systems and cannot be considered, except to a first rough approximation, as constant (Franciosi, 1984; Havner, 1982).

Several other partial results lead to the conclusion that the resolved shear strain distribution is not a suitable set of hardening parameters and that it is hard to avoid taking the dislocation densities directly into account. Starting from the original Taylor (1934) relation,

$$\tau_c = \alpha\mu b\sqrt{\rho} \tag{2.4}$$

where ρ is the total dislocation density, μ the elastic shear modulus and b the Burgers vector intensity (α is a dimensionless parameter of the order of unity), more anisotropic relations may be checked where the individual dislocation densities ρ^h on each slip system (h) are considered and an interaction matrix a^{gh} is introduced, instead of the simple scalar α. Several forms have been proposed for such a relation (e.g. Zarka, 1973; Franciosi, 1978), and some pieces of evidence argue for the following:

$$\tau_c^g = \mu b \sum_h (a^{gh}\rho^h)^{1/2} \tag{2.5}$$

The main aspects of the hardening behaviour of single crystals are fairly well accounted for by such a relation, with an adequate a^{gh} matrix which has to integrate both long-range and short-range interactions between dislocation families. But the main difficulty now lies in the flow rule, since the correlations between the dislocation densities which are efficient with regard to the hardening behaviour are not necessarily those that play the prominent part in the plastic flow. Consequently, mobile dislocations have to be specified and, despite several attempts to resolve this question, in connection with cross-slip behaviour and cell formation, it cannot yet be considered as sufficiently elucidated to allow an explicit transition from dislocation densities to polycrystal macroscopic plastic behaviour.

Accordingly, in the following, the hardening matrix formalism will be adopted in spite of its shortcomings while awaiting further advances towards a firmer description of dislocation mobility leading to a physically consistent (and mechanically tractable) plastic flow rule for isolated single crystals.

2.2. Mechanical Influence of Grain Boundaries

The next step is concerned with the analysis of the mechanical part of grain boundaries in the plasticity of polycrystals. This question may be considered from two distinct points of view. On the one hand, and even if their detailed microstructure is not very well resolved in the general case, grain boundaries are known to interact with lattice dislocations, acting as obstacles to their free motion as well as sources for various relaxation mechanisms, so that they significantly disturb the dislocation motion and the glide pattern, at least in their immediate vicinity. On the other hand, each grain is submitted to long-range interactions with other more or less distant grains, so that an internal stress field adds to the applied one. For that matter, grain boundaries do not act by themselves, but as the interfaces between adjoining grains with different lattice orientations and consequently different plastic glide distributions which generally correspond to an incompatible plastic strain field and associated internal stresses.

Both of these phenomena combine in the actual behaviour of polycrystals and it is not easy to separate them except in very simplified situations. This is the case for bicrystals with a plane boundary, which have been the subject of a number of both experimental and theoretical studies. An extensive discussion of the results of these studies is outside the scope of this chapter, but some of them may be briefly reported in order to point out the main conclusions.

Let us compare the two following situations:

(i) In the first, a bicrystal with a plane boundary (say normal to Ox_2, if an orthogonal cartesian coordinates system x_i is adopted) is pulled according to a tensile axis parallel to the boundary (say Ox_3);
(ii) In the second, two single crystals, each of which is identical with one of the two grains of the foregoing bicrystal, are pulled independently under the same conditions as for the bicrystal.

The general conclusions of this comparison, except for very special situations, are that the bicrystal is harder than each of the single crystals; its stress–strain curve does not exhibit the single glide stage (stage I) of the single crystal curves and then begins with a higher slope; several slip systems are activated almost at the beginning of the plastic flow; moreover, slip heterogeneities are present, with a special predilection of secondary systems for concentrating near the grain boundary, at deformation stages where the single crystals look much more uniformly deformed. Eventually, further observations (Fig. 2) show that the glide pattern is different even from one face of a given grain to the parallel face of the same grain, with a tendency to form distinct areas within the grains, each of which is roughly uniformly deformed, separated by intragranular interfaces, leaning on the trace of the grain boundary on one face or the other (Rey and Zaoui, 1980, 1982).

The first impulse to try to understand such behaviour leads one to plead that the plastic responses of both isolated grains to the applied

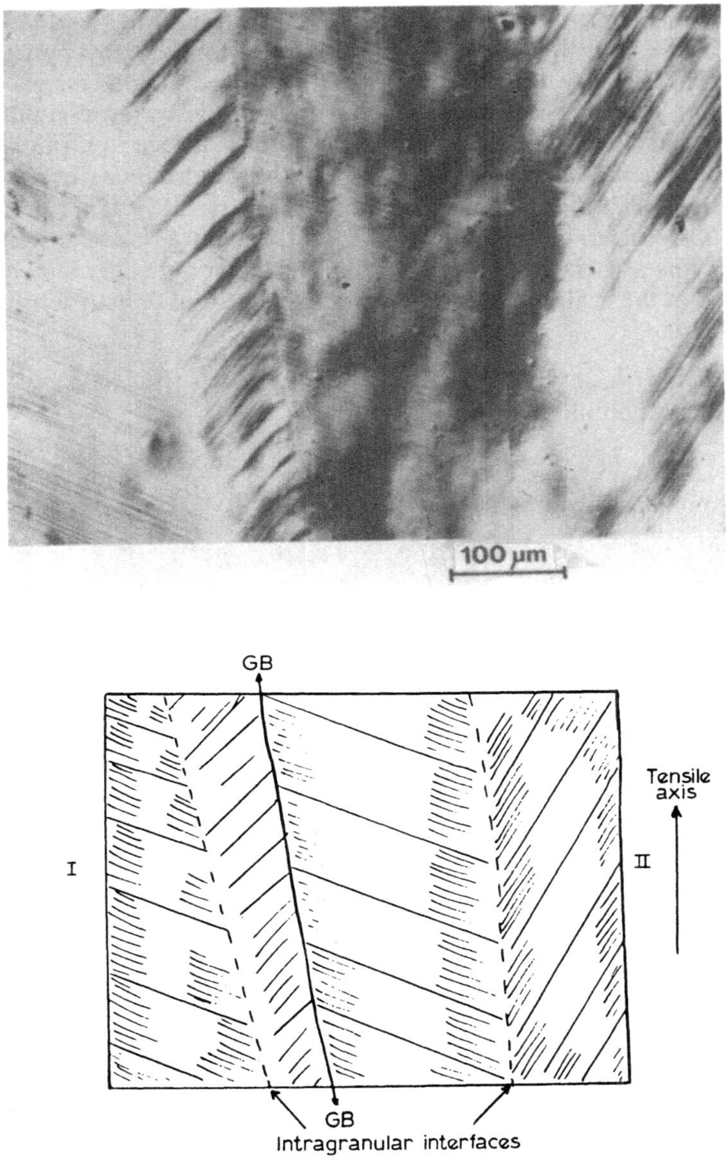

Fig. 2. Intragranular domains, with distinct active slip systems, in copper bicrystals. GB = grain boundary.

stress are incompatible with each other, so that internal stresses develop which are responsible for the activation of secondary slip systems, which is at variance with the situation for isolated single crystals (at low strain). This idea may be made more quantitative by resort either to the physical theory of dislocations, such as dislocation pile-ups (Livingston and Chalmers, 1957), or to the continuous theory of dislocations and plastic incompatibility (Kröner, 1958). If the initial plastic distortions in grains I and II, say $\boldsymbol{\beta}^{\mathrm{pI}}$ and $\boldsymbol{\beta}^{\mathrm{pII}}$, are assumed uniform, the associated plastic incompatibility is only superficial and localized at the grain boundary. With the dislocation density tensor $\boldsymbol{\alpha}$ defined by

$$\alpha_{ij} = -\varepsilon_{ikl}\,\beta^{\mathrm{p}}_{lj,k} \qquad (\boldsymbol{\alpha} = -\mathbf{curl}\,\boldsymbol{\beta}^{\mathrm{p}}) \tag{2.6}$$

or the incompatibility tensor $\boldsymbol{\eta}$, given by

$$\eta_{ij} = -\varepsilon_{ikl}\,\varepsilon_{jmn}\,\varepsilon^{\mathrm{p}}_{lm,kn} \qquad (\boldsymbol{\eta} = -\mathbf{inc}\,\boldsymbol{\varepsilon}^{\mathrm{p}}) \tag{2.7}$$

and correlated with $\boldsymbol{\alpha}$ by

$$\eta_{ij} = \tfrac{1}{2}(\varepsilon_{imn}\,\alpha_{jm,n} + \varepsilon_{jmn}\,\alpha_{im,n}) \tag{2.8}$$

where the comma notation for spatial derivatives and the permutation tensor ε_{ijk} are used together with Einstein's summation convention, one finds easily

$$\alpha_{ij} = -\varepsilon_{i2l}\,\Delta\beta^{\mathrm{p}}_{lj}\,\delta(x_2) = \begin{pmatrix} -\Delta\beta^{\mathrm{p}}_{31} & -\Delta\beta^{\mathrm{p}}_{32} & -\Delta\beta^{\mathrm{p}}_{33} \\ 0 & 0 & 0 \\ \Delta\beta^{\mathrm{p}}_{11} & \Delta\beta^{\mathrm{p}}_{12} & \Delta\beta^{\mathrm{p}}_{13} \end{pmatrix} \delta(x_2) \tag{2.9}$$

and

$$\eta_{ij} = (\varepsilon_{im2}\,\alpha_{jm})_{(ij)}\,\delta'(x_2) = \begin{pmatrix} \Delta\varepsilon^{\mathrm{p}}_{33} & 0 & -\Delta\varepsilon^{\mathrm{p}}_{13} \\ 0 & 0 & 0 \\ -\Delta\varepsilon^{\mathrm{p}}_{13} & 0 & \Delta\varepsilon^{\mathrm{p}}_{11} \end{pmatrix} \delta'(x_2) \tag{2.10}$$

In these formulae, $\Delta\boldsymbol{\beta}^{\mathrm{p}} = \boldsymbol{\beta}^{\mathrm{pII}} - \boldsymbol{\beta}^{\mathrm{pI}}$ and $\delta(x_2)$ and $\delta'(x_2)$ are the Dirac function and its first derivative respectively. For homogeneous and isotropic elasticity, knowledge of $\boldsymbol{\alpha}$ permits calculation of the associated internal stress field from the classical solutions for discrete dislocation walls corresponding to each α_{ij} component and, after a limiting procedure, for continuous distributions. The general stress function tensor method may also be used directly. In any case, we find the internal stress field (Zaoui, 1979):

$$\frac{\sigma_{ij}}{\mu} = \left[-\Delta\varepsilon^{\mathrm{p}}_{ij} + \Delta\varepsilon^{\mathrm{p}}_{2j}\delta_{i2} + \Delta\varepsilon^{\mathrm{p}}_{i2}\delta_{j2} + \frac{\Delta\varepsilon^{\mathrm{p}}_{22}}{1-\nu}(\nu\delta_{ij} - \delta_{i2}\delta_{j2}) + \frac{\nu\Delta\varepsilon^{\mathrm{p}}_{kk}}{1-\nu}(\delta_{i2}\delta_{j2} - \delta_{ij}) \right] \operatorname{sgn} x_2 \qquad (2.11)$$

where $\operatorname{sgn} x_2$ denotes the sign of x_2, δ_{ij} is the Kronecker tensor and ν is the Poisson ratio.

For a plastic deformation of glide type ($\varepsilon^{\mathrm{p}}_{kk} = 0$), it remains simply

$$\left.\begin{aligned} \sigma_{11} &= \frac{\mu}{1-\nu}(\Delta\varepsilon^{\mathrm{p}}_{11} + \nu\Delta\varepsilon^{\mathrm{p}}_{33}) \operatorname{sgn} x_2 \\ \sigma_{33} &= \frac{-\mu}{1-\nu}(\Delta\varepsilon^{\mathrm{p}}_{33} + \nu\Delta\varepsilon^{\mathrm{p}}_{11}) \operatorname{sgn} x_2 \\ \sigma_{13} &= -\mu\Delta\varepsilon^{\mathrm{p}}_{13} \operatorname{sgn} x_2 \\ \sigma_{12} &= \sigma_{22} = \sigma_{23} = 0 \end{aligned}\right\} \qquad (2.12)$$

This solution must be used very cautiously:

(i) First, it corresponds to the case of a perfect fit between the two grains on the boundary. If some grain boundary sliding occurs, the classical compatibility conditions are modified, as soon as this sliding is not uniform (Mussot *et al.*, 1985) and relations (2.12) are no longer valid;
(ii) Secondly, eqns (2.12) give the residual stress field (regardless of edge effects in a finite bicrystal) when the uniform plastic strain fields $\boldsymbol{\varepsilon}^{\mathrm{pI}}$ and $\boldsymbol{\varepsilon}^{\mathrm{pII}}$ are given and 'frozen'. If the bicrystal is plastically flowing, the solution must be given in an incremental form and can be significantly modified (and especially lowered) with regard to eqns (2.12). A fairly good approximation in this case consists in replacing the elastic shear modulus μ in eqns (2.12) by the overall elastic-plastic secant modulus μ^{s} in the case of monotonic radial loading, or both $\boldsymbol{\sigma}$ and $\boldsymbol{\varepsilon}^{\mathrm{p}}$ by the rates $\dot{\boldsymbol{\sigma}}$ and $\dot{\boldsymbol{\varepsilon}}^{\mathrm{p}}$ and μ by the tangent elastic-plastic modulus μ^*. Contingent on these modifications, relations (2.12) lead to good agreement with observations of the active secondary slip systems only in restricted cases, namely when slip inhomogeneities do not occur in the grains.

The problem of slip inhomogeneity cannot be solved only by considering the lattice disorientation across the grain boundary. It also involves the influence of the grain boundary itself, as an obstacle to the easy development of slip when it comes near the boundary. A simplified picture, according to which slip on a given plane takes place by the motion of straight edge dislocations and is slowed down as soon as these dislocations meet the grain boundary, leads to the resulting formation of intragranular plane interfaces, cutting the grain boundary on one of the observed faces and separating plastically deformed and undeformed areas; such interfaces are generally incompatible (even when the bicrystal as a whole may be considered as compatible) and are then associated with specific internal stresses, which obey eqns (2.12) contingent on the above-mentioned modifications and an adequate axes rotation. These internal stresses are responsible for the activation of secondary systems on one or both sides of the interfaces which may rotate and vanish or not according to the nature of the interactions between the systems — both primary and secondary — involved, as they result from latent hardening investigations. Such an interpretation has been corroborated by a number of observations on fcc bicrystals (Rey, 1980).

Even if it is restricted to the case of bicrystals, it is expected to be partly relevant to the plasticity of polycrystals where intragranular slip inhomogeneity is a very common feature which may have an important influence on the overall strain-hardening behaviour as well as on the crystallographic texture development. On the other hand, one would be justified in feeling sceptical about the feasibility of extending such an interpretation, which is already somewhat tedious for simple bicrystals, to the case of polycrystals.

2.3. From the Bicrystal to the Polycrystal

Many other difficulties remain to be overcome on the way to the polycrystal. One of them may be encountered in the case of a tricrystal, with three intersecting plane boundaries. As far as slip inhomogeneity is concerned, a new type of pattern and cause adds to the foregoing ones, due to the effect of the grain triple junction. Observations show that, besides the above-mentioned intragranular 'domains' of the bicrystal type, new interfaces can be formed from the junction so as to split the affected grain into two dihedral domains where different slip systems develop (Fig. 3). This kind of inhomogeneity may be correlated with the very strong internal stress field which prevails in such a tricrystal with uniform plastic strain in each grain; unlike the situation for the bicrystal [eqns (2.12)], it is no longer uniform in the grains. In addition to a regular contribution which depends on the angle from the boundaries, a singular one is included which has a logarithmic variation with the

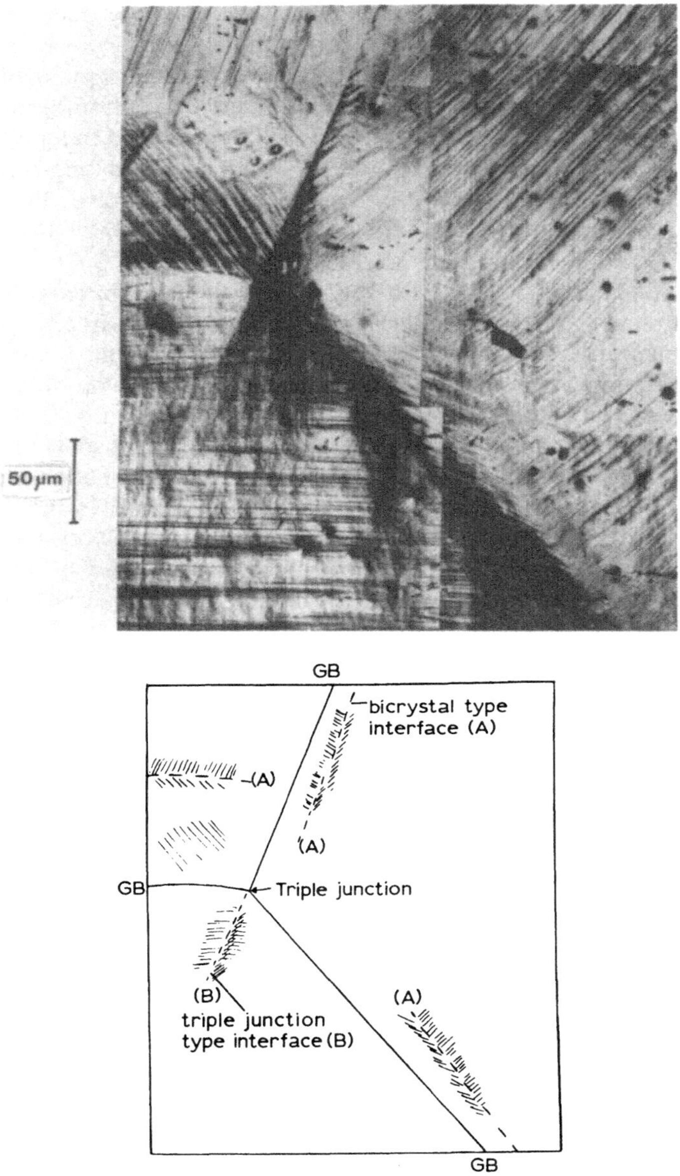

Fig. 3. Two types of intragranular slip inhomogeneities (A and B) observed on copper tricrystals. GB = grain boundary.

distance to the junction. Such a singularity can be relaxed by the formation of new interfaces resting on the junction and limiting differently plastified areas (see Rey *et al.,* 1985); this is a plausible, if not exclusive, reason for the development of such an inhomogeneity.

In any case, besides local effects which are induced in a given grain by the grain boundaries and junctions which bound it, long-range ones must be taken into account which originate in more or less distant plastic incompatibility and internal stress (both inter- and intragranular) sources (Bretheau and Caldemaison, 1981).

Evidently any deterministic treatment of such combined interaction effects seems somewhat hopeless and we must accept, in order to build up constitutive equations for the polycrystal, to hold a number of secondary effects in the background, to be satisfied with only a rough picture of real behaviour and to turn towards statistical methods which would retain the prominent aspects and smooth out most of the contrast of the foregoing phenomena. Such a necessity can be made a virtue if it is appreciated that most of the experimental data available on the plastic behaviour as well as on the microstructure of polycrystals are of a statistical nature; this is the case not only for the macroscopic behaviour itself, which is an average one, but also for the influence of the mean grain size, the description of the crystallographic texture thanks to the texture function, the phase volume fractions for a multiphase material, etc.

Thus we shall soon forget the insight that we have just had into the stress and strain inhomogeneity within the grains of a deformed polycrystal, and, going on towards some 'quasi-physical' modelling of the polycrystal behaviour, consider the aggregate as a random set of fixed homogeneous elements (the 'phases'), each of which will remain, all along the plastic flow, uniformly strained and stressed. The nature of these phases will be discussed in the following but, whether they are individual grains or grain families with the same lattice orientation or whole chemical phases of an actual multiphase material, the main problem remaining now will be concerned with the long-range mechanical interactions between these phases associated with the disparity in their individual mechanical properties.

Nevertheless a general warning must be kept in mind from the foregoing. Since intraphase inhomogeneity is erased, the phase behaviour can no longer be strictly identified with the local behaviour of the constituent material; for instance, the averaged *behaviour* of a given strain (regardless of its mechanical interactions with the other grains of the polycrystal) must more or less differ from the *behaviour* of an isolated, uniformly deformed single crystal, and judicious adjustments of the mechanical characteristics of the single crystal are not only allowed but necessary in order to be consistent with the picture of uniformly deformed phases. Moreover, since this part of the whole modelling procedure is, to a certain extent, physically approximate (if not somewhat arbitrary), the statistical description of the phase spatial distribution cannot be unduly refined (as it can be in the elastic case, where the intraphase behaviour is much better known and described), at the risk of illusive and artificial improvements.

3. MODELLING OF POLYCRYSTAL PLASTIC BEHAVIOUR

Advances in the modelling of polycrystal plastic behaviour have always followed those of elasticity with a more or less important time shift. Whereas Taylor (1938) extended to plastic polycrystals the basic idea developed by Voigt (1889) as much as fifty years earlier, a few years only separated Hershey's (1954) first proposals for a self-consistent modelling of elastic polycrystals from Kröner's (1961) extensions to the case of plasticity. Unfortunately plasticity seems to be losing ground again, since the systematic statistical theory of heterogeneous linear materials has reached full growth within almost twenty years, while a plastic analogue still looks far from conception. That is why in the following, which is concerned with polycrystal plasticity, statistical aspects, if not completely missing, will remain on a quite elementary level. Several attempts will be briefly reported after a description of some general and common principles of modelling procedure and before paying special attention to one of them, namely the self-consistent method, with an illustration by some typical applications and results.

3.1. Some Principles of Modelling Procedure

Regardless of significant differences in the details of their basic assumptions, the various models that have been proposed in the literature obey some general common principles concerning the modelling method and procedure. It can roughly be said that they consist of a combination of three kinds of elements:

— a characterization of the constituent elementary phases
— an averaging procedure
— a localization modelling.

They will be successively briefly discussed.

Constituent Phases

This first step is perhaps the most crucial, since it determines the basic scale of inhomogeneity which is dealt with throughout the calculation. It is concerned with the physical identity of the elementary homogeneous phases, their mechanical characterization as well as the statistical description of their initial distribution in the chosen material. As far as the phase identification is concerned, there is not, of course, a unique answer; it depends on the available structural information and on the expected degree of accuracy of the analysis. For instance, in the case of a two-phase metal, the homogeneous phases may be simply considered as the chemical ones, taken as a whole in turns. If necessary, both chemical phases may be divided into smaller ones by taking the grain lattice orientations into account. But this may still be too coarse for special studies for which the elementary homogeneous phases may have to be individualized as grain families with, at any one time, the same chemical composition, lattice orientation, grain shape and size. Such an analysis may be further refined so as to take inhomogeneities at a lower scale into account, as far as to describe the spatial distribution itself, through correlation functions of more or less high order. Examples of such differences of phase identification will be given in the following (Section 3.3).

The mechanical behaviour of each phase has then to be specified. Even if several arguments against such a simplification have been brought out in the foregoing section, local behaviour only will be considered, with the formal (rate-independent) expression

$$\dot{\boldsymbol{\sigma}} = \mathbf{l}\dot{\boldsymbol{\varepsilon}} \quad \text{or} \quad \dot{\boldsymbol{\varepsilon}} = \mathbf{m}\dot{\boldsymbol{\sigma}} \tag{3.1}$$

where $\mathbf{l}$ and $\mathbf{m}$ are the fourth order tensors of incremental moduli and compliances respectively. Of course, since plasticity is concerned, $\mathbf{l}$ and $\mathbf{m}$ are multi-branched quantities and the constitutive laws (3.1) must be considered together with the yield criterion. Besides the usual parameters involved in plastic laws, all the variables involved in eqns (3.1), and especially $\mathbf{l}$ and $\mathbf{m}$, also depend on the parameters used for the definition of the phases.

Equations (3.1) are easily derived when the phases are directly considered as homogeneous ones, according to the classical formulation of continuum mechanics. Nevertheless, their general structure may be retained even when more refined descriptions are used. For example, for grains that deform by crystallographic glide according to the Schmid law and a hardening matrix rule, we have for active slip systems, with the notations of eqns (2.1) and (2.2),

$$\dot{\tau}^{\mathrm{g}} = \mathbf{R}^{\mathrm{g}}\dot{\boldsymbol{\sigma}} = \sum_{\mathrm{h}} h^{\mathrm{gh}}\dot{\gamma}^{\mathrm{h}} \tag{3.2}$$

This equation can be formally inverted into

$$\dot{\gamma}^{\mathrm{g}} = \sum_{\mathrm{h}} k^{\mathrm{gh}}\,\dot{\tau}^{\mathrm{h}} = \sum_{\mathrm{h}} k^{\mathrm{gh}}\mathbf{R}^{\mathrm{h}}\dot{\boldsymbol{\sigma}} \tag{3.3}$$

so that, according to eqn (2.2),

$$\dot{\boldsymbol{\varepsilon}}^{\mathrm{p}} = \sum_{\mathrm{g}}\sum_{\mathrm{h}} \mathbf{R}^{\mathrm{g}}k^{\mathrm{gh}}\mathbf{R}^{\mathrm{h}}\dot{\boldsymbol{\sigma}} = \mathbf{m}^{\mathrm{p}}\dot{\boldsymbol{\sigma}} \tag{3.4}$$

where $\mathbf{m}^{\mathrm{p}}$ stands for the incremental plastic compliance. Addition of the elastic one formally leads to eqns (3.1), where $\mathbf{m}$ obviously depends on the specific set of slip systems that are active at any step. When lattice rotations are needed, such as in texture development calculations, further analysis also requires the plastic rotation rate tensor $\dot{\boldsymbol{\omega}}^{\mathrm{p}}$, leading to

$$\dot{\boldsymbol{\omega}}^{\mathrm{p}} = \sum_{\mathrm{g}} (\vec{m}^{\mathrm{g}} \otimes \vec{n}^{\mathrm{g}})^{\mathrm{AS}}\,\dot{\gamma}^{\mathrm{g}} \tag{3.5}$$

where the antisymmetrical part of the term between the brackets is used.

In any case, it must be kept in mind that, considering lower scale phenomena and since intraphase inhomogeneities are omitted, the constitutive laws to be considered here are average ones. For instance, if the elementary phase is defined as a grain family with the same lattice orientation in a polycrystal, the hardening matrix must somehow depart from that of an isolated single crystal; this may be done by adjusting the diagonal components of h not on easy glide (stage I) moduli but on multislip (stage II) ones (Berveiller and Zaoui, 1979), since easy single glide is almost ineffective in the grains of a polycrystal. It is even conceivable to integrate into the grain constitutive law itself the grain size dependence of the CRSS, although such a dependence is specific to the polycrystal situation and, of course, not to the single crystal case (Weng, 1983).

Finally, the phase characterization must be completed with the description of the phase distribution in the initial state. Besides very special cases (here excluded) such as periodic media, for which thereby specific and powerful homogenization techniques are relevant, this description is a more or less precise statistical one. According to the particular case, phase volume fractions only may be given (or considered), or one can deal with more detailed correlation data which could specify the characteristics of the space phase distribution. In the first kind of information, special mention may be made of the CODF (the crystal orientation distribution function — say the 'texture function') which specifies the volume fraction of grains with the lattice orientation $\boldsymbol{\Omega}$ in the range $\mathrm{d}\boldsymbol{\Omega}$:

$$\frac{\mathrm{d}V}{V} = f(\boldsymbol{\Omega})\,\mathrm{d}\boldsymbol{\Omega} \tag{3.6}$$

where $\boldsymbol{\Omega}$ denotes lattice orientation parameters (usually the three Euler angles). If other geometrical parameters Y are considered (shape factors, size variables, etc.), more detailed probability density functions $f(\boldsymbol{\Omega}, Y)$ must be considered. If the second kind of description, namely a space dependent one, is adopted, one has to deal with correlation functions of increasing order for the incremental moduli $\mathbf{l}$ (or $\mathbf{m}$). It must be stressed here that one aspect of the difficulties which have to be overcome in the elaboration of a systematic statistical theory for plasticity lies in the fact that such a phase distribution is continuously changing in the course of the plastic flow; this is the case not only because $\mathbf{l}$ and $\mathbf{m}$ are not constant at a given point such as for linear elasticity, but also because of the continuous, and no longer negligible, change of relative distances and orientations between the constituents of the aggregate, so that the correlation functions would have to be modified at every step of any calculation.

Average Relations

Since the macroscopic behaviour must be derived from microscopic data, an average procedure is needed in order to define overall variables from local data. This problem has been discussed in the preceding sections and will not be discussed here in detail. Direct averaging — whether referring to ensemble averages or space averages if the ergodic hypothesis is considered — is supposed to be valid for the derivation of macroscopic stress and strain variables from microscopic ones, namely for the rates

$$\dot{\boldsymbol{\Sigma}} = \langle \dot{\boldsymbol{\sigma}} \rangle \quad \text{and} \quad \dot{\mathbf{E}} = \langle \dot{\boldsymbol{\varepsilon}} \rangle \tag{3.7}$$

For the case of a multiphase material, with the volume fractions c_i for each phase (i), this is simply

$$\dot{\mathbf{\Sigma}} = \sum_i c_i \dot{\sigma}_i \quad \text{and} \quad \dot{\mathbf{E}} = \sum_i c_i \dot{\varepsilon}_i \tag{3.8}$$

The question is now to derive the macroscopic moduli $\mathbf{L}$ (or compliances $\mathbf{M}$) from the local ones $\mathbf{l}$ (or $\mathbf{m}$); of course, this is not only a matter of averaging procedure, a 'localization' analysis is needed.

Localization Modelling

What is involved in this question (which must not be confused, in spite of the same word, with instability analysis) refers to the derivation of local quantities from average ones. Within a given set of statistical data, this procedure is the very modelling process, which differs from one model to the other. Various such procedures will be discussed in the next section. At present they will not be specified other than in their final formal result which may be put in the alternative form

$$\dot{\varepsilon} = \mathbf{A}\dot{\mathbf{E}} \quad \text{or} \quad \dot{\sigma} = \mathbf{B}\dot{\mathbf{\Sigma}} \tag{3.9}$$

Note that such a notation is very condensed since, in the general case, it corresponds to an integro-differential operation whose kernel $\mathbf{A}$ (or $\mathbf{B}$) depends not only on the parameters of the relevant elementary phase ($\mathbf{\Omega}$, Y, ..., or even the position $\vec{r}$ in the case of a spatial description) but also on the overall moduli $\mathbf{L}$ (or $\mathbf{M}$) which are to be derived as the final result. For instance, in the case of an inhomogeneous elastic material, the localization equation coincides with the Lippmann–Schwinger equation (Dederichs and Zeller, 1973)

$$\dot{\varepsilon}_{ij}(\vec{r}) = \dot{E}^0_{ij} + \int \Gamma^0_{ijkl}(\vec{r} - \vec{r}\,')\,\delta L_{klmn}(\vec{r}\,')\,\dot{\varepsilon}_{mn}(\vec{r}\,')\,\mathrm{d}V' \tag{3.10}$$

if $\dot{\mathbf{E}}^0$ is the solution of the homogeneous problem for a reference homogeneous medium with the moduli $\mathbf{L}^0$. Here $\delta\mathbf{L}$ stands for the deviation $(\mathbf{l} - \mathbf{L}^0)$ and $\mathbf{\Gamma}^0$ is derived from the Green tensor of the reference homogenized medium with moduli $\mathbf{L}^0$ by derivation of second order and symmetrization. Notation (3.9) would hold if eqn (3.10) is rewritten in the condensed form

$$\dot{\varepsilon} = \dot{\mathbf{E}}^0 + \mathbf{\Gamma}^0(\mathbf{l} - \mathbf{L}^0)\dot{\varepsilon} \quad \text{or} \quad \dot{\varepsilon} = [\mathbf{I} - \mathbf{\Gamma}^0(\mathbf{l} - \mathbf{L}^0)]^{-1}\dot{\mathbf{E}}^0 \tag{3.11}$$

where $\mathbf{I}$ denotes the unit operator. From eqns (3.7),

$$\dot{\mathbf{E}} = \langle \dot{\boldsymbol{\varepsilon}} \rangle = \langle [\mathbf{I} - \boldsymbol{\Gamma}^0(\mathbf{l} - \mathbf{L}^0)]^{-1} \rangle \dot{\mathbf{E}}^0 \qquad \dot{\boldsymbol{\varepsilon}} = \mathbf{A}\dot{\mathbf{E}}$$

with

$$\mathbf{A} = [\mathbf{I} - \boldsymbol{\Gamma}^0(\mathbf{l} - \mathbf{L}^0)]^{-1} \langle [\mathbf{I} - \boldsymbol{\Gamma}^0(\mathbf{l} - \mathbf{L}^0)]^{-1} \rangle^{-1} \tag{3.12}$$

In this case the integro-differential nature of $\mathbf{A}$ is obvious and its dependence on the unknown $\mathbf{L}$ could be made explicit. This dependence is retained in the simpler case when the space phase distribution is not considered so that $\mathbf{A}$ and $\mathbf{B}$ refer to usual fourth order tensors.

In any case, the integration of eqns (3.1) and (3.9) into (3.7) formally yields the equation for the unknown $\mathbf{L}$ or $\mathbf{M}$ tensors:

$$\begin{aligned} \dot{\boldsymbol{\Sigma}} &= \langle \dot{\boldsymbol{\sigma}} \rangle = \langle \mathbf{l}\dot{\boldsymbol{\varepsilon}} \rangle = \langle \mathbf{l}\mathbf{A} \rangle \dot{\mathbf{E}} \\ \dot{\mathbf{E}} &= \langle \dot{\boldsymbol{\varepsilon}} \rangle = \langle \mathbf{m}\dot{\boldsymbol{\sigma}} \rangle = \langle \mathbf{m}\mathbf{B} \rangle \dot{\boldsymbol{\Sigma}} \end{aligned} \tag{3.13}$$

leading to

$$\mathbf{L} = \langle \mathbf{l}\mathbf{A} \rangle \quad \text{or} \quad \mathbf{M} = \langle \mathbf{m}\mathbf{B} \rangle \tag{3.14}$$

which are integral equations even in simple cases, because of the dependence of $\mathbf{A}$ and $\mathbf{B}$ on $\mathbf{L}$ or $\mathbf{M}$.

Instead of the form (3.9), the localization equation may be derived in a mixed form where $\dot{\boldsymbol{\varepsilon}}, \dot{\boldsymbol{\sigma}}, \dot{\mathbf{E}}$ and $\dot{\boldsymbol{\Sigma}}$ are present at the same time. This will be the case, later on, of the self-consistent scheme whose interaction law is

$$\begin{aligned} \dot{\boldsymbol{\varepsilon}} - \dot{\mathbf{E}} &= -\mathbf{M}^*(\dot{\boldsymbol{\sigma}} - \dot{\boldsymbol{\Sigma}}) \\ \dot{\boldsymbol{\sigma}} - \dot{\boldsymbol{\Sigma}} &= -\mathbf{L}^*(\dot{\boldsymbol{\varepsilon}} - \dot{\mathbf{E}}) \end{aligned} \tag{3.15}$$

where $\mathbf{M}^*$ and $\mathbf{L}^*$ are the usual 'constraint' tensors. Nevertheless, the form (3.15) easily reduces to the form (3.9) with

$$\begin{aligned} \mathbf{A} &= (\mathbf{l} + \mathbf{L}^*)^{-1}(\mathbf{L} + \mathbf{L}^*) \\ \mathbf{B} &= (\mathbf{m} + \mathbf{M}^*)^{-1}(\mathbf{M} + \mathbf{M}^*) \end{aligned} \tag{3.16}$$

Since, when eqns (3.9) are taken into (3.7), the result is that

$$\langle \mathbf{A} \rangle = \langle \mathbf{B} \rangle = \mathbf{I} \tag{3.17}$$

the self-consistent equation is simply obtained by averaging eqns (3.16) and using (3.17), i.e.

$$\begin{aligned} \langle (\mathbf{l} + \mathbf{L}^*)^{-1}(\mathbf{L} + \mathbf{L}^*) \rangle = \mathbf{I} \quad &\text{or} \quad \langle (\mathbf{l} + \mathbf{L}^*)^{-1}(\mathbf{L} - \mathbf{l}) \rangle = 0 \\ \langle (\mathbf{m} + \mathbf{M}^*)^{-1}(\mathbf{M} + \mathbf{M}^*) \rangle = \mathbf{I} \quad &\text{or} \quad \langle (\mathbf{m} + \mathbf{M}^*)^{-1}(\mathbf{M} - \mathbf{m}) \rangle = 0 \end{aligned} \tag{3.18}$$

which can also be written as

$$\begin{aligned} \mathbf{L} &= \langle (\mathbf{l} + \mathbf{L}^*)^{-1} \rangle^{-1} \langle (\mathbf{l} + \mathbf{L}^*)^{-1} \mathbf{l} \rangle \\ \mathbf{M} &= \langle (\mathbf{m} + \mathbf{M}^*)^{-1} \rangle^{-1} \langle (\mathbf{m} + \mathbf{M}^*)^{-1} \mathbf{m} \rangle \end{aligned} \tag{3.19}$$

Note that a complete localization analysis must also be concerned with similar relations for rotations $\boldsymbol{\omega}$, which are basic variables of the plasticity of polycrystals (particularly when texture formation is considered). From a relation such as (3.9), now relating local $\dot{\boldsymbol{\omega}}$ and macroscopic $\dot{\boldsymbol{\Omega}}$ total rotation rate tensors (but also $\dot{\boldsymbol{\varepsilon}}$ in the general case), the total rotation may be divided at every step into its plastic and elastic parts, the latter measuring the lattice rotation which is responsible for the crystallographic texture development (but also for a geometrical source of hardening due to the rotation of slip systems with respect to the applied stresses). Nevertheless, in most usual models which will now be described, local rotations are considered uniform, thus the lattice rotation rate is simply

$$\dot{\boldsymbol{\omega}}^{e} = -\dot{\boldsymbol{\omega}}^{p} + \dot{\boldsymbol{\Omega}} \tag{3.20}$$

where $\dot{\boldsymbol{\Omega}}$ vanishes in a number of practical situations (tensile tests, rolling, etc.) so that the confusion which is not so seldom between plastic and lattice rotations, even if conceptually grievous, may be of no practical importance, except for a sign mistake which 'physical intuition' may easily rectify.

3.2. Space Distribution Independent Modelling

Progress in polycrystal plastic modelling naturally developed on a non-statistical basis, by considering only the crystal orientation distribution (i.e. the phase volume fractions only, if phases are defined as grain families with the same lattice orientation), irrespective of the spatial arrangement of the grains. Even if, as will be argued further, this drastic restriction leads to the fact that none of the models which would be defined within this field can be better (or worse) than the other in any circumstances, each of them has unconditional supporters who would profit by adopting a more statistical point of view.

Initial Models

The Sachs (1928) model may be considered as a precursor, based upon a very specific assumption which makes its statistical meaning somewhat indistinct. Considering a given applied stress state, primary slip systems are defined in each grain as if the stress field were uniform. But these primary systems are then assumed to suffer the same resolved shear stress, equal to the critical one (say τ_0), which leads to a modification of the corresponding stress level, proportionally; averaging these quantities, which differ from one crystal to another, leads to an estimate of the macroscopic yield stress. For instance, in the case of a stretched isotropic fcc polycrystal, the corresponding tensile yield stress is found to be 2·24 times larger than the CRSS, and the hardening slope of the polycrystal $d\sigma/d\varepsilon$ turns out to be $(2{\cdot}24)^2 \simeq 5$ times larger than that of the single crystal ($d\tau/d\gamma$ on the primary system). It must be stressed that the Sachs model does not conform with a uniform stress approach (as is often declared), which deprives it of any extremal character. As a matter of fact, it has not been so extensively used as the Taylor model which, even if getting on for fifty, still looks quite youthful.

According to Taylor's assumption, the elastic part of strain is neglected and the plastic part is uniform throughout the polycrystal. Compatibility requirements are then satisfied, including rotation aspects. According to the compatibility conditions

$$\mathbf{curl}\,\boldsymbol{\beta} = 0 \tag{3.20}$$

which can also be written as

$$\varepsilon_{ljk}\varepsilon_{ki,j} - \omega_{l,i} = 0 \tag{3.21}$$

where the rotation vector ω_i is defined by

$$\omega_i = \tfrac{1}{2}\varepsilon_{ijk}\,\omega_{jk} \tag{3.22}$$

a uniform strain field is only consistent with a uniform (total) rotation field. So the lattice rotation rate just opposes the plastic one, which is a logical addendum to the Taylor model. Since there are a number of possible ways to obtain an arbitrary plastic strain tensor by slip activation (C_N^5, if N is the total number of easy glide systems, considering that five independent conditions are fixed by a traceless symmetrical tensor, except for dependent slip systems sets), the question arises of how to get the best choice of active slip systems and associated resolved shear strain intensity. The solution is given by extremal theorems of standard plasticity (i.e. for materials conforming with Hill's maximum plastic work principle, which is the case for crystals that obey the Schmid law). As a kinematically admissible approximation, Taylor's assumption generally leads to an overestimate of the plastic work, and then, for simple loading paths (tension, shear, . . .), of the associated load parameter corresponding to incipient plastic flow. Consequently, the best Taylor's choice must lead to the lowest upper bound and then make the plastic work a minimum. If the CRSS is equal on all the slip systems and in all the grains, this is equivalent to making the sum of the shear amounts a minimum. That is how to derive, for instance, the upper bound of the tensile flow stress as 3·06 times the CRSS for an isotropic fcc polycrystal (which is, as expected, higher than Sachs's estimate) corresponding to a $(3{\cdot}06)^2 \simeq 9{\cdot}4$ times larger hardening slope for the polycrystal than for the single crystal. An alternative treatment has been built up by Bishop and Hill (1951), which focuses attention on stress aspects (since plastic admissibility of the strain field also implies definite conditions on the associated stress field, even if equilibrium is not needed) and leads to more tractable calculations.

The extremal character of the Taylor model makes it a very useful reference in any plastic analysis of polycrystals. At the same time, the uniform strain assumption indicates that, from a statistical point of view, the Taylor model is quite adapted to extreme structural cases when grains are distributed according to a parallel connection with respect to exterior loading (similarly to the Voigt model for linear elasticity). Likewise, since the strain localization tensor **A** reduces to identity, the overall moduli tensor **L** reduces to the average of the local one **l**; see eqns (3.14), or (3.19) for $\mathbf{M}^* = 0$.

Despite the fact that some ambiguity may remain after the minimization procedure concerning the choice of the set of the active slip systems, and then the value of the plastic rotation (and consequently of the lattice rotation too), the Taylor model has been extensively used for the prediction of crystallographic textures as well as for the analysis of the plastic anisotropy of textured polycrystals (such as the prediction of the Lankford coefficient for rolled sheets). Its main merit lies in the fact that, for the first time, many problems of polycrystal plasticity can be dealt with according to a systematic approach. Nevertheless, its extremal character often leads to a fairly loose quantitative agreement with experimental data (too acute textures, too strong anisotropy, too weak sensitivity with respect to the specific properties of various materials with the same crystallographic symmetry, ...) That is why several attempts have been made to relax its too strong basic assumption. Similar conclusions would surely have to be drawn from use of the dual model of the Taylor one, namely the uniform stress model, for which $\mathbf{B} = \mathbf{I}$ and $\mathbf{M} = \langle \mathbf{m} \rangle$, according to eqns (3.14). But such a static approach, which is more difficult to express in terms of slip systems activity, has been poorly used in the literature, after Batdorf and Budiansky (1949), except here again according to modified versions.

Modified Taylor Models

A straightforward extension of the Taylor model was proposed by Lin (1957) who assumed uniformity of the total strain (i.e. including the elastic part). Thus, the plastic strain is allowed to differ, to a certain extent, from one grain to another. Whereas the overall moduli still obey $\mathbf{L} = \langle \mathbf{l} \rangle$, it may be noticed that, in the case of isotropic elasticity, a relation similar to (3.15), but concerning the plastic strain deviation $(\boldsymbol{\varepsilon}^p - \mathbf{E}^p)$, may be derived:

$$\begin{aligned} \sigma &= \lambda\theta\mathbf{I} + 2\mu(\boldsymbol{\varepsilon} - \boldsymbol{\varepsilon}^p) \\ \boldsymbol{\Sigma} &= \lambda\theta\mathbf{I} + 2\mu(\mathbf{E} - \mathbf{E}^p) \end{aligned} \tag{3.23}$$

and then

$$\sigma - \boldsymbol{\Sigma} = 2\mu(\mathbf{E}^p - \boldsymbol{\varepsilon}^p) \tag{3.24}$$

where λ and μ are the Lamé coefficients and θ denotes the trace of $\boldsymbol{\varepsilon}$ or $\mathbf{E}$ (since $\boldsymbol{\varepsilon}^p$ and $\mathbf{E}^p = \langle \boldsymbol{\varepsilon}^p \rangle$ are traceless tensors). Such an interaction law between internal stress $(\sigma - \boldsymbol{\Sigma})$ and the plastic strain deviation will be used as a reference in the following. Let us already note that, from a practical point of view, eqn (3.24) allows extremely slight plastic strain deviations due to the very high usual values of the elastic shear modulus μ with respect to the limited admissible level of internal stresses.

The major extension of the Taylor model is much more recent (see for instance Honneff and Mecking, 1978). It is referred to as a 'relaxed' Taylor theory, since it only requires uniformity of some of the components of the plastic strain tensor, depending on the problem that is considered. Besides reflecting a natural tendency to 'soften' the classical Taylor model which was admitted to be too 'hard', this modification is based upon the experimental evidence of large changes in the grain shapes after an important oriented plastic flow (such as during rolling). As a matter of fact, compatibility requirements for a flat grains stacking are mainly concerned with the continuity of components of the plastic strain tensor which lie parallel to the larger grain boundaries [see eqn (2.10) for a plane boundary normal to x_2; the incompatibility tensor vanishes if $\Delta\varepsilon^p_{11} = \Delta\varepsilon^p_{13} = \Delta\varepsilon^p_{33} = 0$]. Consequently, adjustment of free strain components according to adequate minimization procedures leads to significant differences with the predictions of the classical Taylor model and to better agreement with experiment. Similar attempts, which cannot be enumerated here (see e.g. Leffers, 1981), have also been developed, on the other hand, in order to raise the lower bound associated with static approaches. Such a purpose of deviating from extreme situations and dealing with more standard ones is satisfied at once by the self-consistent scheme which will be briefly described now.

The Self-consistent Scheme

The essential purpose of this scheme, which was already familiar to solid state physicists when it was extended to elastic and then to plastic inhomogeneous aggregates, was to deliver an approximate estimate of the internal stresses arising in any grain due to the plastic strain misfit with its neighbours. The basis for this approximation consisted in replacing the detailed distribution of plastic strain around a given grain, with the plastic strain ε^p, by an averaged one fixed by the macroscopic value $\mathbf{E}^p$. Assuming isotropic elasticity and a spherical shape for the grain, one is left with Eshelby's inclusion problem, with the slight difference of a uniformly plastified infinite matrix, stressed by $\mathbf{\Sigma}$ at infinity, instead of an unloaded elastic one. Nevertheless, it is easy to change Eshelby's (1961) solution

$$\sigma = -2\mu(1-\beta)\varepsilon^p \tag{3.25}$$

where $\beta = 2(4-5\nu)/[15(1-\nu)]$, into the following (Kröner, 1961):

$$\sigma = \mathbf{\Sigma} + 2\mu(1-\beta)(\mathbf{E}^p - \varepsilon^p) \tag{3.26}$$

Such an 'interaction law' may then be used as a localization one, which may be written in the form (3.15) if stress and strain rates are used, and with

$$\begin{aligned} \mathbf{L}^* &= k\mathbf{I} \\ k &= \frac{2\mu(1-\beta)}{\beta} = \frac{\mu(7-5\nu)}{4-5\nu} \end{aligned} \tag{3.27}$$

Referring to eqn (3.24), it may be concluded that taking account of strain inhomogeneity according to such a procedure only leads to the lowering by a factor of the order of 2 (since $\beta \simeq 1/2$ in current cases) the eigenstress level with respect to Lin's assumption. Such a similarity is corroborated when overall moduli $\mathbf{L}$ are considered. Whereas Lin's assumption leads to $\mathbf{L} = \langle \mathbf{l} \rangle$, the foregoing one yields, according to eqns (3.19) and (3.27),

$$\begin{aligned} \mathbf{L} &= \langle (\mathbf{l} + k\mathbf{I})^{-1} \rangle^{-1} \langle (\mathbf{l} + k\mathbf{I})^{-1}\, \mathbf{l} \rangle \\ &= \langle (\mathbf{l} + k\mathbf{I})^{-1} \rangle^{-1} \langle \mathbf{I} - k(\mathbf{l} + k\mathbf{I})^{-1} \rangle \\ &= \langle (\mathbf{l} + k\mathbf{I})^{-1} \rangle^{-1} - k\mathbf{I} \end{aligned} \tag{3.28}$$

But, since incremental elastic-plastic moduli **L** are currently (i.e. except at the very beginning of the plastic flow) lower than elastic moduli by one or two orders of magnitude, it must be considered that, in a symbolical way, one has $\mathbf{l} \ll k\mathbf{I}$ (as k is about 2 or 3 times the elastic shear modulus μ). A formal development of eqn (3.28) then yields easily

$$\mathbf{L} = \langle \mathbf{l} \rangle - \frac{1}{k}(\langle \mathbf{l}^2 \rangle - \langle \mathbf{l} \rangle^2) + \theta\left(\frac{1}{k^2}\right) \tag{3.29}$$

which indicates that Lin's estimate is very slightly modified by this approach — at least from a numerical point of view.

The foregoing scheme may be criticized as not being fully self-consistent. Self-consistency does not require a uniform plastification of the matrix by $\mathbf{E}^p$, but leads to the assimilation of this matrix with an 'equivalent homogeneous medium', i.e. with a fictitious medium whose behaviour is identified with the overall unknown medium whose derivation is the very purpose of the self-consistent procedure. In other words, the relevant matrix/inclusion problem is not the former one, but the inhomogeneity problem where the inclusion has the moduli **l** and the matrix the unknown moduli **L** (rather than the elastic ones in the foregoing, regardless of the uniform plastic strain $\mathbf{E}^p$). This problem, which has also been analysed by Eshelby for ellipsoidal inclusions, has a general solution which has also the form (3.15), but with a constraint tensor $\mathbf{L}^*$ different from (3.27) and leading to much lower interactions and internal stresses. Whereas the final solution always complies with eqns (3.19), it strongly differs from (3.29) where k has the 'elastic' value (3.27) (Hill, 1965).

Unfortunately the relationship between $\mathbf{L}^*$ and **L** cannot be given in a closed explicit form in the general case. Besides complete numerical treatments (Hutchinson, 1970) or analytical developments for plane problems, which are possible even at finite strain (Iwakuma and Nemat-Nasser, 1984), closed forms are available only in the case of isotropy. This leads to convenient treatments of the plasticity of isotropic polycrystals or to approximate ones when anisotropy is present. Both of these possibilities will be discussed now before evolving the merits and limits of the self-consistent modelling.

3.3. Isotropic Formulation of the Self-consistent Scheme

Besides its own interest as a modelling method, the self-consistent scheme also deserves discussion with more details regarding the formulation because of its specific difficulty. Unlike Taylor's or Lin's models, whose basic assumption allows an individual treatment of each phase since the plastic or total strain tensor is known at each step if the average one is assigned, this is no longer the case owing to the form of the interaction law (3.15) where local and overall variables are coupled. This difficulty may be first discussed in the simpler case of an isotropic two-phase material before dealing with the general case.

Isotropic Plastic Flow of a Two-phase Material

Let us consider a two-phase isotropic material, each phase of which is assumed isotropic and incompressible, with the incremental shear modulus μ_i^* and the volume fraction c_i (i = 1 or 2). The constitutive equation of each phase reduces to

$$'\dot{\sigma}_i = \mathbf{l}_i \dot{\varepsilon}_i \quad \text{with} \quad \mathbf{l}_i = 2\mu_i^* \mathbf{I} \tag{3.30}$$

whereas that of the homogenized composite is

$$'\dot{\mathbf{\Sigma}} = \mathbf{L}\dot{\mathbf{E}} \quad \text{with} \quad \mathbf{L} = 2\mu^* \mathbf{I} \tag{3.31}$$

Here $'\sigma_i$ and $'\mathbf{\Sigma}$ stand for the deviatoric parts of σ_i and $\mathbf{\Sigma}$ respectively.

The average equations are given by (3.8) and the strain localization tensor $\mathbf{A}_i$, such that

$$\dot{\varepsilon}_i = \mathbf{A}_i \dot{\mathbf{E}} \tag{3.32}$$

is obtained straightforwardly from the classical Eshelby solution of the spherical inhomogeneity, namely

$$\mathbf{A}_i = \frac{5\mu^*}{3\mu^* + 2\mu_i^*}\mathbf{I} \tag{3.33}$$

The effective moduli $\mathbf{L}$ are given by eqns (3.14) which reduce to

$$\mathbf{L} = \sum_i c_i \mathbf{l}_i \mathbf{A}_i = \sum_i \frac{10 c_i \mu^* \mu_i^*}{3\mu^* + 2\mu_i^*}\mathbf{I} \tag{3.34}$$

so that the effective modulus μ^* is the solution of the equation

$$5 \sum_i \frac{c_i \mu_i^*}{3\mu^* + 2\mu_i^*} = 1 \tag{3.35}$$

or, equivalently, of

$$5\mu^* \sum_i \frac{c_i}{3\mu^* + 2\mu_i^*} = 1 \tag{3.36}$$

This second-order equation, with only one positive root, in the case of a two-phase material, may clearly be generalized with the same form to a nth order equation in the case of a n-phase composite, but of course such a treatment is tractable only when a few phases are considered. Even if it is a very easy one in this case, it must be admitted that the local variables cannot be derived directly from the overall ones since the localization tensors depend on the effective moduli which are not known at the beginning; this difficulty, which disappears in the Lin–Taylor model, necessarily makes any self-consistent approach more tedious than simpler ones.

Some additional remarks may be made:

(i) This quite simple and, in the isotropic case, fully self-consistent approach may be used, for a two-phase material, almost as easily as the uniform stress or uniform strain approximations, and it is likely to be more adapted to ordinary disordered situations. In spite of its simplicity, it is able to describe the delayed plastification of the harder phase and the associated 'second yield-point' which appears on the experimental stress–strain curves as well as the specific part of the Bauschinger effect which results only from the two-phase inhomogeneity (Berveiller and Zaoui, 1983);

(ii) In the case of monotonic radial loading, the step-by-step integration may be avoided thanks to a finite formulation which deals with 'secant' moduli μ_i^s and μ^s instead of instantaneous ones μ_i^* and μ^*, and with stress and strain tensors instead of their rates; a convenient graphic determination (Fig. 4) may even be developed (see Berveiller and Zaoui, 1981). A similar treatment may also be used for the case of steady state power-law creep, through Hoff's analogy between creep and strain hardening (Chen and Argon, 1979);

(iii) Finally, it may be noticed that, from eqns (3.30)–(3.33), an interaction law in the form (3.15) may be derived as

$$\dot{\sigma}_i = \dot{\mathbf{\Sigma}} + 3\mu^*(\dot{\mathbf{E}} - \dot{\varepsilon}_i) \tag{3.37}$$

This means that the constraint tensor $\mathbf{L}^*$ is

$$\begin{aligned} \mathbf{L}^* &= k^*\mathbf{I} \\ k^* &= 3\mu^* \end{aligned} \tag{3.38}$$

which looks like Kröner's relation (3.27) with $\nu = 1/2$ (due to the assumption of incompressibility), except that the elastic modulus μ is here replaced by the far lower elastic-plastic one μ^*. This point clearly

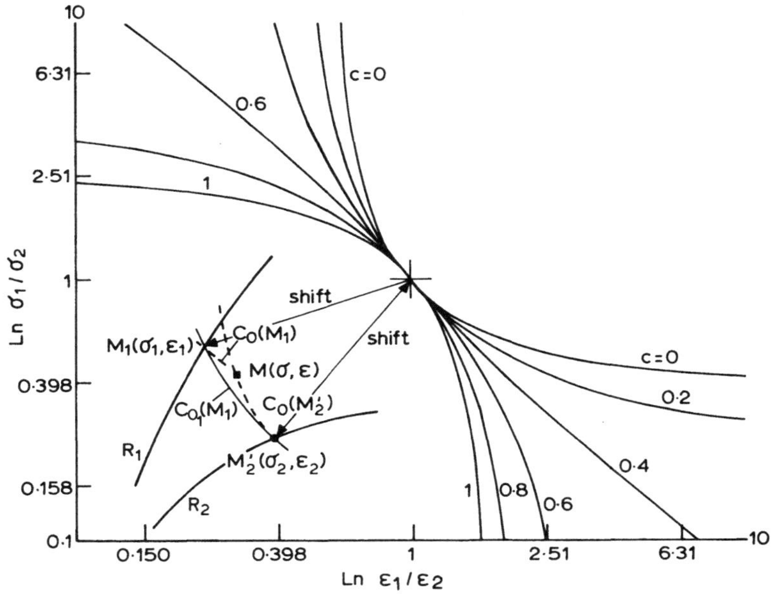

Fig. 4. Graphic determination of the stress–strain curve of a two-phase composite, according to a simplified self-consistent scheme.

expresses the phenomenon of plastic accommodation, i.e. of plastic relaxation of the internal stresses associated with the plastic flow of the composite. In the case of radial monotonic loading, a finite interaction law may be used which has the form (3.26) except that the secant elastic-plastic modulus μ^s must be used, instead of the elastic one.

The convenience of this simple model clearly results not only from the small number of phases but also from the assumption of isotropy. Even if, in the case of ellipsoidal prolate inclusions, Eshelby's general solution could be used, part of the self-consistency of the model would surely be lost owing to the anisotropic resultant behaviour of the composite. Such a treatment could then only be considered as an approximate self-consistent one. The same conclusion holds for the following formulation concerning the plastic behaviour of textured polycrystals.

Plastic Behaviour of Textured Polycrystals

The convenient form (3.28) of the self-consistent equation originates more from the isotropic than from the elastic assumption, concerning the constraint tensor $\mathbf{L}^*$ (3.27). If the unknown $\mathbf{L}$ tensor is, temporarily, considered as an isotropic tensor in the form

$$L_{ijkl} = \mu^* \left(\frac{2\nu^*}{1 - 2\nu^*} \delta_{ij}\delta_{kl} + \delta_{ik}\delta_{jl} + \delta_{il}\delta_{jk} \right) \tag{3.39}$$

where μ^* and ν^* are the instantaneous elastic-plastic shear modulus and Poisson's ratio respectively and δ is the Kronecker tensor, the constraint tensor has an explicit expression given by Eshelby's solution,

$$\mathbf{L}^* = \mathbf{L}(\mathbf{S}^{-1} - \mathbf{I}) \tag{3.40}$$

where $\mathbf{S}$ stands for Eshelby's tensor relating the total and the plastic strain in the basic inclusion problem. For a spherical inclusion, eqn (3.40) reduces to

$$L^*_{ijkl} = \frac{\mu^*(3 - 5\nu^*)}{4 - 5\nu^*} \delta_{ij}\delta_{kl} + \frac{\mu^*(7 - 5\nu^*)}{2(4 - 5\nu)^*} (\delta_{ik}\delta_{jl} + \delta_{il}\delta_{jk}) \tag{3.41}$$

so that the interaction law finally reads, in the case of isotropic elasticity,

$$\dot{\boldsymbol{\sigma}} = \dot{\boldsymbol{\Sigma}} + \alpha\mu(\dot{\mathbf{E}}^{\mathrm{p}} - \dot{\boldsymbol{\varepsilon}}^{\mathrm{p}}) \tag{3.42}$$

where the scalar plastic accommodation function α is given by

$$\alpha = \frac{2\mu^*(7 - 5\nu)}{\mu^*(7 - 5\nu^*) + 2\mu(4 - 5\nu^*)} \tag{3.43}$$

Note that a similar treatment could be performed in the case of monotonic radial loading and using a 'secant' accommodation function (α^{s} say) together with secant moduli μ^{s} (Berveiller and Zaoui, 1979).

The advantage gained with this treatment lies both in its simplicity regarding further calculations and in the right order of magnitude of the intensity of the mechanical reactions between grains, compared with the 'elastic accommodation' assumption in eqn (3.26). As a matter of fact, $\alpha\mu$ in eqn (3.42), as well as $\alpha^{\mathrm{s}}\mu$ in the equivalent equation referring to the 'secant formulation', coincides with $2\mu(1 - \beta)$ in eqn (3.26) in the elastic range only. As soon as some plastic flow occurs, they rapidly drop by one or two orders of magnitude, so that even a constant value of α, of the order 10^{-2}, would lead to meaningful results in a strain range of roughly 10^{-3} to 10^{-1}. The main drawback clearly consists in diminishing the self-consistency of the procedure, except in the case of effective isotropy, since the isotropic estimate (3.40) of the constraint tensor is not consistent with anisotropic overall behaviour. Nevertheless, this does not mean that the final estimate of L according to eqns (3.19) will be isotropic; if an initial texture is taken into account in the average procedure, an anisotropic L tensor will be derived, but this anisotropy will probably be an underestimate of the actual one.

The specific problems of numerical formulation will not be discussed in great detail. What may be stressed is the particular simplicity of the numerical treatment in the case of isotropy. Adapting to this case of a varying α function the original method of Budiansky and Wu (1962), which was developed in connection with Kröner's initial statement (3.26), the interaction law (3.42) may be arranged into

$$\dot{\sigma} + \alpha\mu\dot{\varepsilon}^{\mathrm{p}} = \dot{\Sigma} + \alpha\mu\dot{\mathrm{E}}^{\mathrm{p}} \tag{3.44}$$

which exhibits the uniformity of $(\dot{\sigma} + \alpha\mu\dot{\varepsilon}^{\mathrm{p}})$ throughout the polycrystal at any stage of the calculation. This allows one to deal with each phase in turn, such as in the case of the Lin–Taylor model, as soon as the form of the right member in eqn (3.44) may be anticipated, which is possible in the case of plastic isotropy only. On the contrary, anisotropy requires lengthy iterations or alternative formulations, such as the specification of $\dot{\mathrm{E}}^{\mathrm{p}}$ in eqn (3.44) in terms of the individual phase contributions $\dot{\varepsilon}_i^{\mathrm{p}}$ and the simultaneous resolution of the whole set of equations (Hihi *et al.*, 1985)

$$\dot{\sigma}_i + \alpha\mu\dot{\varepsilon}_i^{\mathrm{p}} - \alpha\mu\sum_j c_j\dot{\varepsilon}_j^{\mathrm{p}} = \dot{\Sigma} \tag{3.45}$$

where the stress rates $\dot{\sigma}_i$ are expressed in terms of $\dot{\varepsilon}_i^{\mathrm{p}}$ thanks to the phase constitutive equations (3.1). Note that, finally, if the crystallographic nature of plastic glide has to be taken into account, the final set of equations has the resolved shear strain rates $\dot{\gamma}^{\mathrm{g}}$ as variables, according to eqn (2.2) which, for a prescribed Σ history, reads

$$c_i\sum_{\mathrm{h}}(\mathrm{h}_i^{\mathrm{gh}} + \alpha\mu\mathbf{R}_i^{\mathrm{g}}\mathbf{R}_i^{\mathrm{h}})\dot{\gamma}_i^{\mathrm{h}} - \alpha\mu\sum_j\sum_{\mathrm{h}} c_i c_j \mathbf{R}_i^{\mathrm{g}}\mathbf{R}_j^{\mathrm{h}}\dot{\gamma}_j^{\mathrm{h}} = c_i\mathbf{R}_i^{\mathrm{g}}\dot{\Sigma} \tag{3.46}$$

where lower indices i and j specify the phases whereas the volume fractions c_i derive from an adequate discretization of the texture function (3.6). (Here a hardening matrix h^{gh} formalism has been adopted.) The step-by-step resolution of eqn (3.46) allows the derivation of $\dot{\varepsilon}_i^{\mathrm{p}}$ and then of $\dot{\mathrm{E}}^{\mathrm{p}}$ as well as the calculation of the lattice rotation rate according to eqn (3.20) with the object of predicting the texture development. (Note that eqn (3.20) could be justified as a rigorous equation, within the self-consistent approach, only in the case of complete isotropy; see Berveiller and Zaoui, 1982.)

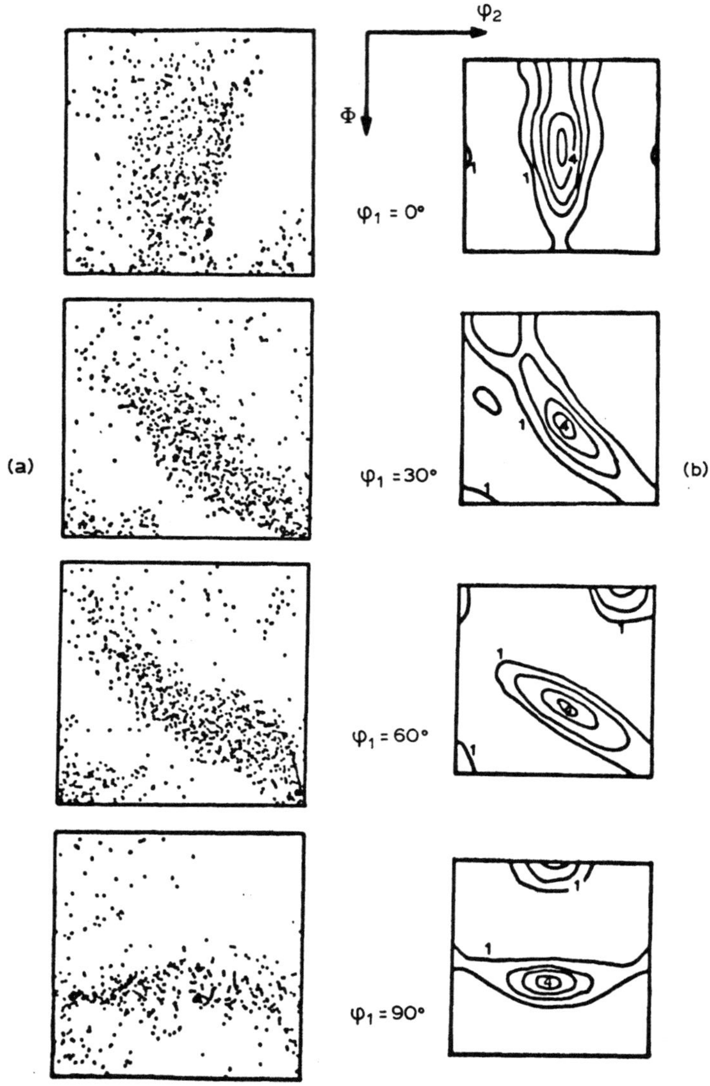

Fig. 5. Simulation of the development of a rolling texture (a) compared with an experimental one (b) (bcc polycrystal).

Figures 5 and 6 show typical results obtained according to such a treatment, compared with experimental data. The isotropic approximation of the constraint tensor **L*** is here proved neither to prevent the resulting overall behaviour from being anisotropic when an initial crystallographic texture is considered nor to allow the calculation of the texture development from an isotropic situation; combination of both of these calculations could be used to simulate the continuous development of an induced plastic anisotropy correlated with the simultaneous evolution of the crystallographic texture. Here the plastic accommodation factor α has been taken as a constant, for the sake of simplicity, in the typical range 10^{-2}–10^{-1}. The hardening matrix was constant too, with two kinds of components (h_2 and h_1 say) according to the nature of the dislocations interaction between the involved slip systems. The ratio $A = h_2/h_1$, larger than unity, is then an intracrystalline anisotropy parameter; moreover, in accordance with the arguments developed in Sections 2.3 and 3.1, the value of h_1, which measures the lower hardening moduli (and especially the self-

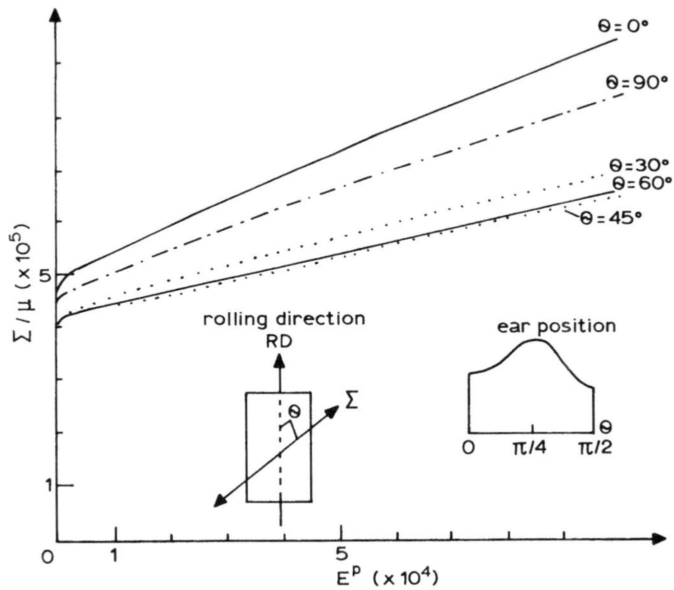

Fig. 6. Simulation of tensile tests on samples cut, at various angles θ, in a cold-rolled sheet (fcc polycrystal).

hardening ones for the diagonal components), has been fitted with the hardening slope of the stage II of fcc single crystals, namely $h_1 \simeq \mu/250$, since plastified grains of a polycrystal rapidly flow according to a multislip regime, contrary to any isolated single crystals. Finally, it may be admitted that the case of a two-phase polycrystal, each phase of which is considered as an aggregate of disoriented grains, can be dealt with as easily as a single phase one; it is only a matter of an increased number of elementary phases (Berveiller and Zaoui, 1983).

4. DISCUSSION AND CONCLUSION

It is clear from the foregoing that none of the reported models can take the phase spatial distribution into account. After the elementary phases have been defined and characterized, the only information which has been (and could be) used concerning the constitution of the aggregate is their volume fractions. According to the statistical systematic theory, and even if such a theory has, up to now, been developed only for linear behaviour, such poor statistical information can only be used to predict, at best, upper and lower bounds for the overall moduli. Since each of these models nevertheless leads to one estimate only for these moduli, it must be concluded that each of them is based upon one definite, but implicit, assumption concerning the phase mutual arrangement. Whereas uniform strain or stress assumptions are known to correspond to very special, highly ordered situations, the self-consistent approach has been proved to conform, in the linear case, with a state of perfect disorder. Even if such a statement cannot be considered as proved in the case of plasticity, it is likely to remain relevant to this situation too, thanks to an adequate definition of the involved correlation functions. Unless satisfied with the opinion that usual polycrystal constitutions are nearer to disorder than to order, one must strive to elaborate new models which could take some more precise statistical information into account.

In these few final comments, exploratory considerations are reported concerning one possible way to progress towards such an objective, starting from the classical self-consistent scheme and extending some of its basic concepts. This way is distinct from a systematic statistical approach which, similarly to the linear case, would try to associate some bounds with each degree of statistical description, but both of these roads could converge and meet in a limit process.

Let us go back to eqn (3.10), which is valid for linear inhomogeneous elasticity but could also be considered as a formal description of the total strain rate field $\dot{\varepsilon}(\vec{r})$ for an elastic-plastic inhomogeneous medium with instantaneous local moduli l, and a reference homogeneous medium with moduli $\mathbf{L}^0$. The 'transition tensor' $\mathbf{T}(\vec{r})$ can be defined by the equation

$$\mathbf{T}(\vec{r})\,\dot{\mathbf{E}}^0 = \delta\mathbf{L}(\vec{r})\,\dot{\varepsilon}(\vec{r}) \tag{4.1}$$

so that, when multiplied by $\delta\mathbf{L}(\vec{r})$, eqn (3.10) yields

$$\mathbf{T}(\vec{r}) = \delta\mathbf{L}(\vec{r}) + \delta\mathbf{L}(\vec{r})\int \mathbf{\Gamma}^0(\vec{r}-\vec{r}\,')\,\mathbf{T}(\vec{r}\,')\,\mathrm{d}V' \tag{4.2}$$

and then

$$\mathbf{T} = (\mathbf{I} - \delta\mathbf{L}\,\mathbf{\Gamma}^0)^{-1}\,\delta\mathbf{L} \tag{4.3}$$

At the same time, we get

$$\dot{\sigma} = (\mathbf{L}^0 + \delta\mathbf{L})\,\dot{\varepsilon} = \mathbf{L}^0\dot{\varepsilon} + \mathbf{T}\dot{\mathbf{E}}^0 \tag{4.4}$$

so that

$$\dot{\mathbf{\Sigma}} = \langle\dot{\sigma}\rangle = \mathbf{L}^0\dot{\mathbf{E}} + \langle\mathbf{T}\rangle\,\dot{\mathbf{E}}^0 = \mathbf{L}\cdot\dot{\mathbf{E}} \tag{4.5}$$

If now the reference medium is taken as the unknown homogenized medium with the moduli $\mathbf{L}$ (i.e. $\mathbf{L}^0 = \mathbf{L}$), this means that $\langle\mathbf{T}\rangle = 0$ or, from eqn (4.3),

$$\langle(\mathbf{I} - \delta\mathbf{L}\,\mathbf{\Gamma})^{-1}\,\delta\mathbf{L}\rangle = 0 \tag{4.6}$$

where $\mathbf{\Gamma}$ refers to the effective medium with moduli $\mathbf{L}$. This is a general implicit equation for the effective moduli L, which can be simplified according to various assumptions, including the crudest one, which leads to $\langle\delta\mathbf{L}\rangle = 0$, as for the Lin–Taylor model. The classical self-consistent scheme clearly refers to elementary situations where one ellipsoidal inhomogeneity is considered at once, which is embedded within an infinite matrix with the moduli $\mathbf{L}$. In this case, $\delta\mathbf{L}$ vanishes everywhere except in the inclusion and the transition tensor is uniform in it, according to Eshelby (1957); the local part of $\mathbf{\Gamma}$ only is operating and the self-consistent equation reduces to

$$\langle(\mathbf{I} - \Delta\mathbf{L}^{\mathrm{i}}\mathbf{\Gamma}^{\mathrm{ii}})^{-1}\,\Delta\mathbf{L}^{\mathrm{i}}\rangle_{(\mathrm{i})} = 0 \tag{4.7}$$

where $\Delta\mathbf{L}^{\mathrm{i}}$ stands for $\delta\mathbf{L}$ in each inclusion (i) and $\mathbf{\Gamma}^{\mathrm{ii}}$ for the average of $\mathbf{\Gamma}$ in it. This equation could easily be identified with (3.19) thanks to the expression of the constraint tensor $\mathbf{L}^*$ in terms of $\mathbf{\Gamma}^{\mathrm{ii}}$ and $\Delta\mathbf{L}^{\mathrm{i}}$.

More complicated elementary situations may now be considered, beginning with two ellipsoidal inhomogeneities (i) and (j) within an infinite matrix with the moduli L. Now $\delta \mathbf{L}$ is non-zero in (i) and (j) only and the transition tensor can be averaged in each of the inclusions:

$$\mathbf{T}^{\mathrm{i}} = \Delta\mathbf{L}^{\mathrm{i}} + \Delta\mathbf{L}^{\mathrm{i}}\boldsymbol{\Gamma}^{\mathrm{ii}}\mathbf{T}^{\mathrm{i}} + \Delta\mathbf{L}^{\mathrm{i}}\boldsymbol{\Gamma}^{\mathrm{ij}}\mathbf{T}^{\mathrm{j}} \tag{4.8}$$

where $\boldsymbol{\Gamma}^{\mathrm{ij}}$ stands for an average of $\boldsymbol{\Gamma}(\vec{r} - \vec{r}\,')$ both for $\vec{r} \in$ (i) and $\vec{r} \in$ (j), so that the long-range part of $\boldsymbol{\Gamma}$ must be considered also. From eqn (4.8) and the similar equation for $\mathbf{T}^{\mathrm{j}}$, the transition tensor may be calculated everywhere, and if the same operation is performed for any pair of inclusions of the medium, the self-consistent equation $\langle \mathbf{T} \rangle = 0$ yields a new estimate of $\mathbf{L}$ according to this two-sites self-consistent scheme. (Note that the results may be given in a closed form for two spherical inclusions in an isotropic matrix; see Berveiller *et al.*, 1981.) Obviously the same procedure may be formally applied to any set of n inclusions, leading to multisite self-consistent schemes of arbitrary order n (Berveiller and Zaoui, 1984).

This is only a general indication of what could be done in order to integrate more statistical information on the space phase distribution into the usual polycrystalline or multiphase modelling. Even if it is still far from being a convenient and completely defined prediction tool, the main differences with the systematic statistical theory may be underlined from now on. Whereas this systematic theory could be able to predict the upper and lower bounds for the estimate of the overall moduli at any order of statistical description, the multisite self-consistent method is expected to give, at the same stage, one estimate only which takes the corresponding information into account and assumes some kind of perfect disorder beyond, concerning the information which is missing at higher orders of statistical description (which conforms with the use of an 'equivalent homogeneous medium' within which definite sets of n inclusions are embedded) (Zaoui, 1984).

In any case, whether such an extension of the classical self-consistent scheme shows, or does not show, an adequate and familiar method for an improved modelling of the behaviour of polycrystals and multiphase materials, further advances in this field strongly depend on the integration of a space-dependent description of the constitution of the aggregates into adequate statistical mechanical treatments. As long as this is not performed, the dispute between Taylor's, Sachs's and Hill's supporters looks vain and purposeless.

REFERENCES

Batdorf, S. B. and Budiansky, B. (1949). A mathematical theory of plasticity based on the concept of slip, NACA TN-1871.

Berveiller, M. and Zaoui, A. (1979). An extension of the self-consistent scheme to plastically flowing polycrystals, *J. Mech. Phys. Solids*, **26**, 325.

Berveiller, M. and Zaoui, A. (1981). A simplified self-consistent scheme for the plasticity of two-phase metals, *Res. Mech. Lett.*, **1**, 119.

Berveiller, M. and Zaoui, A. (1982). Etude de l'anisotropie élastique, plastique et géométrique dans les polycristeaux métalliques. In *Mechanical Behaviour of Anisotropic Solids*, J. P. Boehler (Ed.), CNRS, and Martinus Nyhoff, The Hague, p. 335.

Berveiller, M. and Zaoui, A. (1983). Modelling of the plasticity and the texture development of two-phase metals. In *Deformation of Multi-phase and Particle-containing Materials*, J. B. Bilde-Sørensen, N. Hansen, A. Horsewell, T. Leffers and H. Lilholt (Eds), Risø National Laboratory, Roskilde, Denmark, p. 153.

Berveiller, M. and Zaoui, A. (1984). Modelling of the plastic behaviour of inhomogeneous media, *J. Eng. Mat. Tech.*, **106**, 295.

Berveiller, M., Hihi, A. and Zaoui, A. (1981). Self-consistent schemes for polycrystalline and multiplase materials plasticity. In *Deformation of Polycrystals: Mechanisms and Microstructures*, N. Hansen, A. Horsewell, T. Leffers and H. Lilholt (Eds), Risø National Laboratory, Roskilde, Denmark, p. 145.

Bishop, J. F. W. and Hill, R. (1951). A theory of the plastic distortion of a polycrystalline aggregate under combined stress, *Phil. Mag.*, **422**, 414.

Bretheau, T. and Caldemaison, D. (1981). Test of mechanical interaction models between polycrystal grains by means of local strain measurements. In *Deformation of Polycrystals: Mechanisms and Microstructures*, N. Hansen, A. Horsewell, T. Leffers and H. Lilholt (Eds), Risø National Laboratory, Roskilde, Denmark, p. 157.

Budiansky, B. and Wu, T. T. (1962). Theoretical prediction of plastic strains of polycrystals, Proc. 4th US Nat. Congr. Appl. Mech., p. 1175.

Chen, I. W. and Argon, A. S. (1979). Steady state power law creep in heterogeneous alloys with coarse microstructures, *Acta Metall.*, **27**, 785.

Dederichs, P. H. and Zeller, R. (1973). Variational treatment of the elastic constants of disordered materials, *Z. Physik*, **259**, 103.

Eshelby, J. D. (1957). The determination of the elastic field of an ellipsoidal inclusion and related problems, *Proc. Roy. Soc.* (London), **A241**, 376.

Eshelby, J. D. (1961). Elastic inclusions and inhomogeneities. In *Progress in Solids Mechanics*, Vol. II, I. N. Sneddon and R. Hill (Eds), North-Holland, Amsterdam, p. 87.

Franciosi, P. (1978). Plasticité à froids des monocristaux cfc: étude du durcissement latent, spec. thesis, Université Paris.

Franciosi, P. (1984). Etude théorique et expérimentale du comportement élastoplastique des monocristaux métalliques se déformant par glissement: modélisation pour un chargement complexe quasi statique, doctoral thesis, Université Paris-Nord, Villetaneuse, France.

Franciosi, P. and Zaoui, A. (1982*a*). Multislip in fcc crystals: a theoretical approach compared with experimental data, *Acta Metall.*, **30**, 1627.

Franciosi, P. and Zaoui, A. (1982*b*). Multislip tests on copper crystals: a junction hardening effect, *Acta Metall.,* **30**, 2141.
Franciosi, P., Berveiller, M. and Zaoui, A. (1980). Latent hardening in copper and aluminium single crystals, *Acta Metall.,* **28**, 273.
Havner, K. S. (1982). The theory of finite plastic deformation of crystalline solids. In *Mechanics of Solids,* the Rodney Hill 60th Anniversary Volume, H. G. Hopkins and M. J. Sewell (Eds), Pergamon Press, Oxford, p. 265.
Havner, K. S. and Shalaby, A. H. (1977). A simple mathematical theory of finite distortional latent hardening in single crystals, *Proc. Roy. Soc.* (London), **A358**, 47.
Hershey, A. V. (1954). The elasticity of an isotropic aggregate of anisotropic cubic crystals, *J. Appl. Mech.,* **21**, 236.
Hihi, A., Berveiller, M. and Zaoui, A. (1985). Une nouvelle formulation autocohérente de la plasticité des polycristaux métalliques, *J. Mech. Theor. Appl.,* **4**, 201.
Hill, R. (1965). Continuum micro-mechanics of elastoplastic polycrystals, *J. Mech. Phys. Solids,* **13**, 89.
Hill, R. (1966). Generalized constitutive relations for incremental deformation of metal crystals by multislip, *J. Mech. Phys. Solids,* **14**, 95.
Honneff, M. and Mecking, H. (1978). A method for the determination of the active slip systems and orientation changes during single crystal deformation. In *Textures of Materials,* G. Gottstein and K. Lücke (Eds), Springer, p. 265.
Hutchinson, J. W. (1970). Elastic-plastic behaviour of polycrystalline metals and composites, *Proc. Roy. Soc.* (London), **A319**, 247.
Iwakuma, T. and Nemat-Nasser, S. (1984). Finite elastic-plastic deformation of polycrystalline metals, *Proc. Roy. Soc.* (London), **A394**, 87.
Jackson, P. J. and Bazinski, Z. S. (1967). Latent hardening and the flow stress in copper single crystals, *Can. J. Phys.,* **45**, 707.
Kocks, U. F. (1964). Latent hardening and secondary slip in aluminium and silver, *Trans. Met. Soc. AIME,* **230**, 1160.
Kröner, E. (1958). *Kontinuumstheorie der Versetzungen und Eigenspannungen,* Springer-Verlag, Berlin.
Kröner, E. (1961). Zur plastischen Verformung des Vielkristalls, *Acta Metall.,* **9**, 155.
Leffers, T. (1981). Microstructures and mechanisms of polycrystal deformation at low temperature. In *Deformation of Polycrystals: Mechanisms and Microstructures,* N. Hansen, A. Horsewell, T. Leffers and H. Lilholt (Eds), Risø National Laboratory, Roskilde, Denmark, p. 55.
Lin, T. H. (1957). Analysis of elastic and plastic strains of a fcc crystal, *J. Mech. Phys. Solids,* **5**, 143.
Livingston, J. D. and Chalmers, B. (1957). Multiple slip in bicrystal deformation, *Acta Metall.,* **5**, 322.
Mandel, J. (1965). Généralisation de la théorie de plasticité de W. T. Koiter, *Int. J. Solids Struct.,* **1**, 273.
Mussot, P., Rey, C. and Zaoui, A. (1985). Grain boundary sliding and strain compatibility, *Res. Mech.,* **14**, 69.
Nakada, Y. and Keh, A. S. (1966). Latent hardening in iron single crystals, *Acta Metall.,* **14**, 961.

Peirce, D., Asaro, R. J. and Needleman, A. (1982). An analysis of non-uniform and localized deformation in ductile single crystals, *Acta Metall.,* **30**, 1087.
Rey, C. (1980). Contribution á l'étude de l'influence des interfaces sur la plasticité à froid des métaux purs, doctoral thesis, Université Paris-Nord, Villetaneuse, France.
Rey, C. and Zaoui, A. (1980). Slip heterogeneities in deformed aluminium bicrystals, *Acta Metall.,* **28**, 687.
Rey, C. and Zaoui, A. (1982). Grain boundary effects in deformed bicrystals, *Acta Metall.,* **30**, 523.
Rey, C., Mussot, P., Vroux, A. H. and Zaoui, A. (1985). Effects of interfaces on the plastic behaviour of metallic aggregates, *J. Phys.,* **46** C4, 645.
Sachs, G. (1928). Zur Ableitung einer Fliessbedingung, *Z VDI,* **72**, 734.
Taylor, G. I. (1934). The mechanism of plastic deformation of crystals, *Proc. Roy. Soc.* (London), **A145**, 362.
Taylor, G. I. (1938). Plastic strain in metals, *J. Inst. Metals,* **62**, 307.
Voigt, W. (1889). Über die Beziehung zwischen den beiden Elastizitätskonstanten isotroper Körper, *Wied. Ann.,* **38**, 573.
Weng, G. J. (1979). Kinematic hardening rule in single crystals, *Int. J. Solids Struct.,* **15**, 861.
Weng, G. J. (1983). A micromechanical theory of grain-size dependence in metal plasticity, *J. Mech. Phys. Solids,* **31**, 193.
Zaoui, A. (1979). Aspects fondamentaux de la plasticité des polycristaux métalliques. In *Dislocations et Déformation Plastique,* P. Groh, L. P. Kubin and J. L. Martin (Eds), Editions de Physique, Paris.
Zaoui, A. (1984). Macroscopic plastic behaviour of microinhomogeneous materials. In *Plasticity Today,* A. Sawczuk and G. Bianchi (Eds), Elsevier Applied Science, London, p. 451.
Zarka, J. (1973). Etude du comportement des monocristaux métalliques: application à la traction du monocristal cfc, *J. Mécan.,* **12**, 275.

PART III

PHENOMENOLOGICAL MODELLING

CHAPTER 8

Statistical Modelling

E. KRÖNER
Institut für Theoretische und Angewandte Physik, Universität Stuttgart, and Max-Planck-Institut für Metallforschung, Stuttgart, Federal Republic of Germany

ABSTRACT

Methods of probability theory and statistics are applied to the elasticity of polycrystals. Other properties of polycrystals can be treated in an analogous manner. Statistics permits us to derive not only the overall response, e.g. in terms of effective elastic moduli, but also mean fluctuations of stress and strain, mean stress and strain of grains of given orientation etc. Rigorous results are obtained in the form of various theorems on bounds for effective elastic moduli. These are derived from the classical variational principles of elasticity theory. The most important tools of the statistical part of the theory are the probability density functions and the correlation functions. They are used to provide the information needed to derive the quantities of interest. For the elastic part, Green's functions and modifications thereof have proved to be of great value. All these tools are explained in detail. Only brief remarks concerning the dynamical theory are given. More space would be needed to review, for instance, the present state of wave propagation in polycrystals.

1. INTRODUCTION

1.1. Scope and Content

Almost 100 years ago, Woldemar Voigt (1887) proposed to average the elastic moduli of the crystallites in a macrohomogeneous polycrystal over all orientations and consider the resulting isotropic moduli as effective elastic moduli of the aggregate. Much later, Reuss (1929) proposed doing the same with the elastic compliances. Again much later, Hill (1952) proved that the polycrystal moduli calculated after Voigt and Reuss are upper and lower bounds to the true effective elastic moduli of the polycrystal. The latter are usually defined as the components of a 4th rank tensor connecting the average values $\langle\sigma(\mathbf{r})\rangle$ and $\langle\varepsilon(\mathbf{r})\rangle$ of the stress and strain tensor at some point $\mathbf{r}$:

$$\langle\sigma(\mathbf{r})\rangle = \mathbf{C}^{\text{eff}}\cdot\cdot\langle\varepsilon(\mathbf{r})\rangle \tag{1.1}$$

where the averages are taken over a sufficiently large neighbourhood of the point $\mathbf{r}$. Further progress in the matter of these bounds is due to Mandel and Dantu (1963) and is reviewed in the conclusion to this chapter.

A homogeneous medium with elasticity described by $\mathbf{C}^{\text{eff}}$ is called the *effective medium* (representing the heterogeneous medium) if the two media have also the same shape. One also speaks of *effective medium theory* if eqn (1.1) is applicable. More 'universal conditions [than eqn (1.1)] for assimilation of a heterogeneous material into an effective medium' have been proposed by Huet (1982) who bases his definition on universal conservation laws. This theory is also suited to define nonlinear effective media. For present purposes, however, it seems simpler to use eqn (1.1).

The bounds of Voigt and Reuss are 1st order bounds, i.e. correct in the first power of the deviation of the single crystal moduli from their mean value. In this chapter the deviation is denoted by $\mathbf{c}'$ ($\equiv \mathbf{c} - \langle\mathbf{c}\rangle$). The two theorems are also true if the aggregate is macroanisotropic because of texture. As emphasized by Hill, these results are valid only under a certain statistical hypothesis which applies to many practical cases though, and will also be assumed for a major part of this chapter. As we know today, the hypothesis signifies that the external forces deforming the polycrystal are not correlated with the distribution of elastic moduli over the aggregate.

The bounds of Voigt and Reuss are the best possible if the statistical information is restricted to the volume percentages of the various orientations within the polycrystal. Adding the additional information that the specimen is macroisotropic and the elastic moduli of neighbouring grains are uncorrelated, Hashin and Shtrikman (1963) found narrower bounds which are correct in the 2nd order of $\mathbf{c}'$, thus classified as 2nd order bounds.

The knowledge of the effective elastic moduli has great practical value because their existence implies that the polycrystal behaves macroscopically like a homogeneous elastic medium having the elasticity tensor C^{eff}. Thus boundary value problems of polycrystals can be solved with the methods developed for homogeneous elastic bodies. Of course, it is normally possible to measure these moduli before the specimen is taken into use. On the other hand, important insight can be gained if the values of these moduli can be related to the statistical arrangement of the crystallites, their sizes, shapes, neighbourhoods, etc. This insight could for instance be used to draw conclusions about the internal constitution, texture, grain shapes, etc., of the polycrystal from the knowledge of the monocrystal moduli and the effective moduli measured in simple experiments. Therefore great interest resides in a theory that goes beyond the mentioned bounds and beyond other results such as those of Huber and Schmid (1934), Boas and Schmid (1934), Bruggeman (1934) and others whose formulae are not bounds but approximations that sometimes are good and sometimes not. There are also various methods that give good effective moduli with reasonable effort, although they lack the rigorous foundation that can come only from the stochastic theory, because the problem is one of statistical physics. Such methods have their value but they do not fit into the scope of this chapter. Some of these methods are described by Laws (1980), Ondracek and Schulz (1980), Christensen (1979) and Markov (1981). In these works further references can be found.

Any satisfactory theory of polycrystal behaviour must possess a strong stochastic component, because randomness is the most prominent feature of a polycrystal. It is therefore necessary to combine elasticity theory with probability theory in order to obtain what could be called a statistical theory of elasticity. Having clarified the physical model of a polycrystal in Section 2, its statistical aspects are described in Section 3. Section 4 presents just as much probability theory as is needed later. No previous knowledge of probability is required. Since most of the readers are probably familiar with elasticity theory, no introduction to this field is given, except that in Section 5 on elastic Green's functions which are an indispensable tool for our theory and perhaps less known than the other parts of elasticity theory needed here. The reader who is experienced with Green's functions may skip this section.

With the tools supplied in the earlier sections the systematic theory of effective elastic moduli is developed in Section 6. We show that an exact general solution exists for the problem of deriving the tensor of effective elastic moduli, given the statistical information in terms of correlation functions of all orders up to infinity. Rigorous bound theorems are described that allow us to derive optimum bounds if the information available is more restricted. It is remarkable that the relevant formulae apply just as well to composite materials with random structure. Theoretical predictions and experimental results on polycrystals agree as well as can be expected.

Problems going beyond finding effective elastic moduli are treated in Section 7. Mean fluctuations of stress and strain can be calculated, also mean stress and strain of crystallites of a given orientation. The latter are indispensable for any better theory of polycrystal plasticity where grains of various orientations behave differently. The presentation of this section is relatively brief because the theory is in an early stage. Further development is very desirable.

The dynamical theory, for instance of wave propagation in polycrystals, is more involved by orders of magnitude and touched only briefly in Section 7. The reason for the complications is that inertia forces develop which are correlated with the distribution of elastic moduli. Compared with those of statics, results on dynamics are rather poor in accuracy and mostly refer to composite materials rather than to polycrystals. For recent results see, e.g. Bahr (1977), Bahr and Gambin (1977), Datta (1980), Diener (1980), Ford (1980) and Willis (1980, 1981); for older work see Frisch (1968).

The theory presented in this chapter is basically linear. After reinterpretation of the symbols used it is also the theory of other linear physical properties such as electrical and thermal conductivity, viscosity, dielectricity and diamagnetism. The theory will be useful also for certain nonlinear problems. Polycrystal plasticity has been mentioned already. Many nonlinear problems are of such a nature that they can be solved by iterative solution of linear equations. The present theory clearly applies to such problems. For a basic limitation see McCoy and Fishman (1980).

The purpose of this chapter is to supply the most important tools for handling statistical problems of elasticity theory. Many rigorous results are derived with their help. The great number of basic formulae will be of use to the reader who wants to go on to problems not, or only insufficiently, covered by present theory. It is not the purpose of this chapter to record numerical results or results gained with (sometimes) ingenious but singular methods that are outside the systematics of the statistical approach.

1.2. Notation

Tensors

We apply tensor notation (subscripts) and symbolic notation.

2nd rank tensors are written in lower case Greek letters.

$\sigma = (\sigma_{ij})$ stress tensor
$\varepsilon = (\varepsilon_{ij})$ strain tensor
$\gamma = (\gamma_{ij})$ tensor of displacement Green's function
(δ_{ij}) unit tensor of 2nd rank; Kronecker symbol

4th rank tensors are ordinarily written in lower case italic letters; capital letters are used if the tensor is constant.

$\mathbf{c} = (c_{ijkl})$ tensor of elastic moduli of heterogeneous medium
$\mathbf{C} = (C_{ijkl})$ same for homogeneous medium
$\mathbf{s} = (s_{ijkl})$ tensor of elastic compliances of heterogeneous medium
$\mathbf{S} = (S_{ijkl})$ same for homogeneous medium
$\mathbf{E} = (E_{ijkl})$ tensor solving spherical inclusion problem
$\mathbf{I} = (I_{ijkl})$ unit tensor of 4th rank $= (\delta_{ik}\delta_{jl} + \delta_{il}\delta_{jk})/2$

Integral Operators

These are 4th rank tensors in three-dimensional physical space and 2nd rank tensors in Hilbert space; they are almost always self-adjoint.

Local integral operators. These have kernels arising from 4th rank tensors by multiplication with Dirac's delta function, the unit tensor of Hilbert space. The local operators corresponding to the above-listed 4th rank tensors are symbolically denoted by $\hat{\mathbf{c}}$, $\hat{\mathbf{C}}$, $\hat{\mathbf{s}}$, $\hat{\mathbf{S}}$, $\hat{\mathbf{E}}$; $\hat{\mathbf{I}}$. $\hat{\mathbf{I}}$ is the unit operator; $\hat{\mathbf{E}}$ is also the local part of the modified strain Green's function (see below).

Nonlocal integral operators. These have kernels of nonzero range and are symbolized by capital Greek letters.

$\hat{\mathbf{\Gamma}}$, kernel $\mathbf{\Gamma}_{ijkl}(\mathbf{r}, \mathbf{r}')$ modified strain Green's function
$\hat{\mathbf{\Delta}}$, kernel $\mathbf{\Delta}_{ijkl}(\mathbf{r}, \mathbf{r}')$ modified stress Green's function
$\hat{\mathbf{\Phi}}$, kernel $\mathbf{\Phi}_{ijkl}(\mathbf{r}, \mathbf{r}')$ nonlocal part of $\hat{\mathbf{\Gamma}}$
$\hat{\mathbf{A}}$, kernel $\mathbf{A}_{ijkl}(\mathbf{r}, \mathbf{r}')$ auxiliary operator

The integrations brought about by the integral operators are of convolution type in the wider sense. They always imply also a double scalar multiplication. These convolutions (symbol $*$) are e.g.

$$\hat{\mathbf{C}} * \hat{\mathbf{\Gamma}}, \text{ means } \int C_{ijkl}(\mathbf{r}, \mathbf{r}')\, \Gamma_{klmn}(\mathbf{r}', \mathbf{r}'')\, \mathrm{d}V'$$

$$\hat{\mathbf{\Gamma}} * \boldsymbol{\sigma}, \text{ means } \int \Gamma_{ijkl}(\mathbf{r}, \mathbf{r}')\, \sigma_{kl}(\mathbf{r}')\, \mathrm{d}V'$$

$$\boldsymbol{\sigma} * \hat{\mathbf{\Gamma}}, \text{ means } \int \sigma_{ij}(\mathbf{r})\, \Gamma_{ijkl}(\mathbf{r}, \mathbf{r}')\, \mathrm{d}V$$

$$\boldsymbol{\sigma} * \boldsymbol{\varepsilon} \equiv \boldsymbol{\sigma} * \hat{\mathbf{I}} * \boldsymbol{\varepsilon}, \qquad \text{means } \int \sigma_{ij}(\mathbf{r})\, \varepsilon_{ij}(\mathbf{r})\, \mathrm{d}V$$

Products of type cc are dyadic; they mean $c_{ijkl}c_{mnpq}$. Double scalar products $\mathbf{c} \cdot\cdot\, \mathbf{c}$ mean $c_{ijkl}c_{klmn}$. If not stated otherwise, the summation convention holds.

2. THE PHYSICAL MODEL OF A POLYCRYSTAL

In order to deal with the phenomenological, or macro, behaviour of polycrystalline aggregates, these are often modelled as continua consisting of many coherent subcontinua, also called *grains* or *crystallites,* such that the considered material properties change abruptly when going from one grain to the next. It may be expected that this model works well as long as the material property is inherently located in the interior of the crystallites rather than in their interfaces. Also the latter situation has some physical significance, for instance when inelastic behaviour is caused by grain boundary sliding.

Grain boundaries are *interface defects,* along which the atoms are in irregular positions. These defects have a thickness that in many polycrystals is of the order of one atomic distance, i.e. extremely small compared with those distances that are of interest in macro experiments. Nevertheless the presence of grain boundaries implies the presence of a partial volume in which the cohesive forces between the atoms deviate from those in the interior of the grains. In a sense, the 'grain boundary volume' contains another 'phase' that should be taken into account when it is comparable in size to the volume of the aggregate. Fortunately the volume percentage of the grain boundary decreases quickly with increasing grain size. It is of the order of several 10^{-1} for a mean grain diameter of $10a$, of the order of 10^{-3} for 10^2a, 10^{-5} for 10^3a and 10^{-7} for

10^4a, where a is the atomic spacing. Polycrystals with less than 10^4a grain size are considered extremely fine-grained and occur rarely in practical use. This shows that in most applications grain boundaries need not be taken into account as a separate phase. Grain sizes of order $10a$ or less are sometimes considered in the modelling of amorphous metals; see e.g. Rudee and Howie (1972). The grain boundary must then be considered as a separate phase. For further models of amorphous solids see also the conference proceedings edited by Yonezawa and Ninomiya (1983). In this chapter we shall neglect the finite thickness of the grain boundary and endow our model of a polycrystal with grain boundaries of zero thickness.

Here we are interested in linearly elastic polycrystals. We therefore assume in the model of a polycrystal that its grains are linear elastic continua. It is well known that sound waves in such continua are not dispersed whereas crystals show dispersion at wavelengths of a few atomic distances. Such waves are not phenomenological and will be excluded from our consideration. With this restriction the continuum model of the crystallite reflects its elastic properties with greatest accuracy.

We shall furthermore build into our model of a polycrystal the requirement that the grains have elastic moduli that are constant through the grain so that each grain has its stress–strain law of the general form

$$\sigma_{ij}(\mathbf{r}) = c_{ijkl}\varepsilon_{kl}(\mathbf{r}) \tag{2.1}$$

with constant elastic moduli c_{ijkl}. Of course, the stresses σ_{ij} and strains ε_{ij} will, in general, vary with the position $\mathbf{r}$ in the grain.

The elastic moduli as shown in eqn (2.1) are assumed to be constant in a grain, but to vary abruptly from one grain to another. The values that the c_{ijkl} assume in a particular grain are supposed to depend only on the material and on its lattice orientation described, for instance, by the triple of Eulerian angles taken with respect to some reference frame.

Real crystallites of a given material neither have constant elastic moduli nor do these depend on orientation only. Each grain contains a variety of lattice defects which implies a local change of elastic moduli. If this change is neglected and instead the constant averaged moduli are assigned to the crystallite, the relative deviation of the calculated polycrystal behaviour from that of the real aggregate is well below 10^{-3}, almost a magic accuracy limit both for theories and experiments on polycrystal elasticity.

Well above this limit, namely of the order of a percent, is another effect stemming also from lattice defects, in particular from dislocations. These can bow out even under very small applied stresses, thereby producing an additional reversible strain which usually is classified as anelastic because it proceeds with retardation. This effect leads to an apparent lowering of the elastic moduli of the dislocated crystallite as compared with a defect-free one. A rather complex situation arises because the apparent elastic moduli of two grains with different orientations are no longer connected by the well known tensor transformation law. The best way to escape this difficulty is to use well annealed materials in which the number of defects and the described effect are a minimum. This, however, is not always possible, since deformed materials, which have many defects and therefore lower elastic moduli, are often used in practice thanks to their particular strength properties.

The described effect was explained by Wawra (1979) and is depicted in Fig. 1. It has given rise to misinterpretations of comparisons between theoretical and experimental values of polycrystal moduli. The same is true for another effect of similar magnitude, namely the changing of effective moduli due to texture. It is in fact difficult, if not impossible, to prepare a polycrystalline aggregate that is completely free of texture. The best way for removing texture is through annealing. It seems that an agreement between theory and experiment of 10^{-3} can be reached in the case of well annealed polycrystals whereas, in my estimate, it is restricted to a few percent in the case of cold-worked polycrystals, at least as long as the theory is not especially extended to include the elastic softening by dislocations, and as long as no really accurate methods exist to take into account the present texture. Such methods are usually based on the experimental finding that certain measured moduli depend weakly on the direction in the specimen. For details see e.g. Bradfield and Pursey (1953) and Wawra (1968, 1974).

In the foregoing discussion a model of a polycrystal was introduced that is idealized with respect to the real polycrystal. The model is defined as follows:

(i) The grains of the polycrystal are modelled as linear elastic continua with constant, orientation-dependent elastic moduli.

(ii) The elastic moduli of any two grains of different orientation are connected by the well known tensor transformation law for elastic moduli of monocrystals.

(iii) All grains together constitute the polycrystal. They are modelled as separated by infinitely thin interfaces in the sense of continuum theory.

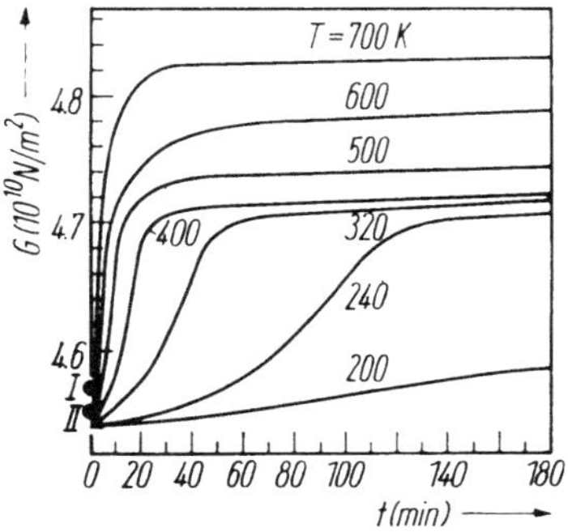

Fig. 1. Shear modulus G of 76% cold-drawn electrolytic copper which subsequently has been annealed at temperatures from 200 K up to 700 K. Measurements were performed at room temperature. Points I and II from Grüneisen. All values of G are corrected for small texture. From Wawra (1979).

3. THE STATISTICAL MODEL OF A POLYCRYSTAL

The foregoing section served to create a physical model which is described by a definite set of physical equations. The equations themselves were not written down, except eqn (2.1), but they are uniquely defined by the statements referring to the linearly elastic behaviour of the grains. Nothing has been said about the arrangement, shape and size of the crystallites. It is easy to imagine, but less easy to prepare, a polycrystal in which the grains form a periodic array. Methods have been developed to deal with such a situation. Best known is perhaps the so-called *homogenization* (see e.g. the book by Bensoussan *et al.*, 1978). This method is appropriate also for treating periodically structured *composites* that play an important role in practice whereas periodic polycrystals have no such significance.

In this chapter the interest is in polycrystals that are highly disordered in their constitution. We shall define later a type of polycrystal called *perfectly disordered.* It is the opposite of the periodic polycrystal. Between the perfectly ordered (= periodic) and the perfectly disordered polycrystal lies a whole range of partially ordered or disordered polycrystals which are accessible to a statistical treatment provided that the disorder is the prevailing feature. These form the subject of our present interest.

The essential concepts and notions of statistics are commonly introduced within the framework of probability theory, a mathematical discipline with axiomatic foundation that has to be omitted here for reasons of space. It is the combination of probability theory and physical theory that leads to the relevant statistical physics or physical statistics. We arrive at such a theory if we now apply probability concepts to the polycrystal. Such concepts are described in a monograph of Beran (1968), for instance, which we recommend to the reader who prefers a more detailed description than can be given here.

As is well known, statistics work better when a larger number of stochastic elements are used in the experiments, the results of which we try to predict. In our case it is the number of crystallites that we assume to be very large, in fact near to infinity, in all directions through the polycrystal. If n is this number, probability theory tells us that the relative deviation of our prediction from the experimental truth will decrease as $1/\sqrt{n}$. The reader may convince himself in an experiment of repeated throwing of a die. In this experiment the expected number of spots is 3·5. The result of the performance is that, with increasing number n of throws, the average number approaches 3·5 with a relative deviation proportional to $1/\sqrt{n}$.

The calculation of expectations, or expectation values, which can be compared with experiments or observations, is the final aim of any statistical theory. As mentioned above, the expectation of any experiment is the mean of the outcomes of many experiments which are alike in a macro sense but differ in the micro details. In the die experiment the microscopic details are the forces and moments exerted on the die during the throwing as well as any microscopic factor that influences the outcome of the experiment. Unlike the case of the die, the statistics of the polycrystal is inherently inside the physical system rather than in the external experimental set-up. This suggests that one performs the same experiment not repeatedly with a single specimen, but rather does the same kind of experiment with many specimens which are prepared alike such that they appear to be macroscopically equal though different in the microscopic details. We expect that the outcomes of these experiments will vary more or less, but in fact very little for a large number of grains. The mean value of the outcomes is equal to the expectation value by definition.

The set of such experiments is called *statistical ensemble* by Beran (1968). The concept goes back to Gibbs who thought of ensembles as large collections of physical systems alike in the above sense. Both forms seem appropriate.

Consider now a very large polycrystal, in fact an aggregate having an 'almost' infinite number of grains over any distance of the smallest phenomenological length, say l, in the problem. Here we speak of numbers so large that also their square roots are 'almost' infinite. In such a case the volume V of the polycrystal can be divided into so-called macrovolume elements ΔV that are 'almost' infinitesimal on the scale of l, but contain nevertheless a huge, 'almost' infinite number of grains. Such an element ΔV is itself a complex statistical object and exhibits some effective material property similar to that exhibited by the polycrystal as a whole. If, for instance, all elements show the same effective elasticity, the elasticity of the polycrystal is called *macrohomogeneous* or, if the arrangement of the grains is disordered rather than ordered, it is called *statistically homogeneous*. Also the term *quasihomogeneous* is used. The elasticity of a polycrystal is called *statistically inhomogeneous* if the effective elasticity varies macroscopically over the elements ΔV.

We have mentioned that statistics does not cover the whole range between the perfectly ordered and the perfectly disordered situation. The question then arises whether there exists any criterion as to which situations can in fact be handled by statistics. To understand this question, consider that the described polycrystal is deformed by external forces distributed over its surface in such a way that the stress and strain would be constant throughout if the polycrystal were a homogeneous elastic body. Performing this experiment over all members of the ensemble (see above) we arrive at ensemble averages of stress and strain that throughout this chapter are denoted by $\langle \sigma_{ij} \rangle$ and $\langle \varepsilon_{ij} \rangle$ respectively.

The question may be raised whether $\langle \sigma_{ij} \rangle$ and $\langle \varepsilon_{ij} \rangle$, when measured as indicated, are independent of the position in the various members of the ensemble. Let $\mathbf{r}^{(1)}$ denote a position in the first polycrystal of the ensemble, $\mathbf{r}^{(2)}$ the *corresponding* position in the second one, etc. All corresponding coordinates of the points $\mathbf{r}^{(1)}, \mathbf{r}^{(2)}, \ldots$ shall have the same numerical values. Thus it suffices to use only one vector, say $\mathbf{r}$, to characterize the positions of corresponding points in all members of the ensemble. Are $\langle \sigma_{ij} \rangle$ and $\langle \varepsilon_{ij} \rangle$ independent of $\mathbf{r}$? There appears to be a good chance of this provided that the number N of members in the ensemble is very large and the grain arrangements are sufficiently irregular so that during the averaging the differences attributable to different points $\mathbf{r}$ are eliminated. Note that a similar expectation is not realistic if the polycrystal is almost periodic, e.g. a slightly distorted chess board pattern.

Recall now the division into macrovolume elements ΔV which show the same effective elasticity in the statistically homogeneous case. It is easy to guess that the mean stress and the mean strain taken over any ΔV are equal in each case, namely equal to the mean stress and mean strain taken over the total volume V. Next consider the macro-inhomogeneous situation in which the averages over ΔV are not equal. This occurs, for instance, if the external forces are applied in a more inhomogeneous manner. The definitions of volume and ensemble averages are e.g. for stress

$$\overline{\sigma_{ij}(\mathbf{r})} \equiv \frac{1}{\Delta V(\mathbf{r})} \int_{\Delta V(\mathbf{r})} \sigma_{ij}(\mathbf{r})\,\mathrm{d}V, \qquad \langle \sigma_{ij}(\mathbf{r}) \rangle \equiv \frac{1}{N} \lim_{N \to \infty} \sum_{N} \sigma_{ij}(\mathbf{r}) \quad (3.1)$$

The averaged functions vary macroscopically with position, i.e. are *smooth* functions contrary to the original random functions that fluctuate on the micro scale. Note that ensemble averaging commutes with differentiation and integration. It is common to use the same letter $\mathbf{r}$ for the position on the micro scale in the random functions and on the macro scale in the averaged functions.

Obviously $\overline{\sigma_{ij}(\mathbf{r})}$ is constant over the considered macroelement. Since ΔV contains an 'almost' infinite number of grains, we expect also that $\langle \sigma_{ij}(\mathbf{r}) \rangle$ is constant in ΔV, just as we expected this in the macro-homogeneous case. Generalizing this idea to arbitrary random functions, we argue that this result is reasonable if and only if for all random functions the equality

$$\text{volume average} = \text{ensemble average} \qquad (3.2)$$

holds true, where the averaging is defined in accordance with eqns (3.1). In fact, the expectation that $\langle \sigma_{ij} \rangle$ does not depend on $\mathbf{r}$ within ΔV implies that $\langle \sigma_{ij}(\mathbf{r}) \rangle$ is also equal to the ensemble average over the volume averages, the fluctuations among which tend to zero with the number of grains in them.

Equation (3.2) is called the *ergodic hypothesis* and has the status of a first principle or postulate in the statistics of *spatially* continuous objects. In problems that are stochastic in *time* rather than in space, a hypothesis is used that equates time average and ensemble average.

Numerous investigations show that real physical systems satisfy the ergodic hypothesis very rarely. Nevertheless the hypothesis works 'well' whenever the situation is 'sufficiently' random. Here the terms 'well' and 'sufficiently' are not precisely defined; they depend on the demands for accuracy. We see that the ergodic hypothesis provides the desired criterion which decides if statistics are applicable or not to a particular problem. It seems that polycrystals, except those grown rather artificially, are disordered enough to satisfy the ergodic hypothesis with an accuracy that is far above that accessible to present experimental techniques. For further details the reader is referred to the extensive literature of statistical mechanics, for example Beran (1968), Fowler (1955), Reif (1965) and Tolman (1938), and also the original work by Mazilu (1981).

We are now ready to introduce the statistical model of a polycrystal under the following assumptions:

(i) All random variables of the model polycrystal obey the ergodic hypothesis defined by eqns (3.1) and (3.2).

(ii) The distribution of elastic moduli over the grains of the model polycrystal is statistically homogeneous. Here a random variable is any variable that is distributed at random, for instance in our applications stress, strain and elastic moduli.

The form of the statistical model implies that we now have a problem of statistics to which all equations and notions of probability theory are applicable. We shall utilize this immediately in the next section.

4. SOME BASIC NOTIONS AND EQUATIONS OF PROBABILITY THEORY

4.1. Probability and Expectation

In the preceding two sections we have introduced a model of the polycrystal that takes into account both elastic and statistical properties. The behaviour of this model crystal is governed by the equations of elasticity theory on the one hand and by those of probability theory on the other hand. Every solution of these equations describes a possible situation of the model crystal. The model is developed in such a way that one may expect it also to represent physical reality rather well.

As mentioned, the main aim of any statistical theory is the derivation of expectation values, given some information about the physical situation. The expectation value is equal to the ensemble average, defined for the example of stress in eqn (3.1). To calculate expectation values we need to know something about probability.

Consider a physical system with stochastic properties and a well defined measurement, here also called an *experiment of chance*, or *trial*.

To have something definite in mind, let the system be a die and the trial the throwing of the die. Each trial leads to the outcome c,* say, which can assume the values 1, 2, 3, 4, 5 or 6, also symbolized by c_i ($i = 1, 2, \ldots, 6$). If we perform n trials we shall observe that the numbers occur with a certain frequency. One may wish to compare these frequencies or, better, the relative frequencies for large numbers of trials. We define the relative *frequency* or *probability* that the outcome of a trial is c as

$$P(c) = \lim_{n\to\infty} \frac{\text{number of trials with outcome } c}{\text{total number } n \text{ of trials}} \tag{4.1}$$

The probability defined in this fashion is also called *a posteriori probability* because first the experiments are performed and then the probability is determined.

If the die is perfectly regular, one may dare to predict already before performing the experiments that the probabilities defined in eqn (4.1) will come out to be 1/6 each. The probability determined by *a priori* reflexions, often of symmetry type, is called *a priori probability.* It was encountered already in Section 3. Whereas it is easy to find out the *a priori* probability of the perfect die, this task is often much more difficult in other cases, depending on the complexity of the statistical situation, as we shall see. If the *a priori* probabilities are correctly found, they coincide numerically with the *a posteriori* ones.

Once the (*a priori* or *a posteriori*) probabilities of all possible trial outcomes are known, all expectation values can be calculated. For the die we have the expectation, often denoted by E, as the average, weighted with the probabilities:

$$E(c) \equiv \langle c \rangle = \sum_{6} p_i c_i = \frac{1}{6}(1 + 2 + 3 + 4 + 5 + 6) = 3{\cdot}5 \tag{4.2}$$

Here p_i is the probability of the outcome of c_i as mentioned before. Note that the appearance of weights is in agreement with the ensemble concept. In fact, the expectation corresponds to a large number of experiments in which the outcomes appear with relative frequencies, i.e. are determined by the probabilities.

*We use the letter c instead of the often used x in order to avoid confusion with the spatial coordinate.

Another expectation of interest is the so-called *variance*

$$E[(c - \langle c \rangle)^2] = \sum_6 p_i (c_i - \langle c \rangle)^2$$

$$= \frac{1}{6}(2{\cdot}5^2 + 1{\cdot}5^2 + 0{\cdot}5^2 + 0{\cdot}5^2 + 1{\cdot}5^2 + 2{\cdot}5^2) \approx 1{\cdot}46 \qquad (4.3)$$

from which follows the *standard deviation,* i.e. the square root of the variance, as 1·21. It is an often used measure of the scattering of measured values.

In the statistical problem of a polycrystal the trial outcomes (e.g. stress, strain or elastic moduli) have a continuous rather than discrete spectrum. In such a situation, probabilities are replaced by probability densities and sums by integrals. Assume that the outcomes have densely lying values c between c_{max} and c_{min}. The probability density $P_1(c)$ is now defined as follows: $P_1(c)\,dc$ is the probability of finding the outcome of the experiment of measuring the quantity in question in the range between c and $c + dc$. Since in eqn (4.1) the probability was defined such that the sum of the probabilities for all outcomes c is unity, meaning certainty, we conclude that

$$\int_{c_{min}}^{c_{max}} P_1(c)\,dc = 1 \qquad (4.4)$$

This equation expresses the certainty that the experiment must have an outcome anywhere between c_{min} and c_{max}.

Knowing $P_1(c)$, the expectation values can be calculated as before in the discrete case. For instance,

$$E(c) = \int_{c_{min}}^{c_{max}} c P_1(c)\,dc \qquad (4.5)$$

and, for any function of c,

$$E[f(c)] = \int_{c_{min}}^{c_{max}} f(c) P_1(c)\,dc \qquad (4.6)$$

4.2. Example: Expectation (Mean) Values of Elastic Moduli

In our applications, c will often symbolize the tensor of elastic moduli and then will be written as $\mathbf{c}$. As an example let us apply eqn (4.5) to the calculation of the averaged tensor in a statistically homogeneous polycrystal. As mentioned, the elastic moduli depend on orientation here symbolized by the letter Ω. By $\mathrm{d}\Omega$ we denote an infinitesimal range of orientation Ω (see below). It is convenient to integrate over all orientations rather than directly over the elastic moduli. To this end we introduce $P(\Omega)$, the probability density for orientations, also called *texture function,* by

$$P_1(\mathrm{c})\,\mathrm{dc} = P_1(\mathrm{c}(\Omega))\frac{\mathrm{d}c(\Omega)}{\mathrm{d}\Omega}\mathrm{d}\Omega \equiv P(\Omega)\,\mathrm{d}\Omega, \quad \int P(\Omega)\,\mathrm{d}\Omega = 1 \tag{4.7}$$

The last equation means that every grain must have an orientation. Ω can be specified in terms of the three Eulerian angles so that each triple θ, ψ, ϕ of these fixes one orientation. The angles are used here with reference to the regular precession of a force-free symmetric gyroscope: θ is the angle between the axes of gyroscope and moment of momentum; $\dot{\psi}$ and $\dot{\phi}$, respectively, are the angular velocities around these axes.

The element of the orientation space is

$$\mathrm{d}\Omega = \sin\theta\,\mathrm{d}\theta\,\mathrm{d}\psi\,\mathrm{d}\phi, \qquad \int \mathrm{d}\Omega = 8\pi^2 \tag{4.8}$$

With the full tensor c_{ijkl} instead of c, eqn (4.5) becomes

$$E(c_{ijkl}) \equiv \langle c_{ijkl}\rangle = \int_0^{\pi} \mathrm{d}\theta \sin\theta \int_0^{2\pi} \mathrm{d}\psi \int_0^{2\pi} \mathrm{d}\phi P(\theta,\psi,\phi)\, c_{ijkl}(\theta,\psi,\phi) \tag{4.9}$$

This equation refers to Cartesian coordinates x_k which are often chosen to coincide with macroscopic symmetry directions of the specimen. Let $x_{k'}$ be Cartesian coordinates which lie in the symmetry directions of a particular crystallite. The elastic moduli referred to these coordinates are denoted by $c_{i'j'k'l'}$ and called the *main constants.* They are related to the c_{ijkl} by the tensor transformation law:

$$c_{ijkl}(\theta,\psi,\phi) = \alpha_i^{i'}\alpha_j^{j'}\alpha_k^{k'}\alpha_l^{l'}\,c_{i'j'k'l'}, \quad \det\{\alpha_i^{i'}\} = 1 \tag{4.10}$$

Here, as throughout the chapter, we use the common summation rule (all tensor indices refer to Cartesian coordinates). The Eulerian angles are measured with respect to the symmetry directions of the crystallite and the α's are the Jacobian matrices of the transformation. They depend on the angles and can be represented as a matrix product of the form

$$(\alpha_i^{i'}) = \begin{pmatrix} \cos\phi & -\sin\phi & 0 \\ \sin\phi & \cos\phi & 0 \\ 0 & 0 & 1 \end{pmatrix} \begin{pmatrix} \cos\psi & -\sin\psi & 0 \\ \sin\psi & \cos\psi & 0 \\ 0 & 0 & 1 \end{pmatrix} \begin{pmatrix} \cos\theta & 0 & \sin\theta \\ 0 & 1 & 0 \\ -\sin\theta & 0 & \cos\theta \end{pmatrix} \quad (4.11)$$

(the order of sequence is important!). If eqns (4.10) and (4.11) are inserted in (4.9) and P is given, the integrals can be calculated.

Since c_{ijkl} contains sin and cos of each angle up to power 4 only, it suffices for the present purpose to know the series development of P up to order 4 in these functions because terms of order higher than 4 do not contribute owing to the orthogonality conditions of the trigonometric functions. As long as the interest is only in the average moduli the function $P(\Omega)$ is sufficiently specified by the coefficients of the mentioned series development. Their values for the various symmetries of the crystallites can be found for instance in the monograph of Bunge (1969). In statistically isotropic aggregates (no texture) all orientations occur with equal frequencies, so that $P(\Omega)$ is constant. In view of the normalization [last eqn (4.7)] of the probability density, $P(\Omega) = 1/8\pi^2$ for macroisotropy. With this the integrations in (4.9) can easily be performed. The somewhat lengthy calculations can be avoided by the observation that any tensor of 4th rank has two linear invariants that do not depend on the orientation. The invariants are c_{iikk} and c_{ikik}. It follows from eqn (4.9) that

$$\langle c_{iikk} \rangle = c_{iikk}, \qquad \langle c_{ikik} \rangle = c_{ikik} \quad (4.12)$$

For a homogeneous and isotropic medium of elastic moduli C_{ijkl} the connexion between its compression (bulk) modulus K, rigidity (shear) modulus G and the invariants of C_{ijkl} is given by

$$\begin{aligned} K &= \tfrac{1}{9} C_{iikk} \\ G &= \tfrac{1}{30}(3C_{ikik} - C_{iikk}) \end{aligned} \quad (4.13)$$

After replacement of C_{ijkl} by $\langle c_{ijkl} \rangle$ these equations give the bulk and shear modulus of a homogeneous and isotropic medium with elastic moduli $\langle c_{ijkl} \rangle$. In this particular case we write K^{V} and G^{V} instead of K and G, in order to indicate that the resulting formulae were first obtained by Voigt (1887). For an aggregate of cubic crystallites with

Voigt's (main) constants c_{11}, c_{12}, c_{44} we conveniently introduce

$$\kappa = (c_{11} + 2c_{12})/3, \quad \mu = c_{44}, \quad \nu = (c_{11} - c_{12})/2 \tag{4.14}$$

It then follows easily from eqns (4.12) and (4.13) that

$$K^{\mathrm{V}} = \kappa, \qquad G^{\mathrm{V}} = (3\mu + 2\nu)/5 \tag{4.15}$$

If the tensor s_{ijkl} of elastic compliances is introduced as the inverse of c_{ijkl},

$$s_{ijkl}\, c_{klmn} = I_{ijmn} \equiv \tfrac{1}{2}(\delta_{im}\delta_{jn} + \delta_{in}\delta_{jm}) \tag{4.16}$$

then all preceding calculations can be repeated with s instead of c. If the result is inverted one obtains the formulae of Reuss (1929):

$$K^{\mathrm{R}} = \kappa, \qquad G^{\mathrm{R}} = 5\mu\nu/(2\mu + 3\nu) \tag{4.17}$$

It was proved by Hill (1952), and follows from the theory developed later, that $K^{\mathrm{V}}, G^{\mathrm{V}}$ and $K^{\mathrm{R}}, G^{\mathrm{R}}$ are respectively upper and lower bounds of the true effective moduli (to be defined later) of macrohomogeneous and macroisotropic elastic moduli. The coincidence of K^{V} and K^{R} holds for aggregates of cubic crystallites only. These are the only crystals in which the compression modulus does not depend on crystallite orientation.

The bounds of Voigt and Reuss are good approximations to the true effective moduli provided that the anisotropy of the crystallites is small. In the case of cubic symmetry the *anisotropy constant* $A \equiv \mu/\nu$ is then close to unity. μ and ν, as introduced in eqns (4.14), are the extreme values for the orientation-dependent shear moduli. Their ratio is shown for some cubic crystals in Table 1 which shows that either μ or ν can have the larger value. It is interesting to see that elements that are close in the periodic table are also close in this table. The tungsten crystal is practically isotropic.

TABLE 1
Anisotropy Constant (at 0 K) for Some Cubic Crystals, Put Together by Ledbetter (1983)

Material	Nb	Cr	V	Mo	W	Al	Pt	Sn
$A = \mu/\nu$	0·52	0·69	0·81	0·90	1·00	1·21	1·48	1·56
Material	Ir	Ta	Ge	Fe	Ni	Pd	Au	Ag
$A = \mu/\nu$	1·57	1·62	1·67	2·34	2·39	2·46	2·84	2·99
Material	Cu	Th	Pb	Cs	K	Rb	Na	Li
$A = \mu/\nu$	3·19	3·45	3·84	7·62	7·63	8·07	8·50	9·24

4.3. Higher Order Probability Densities

Whereas the six probabilities p_i for the outcomes of die experiments form a complete set in the sense that all expectation values can be calculated, this is not so in the case of polycrystal experiments, if only the probability density $P_1(\mathbf{c})$ is known. $P_1(\mathbf{c})$ contains the information about volume percentages in which the various orientations occur. As we have seen, this knowledge enables us to determine bounds, but nothing more. The conclusion is that the macro properties of a polycrystal are determined by more features than the volume percentages of orientation only. These features concern above all the distribution of grain shapes and sizes. It is rather obvious that their description in statistical terms is a complex undertaking.

Let us therefore try to answer a simpler question. Consider a straight line going in a given direction through the polycrystal. What is the average linear dimension of a grain along this line? The answer to this question gives us some statistical information about grain shapes and sizes, which is certainly more than that contained in the probability density $P_1(\mathbf{c})$. We shall see later that the average grain size is not of great interest, because a magnification of the whole specimen, which includes that of the grains, does not affect the material properties. It is however important to know if and how the linear grain dimension varies with the direction in the polycrystal. Which statistical tools can we use to obtain this information?

Let us think of a set of trials through our ensemble so that in each trial two fixed test points $\mathbf{r}_1$ and $\mathbf{r}_2$ are placed in one of the members of the ensemble. These two points also define a direction (the same) in each polycrystal. Assign the number 1 to the outcome that both points lie in the same grain, and 0 to the other outcome. Obviously the ensemble average is the coincidence probability of both test points occurring in one and the same grain. This follows from the definition of probability, eqn (4.1). Let us write $w(\mathbf{r}_1, \mathbf{r}_2)$ for this probability. If the medium is statistically homogeneous, then w depends on $\mathbf{r}_{12} = \mathbf{r}_2 - \mathbf{r}_1$; if it is also statistically isotropic, then $w = w(r_{12})$, $r_{12} = |\mathbf{r}_{12}|$. It is now clear that $w(\mathbf{r}_{12})$ contains important information about the grain shapes, namely if their distribution is anisotropic and, if it is, how much. In fact, it answers the question raised about the dependence on direction of the mean grain dimension.

Let us now repeat the last trial series, but ask in addition for the probability of finding the elastic moduli $\mathbf{c}$ in the range $d\mathbf{c}_1$ at $\mathbf{r}_1$ and in $d\mathbf{c}_2$ at $\mathbf{r}_2$. This probability will be denoted by $P_2(\mathbf{c}_1, \mathbf{c}_2)\, d\mathbf{c}_1\, d\mathbf{c}_2$ and $P_2(\mathbf{c}_1, \mathbf{c}_2)$ is the *two-point probability density* because it refers to simultaneous measurements at two points. If the measurements are made with two coinciding points, $\mathbf{r}_1 = \mathbf{r}_2$, then P_2 reduces to the one-point density P_1.

It is often helpful to decompose P_2 into two parts referring to the coincidence (outcome 1) and non-coincidence (outcome 0) of the two points. Then

$$P_2(\mathbf{c}_1, \mathbf{c}_2) = P_c(\mathbf{c}_1, \mathbf{c}_2|1)\, w(\mathbf{r}_{12}) + P_c(\mathbf{c}_1, \mathbf{c}_2|0)(1 - w(\mathbf{r}_{12})) \quad (4.18)$$

where we read 'given 1' for $|1$. P_c is a conditional probability density of finding $\mathbf{c}_1$ and $\mathbf{c}_2$, given that the two points are in one and the same grain (case 1) or not (case 2). $w(\mathbf{r}_{12})$ is, as before, the probability that the two points are in the same grain.

By eqn (4.18) the purely geometrical probability w is split away from P_2 leaving P_c which contains the material information. Note, first of all, that

$$P_c(\mathbf{c}_1, \mathbf{c}_2|1) = P_1(\mathbf{c}_1)\, \delta(\mathbf{c}_1, \mathbf{c}_2) \quad (4.19)$$

because $|1$ implies that the two points are in the same grain, thus $\mathbf{c}_1 = \mathbf{c}_2$. Note that eqn (4.19) is correctly normalized. The last term in eqn (4.18) vanishes for $\mathbf{r}_1 = \mathbf{r}_2$ but not otherwise. It contains the correlations between the elastic moduli in different grains. If the distribution of elasticity over the grains is uncorrelated, which indicates high disorder, then P_c factors with respect to $\mathbf{c}_1$ and $\mathbf{c}_2$ and we have

$$P_c(\mathbf{c}_1, \mathbf{c}_2|0) = P_1(\mathbf{c}_1)\, P_1(\mathbf{c}_2) \quad (4.20)$$

valid for the non-coincidence case. A more detailed discussion of the factorization problem has been given by Miller (1969).

From these considerations it follows that we can model a polycrystal as an assembly of empty cells constructed according to some statistical description. The grains are then filled with the material property, here elasticity. The filling may be correlated with the grain pattern or not. A correlation arises for instance if some orientations are assigned to more lengthy grains and others to more compact grains. Other correlations arise if, in the progress of filling, notice is taken of what has been done before. Polycrystals that do not show the two types of correlation just described are called *statistically decoupled of grade 2* where the term 'grade 2' refers to the 2-point tests treated in this section.

Everything said here in connexion with the 2-point testing has its analogy for 3-point testing. Three-point probability densities are defined. They contain information about average grain shapes (needle, disc, sphere). For instance, needles oriented in all directions are detected by the 3-point probability density P_3 which in this case has a peak for points $\mathbf{r}_1$, $\mathbf{r}_2$, $\mathbf{r}_3$ lying on a line. It is now obvious that nth-order probability densities can be defined for arbitrary n. They include all densities of order n and describe finer and finer details of the grain arrangement.

According to a well known theorem of probability theory, the full set of probability densities P_n ($n = 1, 2, \ldots, \infty$) contains the complete statistical information about the statistical situation. With their help all desired expectation values can be calculated, if the corresponding integrations are manageable.

Our discussion leaves open the question of finding out the probability densities. We consider this question in the following Section 4.4.

4.4. Correlation Functions

By means of the probability functions another set of functions can be defined. These are called *n-point correlation functions* or correlation functions of order n and denoted here by K_n. They are of extreme importance in the whole of statistical physics. We define them by

$$\begin{aligned} K_1 &\equiv \int c_1 P_1(c_1)\, \mathrm{d}c_1 \\ K_2 &\equiv \int c_1 c_2 P_2(c_1, c_2)\, \mathrm{d}c_1\, \mathrm{d}c_2 \\ K_3 &\equiv \int c_1 c_2 c_3 P_3(c_1, c_2, c_3)\, \mathrm{d}c_1\, \mathrm{d}c_2\, \mathrm{d}c_3 \end{aligned} \tag{4.21}$$

etc., where the interval of integration covers the whole range of c-values and c, for the moment, is an arbitrary random variable to be replaced later by $\mathbf{c}$, the tensor of elastic moduli. Of course, correlation functions can also be formed with stress or strain. It is sometimes an advantage if the correlation functions are not constructed with the original random variable, but rather with its deviation from the mean. We define the deviation as

$$c' = c - \langle c \rangle \tag{4.22}$$

and introduce its correlation functions by

$$\begin{aligned} k_1 &= \int c_1' P_1(c_1)\, \mathrm{d}c_1 \equiv 0 \\ k_2 &= \int c_1' c_2' P_2(c_1, c_2)\, \mathrm{d}c_1\, \mathrm{d}c_2 \equiv \langle c_1' c_2' \rangle \\ k_3 &= \int c_1' c_2' c_3' P_3(c_1, c_2, c_3)\, \mathrm{d}c_1\, \mathrm{d}c_2\, \mathrm{d}c_3 \equiv \langle c_1' c_2' c_3' \rangle \end{aligned} \tag{4.23}$$

It is clear that the K and k correlation functions depend on the position of the points concerned; K_1 is the mean already discussed, while k_1 vanishes because the deviation is defined as a random variable with vanishing mean. For convenience we include K_1 and k_1 in the set of correlation functions. Other authors start the set at $n = 2$. The essential element in forming a correlation function is the relevant probability density. This means that correlation functions and probability densities contain roughly the same statistical information. They contain it exactly, when the whole sets of these functions are considered. There is a difference, however, when the correlation function of order n is compared with the n-point probability function. We see this already in the simplest case of order 1. The mean $\langle c \rangle$ is one single constant, if the medium is statistically homogeneous. If $P_1(c)$ is given instead, also averages $\langle c^2 \rangle$, $\langle c^3 \rangle$, generally $\langle f(c) \rangle$ can be calculated. Besides, if the n-point correlation function is known, the $(n - 1)$-point function does not follow, unlike in the case of densities. Therefore, the n-point probability function contains more information than even the *set* of correlation functions up to order n. If however n approaches infinity, the information content in the two cases becomes more and more equal.

One of the most prominent problems of the theory is to calculate expectation values from the given information. A sharp prediction is only possible if the full statistical information is known and can be processed. This is seldom the case. Having at our disposal only partial statistical information, the best that can be done is to calculate *bounds* of expectation values. It is clear that these bounds are closer when more information has been incorporated.

The foregoing discussion makes it clear that the optimum bounds obtainable from the n-point probability are closer than those obtained from the correlation functions up to order n. Those coming from correlation functions up to order $n + 1$, however, are closer than the first mentioned bounds. Whether one works with probability densities or with correlation functions will therefore be a matter of availability of information and of convenience. It is a fundamental task of the present theory to determine the P's or k's from the given information about the physical system. This task can be extremely difficult. Therefore, often certain models are introduced such as cell structures constructed from a Poisson point distribution by introducing plane cell walls perpendicular to the lines connecting neighbouring Poisson points (see e.g. Miller, 1969, and Benk, 1980). Since the geometrical probabilities are known for the Poisson case, one can (mostly approximately) calculate the desired expectation values, provided that the distribution of elastic moduli over the Poisson cells, also called Voronoi polyhedra, is simple, e.g. decoupled. In the decoupled case the mentioned factorization of the probability densities becomes effective. Besides, there is an important result concerning the k correlation functions in this case. In fact, using eqns (4.18)–(4.20) in order to calculate k_2 from eqns (4.23), we obtain, now with the tensor $\mathbf{c}$ of elastic moduli,

$$k_2(\mathbf{r}_1, \mathbf{r}_2) = \langle \mathbf{c}'^2 \rangle w_2(\mathbf{r}_1, \mathbf{r}_2), \qquad \mathbf{c}'^2 = \mathbf{c}'\mathbf{c}' \tag{4.24}$$

(both c's taken at the same point); the corresponding result for the 3-point correlations is

$$k_3(\mathbf{r}_1, \mathbf{r}_2, \mathbf{r}_3) = \langle \mathbf{c}'^3 \rangle w_3(\mathbf{r}_1, \mathbf{r}_2, \mathbf{r}_3), \qquad \mathbf{c}'^3 = \mathbf{c}'\mathbf{c}'\mathbf{c}' \tag{4.25}$$

and for the 4-point correlations

$$k_4(\mathbf{r}_1, \ldots, \mathbf{r}_4) = \langle \mathbf{c}'^4 \rangle w_4(\mathbf{r}_1, \ldots, \mathbf{r}_4) - 3\langle \mathbf{c}'^2 \rangle^2 (1 - w_4(\mathbf{r}_1, \ldots, \mathbf{r}_4)),$$
$$\mathbf{c}'^4 = \mathbf{c}'\mathbf{c}'\mathbf{c}'\mathbf{c}' \tag{4.26}$$

$\langle \mathbf{c}'^n \rangle$ is called the nth moment of $\mathbf{c}'$ and is constant here. It is easy to express also the higher correlation functions in terms of moments and coincidence probabilities. Only P_1 is needed in order to calculate the moments $\langle \mathbf{c}'^n \rangle$. This means that the main complexity of the problem (see below) arises from the purely geometrical coincidence probabilities. This statement is true only if the geometrical and material statistics are decoupled in the above sense.

Correlation functions are often accessible to experimental measurements. Imagine some device with two electronic feelers that, in fixed relative position $\mathbf{r}_{12}$, scan the polished surface of a polycrystal and, at each moment, register the grain orientation at the points of contact. A connected computer can almost instantly calculate the associated product $\mathbf{c}_1\mathbf{c}_2$ and the average $\langle \mathbf{c}_1\mathbf{c}_2 \rangle$ for the scanning of the whole surface. If the aggregate is statistically homogeneous and if such measurements are repeated for varying $\mathbf{r}_{12}$ and also for other surface orientations, the correlation function $K_2(\mathbf{r}_{12})$ can be found. The same methods work also in the case of 3-point correlations, because three points always lie in a plane. This is not so with four points. However, in many cases it is satisfactory to work with correlations up to order 3 since these contain the most relevant statistical information. Devices based on the ideas described above are in use at the Institut de Recherches de la Sidérurgie Française (IRSID), however with emphasis on phase mixtures rather than on polycrystals (Hersant *et al.,* 1975). Based on original work of Matheron (1967), Serra (1969, 1982) and others, a new branch of science has been developed called 'mathematical morphology' which among others is concerned with the calculation of correlations from geometric-statistical information about the material (see e.g. Jeulin, 1979). It was shown by Benk (1980, Poisson cell materials), Babos (1982, composites with Poisson-distributed circular-cylindrical inclusions) and Chassapis (1983, Poisson-distributed spherical inclusions) that the method works well and allows the calculation of correlation functions up to order 3.

Poisson distributions are of special interest in this context because they describe a very disordered situation. The Poisson cell arrangement is, however, not *perfectly* disordered because the cell walls are plane which implies a principle of order. The plane character of these walls can be detected only by correlation functions of at least 4th order because four is the minimum number of points that do not necessarily lie in a plane. If the grain boundaries are mainly plane, one expects a peak of the 4-point correlation function if the four points $\mathbf{r}_1$, $\mathbf{r}_2$, $\mathbf{r}_3$ and $\mathbf{r}_4$ lie in a plane.

These considerations now show that the correlation functions are not so abstract as might appear when they are first introduced by the formal set of equations (4.21). They are indispensable tools in the statistical theory of heterogeneous and other physical systems. With their description we have completed our introduction to the concepts and notions of statistics that are fundamental not only for the problem of elastic polycrystals but also for many other problems of statistical physics.

5. GREEN'S FUNCTIONS IN ELASTICITY THEORY

In this chapter the interest is in a combination of elasticity and probability theory which together result in a theory of statistical elasticity. So far we have provided the tools of probability theory. Those of elasticity theory are assumed as known to the reader, except one tool of extreme importance. This is the *Green's function,* which appears in our theory in a perhaps not so familiar form. Almost all relevant results of the statistical theory of polycrystal elasticity were obtained using the method of Green's functions. This, however, does not mean that other methods will not be developed in the future. In fact, any method that is helpful in solving boundary value problems should be a possible tool also for statistical problems of elasticity. In this section, the concept and notion of Green's functions is explained corresponding to the needs of this chapter.

5.1. Displacement Green's Function

In classical elasticity theory the Green's function, for our purpose also called the *displacement Green's function,* is introduced in order to write down formal solutions of the two basic boundary value problems (BVP's) of elasticity theory. In the first, or Dirichlet type, boundary value problem the displacement is prescribed over the surface S of the body of volume V, here assumed as simply connected for simplicity. In the second, or Neumann type, boundary value problem, the surface tractions rather than surface displacements are prescribed. There exists also the mixed problem where displacements are given over part of the surface and tractions over the other part. In this introductory treatment we can ignore this complication which would not contribute much to our insight into the statistical aspects of the problem. The equations to be solved for the two boundary value problems are

$$\begin{aligned} D_{ik}(\mathbf{r}, \nabla)\, u_k(\mathbf{r}) &= f_i(\mathbf{r}), && \mathbf{r} \text{ in } V \text{ (both BVP's)} \\ u_i(\mathbf{r}) &= g_i(\mathbf{r}), && \mathbf{r} \text{ on } S \text{ (1st BVP)} \\ n_j(\mathbf{r})\, c_{ijkl}(\mathbf{r})\, \partial_l u_k(\mathbf{r}) &= h_i(\mathbf{r}), && \mathbf{r} \text{ on } S \text{ (2nd BVP)} \end{aligned} \tag{5.1}$$

Here

$$D_{ik}(\mathbf{r}, \nabla) = -\partial_j c_{ijkl}(\mathbf{r})\, \partial_l \tag{5.2}$$

is the elastic differential operator; n_j is the external unit vector normal to the surface, u_k is the required displacement field, f_i represents the external forces per unit volume, h_i those per unit area of the surface and g_i the surface displacements. f_i, g_i and h_i are taken as given.

Consider now two special boundary value problems, namely those when $f_i(\mathbf{r})$ represents a point force of unit strength acting at a point $\mathbf{r}'$, say, in V and when the surface displacements are zero (1st BVP) or the surface tractions are zero except for a part needed to equilibrate the point force (2nd BVP). Write $f_i(\mathbf{r}) = P_m(\mathbf{r})\delta(\mathbf{r} - \mathbf{r}')\delta_{im}$ with the delta function $\delta(\mathbf{r} - \mathbf{r}')$. The solution of this problem can be represented as

$$u_i(\mathbf{r}) = \gamma_{im}(\mathbf{r}, \mathbf{r}')\, P_m(\mathbf{r}') \tag{5.3}$$

where $\gamma_{im}(\mathbf{r}, \mathbf{r}')$ is the Green's function of the chosen boundary value problem. Where an explicit distinction of the two Green's functions is essential, we shall indicate the difference by writing $\gamma^{\rm I}$ and $\gamma^{\rm II}$. Since P_m was assumed to be a unit force, γ_{im} can, for a given i and m, be interpreted as the ith component of the displacement field due to a point force of unit strength acting at point $\mathbf{r}'$ in the m-direction. This interpretation also explains the chosen name *displacement* Green's function. It follows from eqns (5.1) by insertion of the special f_i and from eqn (5.3) that γ_{im} satisfies the equations

$$D_{ik}(\mathbf{r},\nabla)\,\gamma^{\rm I,II}_{km}(\mathbf{r},\mathbf{r}') = \delta(\mathbf{r}-\mathbf{r}')\,\delta_{im}, \qquad \mathbf{r},\mathbf{r}' \text{ in } V \tag{5.4}$$

valid for both boundary value problems, and

$$\gamma^{\rm I}_{im}(\mathbf{r},\mathbf{r}') = 0, \qquad \mathbf{r} \text{ on } S,\ \mathbf{r}' \text{ in } V \tag{5.5}$$

for the first boundary value problem and

$$n_j(\mathbf{r})c_{ijkl}(\mathbf{r})\,\partial_l\gamma^{\rm II}_{km}(\mathbf{r},\mathbf{r}') = -\delta_{im}/S, \qquad \mathbf{r} \text{ on } S,\ \mathbf{r}' \text{ in } V \tag{5.6}$$

for the second boundary value problem. S is the area of the surface. The term $-\delta_{im}/S$ is added on the right-hand side of eqn (5.6) because the body would not be in equilibrium under the action of the point force alone. If eqn (5.6) is integrated over the whole surface, a force equal in size but opposite in sign to P_m results. For details see textbooks on elasticity theory, e.g. Solomon (1968). It is also stated that the displacement field arising from a distribution of volume forces and surface displacements is

$$u_i(\mathbf{r}) = \int_V \gamma^{\rm I}_{ik}(\mathbf{r},\mathbf{r}')f_k(\mathbf{r}')\,dV' + \int_S n_l(\mathbf{r}')c_{klmn}(\mathbf{r}')\,[\partial_{n'}\,\gamma_{mi}(\mathbf{r},\mathbf{r}')]g_k(\mathbf{r}')\,dS' \tag{5.7}$$

and, if instead of surface displacements surface tractions are prescribed,

$$u_i(\mathbf{r}) = \int_V \gamma^{\rm II}_{ik}(\mathbf{r},\mathbf{r}')f_k(\mathbf{r}')\,dV' + \int_S \gamma^{\rm II}_{ik}(\mathbf{r},\mathbf{r}')\,h_k(\mathbf{r}')\,dS' \tag{5.8}$$

Equations (5.7) and (5.8) are formal solutions of the two boundary value problems; 'formal', because finding the two Green's functions is often extremely difficult. From now on we shall consider the second boundary value problem only and, for convenience, omit the superscript.

5.2. Modified (Strain and Stress) Green's Functions

If the stress equilibrium conditions

$$-\partial_i\sigma_{ij} = f_j, \qquad n_i\sigma_{ij} = h_j \tag{5.9}$$

are inserted in eqn (5.8) and the strain

$$\varepsilon_{ij} = (\partial_i u_j + \partial_j u_i)/2 \tag{5.10}$$

is formed, a partial integration gives

$$\varepsilon_{ij}(\mathbf{r}) = \int_V \Gamma_{ijkl}(\mathbf{r}, \mathbf{r}')\, \sigma_{kl}(\mathbf{r}')\, \mathrm{d}V' \tag{5.11}$$

where

$$\Gamma_{ijkl}(\mathbf{r}, \mathbf{r}') \equiv \partial_j \partial_{l'} \gamma_{ik}(\mathbf{r}, \mathbf{r}')|_{(ij)(kl)} \tag{5.12}$$

(ij) and (kl) denote symmetrization in the subscripts concerned. The interesting feature of eqn (5.11) is that it gives the same strain field for different stress fields, provided that these satisfy eqns (5.9). In this respect, eqn (5.11) is different from the inverse of Hooke's law

$$\varepsilon_{ij}(\mathbf{r}) = s_{ijkl}(\mathbf{r})\, \sigma_{kl}(\mathbf{r}) = \int_V s_{ijkl}(\mathbf{r}')\, \delta(\mathbf{r} - \mathbf{r}')\, \mathrm{d}V' \tag{5.13}$$

where we have introduced the integral representation in order to compare (5.13) with eqn (5.11). Since equations of this type occur frequently, it is convenient to use an abridged notation (see Section 1.2). To this end we note that eqns (5.11) and (5.13) are convolutions. Introducing the symbol $*$ for the operation and also suppressing the tensor indices, we rewrite these equations as

$$\varepsilon_{\mathrm{a}} = \hat{\mathbf{\Gamma}} * \sigma \tag{5.14}$$

$$\varepsilon = \hat{\mathbf{s}} * \sigma \tag{5.15}$$

The suffix 'a' added to ε in eqn (5.14) will be explained below. $\hat{\mathbf{\Gamma}}$ and $\hat{\mathbf{s}}$ have the status of integral operators with kernels $\Gamma_{ijkl}(\mathbf{r}, \mathbf{r}')$ and $s_{ijkl}(\mathbf{r}')\delta(\mathbf{r}, \mathbf{r}')$ respectively. Both are self-adjoint operators, e.g.

$$\Gamma_{ijkl}(\mathbf{r}, \mathbf{r}') = \Gamma_{klij}(\mathbf{r}', \mathbf{r}) \tag{5.16}$$

Equation (5.16) follows, because γ_{ik} is self-adjoint:†

$$\gamma_{ik}(\mathbf{r}, \mathbf{r}') = \gamma_{ki}(\mathbf{r}', \mathbf{r}) \tag{5.17}$$

†Not only is γ self-adjoint, it is also symmetric in i, k and in $\mathbf{r}, \mathbf{r}'$, because the governing equations (5.4)–(5.6) have this symmetry. The latter symmetries are lost if these equations are replaced by those of non-local (linear) elasticity theory. In that case the compliance operator $\hat{\mathbf{s}}$ has no longer the simple kernel as above but is, like $\hat{\mathbf{\Gamma}}$, a more general function of $\mathbf{r}$, and $\mathbf{r}'$. Equations (5.14) and (5.15) remain valid. $\hat{\mathbf{\Gamma}}$ and $\hat{\mathbf{s}}$ are still self-adjoint, a property that is important in variational principles.

The difference between eqns (5.14) and (5.15) is of basic importance. The point is that stresses can exist in the absence of external forces, i.e. when f_i and h_j in eqn (5.9) vanish. Such stresses are called *internal* or *eigen* stresses and the strains are found from them by application of the inverse Hooke's law.

Equation (5.15) includes incompatible strains which are not derivable from a displacement field. In other words, neither eqn (5.10) nor eqn (5.14) which was derived by use of (5.10) is appropriate for the situation of incompatible strain, whereas eqn (5.15) is equally applicable for compatible *and* incompatible strain.† With eqn (5.15) also the direct Hooke's law

$$\boldsymbol{\sigma} = \hat{\mathbf{c}} * \boldsymbol{\varepsilon} \tag{5.18}$$

applies to compatible and incompatible strain.

Let us introduce the decomposition

$$\boldsymbol{\varepsilon} = \boldsymbol{\varepsilon}_{\mathrm{a}} + \boldsymbol{\varepsilon}_{\mathrm{b}} \tag{5.19}$$

of the resulting elastic strain $\boldsymbol{\varepsilon}$ into compatible strain $\boldsymbol{\varepsilon}_{\mathrm{a}}$ and incompatible strain $\boldsymbol{\varepsilon}_{\mathrm{b}}$. With eqn (5.18), eqn (5.14) becomes

$$\boldsymbol{\varepsilon}_{\mathrm{a}} = \hat{\boldsymbol{\Gamma}} * \hat{\mathbf{c}} * \boldsymbol{\varepsilon} \tag{5.20}$$

It follows from our derivation that $\boldsymbol{\varepsilon}_{\mathrm{a}}$ vanishes if $\boldsymbol{\varepsilon}$ on the right-hand side is purely incompatible. In other words, $\hat{\boldsymbol{\Gamma}} * \hat{\mathbf{c}}$ acts as a projection operator and annihilates incompatible strain. Projection operators of this kind have been investigated by Kunin (1964, 1983) and by Simmons and Bullough (1970).

Insert now eqn (5.18) into eqns (5.14) and (5.15); let $\hat{\mathbf{c}}$ operate on both equations and substract the upper from the lower one. Using also eqn (5.19) the result is

$$\boldsymbol{\sigma}_{\mathrm{b}} = \hat{\mathbf{c}} * \boldsymbol{\varepsilon}_{\mathrm{b}} = \hat{\mathbf{c}} * (\hat{\mathbf{s}} - \hat{\boldsymbol{\Gamma}}) * \hat{\mathbf{c}} * \boldsymbol{\varepsilon} \tag{5.21}$$

or

$$\boldsymbol{\sigma}_{\mathrm{b}} = \hat{\boldsymbol{\Delta}} * \boldsymbol{\varepsilon} \tag{5.22}$$

where

$$\hat{\boldsymbol{\Delta}} \equiv \hat{\mathbf{c}} - \hat{\mathbf{c}} * \hat{\boldsymbol{\Gamma}} * \hat{\mathbf{c}} \tag{5.23}$$

†The argument for this is: Whether a strain is compatible or not is decided by its derivatives rather than by the local strain. Derivatives do not occur in Hooke's law.

is known together with $\hat{\mathbf{\Gamma}}$; σ_b describes the internal stress state. From eqns (5.18) and (5.22) it follows that $\mathbf{\Delta} * \hat{\mathrm{s}}$ has the status of a projection operator which annihilates 'load' stresses $\boldsymbol{\sigma}_a = \boldsymbol{\sigma} - \boldsymbol{\sigma}_b$.

Both $\hat{\mathbf{\Gamma}}$ and $\hat{\mathbf{\Delta}}$ are fundamental quantities of the theory. In particular, $\hat{\mathbf{\Gamma}}$ will be used to find upper bounds to effective elastic moduli, whereas lower bounds will be derived by means of $\hat{\mathbf{\Delta}}$. $\hat{\mathbf{\Gamma}}$ and $\hat{\mathbf{\Delta}}$ have been called *strain* and *stress Green's functions* by Gubernatis and Krumhansl (1975).

There are interesting results also for the analogous modified Green's functions of the first boundary value problem, $\hat{\mathbf{\Gamma}}^{\mathrm{I}}$ and $\hat{\mathbf{\Delta}}^{\mathrm{I}}$ (see Kröner, 1974). Here we mention only a set of relations that hold true if the considered (locally) elastic medium is homogeneous with tensors of moduli **C** and compliances **S**:†

$$\int_V \Gamma^{\mathrm{I}}_{ijkl}(\mathbf{r}, \mathbf{r}')\,\mathrm{d}V' = S_{ijkl}, \qquad \int_V \Gamma^{\mathrm{II}}_{ijkl}(\mathbf{r}, \mathbf{r}')\,\mathrm{d}V' = 0 \tag{5.24}$$

$$\int_V \Delta^{\mathrm{I}}_{ijkl}(\mathbf{r}, \mathbf{r}')\,\mathrm{d}V' = 0, \qquad \int_V \Delta^{\mathrm{II}}_{ijkl}(\mathbf{r}, \mathbf{r}')\,\mathrm{d}V' = C_{ijkl} \tag{5.25}$$

The first integral, for instance, is verified by taking $\boldsymbol{\sigma}$ = constant in eqns (5.14) and (5.15). Then the related strain $\boldsymbol{\varepsilon}$ is compatible, equal to $\boldsymbol{\varepsilon}_a$. Integration over V of eqns (5.14) and (5.15) now leads to the same result, i.e. to the first eqn (5.24).

The modified Green's functions have interesting implications also in ordinary elasticity theory. For instance, it is often easy to find stresses that satisfy the stress equations (5.9). If these stresses are inserted in eqn (5.11), the (second) boundary value problem is solved by the resulting strains. For further applications see Section 5.4.

5.3. Some Explicit Results on Green's Functions of Homogeneous Media

Both ordinary and modified Green's functions have singularities at $\mathbf{r} = \mathbf{r}'$. Although some caution is necessary, eqns (5.24) and (5.25) show that the singularities are relatively well behaved. In the homogeneous infinite medium $\boldsymbol{\gamma}$ depends on r and $\mathbf{r}'$ in the combination $\mathbf{r} - \mathbf{r}'$. This suggests representing $\boldsymbol{\gamma}$, now denoted by $\boldsymbol{\gamma}^{\infty}$, as a Fourier integral (method of Zeilon, 1910/11). It is easily verified that the function

†Capital letters **C** and **S** stand for elastic tensors that refer to homogeneous media.

$$\gamma_{ik}^{\infty}(\mathbf{r}-\mathbf{r}') = -\frac{1}{8\pi^3}\int_{V_\mathbf{k}} D_{ik}^{-1}(\mathbf{k})\, e^{i\mathbf{k}\cdot(\mathbf{r}-\mathbf{r}')} \mathrm{d}V_\mathbf{k} \tag{5.26}$$

with

$$D_{ik}(\mathbf{k}) = -C_{ijkl}k_jk_l \tag{5.27}$$

satisfies eqn (5.4) [cf. eqn (5.2)]. The integration in eqn (5.26) extends over the (infinite) volume $V_\mathbf{k}$ of the $\mathbf{k}$ (Fourier) space. For isotropic media the elastic tensor has the form

$$C_{ijkl} = \lambda\delta_{ij}\delta_{kl} + 2\mu I_{ijkl} \tag{5.28}$$

with λ,μ as Lamé's constants and I_{ijkl} as defined in eqn (4.16). The integration in eqn (5.26) for the isotropic case yields

$$\gamma_{ij}^{\infty}(\mathbf{r}-\mathbf{r}') = \frac{1}{8\pi\mu}\left(\frac{\lambda+3\mu}{\lambda+2\mu}\delta_{ij} + \frac{\lambda+\mu}{\lambda+2\mu}e_ie_j\right)\frac{1}{|\mathbf{r}-\mathbf{r}'|} \tag{5.29}$$

or, if λ,μ are replaced by the set κ,μ (compression and shear modulus),

$$\gamma_{ij}^{\infty}(\mathbf{r}-\mathbf{r}') = \frac{1}{8\pi\mu}\left(\frac{3\kappa+7\mu}{3\kappa+4\mu}\delta_{ij} + \frac{3\kappa+\mu}{3\kappa+4\mu}e_ie_j\right)\frac{1}{|\mathbf{r}-\mathbf{r}'|} \tag{5.30}$$

where e_i are the Cartesian components of the unit vector $\mathbf{e}$ pointing from $\mathbf{r}'$ to $\mathbf{r}$.

In the case of elastic anisotropy, the integration in (5.26) cannot be performed in closed form, except at transverse isotropy. Many methods have been developed in order to find approximations. Useful information about this is found in the monographs of Bacon *et al.* (1978) and of Mura (1982).

If Γ_{ijkl} is formed in analogy to the prescription (5.12) and the result integrated over the infinite volume V_∞, one obtains with eqn (5.26)

$$\begin{aligned} E_{ijkl} &\equiv \int_{V_\infty} \Gamma_{ijkl}^{\infty}(\mathbf{r}-\mathbf{r}')\,\mathrm{d}V' \\ &= \frac{1}{8\pi^3}\int_{V_\infty} \mathrm{d}V'\, e^{i\mathbf{k}\cdot(\mathbf{r}-\mathbf{r}')}\int_{V_\mathbf{k}} \mathrm{d}V_\mathbf{k}\,[D_{jl}^{-1}(\mathbf{k})\,k_ik_k]_{(ij)(kl)} \end{aligned} \tag{5.31}$$

The first integral on the right-hand side is equal to $8\pi^3\delta(\mathbf{k}) = 2\pi^2\delta(k)/k^2$, so that

$$E_{ijkl} = \frac{1}{4\pi}\int_0^\infty \mathrm{d}k\,\delta(k)\int_0^\pi \mathrm{d}\Theta\,\sin\Theta\int_0^{2\pi}\mathrm{d}\Phi\,[D_{jl}^{-1}(\mathbf{k})\,k_ik_k]_{(ij)(kl)} \tag{5.32}$$

Here k, Θ, Φ are the spherical coordinates of the $\mathbf{k}$ space, and the expression in the brackets does not depend on k [see eqn (5.27)]. Thus

$$E_{ijkl} = \frac{1}{4\pi}\int_0^{\pi} \mathrm{d}\Theta \sin\Theta \int_0^{2\pi} \mathrm{d}\Phi \, [D_{jl}^{-1}(\mathbf{k})\, k_i k_k]_{(ij)(kl)} \qquad (5.33)$$

For the isotropic case the integration yields

$$E_{ijkl} = \frac{1}{15\mu}\left(-\frac{3\kappa+\mu}{3\kappa+4\mu}\delta_{ij}\delta_{kl} + 9\frac{\kappa+2\mu}{3\kappa+4\mu}I_{ijkl}\right) \qquad (5.34)$$

Obviously this is not equal to S_{ijkl} as could have been surmised because the integrals in (5.24) and (5.25) are shape-independent. The point is that the Green's function (5.30) does not cover the case σ = constant used in the derivation of (5.24). Thus eqn (5.30) is the correct Green's function only for states that decay to infinity. The Green's function which includes also stress and strain states that are constant to infinity is obtained from that of finite media by a limiting process.

The 'completed' Green's function thus obtained does not depend on the shape of the medium and obeys eqns (5.24). The first statement is obvious from the derivation of these equations. The reader may convince himself of the correctness of the second statement by using the Green's function of a sphere (see e.g. Solomon, 1968). He can then easily check that the addition to $\hat{\Gamma}^{\infty}$ is $(\hat{\mathrm{S}} - \hat{\mathrm{E}})/V$ as it should be to obtain a shape-independent result.

We mention in passing that the integral in eqns (5.26) can be found in closed form for hexagonal (= transversely isotropic) symmetry as well, but not for other crystal symmetries (Lifshitz and Rosentsveig, 1947; Kröner, 1953).†

The modified Green's function $\hat{\Gamma}^{\infty}$, without the mentioned addition, is a central quantity of the theory. It is instructive to consider it also as a tensor in Hilbert space with unity $\delta(\mathbf{r}-\mathbf{r}')$. From this view the integral in the first eqn (5.24) is the trace of $\hat{\Gamma}^{\mathrm{I}}$. Decomposing $\hat{\Gamma}^{\infty}$ in trace and deviator gives

$$\hat{\Gamma}^{\infty} = \hat{\mathrm{E}} + \hat{\Phi} \qquad (5.35)$$

where $\hat{\mathrm{E}}$ has the kernel $E_{ijkl}\delta(\mathbf{r}-\mathbf{r}')$ and

$$\int_{V_\infty} \Phi_{ijkl}(\mathbf{r}-\mathbf{r}')\,\mathrm{d}V' = 0 \qquad (5.36)$$

†Kröner's result is somewhat more explicit. The right-hand side of his formula [eqn (6)] has to be multiplied by $-1/c_{11}c_{44}c_{66}$.

These equations are true both for isotropic and for anisotropic media. For the latter, E_{ijkl} does not have the simple form (5.34). In the isotropic case, $\hat{\mathbf{\Gamma}}^{\infty}$ is easily calculated from γ^{∞}; the kernel of $\hat{\mathbf{\Phi}}$ follows in view of eqn (5.35) (Kröner and Koch, 1976; Mazilu, 1981):

$$\Phi_{ijkl}(\mathbf{r}-\mathbf{r}') = \frac{1}{8\pi\mu(3\kappa+4\mu)} \cdot \frac{1}{|\mathbf{r}-\mathbf{r}'|^3}\Big[6\mu I_{ijkl} - (3\kappa+\mu) \times (\delta_{ij}\delta_{kl} - 3e_ie_j\delta_{kl} - 3e_ke_l\delta_{ij} - 15e_ie_je_ke_l) + 6(3\kappa-2\mu)(e_ie_k\delta_{jl})_{(ij)(kl)}\Big] \tag{5.37}$$

This result shows that the kernel of $\hat{\mathbf{\Phi}}$ is of the form

$$\Phi_{ijkl}(\mathbf{r}-\mathbf{r}') = \frac{1}{|\mathbf{r}-\mathbf{r}'|^3}\phi_{ijkl}(\theta,\phi) \tag{5.38}$$

where θ,ϕ are the directional angles of the, already mentioned, vector $\mathbf{e} = (\mathbf{r}-\mathbf{r}')/|\mathbf{r}-\mathbf{r}'|$. It follows from the homogeneity in the k's of the tensor $\mathbf{D}$, eqn (5.27), that the form (5.38) is valid also for anisotropic media.

5.4. Relation to the Elastic Inclusion Problem

The tensor E_{ijkl} is closely related to a tensor denoted by S_{ijkl} in Eshelby's (1957) work on ellipsoidal inclusions. To see this, cut out a simply connected, not necessarily ellipsoidal, region (called an *inclusion*) from, in general, an anisotropic infinite stress-free elastic medium. Give the inclusion a constant stress-free strain, say ε^*_{kl}. For brevity, ε^*_{kl} will be called the *misfit tensor* or simply *misfit*. Bring the inclusion back to its original form by an elastic strain, insert it in its former position, glue it along the interface and then relax. The total strain from the initial to the final state shall be denoted by $\varepsilon^{\mathrm{T}}_{ij}$ and named *repression* because it is a measure of how far the environment of the inclusion is pushed away by the non-fitting inclusion. According to Eshelby, $\varepsilon^{\mathrm{T}}_{ij}$ is constant in the ellipsoid and related to ε^*_{kl} by

$$\varepsilon^{\mathrm{T}}_{ij} = S^{\mathrm{Esh}}_{ijkl}\,\varepsilon^*_{kl} \tag{5.39}$$

The stress follows from

$$\sigma_{ij} = C_{ijkl}(\varepsilon^{\mathrm{T}}_{kl} - \varepsilon^*_{kl}) \tag{5.40}$$

Thus the tensor $\mathbf{S}^{\mathrm{Esh}}$ (not the inverse of $\mathbf{C}$!) solves the elastic inclusion problem for the ellipsoid. As shown by Eshelby this solution also allows us easily to find the solution of another type of inclusion problem in which the inserted ellipsoid has elastic moduli different from those of the matrix.

From the results of the preceding Section 5.3 the inclusion problem can be solved as follows. $\varepsilon_{ij}^{\mathrm{T}}$ is the elastic relaxation strain from the state immediately after reinsertion and glueing, now named the unrelaxed state, to the final relaxed state. To maintain the unrelaxed state, interface tractions must be applied from outside. They follow from the first (!) eqn (5.9), if σ_{ij} there is taken as $-C_{ijkl}\varepsilon_{kl}^{*}$ inside the inclusion and zero outside. Recall that the second eqn (5.9) refers to the *external* surface where $h_j = 0$.

To permit the relaxation process, interface tractions that annihilate the first ones must be applied. We are interested in the elastic strain produced by these tractions. Since they are equal in size but opposite in sign to those initially applied, we conclude that stresses that are $+C_{ijkl}\varepsilon_{kl}^{*}$ in the inclusion and zero outside satisfy the stress equilibrium conditions (5.9). With the corresponding interpretation of eqn (5.11) we obtain

$$\varepsilon_{ij}^{\mathrm{T}}(\mathbf{r}) = \int_{\mathrm{incl}} \Gamma_{ijkl}^{\infty}(\mathbf{r}-\mathbf{r}')\,\mathrm{d}V'\,C_{klmn}\varepsilon_{mn}^{*} \tag{5.41}$$

where the integration extends over the volume of the inclusion, because only there does ε_{mn}^{*} not vanish. For this reason, the integral is not equal to that of (5.31) but will, in general, depend on $\mathbf{r}$. Averaging eqn (5.41) over the inclusion volume gives

$$\overline{\varepsilon_{ij}^{\mathrm{T}}}^{\,\mathrm{incl}} = \frac{1}{V_{\mathrm{incl}}}\iint_{\mathrm{incl}} \Gamma_{ijkl}^{\infty}(\mathbf{r}-\mathbf{r}')\,\mathrm{d}V\,\mathrm{d}V'\,C_{klmn}\,\varepsilon_{mn}^{*} \tag{5.42}$$

We know from Eshelby's work that for ellipsoidal inclusions $\varepsilon_{ij}^{\mathrm{T}}$ is constant inside the inclusions so that for these the average bar in eqn (5.42) can be omitted. By comparison with eqn (5.39) we find

$$S_{ijkl}^{\mathrm{Esh}} = \frac{1}{V_{\mathrm{incl}}}\iint_{\mathrm{incl}} \Gamma_{ijkl}^{\infty}(\mathbf{r}-\mathbf{r}')\,\mathrm{d}V\,\mathrm{d}V'\,C_{ijkl} \tag{5.43}$$

Thus the ellipsoidal inclusion problem is solved by the double integral (5.43). This integral can also be applied to non-ellipsoidal inclusions; it then gives the average total strain. Reproducing unpublished work of the present author, Kneer (1964, 1965) gave the following more explicit expression which follows by use of eqn (5.26):†

† In Kneer's work the parentheses (ij), (kl), meaning symmetrization, are lacking owing to a misprint.

$$\frac{1}{V_{\text{incl}}}\iint\limits_{\text{incl}} \Gamma^{\infty}_{ijkl}(\mathbf{r}-\mathbf{r}')\,\mathrm{d}V\,\mathrm{d}V' = \int_0^{\pi} \mathrm{d}\Theta \sin\Theta$$

$$\int_0^{2\pi} \times\, \mathrm{d}\Phi\, D^{-1}_{jl}(\mathbf{k})\,k_i\,k_k]_{(ij)(kl)} A(\Theta,\Phi) \qquad (5.44)$$

$$A(\Theta,\Phi) = \frac{1}{8\pi^3 V_{\text{incl}}}\int_0^{\infty} \mathrm{d}k\, k^2 \iint\limits_{\text{incl}} \mathrm{d}V\,\mathrm{d}V'\, \mathrm{e}^{i\mathbf{k}\cdot(\mathbf{r}-\mathbf{r}')} \qquad (5.45)$$

Obviously A is a geometrical factor independent of the elastic moduli. For a sphere, $A = 1/4\pi$, so that by comparison with eqn (5.33)

$$\frac{1}{V_{\text{sphere}}}\iint\limits_{\text{sphere}} \Gamma^{\infty}_{ijkl}(\mathbf{r}-\mathbf{r}')\,\mathrm{d}V\,\mathrm{d}V' = E_{ijkl} \qquad (5.46)$$

This equation establishes the relationship between the general theory of the Green's function and that of the inclusion problem. In fact, we can now interpret the tensor $E_{ijkl}C_{klmn}$ as the repression of a misfit of unit strength in the spherical inclusion problem.

A computer program for the double integral has been given by Ghahremani (1977). For other treatments and, partly, more general theorems of the inclusion problem see e.g. Kunin and Sosnina (1971), Kinoshita and Mura (1971), Walpole (1977) and Korringa *et al.* (1978).

6. STATISTICAL THEORY OF EFFECTIVE ELASTIC MODULI

6.1. Effective Elastic Moduli

In the preceding sections those tools of probability and elasticity theory have been supplied that are indispensable for the treatment of probabilistic problems of elasticity theory. Up to now the most extensively investigated such problem is that of finding effective elastic moduli of macrohomogeneous polycrystals, given the moduli of the crystallites and some statistical information about the arrangement of the grains. The effective elastic moduli have the status of expectations. We mention in passing a recent proposal by Pompe *et al.* (1979) to treat our problem in the language of information theory. Although this might become an interesting alternative, we shall not discuss it here. It is unlikely that this approach will become a serious competitor of the statistical approach treated here.

The concept of effective elastic moduli is very old, in fact of the last century. Obviously one cannot obtain information as to whether some specimen is polycrystalline or homogeneous by purely phenomenological measurements on its elasticity. The results of such measurements are at best local volume (or surface) averages of stress and strain which owing to the ergodic hypothesis are equal to ensemble averages. It is then possible to write a stress–strain law in the form

$$\langle \sigma_{ij} \rangle = \langle c_{ijkl} \varepsilon_{kl} \rangle \equiv C^{\text{eff}}_{ijkl} \langle \varepsilon_{kl} \rangle \tag{6.1}$$

where C^{eff}_{ijkl} form the 4th rank tensor of effective elastic moduli defined by eqn (6.1). These do not depend on position in the macrohomogeneous situation considered here, contrary to σ_{ij}, ε_{kl} which may depend on $\mathbf{r}$ and c_{ijkl} which always depends on $\mathbf{r}$. As we shall see, it is convenient to replace the tensor $\mathbf{C}^{\text{eff}}$ by the related *operator* $\widehat{\mathbf{C}}^{\text{eff}}$ with kernel $C^{\text{eff}}_{ijkl}\,\delta(\mathbf{r} - \mathbf{r}')$. Equation (6.1) then reads, in our abridged notation,

$$\langle \boldsymbol{\sigma} \rangle = \langle \mathbf{c} * \boldsymbol{\varepsilon} \rangle = \widehat{\mathbf{C}} * \langle \boldsymbol{\varepsilon} \rangle \tag{6.2}$$

Let us return to eqn (5.8) and apply it to the heterogeneous polycrystal. Take the ensemble average and interchange averaging with integrating. Provided that the external forces are not correlated with the Green's function, which implies that they are not correlated with the distribution of elastic moduli in the aggregate, the average of the products is equal to the product of the averages. Introducing eqns (5.9) and (5.10) and observing that averaging commutes also with differentiation, one obtains, in analogy to eqn (5.14),

$$\langle \boldsymbol{\varepsilon}_{\text{a}} \rangle = \langle \widehat{\boldsymbol{\Gamma}} \rangle * \langle \boldsymbol{\sigma} \rangle \tag{6.3}$$

Comparing this and the inverse of the elasticity law, namely

$$\langle \boldsymbol{\varepsilon} \rangle = \widehat{\mathbf{S}}^{\text{eff}} * \langle \boldsymbol{\sigma} \rangle, \qquad \widehat{\mathbf{S}}^{\text{eff}} \equiv (\widehat{\mathbf{C}}^{\text{eff}})^{-1} \tag{6.4}$$

with eqns (5.14) and (5.15) shows that $\langle \widehat{\boldsymbol{\Gamma}} \rangle$ is the modified Green's function of a homogeneous medium with elastic operator $\widehat{\mathbf{C}}^{\text{eff}}$. This is a first fundamental result of our theory. It is valid under the assumption that the external forces are applied in a manner uncorrelated with the distribution of $\widehat{\mathbf{c}}$.

What is the physical implication of this limitation? It is clear that there are no correlations of the mentioned type if the external forces vary smoothly in the volume and on the surface of the polycrystal. This will be the case in many practical applications. Volume forces that are clearly correlated with elastic moduli are gravitational forces acting on some composite material in which, unlike in polycrystals, the mass density varies from one component to another. Inertia forces arising for instance during wave propagation depend on density and velocity and are therefore correlated with **c** both in composites and in polycrystals. This observation explains why dynamical problems of statistical elasticity are so much more complex than static ones.

No particular problem arises, however, if smooth surface displacements are applied to the polycrystal, although the surface forces h_j needed in order to produce these displacements are correlated with $\widehat{\mathbf{c}}$. To see the point,† decompose h_j into mean $\langle h_j \rangle$ and deviation h_j' in eqn (5.8). The question is then whether or not h_j contributes to the mean displacement, i.e. whether

$$\langle u_i(\mathbf{r}) \rangle = \int_S \langle \gamma_{ik}^{\mathrm{II}}(\mathbf{r}, \mathbf{r}')\, h_k'(\mathbf{r}') \rangle \,\mathrm{d}S' \tag{6.5}$$

vanishes or not. Obviously the brackets define a sort of correlation from point **r** to point **r**′. Assume that the range of these correlations is rather short, say of order *d*, where *d* might be, for instance, the mean grain diameter. Then for any **r** whose distance from the surface is greater than *d*, the brackets can be factorized so that $\langle h_k' \rangle$ (= 0) appears and the integral vanishes. The same procedure is not possible if **r** is within a boundary layer of thickness less than *d*. If the ratio d/l, with l a macrolength, is sufficiently small, we can ignore the boundary layer. We then need not distinguish between a strain state produced by smooth surface displacements and smooth surface tractions. If, however, the mentioned correlations extend over larger (finite) distances, the boundary layer becomes thicker and the situation elastically non-local on the macro scale, even when it is local on the micro scale. This observation offers an interesting way to construct elastic media, which are non-local over larger distances. So far, the statistical theory has not been applied much to such situations, in which $\widehat{\mathbf{C}}^{\mathrm{eff}}$ would have a kernel that does not depend on **r** and **r**′ like a delta function (see, however, Beran and McCoy, 1970; Levin, 1971; Mazilu, 1972; Diener *et al.*, 1981). In this chapter we restrict ourselves to situations that are local also on the macro scale and to those in which the external forces are uncorrelated with $\widehat{\mathbf{c}}$. The problem of 'correlated' forces in local elasticity theory is treated in Section 7.1.

†The following arguments are related to those given by Mandel and Dantu (1963).

6.2. The Formal Solution

Since we are interested in macro experiments we exclude also external forces that fluctuate on a small scale, though uncorrelatedly with $\hat{\mathbf{c}}$. In other words, we consider (static) boundary value problems in which volume forces as well as surface displacements or surface tractions are prescribed as smooth functions of position. At the end of Section 5.2 we explained how the 2nd boundary value problem can be solved by means of the modified Green's function $\hat{\mathbf{\Gamma}}$. The argument can immediately be applied to the present situation: if $\langle\hat{\mathbf{\Gamma}}\rangle$ and any $\langle\boldsymbol{\sigma}\rangle$ satisfying the averaged equilibrium conditions are known, the problem is solved by eqn (6.3).

Proceeding in this way may often not be the most convenient method. If the $\hat{\mathbf{C}}^{\text{eff}}$ belonging to $\langle\hat{\mathbf{\Gamma}}\rangle$ were known, one could utilize all methods developed for boundary value problems of homogeneous bodies. To find $\hat{\mathbf{C}}^{\text{eff}}$ the equation analogous to the first eqn (5.24) can be used:

$$S^{\text{eff}}_{ijkl} = \int_{V_\infty} \langle \Gamma_{ijkl}(\mathbf{r}-\mathbf{r}')\rangle \, \mathrm{d}V' \tag{6.6}$$

Here we need the 'completed' modified Green's function of the infinite medium as described in Section 5.3. Although this is probably a good procedure, it has become common to derive $\hat{\mathbf{C}}^{\text{eff}}$ in a different way. To this end we introduce the concept of a homogeneous comparison medium with operator $\hat{\mathbf{C}}^\circ$ and modified Green's function $\hat{\mathbf{\Gamma}}^\circ$. This medium shall have the same shape as the real medium, which is the polycrystal with operator $\hat{\mathbf{c}}$ and Green's function $\hat{\mathbf{\Gamma}}$. Both $\hat{\mathbf{\Gamma}} * \hat{\mathbf{c}}$ and $\hat{\mathbf{\Gamma}}^\circ * \hat{\mathbf{C}}^\circ$ are projection operators. Therefore we deduce from eqn (5.20) that

$$\boldsymbol{\varepsilon}_{\text{a}} \equiv \hat{\mathbf{\Gamma}}^\circ * \hat{\mathbf{C}}^\circ * \boldsymbol{\varepsilon}, \qquad \boldsymbol{\varepsilon}_{\text{a}} = \hat{\mathbf{\Gamma}} * \hat{\mathbf{c}} * \boldsymbol{\varepsilon} = \hat{\mathbf{\Gamma}}^\circ * \hat{\mathbf{C}}^\circ * \hat{\mathbf{\Gamma}} * \hat{\mathbf{c}} * \boldsymbol{\varepsilon} \tag{6.7}$$

Because this is true for any $\boldsymbol{\varepsilon}$, we have

$$\hat{\mathbf{\Gamma}} * \hat{\mathbf{c}} = \hat{\mathbf{\Gamma}}^\circ * \hat{\mathbf{C}}^\circ * \hat{\mathbf{\Gamma}} * \hat{\mathbf{c}} \tag{6.8}$$

Since $\hat{\mathbf{c}}$ (not $\hat{\mathbf{\Gamma}}$!) has an inverse we can cancel $\hat{\mathbf{c}}$ so that

$$\hat{\mathbf{\Gamma}} = \hat{\mathbf{\Gamma}}^\circ * \hat{\mathbf{C}}^\circ * \hat{\mathbf{\Gamma}} \tag{6.9}$$

This integral equation applies to the case $\hat{\mathbf{\Gamma}}^\circ = \hat{\mathbf{\Gamma}}$ too, so any modified Green's function obeys the equation

$$\hat{\mathbf{\Gamma}} = \hat{\mathbf{\Gamma}} * \hat{\mathbf{c}} * \hat{\mathbf{\Gamma}} \tag{6.10}$$

Since $\hat{\Gamma}$ has no inverse it is not possible to cancel one $\hat{\Gamma}$ on either side so that $\hat{\Gamma}$ is not the operator inverse to $\hat{\mathbf{c}}$, as it would be if $\hat{\Gamma}$ were a true Green's function. Interchanging the role of $\hat{\mathbf{c}}$ and $\hat{\mathbf{C}}^\circ$ in eqn (6.9) yields

$$\hat{\Gamma}^\circ = \hat{\Gamma} * \hat{\mathbf{c}} * \hat{\Gamma}^\circ = \hat{\Gamma}^\circ * \hat{\mathbf{c}} * \hat{\Gamma} \tag{6.11}$$

The latter equation holds because the $\hat{\mathbf{c}}$'s and $\hat{\Gamma}$'s are self-adjoint. Subtracting eqn (6.11) from (6.9) leads to

$$\hat{\Gamma} - \hat{\Gamma}^\circ = -\hat{\Gamma}^\circ * \delta\hat{\mathbf{c}}^\circ * \hat{\Gamma}, \qquad \delta\hat{\mathbf{c}}^\circ \equiv \mathbf{c} - \mathbf{C}^\circ \tag{6.12}$$

or

$$\hat{\Gamma} = \hat{\mathbf{A}}^\circ * \hat{\Gamma}^\circ, \qquad \hat{\mathbf{A}}^\circ \equiv (\hat{\mathbf{I}} + \hat{\Gamma}^\circ * \delta\hat{\mathbf{c}}^\circ)^{-1} \tag{6.13}$$

Equation (6.12) is an integral equation which allows us to calculate the usually complex modified Green's function $\hat{\Gamma}$ in terms of the simpler function $\hat{\Gamma}^\circ$ of the comparison medium, given the operators $\hat{\mathbf{c}}$ and $\hat{\mathbf{C}}^\circ$ of the elastic moduli. Equations similar to (6.12) are known in quantum mechanical scattering theory as the Lippmann–Schwinger (also Dyson) equation. We shall therefore refer to it as the LSD equation. It was derived in the present context by Kröner (1967, slightly erroneously) and, independently, by Eimer (1967, 1968), Zeller and Dederichs (1973) and Korringa (1973). Its great importance lies in the possibility for calculating complicated Green's functions in terms of simpler ones. The concept underlying eqn (6.12) has been applied to general heterogeneous media by Rieder (1962).

Equation (6.13) represents the formal solution of the LSD equation. Here the operator $\hat{\mathbf{A}}^\circ$ is introduced formally as the inverse of the operator $\hat{\mathbf{I}} + \hat{\Gamma}^\circ * \delta\hat{\mathbf{c}}^\circ$. This inverse is explained as the related formal geometric progression and therefore has a meaning only if this series converges. This problem has been investigated by Gambin and Kröner (1981) with the result that for polycrystals $\hat{\mathbf{C}}^\circ$ can always be chosen such that the series converges. In composites the convergence breaks down if the elastic moduli of one of the constituents are zero (porous media) or infinite. Related convergence theorems are due to Mazilu (1976).

Let $\boldsymbol{\varepsilon}^\circ$ be the strain field produced in the comparison medium by those external forces that cause the strain field $\boldsymbol{\varepsilon}$ in the polycrystal. We have $\boldsymbol{\varepsilon} = \hat{\Gamma} * \boldsymbol{\sigma}$, $\boldsymbol{\varepsilon}^\circ = \hat{\Gamma}^\circ * \boldsymbol{\sigma}$ (why?) so that also

$$\boldsymbol{\varepsilon} = \hat{\mathbf{A}}^\circ * \boldsymbol{\varepsilon}^\circ \tag{6.14}$$

holds true. If this is inserted into the equation (6.2) defining $\hat{\mathbf{C}}^{\text{eff}}$ we obtain, after cancelling $\boldsymbol{\varepsilon}^\circ$,

$$\hat{\mathbf{C}}^{\text{eff}} = \langle \mathbf{c} * \hat{\mathbf{A}}^\circ \rangle * \langle \hat{\mathbf{A}}^\circ \rangle^{-1} \tag{6.15}$$

$\hat{\mathbf{A}}^\circ$ is defined in eqn (6.13). Equation (6.15), first given by Eimer (1967, 1968), represents the formal solution of the problem posed initially, namely to calculate the effective elastic moduli, given the moduli of the crystallites and some statistical information about their distribution.

If $\hat{\mathbf{A}}^\circ$ in eqn (6.15) is written in the form of the mentioned geometric progression, then also $\hat{\mathbf{C}}^{\text{eff}}$ appears as a series, which can be brought into the following form (Kröner, 1977):

$$\begin{aligned}\hat{\mathbf{C}}^{\text{eff}} = {} & \hat{\mathbf{C}}^\circ + \langle \delta\hat{\mathbf{c}}^\circ \rangle - \langle \hat{\mathbf{c}}' * \hat{\mathbf{\Gamma}}^\circ * \hat{\mathbf{c}}' \rangle + \langle \hat{\mathbf{c}}' * \hat{\mathbf{\Gamma}}^\circ * \delta\hat{\mathbf{c}} * \hat{\mathbf{\Gamma}}^\circ * \hat{\mathbf{c}}' \rangle \\ & - \langle (\hat{\mathbf{c}}' * \hat{\mathbf{\Gamma}}^\circ * \delta\hat{\mathbf{c}}^\circ)' * \hat{\mathbf{\Gamma}}^\circ * (\delta\hat{\mathbf{c}}^\circ * \hat{\mathbf{\Gamma}}^\circ * \hat{\mathbf{c}}')' \rangle \\ & + \langle (\hat{\mathbf{c}}' * \hat{\mathbf{\Gamma}}^\circ * \delta\hat{\mathbf{c}}^\circ)' * \hat{\mathbf{\Gamma}}^\circ * \delta\hat{\mathbf{c}}^\circ * \hat{\mathbf{\Gamma}}^\circ * (\delta\hat{\mathbf{c}}^\circ * \hat{\mathbf{\Gamma}}^\circ * \hat{\mathbf{c}}')' \rangle \ldots\end{aligned} \tag{6.16}$$

The primes, including those applied to parentheses, denote deviations from the mean. Therefore each $\hat{\mathbf{\Gamma}}^\circ$ is involved in at least one convolution with a deviation. We discussed in the preceding Section 6.1 that in the case of macrolocal elasticity $\hat{\mathbf{C}}^{\text{eff}}$ is independent of the shape and size of the body which therefore can be taken as infinite. It was also clear that the 'completed' Green's function of the infinite medium must be taken if $\hat{\mathbf{C}}^{\text{eff}}$ is calculated via eqn (6.6). This is not so if eqn (6.16) is used. In fact, it was argued in Section 5 that the 'completing part' of the modified Green's function is constant so that its convolution with a deviation vanishes. For this reason we can use the (non-completed) Green's function of the infinite medium in eqn (6.16). This has also been shown by Dederichs and Zeller (1972) by a different argument. If the polycrystal is macroisotropic, we can take the modified Green's function given by eqns (5.35), (5.34) and (5.37).

6.3. Bounds for Effective Elastic Moduli

Equation (6.16) represents the formal solution of our problem in terms of a series of multiple integrals of increasing order. In fact, it is closely related to the Neumann series which solves the LSD equation. As is evident, eqn (6.16) is also a series in terms of correlation functions in $\hat{\mathbf{c}}$ of increasing order. This becomes clear if $\delta\hat{\mathbf{c}}^\circ$ is replaced by

$$\delta\hat{\mathbf{c}}^\circ = \hat{\mathbf{c}} - \hat{\mathbf{C}}^\circ = \hat{\mathbf{c}}' + \delta\hat{\mathbf{C}}^\circ, \qquad \delta\hat{\mathbf{C}}^\circ = \langle \hat{\mathbf{c}} \rangle - \hat{\mathbf{C}}^\circ \tag{6.17}$$

Both $\delta\widehat{\mathbf{c}}^{\circ}$ and $\widehat{\boldsymbol{\Gamma}}^{\circ}$ are smooth functions which can be set outside the averaging brackets so that the correlation functions remain. The more of these that are known the further one can go in the processing of the integrations. This suggests that bounds to effective moduli can be obtained if the series in eqn (6.16) is truncated after a term of order n. Obviously only the terms up to order 5 have been written explicitly in eqn (6.16). There exist two general bound theorems in this relation; one due to Dederichs and Zeller (1973) yields the bounds of odd order, the other due to Kröner (1977) those of even order. Let us describe briefly how such bound theorems can be established.

For simplicity, consider the set of upper bounds alone. As in other parts of physics, such bounds are derived from a variational principle, in our case that of minimum elastic energy (Beran and Molyneux, 1966; Dederichs and Zeller, 1973; for other useful formulations see also Hashin, 1965, and Willis, 1983).

In our present notation, the classical minimum energy principle states that for the equilibrium state $\boldsymbol{\varepsilon}$ of an elastic body, here taken as any realization of the real polycrystal,

$$\delta W \equiv \delta(\boldsymbol{\varepsilon} * \widehat{\mathbf{c}} * \boldsymbol{\varepsilon} - 2\boldsymbol{\sigma}_0 * \boldsymbol{\varepsilon}) = 0, \qquad \operatorname{div} \boldsymbol{\sigma}_0 + \mathbf{f} = 0 \quad (6.18)$$

where $\boldsymbol{\sigma}_0$ satisfies the traction boundary conditions. In the form here given with the *operator* $\widehat{\mathbf{c}}$ the principle is valid even for non-local elasticity. The variation is to be performed in such a way that prescribed displacements are not varied† and that with $\boldsymbol{\varepsilon}$ also $\boldsymbol{\varepsilon} + \delta\boldsymbol{\varepsilon}$ is a compatible strain. If eqn (6.18) is averaged over the ensemble we obtain the form

$$\delta\langle W\rangle = \delta(\langle \boldsymbol{\varepsilon} * \widehat{\mathbf{c}} * \boldsymbol{\varepsilon}\rangle - 2\boldsymbol{\sigma}_0 * \langle \boldsymbol{\varepsilon}\rangle) = 0 \qquad (6.19)$$

of the principle which now refers to the whole ensemble. Note that $\langle \delta W\rangle = \delta\langle W\rangle$. Decomposing $\boldsymbol{\varepsilon}$ into mean $\langle\boldsymbol{\varepsilon}\rangle$ and deviation $\boldsymbol{\varepsilon}'$ we have $\delta\boldsymbol{\varepsilon} = \delta\langle\boldsymbol{\varepsilon}\rangle + \delta\boldsymbol{\varepsilon}'$ where $\langle\delta\boldsymbol{\varepsilon}'\rangle = 0$. Since $\langle\boldsymbol{\varepsilon}\rangle$ and $\boldsymbol{\varepsilon}'$ are independent we can rewrite eqn (6.19) as

$$\frac{\delta\langle W\rangle}{\delta\langle\boldsymbol{\varepsilon}\rangle} \cdot\cdot\; \delta\langle\boldsymbol{\varepsilon}\rangle + \frac{\delta\langle W\rangle}{\delta\boldsymbol{\varepsilon}'} \cdot\cdot\; \delta\boldsymbol{\varepsilon}' = 0 \qquad (6.20)$$

†This condition will not be considered any further, i.e. we assume for simplicity that only tractions are described on the surface. This is no restriction as far as the calculation of local effective moduli is concerned. In fact, these are independent of the particularly chosen boundary value problem. From our discussion on boundary layers we expect that the situation is different if the effective elastic moduli are non-local.

where we have used the variational derivatives. The principle (6.18) is satisfied if simultaneously

$$\frac{\delta\langle W\rangle}{\delta\langle\varepsilon\rangle} = 0, \qquad \frac{\delta\langle W\rangle}{\delta\varepsilon'} = 0 \tag{6.21}$$

With $\delta\langle\varepsilon\rangle = 0$ in the second eqn (6.21) this becomes equivalent to the variational principle

$$\delta\langle\varepsilon * \hat{c} * \varepsilon\rangle = 0, \qquad \delta\langle\varepsilon\rangle = 0 \tag{6.22}$$

because the second term in eqn (6.19) now drops out. Equation (6.22) implies

$$\langle\varepsilon * \hat{c} * \varepsilon\rangle \leqslant \langle\varepsilon^* * \hat{c} * \varepsilon^*\rangle, \qquad \langle\varepsilon^*\rangle = \langle\varepsilon\rangle \tag{6.23}$$

where the trial function ε^* is a compatible strain. But

$$\langle\varepsilon * \hat{c} * \varepsilon\rangle = \langle\varepsilon\rangle * \hat{C}^{\text{eff}} * \langle\varepsilon\rangle \tag{6.24}$$

as follows from the definition (6.1) provided that the external forces are not correlated with the moduli. Thus we obtain Dederichs and Zeller's (1973) variational inequality, there derived by different reasoning:

$$\langle\varepsilon\rangle * \hat{C} * \langle\varepsilon\rangle \leqslant \langle\varepsilon^* * \hat{c} * \varepsilon^*\rangle, \qquad \langle\varepsilon^*\rangle = \langle\varepsilon\rangle \tag{6.25}$$

A comparison of eqns (6.1) and (6.24) shows that

$$\langle\varepsilon * \hat{c} * \varepsilon\rangle = \langle\varepsilon\rangle * \langle\hat{c} * \varepsilon\rangle \tag{6.26}$$

This is sometimes called the condition of consistency, because it ensures that eqns (6.1) and (6.24) can be used as alternative definitions of $\hat{C}^{\text{eff}}$. Obviously this possibility is a prerequisite that the effective medium, described by $\hat{C}^{\text{eff}}$, behaves like an ordinary elastic medium.

Whereas the principle (6.25) leads to upper bounds for $\hat{C}^{\text{eff}}$, upper bounds for $\hat{S}^{\text{eff}} = (\hat{C}^{\text{eff}})^{-1}$ can be derived from the principle of complementary energy. By analogous reasoning one obtains

$$\langle\sigma\rangle * \hat{S}^{\text{eff}} * \langle\sigma\rangle \leqslant \langle\sigma^* * \hat{s} * \sigma^*\rangle, \langle\sigma^*\rangle = \langle\sigma\rangle \tag{6.27}$$

Observe the complete symmetry in the formulations leading to moduli and compliances. This is the reason that we can restrict ourselves to treating the moduli problem alone.

The two conditions on the trial function ε^*, namely that it is a compatible strain with $\langle\varepsilon^*\rangle = \langle\varepsilon\rangle$, are satisfied by the *ansatz*

$$\varepsilon^* = (\hat{I} + P\hat{\Gamma}^\circ * \hat{T}) * \langle\varepsilon\rangle \equiv \hat{\Sigma} * \varepsilon \tag{6.28}$$

where T is an arbitrary operator and P is the so-called deviation (projection) operator introduced by Beran and McCoy (1970). It produces the deviation of any quantity to its right-hand side (the adjoint $P^\dagger$ does the same to any quantity to its left-hand side). Thus the presence of P guarantees that $\langle \boldsymbol{\varepsilon}^* \rangle = \langle \boldsymbol{\varepsilon} \rangle$. $\hat{\boldsymbol{\Gamma}}^\circ$ transforms $\hat{\mathbf{T}} * \langle \varepsilon \rangle$ into a compatible strain [cf. eqn (5.20)] so that eqn (6.28) satisfies all conditions on the trial strain. Since $\hat{\mathbf{T}}$ is arbitrary, eqn (6.28) is the most general trial function. It is convenient and non-restrictive to use the modified Green's function $\hat{\boldsymbol{\Gamma}}^\circ$ of a homogeneous comparison medium. With eqn (6.28) the inequality (6.25) reduces to

$$\langle \boldsymbol{\varepsilon} * \mathbf{C}^{\text{eff}} * \boldsymbol{\varepsilon} \rangle \leqslant \langle \boldsymbol{\varepsilon} \rangle * \hat{\boldsymbol{\Sigma}}^\dagger * \mathbf{c} * \hat{\boldsymbol{\Sigma}} * \langle \boldsymbol{\varepsilon} \rangle \tag{6.29}$$

where $\hat{\boldsymbol{\Sigma}}^\dagger$ is the adjoint to $\hat{\boldsymbol{\Sigma}}$. The inequality (6.29) is equivalent to the operator inequality

$$\hat{\mathbf{C}}^{\text{eff}} \leqslant \langle \hat{\boldsymbol{\Sigma}} * \hat{\mathbf{c}} * \boldsymbol{\Sigma} \rangle \tag{6.30}$$

which was used by Dederichs and Zeller (1973) and by Kröner (1977) in order to derive the mentioned bound theorems.

Before utilizing the last result let us say a word on the first eqn (6.21). With eqn (6.24), eqn (6.19) becomes

$$\delta(\langle \boldsymbol{\varepsilon} \rangle * \hat{\mathbf{C}}^{\text{eff}} * \langle \boldsymbol{\varepsilon} \rangle - 2\sigma_0 * \langle \boldsymbol{\varepsilon} \rangle) = 0 \tag{6.31}$$

where the variation goes with respect to $\langle \boldsymbol{\varepsilon} \rangle$. Thus eqn (6.31) is the variational principle of the effective medium with already determined $\hat{\mathbf{C}}^{\text{eff}}$. It can be used to solve boundary value problems of this medium. This, however, is not a topic of the present chapter.

Since inequalities of the kind (6.30) occur frequently in connexion with variational principles we explain what is meant if an operator or tensor is called 'smaller' than another one. Consider two 4th rank tensors **a** and **b** with the symmetry of the elasticity tensor. We write symbolically $\mathbf{a} < \mathbf{b}$ or $\mathbf{a} - \mathbf{b} < 0$, or also $\hat{\mathbf{a}} - \hat{\mathbf{b}} < 0$, if '**a** is smaller than **b**'. By this we mean that all quadratic forms of $\mathbf{a} - \mathbf{b}$ are negative-definite. These forms can be written in Voigt's notations, e.g. as $\Sigma(a_{ij} - b_{ij})\varepsilon_i\varepsilon_j$ $(i, j = 1, 2, \ldots, 6)$. It is then easily derived that $\mathbf{a} \leqslant \mathbf{b}$ implies

$$a_{ii} \leqslant b_{ii}, \qquad a_{ii}b_{jj} - a_{ij}^2 \leqslant b_{ii}b_{jj} - b_{ij}^2 \qquad \text{(no summation)} \tag{6.32}$$

The first 6 relations above are found if the special $\varepsilon_i \neq 0$, $\varepsilon_{j \neq i} = 0$ is examined, whereas the second 15 relations follow from the consideration of the case $\varepsilon_i \neq 0$, $\varepsilon_j \neq 0$, $\varepsilon_k = 0$ $(i \neq k \neq j)$. Equations (6.32) are 21 relations, in accord with the number of independent components of the most general tensor of elastic moduli.

A particularly useful set of trial functions is obtained in the form

$$\hat{\mathbf{I}} - P\hat{\mathbf{\Gamma}}^\circ * \delta\hat{\mathbf{c}}^\circ + (P\hat{\mathbf{\Gamma}}^\circ * \delta\hat{\mathbf{c}}^\circ)^2 \ldots + (-1)^m (P\hat{\mathbf{\Gamma}}^\circ * \delta\hat{\mathbf{c}}^\circ)^m \tag{6.33}$$

if, for example,

$$\hat{\mathbf{a}}^2 \equiv \hat{\mathbf{a}} * \hat{\mathbf{a}}, \qquad \hat{\mathbf{a}}^3 \equiv \hat{\mathbf{a}} * \hat{\mathbf{a}} * \hat{\mathbf{a}} \qquad \text{etc.}$$

$$P\hat{\mathbf{\Gamma}}^\circ * \delta\hat{\mathbf{c}}^\circ * P\hat{\mathbf{\Gamma}}^\circ * \delta\hat{\mathbf{c}}^\circ = P(\hat{\mathbf{\Gamma}}^\circ * \delta\hat{\mathbf{c}}^\circ * P\hat{\mathbf{\Gamma}}^\circ * \delta\hat{\mathbf{c}}^\circ) \qquad \text{etc.}$$

It is shown by Kröner (1977) that the trial function (6.33) gives the correct $\hat{\mathbf{C}}^{\text{eff}}$ in the limit $m \to \infty$ for any choice of (homogeneous) $\hat{\mathbf{C}}^\circ$ provided that the series converges. For finite m it reproduces the series (6.16) up to the term of order $n = 2m + 1$. Thus one obtains upper bounds of odd order n, if the series (6.16) is truncated after any term of odd order. This is the bound theorem of Dederichs and Zeller (1973). We denote these bounds by a superscript $+n'$, $\hat{\mathbf{C}}^{(+n')}$, where the prime is added in order to distinguish these bounds from the optimum bounds to be discussed shortly. The + sign indicates that it is an *upper* bound.

To all bounds of the operators correspond bounds of the tensors. The bound $\mathbf{C}^{(+1')}$ is that of Voigt. In fact,

$$\mathbf{C}^{(+1')} = \mathbf{C}^\circ + \langle \delta\mathbf{c}^\circ \rangle = \mathbf{C}^\circ + \langle \mathbf{c} - \mathbf{C}^\circ \rangle = \langle \mathbf{c} \rangle \tag{6.34}$$

The bound of Reuss is

$$\mathbf{C}^{(-1')} = (\mathbf{S}^{(+1')})^{-1} = \langle \mathbf{s} \rangle^{-1} \tag{6.35}$$

The bound theorem of Kröner (1977) is a little bit more delicate. Upper bounds of even order are obtained by truncating the series (6.16) provided that $\mathbf{C}^\circ$ is chosen as $\mathbf{C}^{(+0)}$, the upper bound of the 0th order, or larger. Since the upper bounds shift toward higher values if $\mathbf{C}^\circ$ is taken larger than $\mathbf{C}^{(+0)}$, we shall always use $\mathbf{C}^\circ = \mathbf{C}^{(+0)}$ for the upper bounds of even order. The existence of a bound (upper and lower) of 0th order has to do with the fact that the average of any random function lies between its maximum and minimum value. In our case we write

$$\mathbf{C}^{(+0)} \geqslant \mathbf{c} \geqslant \mathbf{C}^{(-0)} \equiv (\mathbf{S}^{(+0)})^{-1} \tag{6.36}$$

where $\mathbf{C}^{(+0)}$ and $\mathbf{C}^{(-0)}$ are defined as the largest and smallest quantities for which (6.36) is true.

The calculation of $\mathbf{C}^{(\pm 0)}$, given $\mathbf{c}$, is not difficult if one looks for isotropic tensors. For instance, consider a macroisotropic aggregate of cubic crystallites with main constants κ, μ, ν. As mentioned in Section 4.1, κ does not depend on orientation. μ and ν are the two extreme values which the shear moduli can assume. This can easily be verified if the tensor transformation law for cubic symmetry is written down. Thus $\mathbf{C}^{(+0)}$ is the larger, $\mathbf{C}^{(-0)}$ the smaller of the two isotropic tensors defined by the pairs of constants (κ, μ) and (κ, ν). Of course, the lower bounds of even order for $\mathbf{C}^{\text{eff}}$ are obtained from the $\hat{\mathrm{s}}$ series, analogous to (6.16), by use of $\mathbf{S}^{(+0)} = (\mathbf{C}^{(-0)})^{-1}$.

6.4. Optimum Bounds

In the preceding Section 6.3 we have discussed sets of bounds that become closer and closer the more correlation functions are included. If all of them are included and the relevant multiple integrals calculated, one obtains coinciding bounds, equal to the true expectation of the effective elastic moduli, provided that the series (6.16) and the corresponding series in $\hat{\mathrm{s}}$ converge. Otherwise the trial function taken from (6.33) still yields bounds but these diverge with increasing order.

The normal situation will be that one is content with bounds of low order, mostly those for $n \leqslant 3$. In this case the question arises of bounds which are optimum with respect to the restricted statistical information, here given in terms of correlation functions up to order n. Such bounds will be denoted by $\mathbf{C}^{(\pm n)}$. They are derived following a method which in the present context was originally proposed by Beran and Molyneux (1966) and later resumed by Dederichs and Zeller (1973). So far, the trial function leading to the nth order bound was taken from (6.33). However, this was not the most general trial function relating to correlation functions of order $\leqslant n$. The most general trial function of this type contains powers of $\hat{\mathbf{\Gamma}}^\circ * \hat{\mathrm{c}}$ up to $m = (n-1)/2$. It has the form

$$(\hat{\mathbf{I}} + P\hat{\mathbf{\Gamma}}^\circ * \hat{\mathrm{T}}) * \langle \varepsilon \rangle = \{\hat{\mathbf{I}} + P[\hat{\mathbf{\Gamma}}^\circ * \hat{\mathrm{c}} * \hat{\mathbf{\Lambda}}_1 + (\hat{\mathbf{\Gamma}}^\circ * \hat{\mathrm{c}})^2 * \hat{\mathbf{\Lambda}}_2 + \ldots + (\hat{\mathbf{\Gamma}}^\circ * \hat{\mathrm{c}})^m * \hat{\mathbf{\Lambda}}_m]\} * \langle \varepsilon \rangle \tag{6.37}$$

where the multipliers $\hat{\mathbf{\Lambda}}_\nu$ $(\nu = 1, 2, \ldots, m)$ are non-random because we consider a power series in the random functions. This trial function contains as a special case our former choice (6.33). Using (6.37) in the inequality (6.30) we determine the $\hat{\mathbf{\Lambda}}$'s so that $\langle \hat{\mathbf{\Sigma}} * \hat{\mathrm{c}} * \hat{\mathbf{\Sigma}} \rangle$ becomes a minimum. The lower bound $\mathbf{C}^{(-n)}$ is derived in an analogous way. The bounds $\mathbf{C}^{(\pm n)}$ are optimum because (6.37) is the most general function that contains the powers of $\hat{\mathbf{\Gamma}}^\circ * \hat{\mathrm{c}}$ up to the degree m, thus incorporating in an optimum manner the correlation functions up to order $n = 2m + 1$. This is so because the operators $\hat{\mathbf{\Lambda}}$ are determined in optimum fashion. They depend on the $\mathbf{C}^\circ$ chosen for the construction of $\mathbf{\Gamma}^\circ$, as we show using the example with $n = 3$. For this example set

$\hat{\boldsymbol{\Gamma}}^\circ * \hat{\mathbf{T}} = -\hat{\boldsymbol{\Gamma}}^\circ * \hat{\mathbf{c}} * \hat{\boldsymbol{\Lambda}}$. A simple calculation gives

$$\hat{\mathbf{C}}^{(+3)} = \langle\hat{\mathbf{c}}\rangle - \langle\hat{\mathbf{c}}' * \hat{\boldsymbol{\Gamma}}^\circ * \hat{\mathbf{c}}'\rangle * \hat{\boldsymbol{\Lambda}} \tag{6.38}$$

$$\begin{aligned}\hat{\boldsymbol{\Lambda}} &= \langle\hat{\mathbf{c}}' * \hat{\boldsymbol{\Gamma}}^\circ * \hat{\mathbf{c}} * \hat{\boldsymbol{\Gamma}}^\circ * \hat{\mathbf{c}}'\rangle^{-1} * \langle\hat{\mathbf{c}}' * \hat{\boldsymbol{\Gamma}}^\circ * \hat{\mathbf{c}}'\rangle \\ &= \langle\hat{\mathbf{c}}' * \hat{\boldsymbol{\Gamma}}^\circ * \hat{\mathbf{c}}' + \hat{\mathbf{c}}' * \hat{\boldsymbol{\Gamma}}^\circ * \delta\hat{\mathbf{c}}^\circ * \hat{\boldsymbol{\Gamma}}^\circ * \hat{\mathbf{c}}'\rangle^{-1} * \langle\hat{\mathbf{c}}' * \boldsymbol{\Gamma}^\circ * \hat{\mathbf{c}}'\rangle\end{aligned} \tag{6.39}$$

where we have used eqn (6.10) to derive the last equation. $\hat{\boldsymbol{\Lambda}}$ depends on $\hat{\mathbf{C}}^\circ$ in such a fashion that $\hat{\mathbf{C}}^{(+3)}$ is independent of $\hat{\mathbf{C}}^\circ$. In order to see this, change $\hat{\mathbf{C}}^\circ$, which occurs implicitly in eqns (6.38) and (6.39), by $\delta\hat{\mathbf{C}}^\circ$. Let $\delta\hat{\boldsymbol{\Gamma}}^\circ = \hat{\boldsymbol{\Gamma}} - \hat{\boldsymbol{\Gamma}}^\circ$ be the corresponding change of $\hat{\boldsymbol{\Gamma}}^\circ$. From the LSD equation (6.12) we obtain, for infinitesimal $\delta\hat{\boldsymbol{\Gamma}}^\circ$,

$$\delta\hat{\boldsymbol{\Gamma}}^\circ = -\hat{\boldsymbol{\Gamma}}^\circ * \delta\hat{\mathbf{C}}^\circ * \boldsymbol{\Gamma}^\circ \tag{6.40}$$

where $\hat{\boldsymbol{\Gamma}}^\circ$ is formed with $\hat{\mathbf{C}}^\circ$. Using this and the fact that $\hat{\boldsymbol{\Gamma}}^\circ$ and $\hat{\mathbf{c}}'$ are self-adjoint, it needs only a routine calculation to show that $\hat{\mathbf{C}}^{(+3)}$ is independent of $\hat{\mathbf{C}}^\circ$. This result is of great interest because it allows us to take a convenient choice of $\hat{\mathbf{C}}^\circ$. Such a choice is for instance $\hat{\mathbf{C}}^\circ = \langle\mathbf{c}\rangle$ which leads to $\delta\hat{\mathbf{c}} = \hat{\mathbf{c}}'$ in eqn (6.39) so that $\hat{\mathbf{C}}^{(+3)}$ appears composed of $\langle\hat{\mathbf{c}}\rangle$, $\langle\hat{\mathbf{c}}' * \hat{\boldsymbol{\Gamma}}^\circ * \hat{\mathbf{c}}'\rangle$ and $\langle\hat{\mathbf{c}}' * \hat{\boldsymbol{\Gamma}}^\circ * \hat{\mathbf{c}}' * \hat{\boldsymbol{\Gamma}}^\circ * \hat{\mathbf{c}}'\rangle$. Even more important is the observation that an isotropic $\hat{\mathbf{C}}^\circ$ can be chosen even if the polycrystal is statistically anisotropic (case of texture) when also $\langle\hat{\mathbf{c}}\rangle$ is anisotropic.

Also the optimum bound of 2nd order can be found with the *ansatz* $\hat{\boldsymbol{\Gamma}}^\circ * \hat{\mathbf{T}} = -\hat{\boldsymbol{\Gamma}}^\circ * \hat{\mathbf{c}} * \hat{\boldsymbol{\Lambda}}$ (Kröner, 1977). The result is

$$\mathbf{C}^{(+2)} = \langle\hat{\mathbf{c}}\rangle - \langle\hat{\mathbf{c}}' * \hat{\boldsymbol{\Gamma}}^\circ * \hat{\mathbf{c}}'\rangle, \qquad \hat{\boldsymbol{\Gamma}}^\circ \text{ formed with } \mathbf{C}^{(+0)} \tag{6.41}$$

This expression is not a bound if $\hat{\boldsymbol{\Gamma}}^\circ$ is formed with $\langle\hat{\mathbf{c}}\rangle$. Since we are interested only in macrolocal elasticity, the operators $\hat{\mathbf{C}}^{(\pm n)}$, $\langle\hat{\mathbf{c}}\rangle$, $\langle\hat{\mathbf{c}}' * \hat{\boldsymbol{\Gamma}}^\circ * \hat{\mathbf{c}}'\rangle$ and $\langle\hat{\mathbf{c}}' * \hat{\boldsymbol{\Gamma}}^\circ * \hat{\mathbf{c}}' * \hat{\boldsymbol{\Gamma}}^\circ * \hat{\mathbf{c}}'\rangle$ have a delta function kernel. It is then convenient to replace them by the corresponding 4th rank tensors. That means that $\hat{\mathbf{C}}$, $\hat{\mathbf{c}}$ and $\hat{\mathbf{c}}'$ are replaced by the relevant tensors $\mathbf{C}$, $\mathbf{c}$ and $\mathbf{c}'$. Only $\hat{\boldsymbol{\Gamma}}^\circ$ does not have a delta function kernel and therefore always implies an integration. Using the kernel $\Gamma^\circ_{ijkl}(\mathbf{r}, \mathbf{r}')$ of $\hat{\boldsymbol{\Gamma}}^\circ$ we find for the tensor representation of $\langle\hat{\mathbf{c}}' * \hat{\boldsymbol{\Gamma}}^\circ * \hat{\mathbf{c}}'\rangle$

$$\begin{aligned}&\left\langle c'_{ijmn}(\mathbf{r}_1) \int_{V_\infty} \Gamma^\circ_{mnpq}(\mathbf{r}_1 - \mathbf{r}_2)\, c'_{pqkl}(\mathbf{r}_2)\right\rangle \mathrm{d}V_2 \\ &= \int_{V_\infty} \Gamma^\circ_{mnpq}(\mathbf{r}_1 - \mathbf{r}_2) \left\langle c'_{ijmn}(\mathbf{r}_1)\, c'_{pqkl}(\mathbf{r}_2)\right\rangle \mathrm{d}V_2\end{aligned} \tag{6.42}$$

The 2-point correlation function appearing on the right-hand side depends only on $\mathbf{r}_1 - \mathbf{r}_2$ because we are dealing with a statistically homogeneous situation. This implies that the integral in eqn (6.42) is constant. Similarly, the tensor representation of $\langle \hat{\mathbf{c}}' * \hat{\mathbf{\Gamma}}^\circ * \hat{\mathbf{c}}' * \hat{\mathbf{\Gamma}}^\circ * \hat{\mathbf{c}}' \rangle$ is

$$\iint_{V_\infty} \Gamma^\circ_{mnpq}(\mathbf{r}_1 - \mathbf{r}_2)\, \Gamma^\circ_{rsuv}(\mathbf{r}_2 - \mathbf{r}_3) \langle c'_{ijmn}(\mathbf{r}_1)\, c'_{pqrs}(\mathbf{r}_2)\, c'_{uvkl}(\mathbf{r}_3)\rangle \, \mathrm{d}V_2\, \mathrm{d}V_3 \qquad (6.43)$$

Also this integral is constant. Equations (6.42) and (6.43) represent those integrals that must be calculated in order to obtain the optimum upper bounds of 2nd and 3rd order. If in these formulae Γ and c' are replaced by Δ and s', one obtains the optimum lower bounds of 2nd and 3rd order. Optimum bounds of order higher than 3 can be derived in an analogous manner. This, however, has not yet been done.

6.5. Perfect and Graded Disorder

The bounds discussed in the preceding Section 6.4 were optimum with respect to statistical information given in terms of correlation functions. As discussed in Section 4, it is also possible to provide statistical information in terms of probability densities P_n. 'P_n given' implies knowledge of all probability densities of order $\leqslant n$. Thus the problem arises that one has to find optimum bounds belonging to P_n. The derivation of these bounds is one of the unsolved problems of our theory. Leaving this aside we rewrite the general formal solution, eqn (6.15), in the form

$$\hat{\mathbf{C}}^{\mathrm{eff}} = \hat{\mathbf{C}}^\circ + \langle \delta\hat{\mathbf{c}} * \hat{\mathbf{A}}^\circ \rangle * \langle \hat{\mathbf{A}}^\circ \rangle^{-1}, \qquad \hat{\mathbf{A}}^\circ \equiv (\hat{\mathbf{I}} + \hat{\mathbf{\Gamma}}^\circ * \delta\hat{\mathbf{c}}^\circ)^{-1} \qquad (6.44)$$

Introduce the so-called transition operator $\hat{\mathbf{T}}$ by

$$\hat{\mathbf{T}} = \delta\hat{\mathbf{c}}^\circ * \mathbf{A}^\circ, \qquad \hat{\mathbf{I}} - \hat{\mathbf{\Gamma}}^\circ * \hat{\mathbf{T}} = \hat{\mathbf{A}}^\circ \qquad (6.45)$$

Using this, eqn (6.44) becomes

$$\hat{\mathbf{C}}^{\mathrm{eff}} = \hat{\mathbf{C}}^\circ + \langle \hat{\mathbf{T}} \rangle * (\hat{\mathbf{I}} - \hat{\mathbf{\Gamma}}^\circ * \hat{\mathbf{T}})^{-1} \qquad (6.46)$$

as first derived by Dederichs and Zeller (1973). Taking $\hat{\mathbf{\Gamma}}^\circ = \hat{\mathbf{E}} + \hat{\mathbf{\Phi}}$ [eqn (5.35)], Eberhardt (1980) and Gairola (1980) find, after a simple calculation,

$$\hat{\mathbf{T}} = \hat{\mathbf{t}} * (\hat{\mathbf{I}} + \hat{\mathbf{\Phi}} * \hat{\mathbf{t}})^{-1}, \qquad \hat{\mathbf{t}} \equiv \delta\hat{\mathbf{c}}^\circ * (\hat{\mathbf{I}} + \hat{\mathbf{E}} * \delta\hat{\mathbf{c}})^{-1} \qquad (6.47)$$

an important prerequisite for what follows.

Because the operator $\hat{\mathbf{E}}$ has a delta kernel, $\hat{\mathbf{t}}$ refers to one point and is therefore called a single-site transition operator. According to eqn (6.46) $\hat{\mathbf{C}}^{\text{eff}}$ is determined by $\hat{\mathbf{\Gamma}}^{\circ}$ and $\langle\hat{\mathbf{T}}\rangle$ where, using eqn (6.47),

$$\langle\hat{\mathbf{T}}\rangle = \langle\hat{\mathbf{t}}\rangle - \langle\hat{\mathbf{t}} * \hat{\mathbf{\Phi}} * \hat{\mathbf{t}}\rangle + \langle\hat{\mathbf{t}} * \hat{\mathbf{\Phi}} * \hat{\mathbf{t}} * \hat{\mathbf{\Phi}} * \hat{\mathbf{t}}\rangle \ldots \tag{6.48}$$

Since the choice of $\hat{\mathbf{C}}^{\circ}$ is arbitrary we can set it equal to $\hat{\mathbf{C}}^{\text{eff}}$ so that

$$\langle\hat{\mathbf{T}}\rangle = 0 \qquad (\text{for } \hat{\mathbf{C}}^{\circ} = \hat{\mathbf{C}}^{\text{eff}}) \tag{6.49}$$

Let us take this choice in eqn (6.48). We now ask if the distribution of elastic moduli in a polycrystal can be such that

$$\langle\mathbf{t}\rangle = \langle\delta\mathbf{c}^{\circ}(\mathbf{I} + \mathbf{E}\cdot\cdot\,\delta\mathbf{c})^{-1}\rangle = 0 \tag{6.50}$$

Here we have replaced the operator equation by the relevant tensor equation which is simple because all operators in eqn (6.48) have a delta kernel. Obviously no more integration occurs in eqn (6.50) which can be considered as an equation for the tensor $\mathbf{C}^{\circ}$. In this special application $\mathbf{C}^{\circ}$ is also called *self-consistent* and denoted by $\mathbf{C}^{\text{sc}}$. Solving eqn (6.50) for $\mathbf{C}^{\text{sc}}$ yields

$$\mathbf{C}^{\text{sc}} = \langle\mathbf{c}(\mathbf{I} + \mathbf{E}\cdot\cdot\,\delta\mathbf{c})^{-1}\rangle\cdot\cdot\langle(\mathbf{I} + \mathbf{E}\cdot\cdot\,\delta\mathbf{c})^{-1}\rangle^{-1}, \qquad \mathbf{C}^{\circ} = \mathbf{C}^{\text{sc}} \tag{6.51}$$

which also follows from eqn (6.44) if $\hat{\mathbf{\Gamma}}^{\circ}$ is there replaced by the local part of $\hat{\mathbf{\Gamma}}^{\circ}$. The self-consistent moduli were introduced by Hershey (1954) for aggregates of cubic crystallites and by Kröner (1958) for the general case; the form given here is close to a result of Walpole (1966). Equation (6.51) is also good for anisotropic situations (texture!).

We have not yet answered the question raised above. In fact, $\langle\mathbf{t}\rangle = 0$ implies $\mathbf{t} = \mathbf{t}'$, so that $\mathbf{C}^{\text{sc}}$ is a solution of our basic problem [cf. eqns (6.46) and (6.48)] if, and only if,

$$\langle\hat{\mathbf{t}}' * \hat{\mathbf{\Phi}} * \hat{\mathbf{t}}'\rangle - \langle\hat{\mathbf{t}}' * \hat{\mathbf{\Phi}} * \hat{\mathbf{t}}' * \hat{\mathbf{\Phi}} * \hat{\mathbf{t}}'\rangle \ldots = 0 \tag{6.52}$$

($\hat{\mathbf{\Phi}}$ formed with $\hat{\mathbf{C}}^{\text{sc}}$). Let us write $\langle\hat{\mathbf{t}}' * (\hat{\mathbf{\Phi}} * \hat{\mathbf{t}}')^{p}\rangle, p = n - 1$, for the term of nth order in eqn (6.52). This equation is satisfied if

$$\langle\hat{\mathbf{t}}' * (\hat{\mathbf{\Phi}} * \hat{\mathbf{t}}')^{p}\rangle = 0, \qquad p = 1, 2, \ldots, \infty \tag{6.53}$$

Note that all correlation functions of $\mathbf{t}'$ of order $n \geqslant 2$ appear in eqn (6.52). Together with $\langle\mathbf{t}\rangle$ they contain the same information as $\langle\mathbf{c}\rangle$ plus the correlation functions of $\mathbf{c}'$, that is the whole statistical information. It is likely, but has not been proved yet, that eqn (6.53) is the only realization of eqn (6.52).

Is it possible that eqn (6.52) is satisfied in a realistic situation? The following argument shows that it is. It follows, from the results of Hill (1952) and from our discussion of optimum bounds, that the bounds $\mathbf{C}^{(\pm 1)}$ of Voigt and Reuss are optimum. This implies that, depending on the grain statistics, effective elastic moduli can be anywhere between $\mathbf{C}^{(+1)}$ and $\mathbf{C}^{(-1)}$. These findings are discussed in more detail by Kröner (1980). In the range between $\mathbf{C}^{(+1)}$ and $\mathbf{C}^{(-1)}$ lies also $\mathbf{C}^{sc}$, for which $\mathbf{C}^{(+1)}$ and $\mathbf{C}^{(-1)}$ are upper and lower bounds. In fact, if eqn (6.51) is written in the form of the series (6.16) with $\hat{\mathbf{\Gamma}}$ replaced with $\hat{\mathbf{E}}$, then $\hat{\mathbf{C}}^{sc} \leqslant \hat{\mathbf{C}}^{(+1)}$ in view of the convergence of this series. Similarly $\hat{\mathbf{C}}^{sc} \geqslant \hat{\mathbf{C}}^{(-1)}$. This means that aggregates can exist for which eqn (6.52) is strictly satisfied. In this case, only the autocorrelation part (coinciding points $\mathbf{r}_1$, $\mathbf{r}_2$, etc.) of the correlation functions contributes. In other words, correlations of finite range do not exist in such an aggregate which therefore has the highest imaginable degree of disorder. In fact, its constitution does not reveal any principle of order and is therefore called *perfectly disordered.* It is necessarily statistically decoupled (Section 4.3) but has in addition the property that no distinguished grain shape exists. In fact, we had argued before that, for instance, lengthy grains in statistical orientation can be detected by 3-point correlation functions. Such correlations would give a contribution to the left-hand side of eqn (6.52).

The preceding discussion shows that the self-consistent elastic moduli can represent the rigorous solution of the statistical problem, namely when the aggregate is perfectly disordered in the described sense. Experimental tests have shown good agreement in all cases where the polycrystals were prepared with the intention of making them as disordered as possible. Obviously perfect disorder is a 'white noise' type of situation, encountered in other parts of statistical physics. Self-consistent moduli are often less good in the case of composites. For instance, matrix-inclusion types of composites are never perfectly disordered because already the statement that one phase is the matrix whereas the other forms the inclusions implies a principle of order.

It is interesting to recall that the self-consistent moduli were introduced originally almost without reference to statistics. It was assumed that the average grain shape is a sphere, i.e. no direction in the aggregate is distinguished by the grain statistics. A model was then set up in which the average grain, the sphere, was considered as an inclusion with lattice-orientation-dependent elastic moduli, embedded in an environment described as homogeneously elastic with moduli equal to the effective elastic moduli. Self-consistence comes in by the requirement that the apparent elastic moduli of the inclusion as calculated by solving the boundary value problem, when averaged over all lattice orientations, should equal the effective elastic moduli. Of course, the coincidence of the solution and that described above by eqn (6.51) comes from the discussed double role of E, being both the solution of the spherical inclusion problem and the local part of the modified Green's function.

For macroisotropic aggregates of cubic crystallites the self-consistent shear modulus, say G, follows from the cubic equation

$$G^3 + (\alpha_1 - \alpha_2)G^2 + (\beta_1 - \beta_2)G - \gamma = 0 \tag{6.54}$$

with κ, μ, ν as in eqn (4.14) and

$$\begin{aligned} \alpha_1 &= 3(15\kappa + 8\mu + 12\nu)/40 \\ \alpha_2 &= (3\mu + 2\nu)/5 \\ \beta_1 &= 3\kappa(2\mu + 3\nu)/20 \\ \beta_2 &= 3(6\kappa\nu + 9\kappa\mu + 20\mu\nu)/40 \\ \gamma &= 3\kappa\mu\nu/4 \end{aligned} \tag{6.55}$$

According to Gairola and Kröner (1981) eqn (6.54) is equivalent to

$$G = \frac{\alpha_2 G^2 + \beta_2 G + \gamma}{G^2 + \alpha_1 G + \beta_1} \tag{6.56}$$

It results from eqn (6.51) for macrohomogeneous and macroisotropic media. If G is set equal to ∞ and 0 on the right-hand side, one obtains on the left-hand side $G^{(+1^*)}$ and $G^{(-1^*)}$, the bounds of Voigt and Reuss.† Using these bounds on the right-hand side yields $G^{(\pm 3^*)}$, and in the next step $G^{(\pm 5^*)}$ etc. If we start with $G^{(\pm 0)}$ on the right-hand side we reproduce $G^{(\pm 2^*)}$ which turn out to be the celebrated bounds of Hashin and Shtrikman (1963). The next step leads to $G^{(\pm 4^*)}$ etc. Usually the sequence $G^{(\pm n^*)}$ converges quickly, leading to the self-consistent modulus G.

The $G^{(\pm n^*)}$ are optimum bounds for perfectly disordered, macrohomogeneous and macroisotropic media. Obviously eqn (6.56) can be considered as a recursion formula that allows us to calculate bounds of

† We use n^* rather than n or n' in order to characterize the set of bounds now under discussion. These bounds belong to rather special statistical information.

order n^* if those of order $n^* - 2$ are given. The same is true for the more general equation (6.51) which we rewrite as

$$\mathbf{C}^{(\pm n^*)} = \langle \mathbf{c} \,..\, (\mathbf{I} + \mathbf{E}^{(l)} \,..\, \delta\mathbf{c}^{(l)})^{-1} \rangle \,..\, \langle (\mathbf{I} + \mathbf{E}^{(l)} \,..\, \delta\mathbf{c}^{(l)})^{-1} \rangle^{-1},$$
$$l \equiv \pm n^* \mp 2 \qquad (6.57)$$

in order to emphasize the recursion character. Equation (6.57) offers a convenient way to calculate numerically the self-consistent moduli by iteration.

The self-consistent method has also been applied to problems with texture, e.g. by Kneer (1964, 1965), Morris (1970) and others.

It has been mentioned that it is often difficult to obtain statistical information in terms of correlation functions. In particular, those of order higher than 3 are seldom available. Therefore it is convenient to introduce a classification scheme (n_1, n_2, n_3) for the statistical properties homogeneity, isotropy and disorder (in this sequence). We say that a polycrystal is of *grade n* in one of these properties if this one is reflected by the correlation functions of the order up to n. For instance the scheme (2, 2, 2) implies that nothing is known of correlation functions of order >2. The occurrence of disorder of grade n_3 always has the consequence that $\hat{\mathbf{\Gamma}}^\circ$ is replaced by $\hat{\mathbf{E}}$ in the integrals containing correlation functions up to order n_3. Thus the optimum upper bound of the (2, 2, 2) type materials is

$$\hat{\mathbf{C}}^{(2)} = \langle \mathbf{c} \rangle - \langle \mathbf{c}' \,..\, \mathbf{E} \,..\, \mathbf{c}' \rangle, \qquad \mathbf{E} \text{ formed with } \mathbf{C}^{(+0)} \qquad (6.58)$$

[cf. eqn (6.41)]. Of course, $\mathbf{C}^{(2)}$ is also an upper bound to the effective moduli of the perfectly disordered material.

The upper Hashin–Shtrikman bound is obtained if, in eqn (6.51), $\mathbf{C}^\circ$ is taken as $\mathbf{C}^{(+0)}$. In order to calculate it for the macroisotropic case, the statistical information according to the classification $(\infty, \infty, 2)$ is needed. The ∞ symbols mean that the specimen must be macrohomogeneous and macroisotropic in all correlation functions, whereas the disorder is confined to 2nd order only. It is the additional information that allows us to calculate all terms $\langle \mathbf{c} \,..\, \mathbf{E} \,..\, \mathbf{c} \,..\, \mathbf{E} \,..\, \mathbf{c} \rangle$ etc. which appear if eqn (6.51) is given in series representation. It is not surprising that the bounds $\mathbf{C}^{(\pm 2^*)}$ thus obtained are (slightly) closer than the optimum bounds $\mathbf{C}^{(\pm 2)}$ derived in the case (2, 2, 2). Recall that we have used n for the optimum bound order if the statistical information is restricted to correlation functions up to order n.

In the quantum mechanical scattering theory, the $\hat{\mathbf{T}}$ operator is used to develop a most powerful scheme called *coherent potential approxi-*

mation (CPA; Soven, 1967). Eberhardt (1973, 1980), Hori (1973, 1975/77), Hori and Yonezawa (1974/75) and Gubernatis and Krumhansl (1975) studied the connexions in detail. Accordingly, the self-consistent method corresponds to the so-called single-site (1-site) CPA. It is suggestive to say that $\langle \mathbf{t} \rangle - \langle \mathbf{t} * \hat{\mathbf{\Phi}} * \mathbf{t} \rangle = 0$ would correspond to the 2-site CPA etc. Not everything is clear, however.

6.6. Brief Remarks on Experimental Results

Although it is true that the physical relevance of a theory is finally decided by experiment or observation, one has to be cautious when judging announcements about successful testing of theories in our field. In fact, so much data on elastic moduli of polycrystals has been compiled over the decades that almost any theory can be 'confirmed' if the comparisons are made without care. Many descriptions of measurements lack a reliable characterization of the preparation or internal state of the specimen. Such measurements are usually worthless for testing theories in our field. Some reason has been given already in Section 2. A general guide should be the requirement that the experiment ought to refer to the situation described by the theory. In cases where sufficient care was taken the theory was confirmed with inaccuracy of well below 1%.

Let us give two examples of what we hope was good experimental testing. Using well annealed specimens (see Section 1) of copper polycrystals at room temperature, Bradfield (1964) finds $0{\cdot}483 \times 10^{11}\ \mathrm{N\ m^{-2}}$ for the effective shear modulus of well annealed specimens. Using eqn (6.56), we find the bounds $G^{(\pm n^*)}$ and the effective shear modulus G as given in Table 2 which shows not only outstanding agreement

TABLE 2

Optimal Bounds of 0th to 3rd Order of Shear Modulus of Macroisotropic Aggregates of Cubic Crystallites of Copper. Monocrystal constants from Ledbetter (1980), with a typographical correction for c_{44}

c_{11}	c_{12}	c_{44}	$G^{(\pm 0)}$	$G^{(\pm 1)}$	$G^{(\pm 2^*)}$	$G^{(\pm 3^*)}$	G^{sc}	G^{exp}
			0·754 0	0·546 8	0·495 1	0·483 2		
1·696	1·224	0·754 0					0·482 5	0·483
			0·236 0	0·401 5	0·460 6	0·477 0		

$G^{(\pm 0)}$ higher and lower of shear moduli $(c_{11} - c_{12})/2$ and c_{44} $(= G^{(\pm 0')} = G^{(\pm 0^*)})$.
$G^{(\pm 1)}$ bounds of Voigt and Reuss $(= G^{(\pm 1')} = G^{(\pm 1^*)})$.
$G^{(\pm 2^*)}$ bounds of Hashin and Shtrikman.
G^{sc} self-consistent modulus, calculated from eqn (6.56).
G^{exp} experimental value from Bradfield (1964).

between the theoretical and experimental G but also gives an impression of the convergence of the bounds.

Wawra *et al.* (1977) show that the effective shear moduli of more than 60 metal and alloy polycrystals lie more or less uniformly distributed within the (scaled) 3rd order bounds calculated from eqn (6.51). It is clear that most of the points would lie outside the 4th order bounds if these were calculated. The specimens had been taken from the literature and from their own measurements such as to ensure the greatest possible disorder and isotropy. The result shows that, despite this endeavour, it was not possible to reproduce a disorder greater than of grade 3. It also appears doubtful if specimens can be prepared with reasonable effort so that they are macrohomogeneous and macro-isotropic of grade higher than 3.

7. EXTENSIONS AND CONCLUSION

7.1. External Forces Correlated with Elastic Moduli

The preceding Section 6 was devoted mainly to the effective elastic moduli. These are quantities of extreme importance because knowledge of them permits us to solve all boundary value problems of polycrystals by the well established methods in use for homogeneous media. This applies also to variational approaches since it was assumed in the theory of Section 6, eqn (6.24), that the energy $\langle \varepsilon * \hat{c} * \varepsilon \rangle$ can be set equal to the energy $\langle \varepsilon \rangle * \hat{C}^{\text{eff}} * \langle \varepsilon \rangle$ formed with the effective moduli and the ensemble averaged strains. We gave realistic examples in which such an assumption is not possible.

Consider an arbitrary stress–strain state of such a more general situation. Then, under the validity of the ergodic hypothesis, the elastic energy of the medium is

$$U = \tfrac{1}{2}\int_V \sigma_{ij}\,\varepsilon_{ij}\,\mathrm{d}V = \langle U \rangle \doteq \tfrac{1}{2}\int_V \langle \sigma_{ij}\,\varepsilon_{ij} \rangle\,\mathrm{d}V \equiv \frac{V}{2}\langle \sigma * \varepsilon \rangle \tag{7.1}$$

In the following we shall make repeated use of the formulae

$$\langle ab \rangle = \langle a \rangle \langle b \rangle + \langle a'b' \rangle, \qquad \langle ab' \rangle = \langle a'b \rangle = \langle a'b' \rangle \tag{7.2}$$

where the prime denotes deviation and a, b are any random functions or tensors. We thus have

$$\langle \sigma * \varepsilon \rangle = \langle \sigma \rangle * \langle \varepsilon \rangle + \langle \sigma' * \varepsilon' \rangle \tag{7.3}$$

where, unlike in the preceding section, $\langle \sigma' * \varepsilon' \rangle \neq 0$ so that the definitions (6.1) and (6.24) of $\widehat{\mathbf{C}}^{\text{eff}}$ are not consistent with each other and, therefore, $\widehat{\mathbf{C}}^{\text{eff}}$ has no relevance to our present problem. Equation (7.3) occurs in the work of Mandel and Dantu (1963). Following Kröner (1973), we write, with $\widehat{\mathrm{s}} * \widehat{\mathrm{c}} = \widehat{\mathbf{I}}$, the unit operator,

$$\begin{aligned} \langle \sigma' * \varepsilon' \rangle &\equiv \langle \sigma' * \widehat{\mathbf{I}} * \varepsilon' \rangle = \langle \sigma' * \widehat{\mathrm{s}} * \widehat{\mathrm{c}} * \varepsilon' \rangle \\ &= \langle \sigma' * \widehat{\mathrm{s}} \rangle * \langle \widehat{\mathrm{c}} * \varepsilon' \rangle + \langle (\sigma' * \widehat{\mathrm{s}})' * (\mathrm{c} * \varepsilon')' \rangle \end{aligned} \tag{7.4}$$

$$\begin{aligned} &\langle (\sigma' * \widehat{\mathrm{s}})' * (\widehat{\mathrm{c}} * \varepsilon')' \rangle = \langle (\sigma' * \widehat{\mathrm{s}})' * \widehat{\mathrm{s}} * \widehat{\mathrm{c}} * (\widehat{\mathrm{c}} * \varepsilon')' \rangle \\ &= \langle (\sigma' * \widehat{\mathrm{s}})' * \widehat{\mathrm{s}} \rangle * \langle \widehat{\mathrm{c}} * (\widehat{\mathrm{c}} * \varepsilon')' \rangle + \langle ((\sigma' * \widehat{\mathrm{s}})' * \widehat{\mathrm{s}})' * (\widehat{\mathrm{c}} * (\widehat{\mathrm{c}} * \varepsilon')')\rangle \end{aligned} \tag{7.5}$$

Continuing like this one obtains the elastic energy

$$\begin{aligned} U = \frac{V}{2}\Big[&\langle \sigma \rangle * \langle \varepsilon \rangle + \langle \sigma' * \widehat{\mathrm{s}} \rangle ** \langle \widehat{\mathrm{c}} * \varepsilon' \rangle \\ &+ \langle (\sigma' * \widehat{\mathrm{s}})' * \mathrm{s}' \rangle *** \langle \widehat{\mathrm{c}}' * (\widehat{\mathrm{c}} * \varepsilon')' \rangle + \ldots \Big] \end{aligned} \tag{7.6}$$

A somewhat different version of the theory, introducing couple stresses of various order, has been given by Levin and Nikolaievski (1978).

In eqn (7.6) the elastic energy consists of an infinite number of contributions. The leading term depends on average stress and strain only, whereas the other terms depend on how stress and strain are correlated with compliances and moduli. In fact, in eqn (7.6) the brackets containing elastic constants and their deviations are so-called *mixed correlation functions.* Note that σ and ε appear linear within these brackets because we consider linear elasticity.

Equation (7.6) implies that the macro state is completely known only if, besides $\langle \sigma \rangle$ and $\langle \varepsilon \rangle$, the correlation functions $\langle \sigma_1' \,..\, \mathrm{s}_2' \rangle$, $\langle \mathrm{c}_1' \,..\, \varepsilon_2' \rangle$, $\langle (\sigma_1' \,..\, \mathrm{s}_2)' \,..\, \mathrm{s}_3' \rangle$, $\langle (\mathrm{c}_1' \,..\, (\mathrm{c}_2 \,..\, \varepsilon_3')' \rangle$ etc. are also known. Here the suffixes 1, 2, 3, etc. denote dependence on $\mathbf{r}_1$, $\mathbf{r}_2$, $\mathbf{r}_3$ etc. Of course, eqn (7.6) is the elastic energy also in problems of wave propagation. A more detailed analysis shows that there exists a main wave characterized by $\langle \sigma \rangle$ which, in a sort of dynamic equilibrium, is accompanied by *correlation waves* determined by the manner in which the *correlation stresses* $\langle \sigma_1' \, \mathrm{s}_2' \rangle$, $\langle (\sigma_1' \, \mathrm{s}_2)' \, \mathrm{s}_3' \rangle$ etc. depend on position and time. This is a fascinating field that still is in its infancy.

7.2. Further Results

The whole statistical information about the polycrystal is contained in the entity of mixed correlation functions between $\boldsymbol{\sigma}$, or $\boldsymbol{\varepsilon}$, or $\mathbf{u}$, the displacement field, on the one hand, and $\mathbf{c}$ and $\mathbf{f}$ on the other hand, where, for simplicity, $\mathbf{f}$ shall include volume and surface forces. We now show how some correlation functions can be calculated, given the correlation functions of $\mathbf{c}$ (assuming $\mathbf{f}$ uncorrelated with $\mathbf{c}$).

Of great interest is the correlation function $\langle\boldsymbol{\sigma}_1\,\boldsymbol{\sigma}_2\rangle$ and especially the *autocorrelation function* $\langle\boldsymbol{\sigma}_1\,\boldsymbol{\sigma}_1\rangle \equiv \langle\boldsymbol{\sigma}^2\rangle$ where both $\boldsymbol{\sigma}$'s refer to the same point $\mathbf{r}_1$. Assume that $\mathbf{C}^{\text{eff}}$ is already known. Then it is convenient to take $\mathbf{C}^\circ = \mathbf{C}^{\text{eff}}$, so that eqn (6.14), $\boldsymbol{\varepsilon} = \widehat{\mathbf{A}}^\circ * \boldsymbol{\varepsilon}_0$, becomes $\boldsymbol{\varepsilon} = \widehat{\mathbf{A}} * \langle\boldsymbol{\varepsilon}\rangle$, $\widehat{\mathbf{A}}$ formed with $\mathbf{C}^{\text{eff}}$. We have for instance

$$\langle\sigma_{ij}(\mathbf{r}_1)\,\sigma_{kl}(\mathbf{r}_2)\rangle = \langle c_{ijmn}(\mathbf{r}_1)\,\varepsilon_{mn}(\mathbf{r}_1)\,c_{klpq}(\mathbf{r}_2)\,\varepsilon_{pq}(\mathbf{r}_2)\rangle$$

$$= \langle c_{ijmn}(\mathbf{r}_1)\int_{V_\infty} \mathrm{d}V_3\,\mathbf{A}_{mnrs}(\mathbf{r}_1,\mathbf{r}_3)\,c_{klpq}(\mathbf{r}_2)\int_{V_\infty} \mathrm{d}V_4\,\mathbf{A}_{pquv}(\mathbf{r}_2,\mathbf{r}_4)\rangle\langle\varepsilon_{rs}(\mathbf{r}_3)\rangle\langle\varepsilon_{uv}(\mathbf{r}_4)\rangle \tag{7.7}$$

or in abridged notation

$$\begin{aligned}\langle\boldsymbol{\sigma}_1\boldsymbol{\sigma}_2\rangle &= \langle\widehat{\mathbf{c}} * \widehat{\mathbf{A}} * \langle\boldsymbol{\varepsilon}\rangle\,\widehat{\mathbf{c}} * \widehat{\mathbf{A}} * \langle\boldsymbol{\varepsilon}\rangle\rangle \\ &= \langle\boldsymbol{\varepsilon}\rangle * \langle\widehat{\mathbf{A}}^\dagger * \widehat{\mathbf{c}}\widehat{\mathbf{c}} * \widehat{\mathbf{A}}\rangle * \langle\boldsymbol{\varepsilon}\rangle\end{aligned} \tag{7.8}$$

If $\widehat{\mathbf{A}}$ is replaced by the relevant geometric progression in $\widehat{\boldsymbol{\Gamma}}^\circ * \delta\widehat{\mathbf{c}}^\circ$ we obtain $\langle\boldsymbol{\sigma}_1\boldsymbol{\sigma}_2\rangle$ in terms of the correlation functions of $\mathbf{c}$ up to infinite order. The autocorrelation function (variance) $\langle\boldsymbol{\sigma}^2\rangle$ follows if eqn (7.7) is multiplied by $\delta(\mathbf{r}_1 - \mathbf{r}_2)$ and integrated over $\mathbf{r}_2$.

The variance of $\boldsymbol{\sigma}$ is a measure of the mean fluctuations of the stresses and is therefore of particular interest.

Similarly also

$$\langle\boldsymbol{\varepsilon}_1\,\boldsymbol{\varepsilon}_2\rangle = \langle\boldsymbol{\varepsilon}\rangle * \langle\widehat{\mathbf{A}}^\dagger\,\widehat{\mathbf{A}}\rangle * \langle\boldsymbol{\varepsilon}\rangle \tag{7.9}$$

In a polycrystal it will often be desirable to know not only the average stress and mean fluctuations, but also the stress averaged over those grains that lie in an orientation range $\mathrm{d}\boldsymbol{\Omega}$ around $\boldsymbol{\Omega}$. This average stress is *conditional;* we write

$$\boldsymbol{\sigma}_\Omega \equiv \langle\boldsymbol{\sigma}|\boldsymbol{\Omega}\rangle = \langle\boldsymbol{\sigma}(\mathbf{r})\,|\,\mathbf{r}\text{ in grains of }\boldsymbol{\Omega}\rangle \tag{7.10}$$

$\boldsymbol{\sigma}_\Omega$ is important, for instance, for the plasticity of polycrystals, because the onset of plastic flow in a crystallite of orientation $\boldsymbol{\Omega}$ depends on the resulting stress acting in this crystallite. A satisfactory theory of polycrystal plasticity must take into account the orientation dependence of this stress.

We start with the former equation

$$\sigma = \hat{c} * \hat{\Gamma} * \sigma \tag{7.11}$$

If the external forces are non-random, i.e. $\mathbf{f} = \langle \mathbf{f} \rangle$, we can replace $\boldsymbol{\sigma}$ by $\langle \boldsymbol{\sigma} \rangle$ on the right-hand side of eqn (7.11) (not on the left-hand side!). Taking the conditional average of eqn (7.11) gives us

$$\sigma_\Omega = \langle \hat{c} * \hat{\Gamma} | \Omega \rangle * \sigma \tag{7.12}$$

If $\hat{\Gamma}$ is replaced by the Neumann series in $\delta\hat{c}^\circ$, the conditional correlation functions

$$\langle \delta\hat{c}_1^\circ | \Omega \rangle, \qquad \langle \delta\hat{c}_1^\circ \delta\hat{c}_2^\circ | \Omega \rangle, \qquad \langle \delta\hat{c}_1^\circ \delta\hat{c}_2^\circ \delta\hat{c}_3^\circ | \Omega \rangle, \ \ldots \tag{7.13}$$

occur. They are to be formed with the conditional probabilities

$$P_1(\hat{c}_1 | \Omega), \qquad P_2(\hat{c}_1, \hat{c}_2 | \Omega), \ \ldots \tag{7.14}$$

Since the condition is that we find $\hat{c}_\Omega \equiv \hat{c}(\Omega)$ at $\mathbf{r}_1$, the densities are

$$\begin{aligned} P_1(\hat{c}_1 | \Omega) &= \delta(\hat{c}_1 - \hat{c}_\Omega) \\ P_2(\hat{c}_1, \hat{c}_2 | \Omega) &= \delta(\hat{c}_1 - \hat{c}_\Omega)\, P_1(\hat{c}_2) \end{aligned} \tag{7.15}$$

etc., so that

$$\begin{aligned} \langle \delta\hat{c}_1^\circ | \Omega \rangle &= \hat{c}_\Omega - \hat{C}^\circ \\ \langle \delta\hat{c}_1^\circ \delta\hat{c}_2^\circ | \Omega \rangle &= (\hat{c}_\Omega - \hat{C}^\circ)\langle \delta\hat{c}_2^\circ \rangle \\ \langle \delta\hat{c}_1^\circ \delta\hat{c}_2^\circ \delta\hat{c}_3^\circ | \Omega \rangle &= (\hat{c}_\Omega - \hat{C}^\circ)\langle \delta\hat{c}_2^\circ \delta\hat{c}_3^\circ \rangle \end{aligned} \tag{7.16}$$

etc. These results show that we need the same statistical information required already for the calculation of $\mathbf{C}^{\text{eff}}$.

If $\langle \boldsymbol{\sigma} \rangle$ is taken constant in eqn (7.12) and replaced by $\hat{\mathbf{C}}^{\text{eff}} * \langle \boldsymbol{\varepsilon} \rangle$, then it makes sense to write

$$\sigma_\Omega = \hat{C}_\Omega^{\text{eff}} * \langle \sigma \rangle, \qquad \hat{C}_\Omega^{\text{eff}} = \langle \hat{c} * \hat{\Gamma} | \Omega \rangle * \hat{C}^{\text{eff}} \tag{7.17}$$

where the $\hat{\mathbf{C}}_\Omega^{\text{eff}}$ represent partial (with respect to orientation) effective moduli. Of course the variance of $\boldsymbol{\sigma}_\Omega$ and other quantities can also be calculated, once the correlation functions are known.

In this chapter we have been concerned with stress–strain states that are produced by externally applied loads. If a polycrystal is deformed plastically and then the loads are removed, the body remains in an internal stress state, or eigenstress state. The sources of these stresses are not external forces but rather incompatibilities which are a measure of the heterogeneity of the plastic flow. The cause of this heterogeneity is the fact that there exist more or less favourably oriented grains, so that the onset of flow varies with orientation. We therefore expect that the incompatibilities will be correlated with the orientations and thus with the elastic moduli. This creates a situation which is as complex as that where external forces are correlated correspondingly. Investigations of these phenomena are very desirable, particularly in view of a later theory of plasticity.

7.3. Conclusion

After the preparation in Sections 1–5, results of the statistical elasticity theory of polycrystals were presented in Sections 6 and 7. Section 6 contains what today is often called the *effective medium theory,* whereas Section 7 treats more general situations in polycrystals. Naturally, the effective medium theory, dealing with the simpler physical situation, is better developed and has produced many rigorous results. In particular, exact expressions for the effective elastic moduli are derived, assuming that complete statistical information about the polycrystal is given, and rigorous (often optimum) bounds for these moduli are set up, when the statistical information is limited. Two conditions must be satisfied if a problem is to belong to the effective medium theory. Firstly, the ergodic hypothesis must hold, which implies a sufficiently stochastic constitution of the polycrystal. This prerequisite will usually be satisfied. Secondly, the distribution of the forces including inertia forces acting on the body must be uncorrelated with that of the elastic moduli. Thus dependence not only on the structure of the polycrystal but also on the nature of the problem itself determines whether or not it can be treated within the effective medium theory.

A formal rigorous expression for the effective elastic moduli, which exists independently of the particular problem as an important characterization of the polycrystal, is that of eqns (6.15) and (6.16). The latter is a Neumann type series in terms of multiple integrals and correlation functions containing the statistical information in a systematic way. There is one particular class, idealized as often in physics, of polycrystals in which the grain arrangement and distribution of elastic moduli is such that the correlation functions permit all indicated integrations, elastic anisotropy (texture) included. These polycrystals, classified as perfectly disordered, are in a white noise situation characterized by a complete lack of any principle of order.

Thoroughly prepared polycrystals can approach such a situation.

Also, bounds for effective elastic moduli have a rigorous meaning. They are classified according to their order n. A bound of order n deviates from the true effective moduli by the power c'^{n+1} of the deviation from the mean of the elastic moduli. Optimum bounds are of great value. The term 'optimum' has a relative meaning. It is always understood with respect to the statistical information incorporated. Closer optimum bounds are not generally 'better' than wider optimum bounds because the usefulness of particular bounds depends on the problem and the information. The 1st order bounds of Voigt and Reuss will always be especially valuable.

It was shown in Section 7.1 why dynamic problems are much more involved than static ones. The problem of wave propagation resembles somewhat that of turbulence. In both situations a dynamic equilibrium plays an important role. The present state of the theory is rather poor. Although it seems possible to arrive at an apparently relatively simple formulation in terms of space-time dependent Green's functions — which has not been shown here — any actual calculation of some accuracy is extremely complex. There are good prospects for a method already considered by Willis (1981) which starts from the extension to dynamics of the principle of minimum energy.

From the statistical information (correlation functions) needed to calculate effective elastic moduli, other quantities of interest can also be derived. Examples are the variance, or standard deviation, of stress and strain, the mean stress and strain and their variances in all grains having a particular lattice orientation, correlations between stresses and elastic moduli (not shown) etc. Some of these results will be indispensable for a later theory of polycrystal plasticity.

Many of the formulae given here apply just as well to disordered composites whereas many results of the theory of composites are not applicable to polycrystals. Literature on composites has been omitted almost completely, although there are results that are relevant to polycrystals too. An extensive bibliography can be found in the survey article of McCoy (1981) which in itself is recommended for further reading.

A chapter of this length can never cover all important works in a field as broad as ours. Contributions of many researchers have been incorporated, but probably the present author has not succeeded enough in limiting reference to his own contributions, so he apologizes to all those who might feel that their results are not adequately represented.

The final remark relates to Professor Mandel's contribution to our field. Important insights were gained in the work of Mandel and Dantu (1963) where Mandel was responsible for the theoretical part. Here he considered a macrohomogeneous and macroisotropic solid consisting of a mixture of arbitrarily many elastic isotropic phases. He noticed that, as a consequence of Hill's theorem about 1st order bounds, the effective compression and shear modulus follow from a 'rule of mixture', whereas the same is not true for Young's modulus. The 1st order upper bound is, in fact, not $\langle e \rangle$, if e is Young's modulus of the constituents, but a more complex expression lying slightly above $\langle e \rangle$. Mandel also discusses the narrower bounds of Hashin and Shtrikman and finds that in 49 tests with very different materials the effective Young's moduli are always within Mandel's (1st order) bounds, but sometimes outside the bounds of Hashin and Shtrikman. This result shows that in these cases the conditions of experiment and Hashin and Shtrikman's theory were not identical. Mandel also gave a justification for Hill's statistical hypotheses $\langle \sigma * \varepsilon \rangle = \langle \sigma \rangle * \langle \varepsilon \rangle$. Although he did not consider volume forces in his proof, which therefore has somewhat restricted application, this was an important step forward.

In his later work, Mandel was fascinated by the properties of plasticity and viscoplasticity. He nevertheless expressed his continuing interest in the elasticity of polycrystals and composites in a letter sent to me in 1978. It is with adoration and admiration that I dedicate this chapter to the great scientist and man Jean Mandel.

ACKNOWLEDGEMENTS

I am very grateful to Professor Hélène Lanchon (Nancy), Dr H. M. Ledbetter (Boulder, Colorado) and Dr B. K. D. Gairola (Stuttgart) for numerous engaging discussions, and to Ms Karin Möss (Stuttgart) for the careful typing of the manuscript. My own work in the field discussed in this chapter has been supported by the Volkswagenstiftung, Land Niedersachsen.

REFERENCES

Babos, G. (1982). Zur Theorie der Eigenschaften statistisch aufgebauter Kompositen, thesis Fak. Physik, Universität Stuttgart, available on request.

Bacon, D. J., Barnett, D. M. and Scattergood, R. O. (1978). Anisotropic continuum theory of lattice defects. In *Progr. Mat. Sci.*, B. Chalmers, J. W. Christian and T. B. Massalski (Eds), Pergamon, New York, p. 51.
Bahr, U. (1977). Einfluss des Randes auf das effektive Verhalten heterogener Körper: Anwendung auf Rayleigh-Wellen, *Ann. Phys.*, **7**, 267.
Bahr, U. and Gambin, B. (1977). Scattering of a wave from a heterogeneous material, *Arch. Mech. Stos.*, **29**, 769.
Benk, S. (1980). Zur Theorie der Eigenschaften statistisch aufgebauter Medien, Thesis Fak. Physik, Universität Stuttgart, available on request.
Benk, S. and Kröner, E. (1979). Effective properties of random media: new rigorous results, *Proc. Phys. Symp. Dacca*, Bangladesh Phys. Soc. edn, Dacca, p. 55.
Bensoussan, A., Lions, J. L. and Papanicolaou, G. (1978). Asymptotic analysis for periodic structures. In *Studies in Mathematics and its Applications*, North-Holland, Amsterdam.
Beran, M. J. (1968). *Statistical Continuum Theories*, Interscience, New York.
Beran, M. and McCoy, J. J. (1970). Mean field variation in random media, *Quart. Appl. Math.*, **28**, 245.
Beran, M. and Molyneux, J. (1966). Use of classical variational principles to determine bounds for the effective bulk modulus in heterogeneous media, *Quart. Appl. Math.*, **24**, 2.
Boas, W. and Schmid, E. (1934). Zur Berechnung physikalischer Konstanten quasiisotroper Vielkristalle, *Helv. Phys. Acta*, **7**, 628.
Bradfield, G. (1964). Use in industry of elasticity measurements with the help of mechanical vibrations, *Notes on Applied Science*, No. 30, National Phys. Lab., Teddington, England.
Bradfield, G. and Pursey (1953). The role of preferred orientation in elasticity investigations, *Phil. Mag.*, **44**, 437.
Bruggeman, D. A. G. (1934). Berechnung der elastischen Moduln für die Verschiedenen Texturen der Regulären Metalle, *Z. Physik*, **92**, 561.
Bunge, H. J. (1969). *Mathematische Methoden der Texturanalyse*, Akademie-Verlag, Berlin.
Chassapis, D. (1983). Effektive Wärmeleitfähigkeit von Verbundwerkstoffen, Thesis Fak. Physik, Universität Stuttgart, available on request.
Christensen, R. M. (1979). *Mechanics of Composite Materials*, Wiley, New York.
Datta, S. K. (1980). Wave propagation in the presence of a random distribution of inclusions, *Proc. Symp. Continuum Models of Discrete Systems 3* (Freudenstadt), E. Kröner and K.-H. Anthony (Eds), Univ. of Waterloo Press, Waterloo, Canada, p. 565.
Dederichs, P. H. and Zeller, R. (1972). Elastische Konstanten von Vielkristallen, KFA-Jül-Rep., JÜL-877-FF.
Dederichs, P. and Zeller, R. (1973). Variational treatment of the elastic constants of disordered materials, *Z. Physik*, **259**, 103.
Diener, G. (1980). A derivation of the radiation transfer theory for random media, *Physica*, **106A**, 398.
Diener, G., Raabe, C. and Weissbarth, J. (1981). Bounds for the nonlocal effective properties of random media, *J. Mech. Phys. Solids*, **29**, 181.

Eberhardt, O. (1973). Klassische und quantenmechanische Theorie ungeordneter Medien, Thesis Fak. Physik, Universität Stuttgart, available on request.
Eberhardt, O. (1980). The coherent potential approximation in random field theories, *Proc. Symp. Continuum Models of Discrete Systems 3* (Freudenstadt), E. Kröner and K.-H. Anthony (Eds), Univ. of Waterloo Press, Waterloo, Canada, p. 415.
Eimer, C. (1967). Stresses in multi-phase media, *Arch. Mech. Stos.,* **19**, 521.
Eimer, C. (1968). The boundary effect in elastic multiphase bodies, *Arch. Mech. Stos.,* **20**, 87.
Eshelby, J. D. (1957). The determination of the elastic field of an ellipsoidal inclusion, and related problems, *Proc. Roy. Soc.* (Lond.), **A241**, 376.
Ford, G. W. (1980). Multiple scattering of waves and particles, *Proc. Symp. Continuum Models of Discrete Systems 3* (Freudenstadt), E. Kröner and K.-H. Anthony (Eds), Univ. of Waterloo Press, Waterloo, Canada, p. 583.
Fowler, R. H. (1955). *Statistical Mechanics,* 2nd edn, Cambridge University Press.
Frisch, U. (1968). Wave propagation in random media. In *Probabilistic Methods in Applied Mathematics,* Vol. 1, A. T. Bharucha-Reid (Ed.), Academic Press, New York, p. 76.
Gairola, B. K. D. (1980). Unified method of estimating bounds and self-consistent values for effective moduli of anisotropic polycrystals, *Proc. Symp. Continuum Models of Discrete Systems 3* (Freudenstadt), E. Kröner and K.-H. Anthony (Eds), Univ. of Waterloo Press, Waterloo, Canada, p. 387.
Gairola, B. K. D. and Kröner, E. (1981). A simple formula for calculating the bounds and the self-consistent value of the shear modulus of a polycrystalline aggregate of cubic crystals, *Int. J. Eng. Sci.,* **19**, 865.
Gambin, B. and Kröner, E. (1981). Convergence problems in the theory of random elastic media, *Int. J. Eng. Sci.,* **19**, 313.
Ghahremani, F. (1977). Numerical evaluation of the stresses and strains in ellipsoidal inclusions in an anisotropic elastic material, *Mech. Res. Comm.,* **4**, 89.
Gubernatis, J. E. and Krumhansl, J. A. (1975). Macroscopic engineering properties of polycrystalline materials: elastic properties, *J. Appl. Phys.,* **46**, 1875.
Hashin, Z. (1965). Elasticity of random media, *Trans. Soc. Rheol.,* **9**(1), 381.
Hashin, Z. and Shtrikman, S. (1963). A variational approach to the theory of the elastic behaviour of multiphase materials, *J. Mech. Phys. Solids,* **11**, 127.
Hersant, T., Jeulin, D. and Parnière, P. (1975). Basic notations of mathematical morphology used in quantitative metallography, Parts 1–5, Reports Institut de Recherches de la Sidérurgie Française (IRSID).
Hershey, A. V. (1954). The elasticity of an isotropic aggregate of anisotropic cubic crystals, *Trans. ASME, J. Appl. Mech.,* **21**, 236.
Hill, R. (1952). The elastic behaviour of a crystalline aggregate, *Proc. Phys. Soc.* (Lond.), **A65**, 349.
Hori, M. (1973). Statistical theory of effective electrical, thermal, and magnetic properties of random heterogeneous materials, Parts I and II, *J. Math. Phys.,* **14**, 514, 1942.
Hori, M. (1975/77). Statistical theory of effective electrical, thermal, and magnetic properties of random heterogeneous materials, Parts VI and VII, *J. Math. Phys.,* **16**, 1777; **18**, 487.

Hori, M. and Yonezawa, F. (1974/75). Statistical theory of effective electrical, thermal, and magnetic properties of random heterogeneous materials, Parts III–V, *J. Math. Phys.*, **15**, 2177; **16**, 352, 365.
Huber, A. and Schmid, E. (1934). Bestimmung der elastischen Eigenschaften quasiisotroper Vielkristalle durch Mittelung, *Helv. Phys. Acta*, **7**, 620.
Huet, C. (1982). Universal conditions for assimilation of heterogeneous material to an effective continuum, *Mech. Res. Comm.*, **9**, 165.
Jeulin, D. (1979). Morphologie mathématique et propriétés physiques des agglomères de minerais de fer et du coke métallurgique, thèse Ecole Nat. Sup. des Mines de Paris.
Kinoshita, N. and Mura, T. (1971). Elastic fields of inclusions in anisotropic media, *Phys. Stat. Sol.* (a), **5**, 759.
Kneer, G. (1964). Zur Elastizität vielkristalliner Aggregate mit und ohne Textur, Diss. Fak. Natur- und Geisteswissenschaften, Bergakademie Clausthal (TH), Clausthal-Zellerfeld.
Kneer, G. (1965). Über die Berechnung der Elastizitätsmoduln vielkristalliner Aggregate mit Textur, *Phys. Stat. Sol.*, **9**, 825.
Korringa, J. (1973). Theory of elastic constants of heterogeneous media, *J. Math. Phys.*, **14**, 509.
Korringa, J., Lin, I.-H. and Mills, R. L. (1978). General theorems about homogeneous ellipsoidal inclusions, *Am. J. Phys.*, **46**, 517.
Kröner, E. (1953). Das Fundamentalintegral der Anisotropen Elastischen Differentialgleichungen, *Z. Physik*, **136**, 402.
Kröner, E. (1958). Berechnung der Elastischen Konstanten des Vielkristalls aus den Konstanten des Einkristalls, *Z. Physik*, **151**, 504.
Kröner, E. (1967). Elastic moduli of perfectly disordered composite materials, *J. Mech. Phys. Solids*, **15**, 319.
Kröner, E. (1973). Zur klassischen Theorie statistisch aufgebauter Festkörper, *Int. J. Eng. Sci.*, **11**, 171.
Kröner, E. (1974). On the physics and mathematics of self-stresses. In *Topics in Applied Continuum Mechanics*, J. L. Zeman and F. Ziegler (Eds), Springer-Verlag, Wien, p. 22.
Kröner, E. (1977). Bounds for effective elastic moduli of disordered materials, *J. Mech. Phys. Solids*, **25**, 137.
Kröner, E. and Koch, H. (1976). Effective properties of disordered materials, *SM Archives*, **1**, 183.
Kunin, I. A. (1964). Green's tensor for anisotropic elastic medium with internal stress sources, *Dokl. Akad. Nauk SSSR*, **157**, 1320.
Kunin, I. A. (1983). *Elastic Media with Microstructure*, Part II, Springer Series in Solid-State Sciences, Vol. 44, Springer-Verlag, Heidelberg.
Kunin, I. A. and Sosnina, E. G. (1971). An ellipsoidal inhomogeneity in an elastic medium, *J. Appl. Math. Mech.* (USSR), **37**, 501.
Laws, N. (1980). The elastic response of composite materials, *Physics of Modern Materials*, Int. Atomic Energy Agency, Vienna, p. 465.
Ledbetter, H. M. (1980). Sound velocities and elastic-constant averaging for polycrystalline copper, *J. Phys.* (D), *Appl. Phys.*, **13**, 1879. The values of elastic moduli of single crystals given there have been slightly changed after personal communication with Dr Ledbetter.
Ledbetter, H. M. (1983). Personal communication.

Levin, V. M. (1971). The relation between mathematical expectations of stress and strain tensors in elastic microheterogeneous media, *J. Appl. Math. Mech.*, **35**, 694 (translated from Russian).

Levin, V. M. and Nikolaievski, V. N. (1978). On a general theory of composite materials and micro-inhomogeneous elastic media, *Engng Trans.* (Poland), **26**, 115.

Lifshitz, I. M. and Rosentsveig, L. N. (1947). Green's tensor for anisotropic unbounded elastic medium, *Zh. Exp. Teor. Fiz.* (USSR), **17**, 783.

Mandel, J. and Dantu, P. (1963). Contribution à l'étude théorique et expérimentale du coefficient d'élasticité d'un milieu hétérogène mais statistiquement homogène, *Ann. Ponts Chaussées* (Paris), **133**, 115.

Markov, K. Z. (1981). One-particle approximations in mechanics of composite materials, *Proc. Symp. Continuum Models of Discrete Systems 4* (Stockholm), O. Brulin and R. K. T. Hsieh (Eds), North-Holland, Amsterdam.

Matheron, G. (1967). *Eléments pour une Théorie des Milieux Poreux,* Masson, Paris.

Mazilu, P. (1972). On the theory of linear elasticity in statistically homogeneous media, *Rev. Roum. Mat. Pur. Appl.*, **17**, 261.

Mazilu, P. (1976). On the theory of linear elasticity in statistically homogeneous media, *Arch. Mech. Stos.* (Poland), **28**, 517.

Mazilu, P. (1981). Criteria obtained from the small volumes theory for choosing adequate average operator in statistical continuum mechanics, *Proc. Symp. Continuum Models of Discrete Systems 4* (Stockholm), O. Brulin and R. K. T. Hsieh (Eds), North-Holland, Amsterdam, p. 35.

McCoy, J. J. (1981). Macroscopic response of continua with random microstructure. In *Mechanics Today,* Vol. 6, S. Nemat-Nasser (Ed), Pergamon, New York, p. 1.

McCoy, J. J. and Fishman, L. (1980). A fundamental limitation on effective property theories for nonlinear statistical continua, *Proc. Symp. Continuum Models of Discrete Systems 3* (Freudenstadt), E. Kröner and K.-H. Anthony (Eds), Univ. of Waterloo Press, Waterloo, Canada, p. 375.

Miller, M. N. (1969*a*). Bounds for effective electrical, thermal and magnetic properties of heterogeneous materials, *J. Math. Phys.*, **10**, 1988.

Miller, M. N. (1969*b*). Bounds for effective bulk modulus of heterogeneous materials, *J. Math. Phys.*, **10**, 2005.

Morris, P. (1970). Elastic constants of polycrystals, *Int. J. Eng. Sci.*, **8**, 49.

Mura, T. (1982). *Micromechanics of Defects in Solids,* Nijhoff, The Hague.

Ondracek, G. and Schulz, B. (1980). The microstructure-field property relationship of multiphase materials, *Proc. Symp. Continuum Models of Discrete Systems 3* (Freudenstadt), E. Kröner and K.-H. Anthony (Eds), Univ. of Waterloo Press, Waterloo, Canada, p. 499.

Pompe, W., Vojta, G. and Kreher, W. (1979). Statistical continuum theory of heterogeneous solid media: information theory approach, *Phys. Stat. Sol.* (b), **95**, 561.

Reif, F. (1965). *Fundamentals of Statistical and Thermal Physics,* McGraw-Hill, New York.

Reuss, A. (1929). Berechnung der Fliessgrenze von Mischkristallen auf Grund der Plastizitätsbedingung für Einkristalle, *Z. Ang. Math. Mech.,* **9**, 49.

Rieder, G. (1962). Iterationsverfahren und Operatorgleichungen in der Elastizitätstheorie, *Abh. Braunschw. Wiss. Ges.*, **14**(2), 109.
Rudee, M. L. and Howie, A. (1972). The structure of amorphous Si and Ge, *Phil. Mag.*, **25**, 1001.
Serra, J. (1969). *Introduction à la Morphologie Mathématique*, Ecole des Mines de Paris.
Serra, J. (1982). *Image Analysis and Mathematical Morphology*, Academic Press, New York.
Simmons, J. A. and Bullough, R. (1970). Internal stress and the incompatibility problem in anisotropic elasticity, *Proc. Conf. Fundamental Aspects of Dislocation Theory* (Gaithersburg), J. A. Simmons, R. de Wit and R. Bullough (Eds), Nat. Bur. Stand. U.S., Spec. Publ. 317, Vol. 1, Washington, p. 89.
Solomon, L. (1968). *Elasticité Linéaire*, Masson, Paris.
Soven, P. (1967). Coherent-potential model of substitutional disordered alloys, *Phys. Rev.*, **156**, 809.
Tolman, R. C. (1938). *The Principles of Statistical Mechanics*, Oxford University Press.
Voigt, W. (1887). Theoretische Studien über die Elastizitätsverhältnisse der Krystalle, *Abh. Kgl. Ges. Wiss. Göttingen, Math. Kl.*, **34**, 1, in part, pp 47 ff.
Walpole, L. J. (1966). On bounds of the overall elastic moduli of inhomogeneous systems-I and II, *J. Mech. Phys. Solids*, **14**, 151, 289.
Walpole, L. J. (1977). The determination of the elastic field of an ellipsoidal inclusion in an anisotropic medium, *Math. Proc. Camb. Phil. Soc.*, **81**, 283.
Wawra, H. H. (1968). Über die Abhängigkeit des Elastizitäts- und Torsionsmoduls gezogener Kupfer- und α-Messingstangen von der Anlasstemperatur, *Z. Metallkunde*, **59**, 268.
Wawra, H. H. (1974). Wesentliche Erweiterung des Anwendungsbereichs von Elastomat II zur Messung Elastischer Grössen, *Z. Metallkunde*, **65**, 579.
Wawra, H. H. (1979). On the role of Grüneisen's elastic parameters in the theory of polycrystal elasticity, *Phys. Stat. Sol.* (a), **54**, 291.
Wawra, H. H., Koch, H. and Kröner, E. (1977). Comparison between experimental values and theoretical bounds for the E- and G-modulus of aggregates of cubic crystallites, *J. Mater. Sci. Soc. Japan*, **14**, 131.
Willis, J. (1980). A polarization approach to the scattering of elastic waves. I, Scattering by a single inclusion; II, Multiple scattering from inclusions, *J. Mech. Phys. Solids*, **28**, 287, 307.
Willis, J. R. (1981). Variational principles for dynamic problems for inhomogeneous elastic media, *Wave Motion*, **3**, 1.
Willis, J. R. (1983). The overall elastic response of composite materials, *Trans. ASME, J. Appl. Mech.*, **50**, 1202.
Yonezawa, F. and Ninomiya, T. (1983). *Topological Disorder in Condensed Matter*, Springer Series in Solid-State Sciences, Vol. 46, Springer-Verlag, Heidelberg.
Zeilon, N. (1910/11). Das Fundamentalintegral der allgemeinen partiellen linearen Differentialgleichung mit konstanten Koeffizienten, *Arkiv Mat., Astron., Fys.*, **6**, No. 38, 1.
Zeller, R. and Dederichs, P. H. (1973). Elastic constants of polycrystals, *Phys. Stat. Sol.* (b), **55**, 831.

CHAPTER 9

Phenomenological Constitutive Models for Metals

ZENON MRÓZ
Institute of Fundamental Technological Research, Warsaw, Poland

ABSTRACT

In this chapter the hardening rules for metals in the range of small strains are discussed. The simple hardening rules, such as isotropic, kinematic or mixed hardening models, involve one scalar and/or one tensor state variable whose evolution rule is expressed in terms of the plastic strain rate. The composite hardening rules involve more state variables and multiple loading surfaces for which evolution rules depend on a previous deformation history. Such composite rules can be formulated by using either the structural approach or the representative variable approach. Whereas the structural approach requires numerous state variables and material parameters, the representative variable approach offers a much simpler material description. The creep hardening rules for viscoplastic deformation processes are discussed at the end of the chapter.

1. INTRODUCTION

The present chapter is devoted to a discussion of the phenomenological constitutive models that are most widely used in engineering applications. In this approach, only the most fundamental aspects of inelastic deformation are retained and the details at the microstructural level are neglected. The microstructural changes are roughly described by proper evolution rules for state variables and proper selection of material parameters occurring in constitutive relations.

Depending on the type of loading, different constitutive equations should be applied in order to simulate the deformation process. Thus, for monotonic loading at room temperature, the details of unloading and reverse plasticity are unimportant, whereas for varying or cyclic loads the proper modelling of evolution of hardening due to repetitive action of loads, demonstrated through varying hysteretic loops, is important. At elevated temperatures and also for some materials at room temperature the creep effects cannot be neglected and the constitutive model should account for viscous deformation.

In this chapter we shall first discuss simple hardening models involving only a single loading surface and one or two state variables. In Section 3 some more complex hardening rules will be discussed that are aimed at a description of hardening effects during variable loading. Such more complex hardening rules usually require a set of loading surfaces in the stress or strain spaces that allow incorporation of sufficient memory of particular loading events. Structural and representative variable approaches will be presented and compared. In Section 4 the creep hardening rules are briefly discussed within the state variable theory.

2. SIMPLE HARDENING RULES

2.1. Isotropic, Kinematic and Combined Hardening Models

During the quasi-static deformation process at room temperature the viscous effects are usually neglected and the plastic strains may be regarded as occurring instantaneously with applied loads. The plastic anisotropy developing during the deformation process is mainly due to two physical effects: (i) occurrence of microstresses resulting from inhomogeneity of plastic deformation on both crystalline and dislocation levels, and (ii) formation of texture resulting from rotation and elongation of crystalline grains. Whereas the second effect is typical for large plastic deformation, the oriented structure of microstresses is inherently related to any plastic strain. On the phenomenological level the microstresses affect the stress–strain curve during loading and reverse loading by changing directional stiffness of the representative element or the conventional yield limit after plastic prestrain (Bauschinger effect in the uniaxial case).

A perfectly plastic model is characterized by the yield condition depending only on the stress state, $F(\sigma_{ij}) = 0$. During deformation this yield condition does not vary and there is no change of the material state. For the uniaxial case the corresponding stress–strain curve is shown in Fig. 1(a).

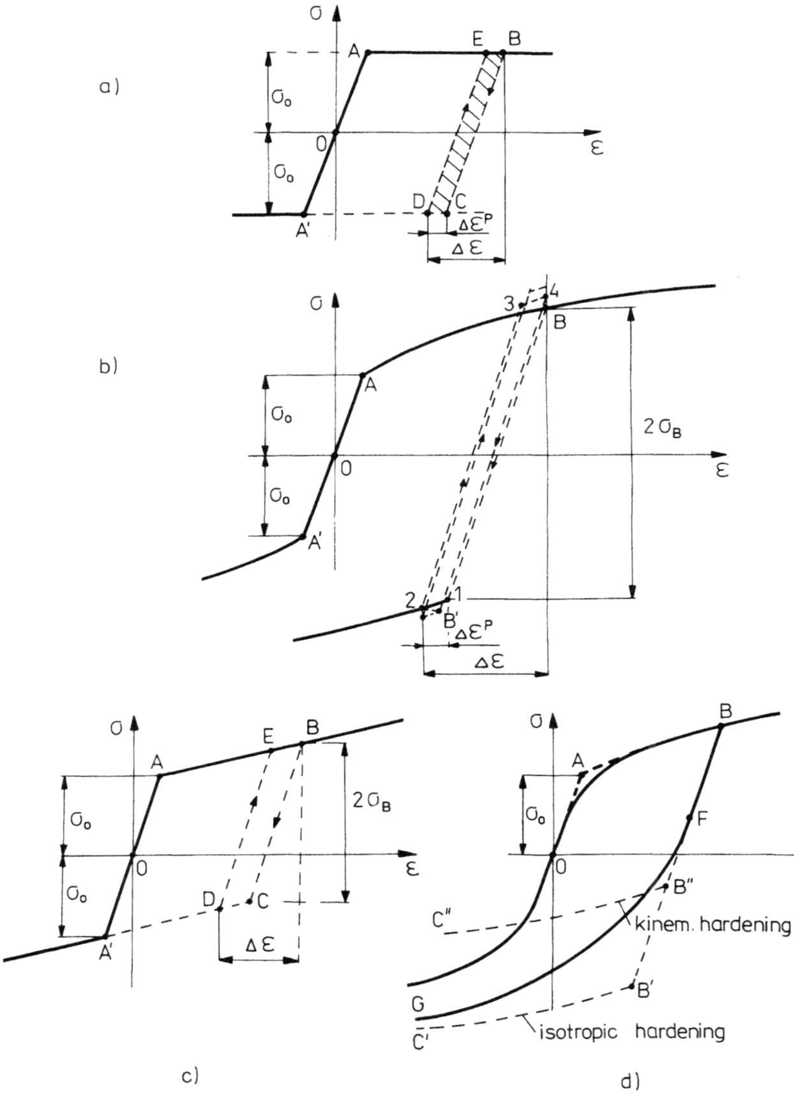

Fig. 1. Simple models of hardening: (a) perfectly plastic material; (b) isotropic hardening model; (c) kinematic hardening model; (d) approximation of the actual stress–strain curve by kinematic and isotropic hardening models.

A simple account of hardening can be made by assuming that the yield condition depends on a scalar variable λ monotonically growing with the plastic strain, for instance

$$\lambda_1 = \int_0^t \sigma_{ij}\dot{\varepsilon}^{\mathrm{p}}_{ij}\,\mathrm{d}t = \int_0^t (\boldsymbol{\sigma}\cdot\dot{\boldsymbol{\varepsilon}}^{\mathrm{p}})\,\mathrm{d}t \tag{2.1}$$

$$\lambda_2 = \int_0^t (\dot{\varepsilon}^{\mathrm{p}}_{ij}\dot{\varepsilon}^{\mathrm{p}}_{ij})^{1/2}\,\mathrm{d}t = \int (\boldsymbol{\varepsilon}^{\mathrm{p}}\cdot\dot{\boldsymbol{\varepsilon}}^{\mathrm{p}})^{1/2}\,\mathrm{d}t \tag{2.2}$$

and the yield condition is

$$F = f(\sigma_{ij}, \lambda) = 0, \qquad \frac{\partial F}{\partial \lambda} \leqslant 0 \tag{2.3}$$

or in a more particular form

$$F = f(\sigma_{ij}) - \sigma_{\mathrm{p}}(\lambda) = 0, \qquad \frac{\mathrm{d}\sigma_{\mathrm{p}}}{\mathrm{d}\lambda} \geqslant 0 \tag{2.4}$$

Since no preferred orientation is induced due to plastic straining, the initially isotropic material stays isotropic after arbitrary deformation. Thus eqns (2.3) and (2.4) combined with (2.1) and (2.2) constitute the *isotropic hardening model.* In particular, the form (2.4) implies that the initial yield surface (for which $\sigma_{\mathrm{p}}(0) = \sigma_{\mathrm{p}0}$ where $\sigma_{\mathrm{p}0}$ is the initial yield limit) preserves its shape and varies only in size in the course of plastic straining. Figure 1(b) presents the predicted material response during tension followed by unloading and compression. Assuming identical material response with respect to change of sign of stress, it is seen that $\sigma_{\mathrm{B}} = |\sigma_{\mathrm{B}'}| > \sigma_0$ when $\mathrm{d}\sigma_{\mathrm{p}}/\mathrm{d}\lambda > 0$. Any cyclic loading of prescribed stress amplitude between B and B′ will thus correspond to elastic response. On the other hand, for specified strain amplitude $\Delta\varepsilon$ the elastic state is reached asymptotically after an infinite number of cycles 1–2–3–4– … ; cf. Fig. 1(b). When σ_{p} tends to its limiting value $\sigma_{\mathrm{p}} \to \sigma_{\mathrm{u}}$ for $\lambda \to \infty$, the material becomes asymptotically perfectly plastic. It is seen that this prediction of the isotropic hardening model limits its applicability to monotonic loading since it is not able to simulate hysteretic phenomena occurring after unloading and reloading between B and B′.

In order to describe the Bauschinger effect and anisotropy associated with the residual microstress field, let us assume the yield condition in the form

$$F = f(\boldsymbol{\sigma} - \boldsymbol{\alpha}) - \sigma_{\mathrm{p}}(\lambda) = 0 \tag{2.5}$$

that is depending on two state variables $\boldsymbol{\alpha}$ and λ. When $\sigma_{\mathrm{p}}(\lambda) = \text{const}$ the yield surface undergoes a rigid-body translation which corresponds to the case of kinematic hardening, shown in Fig. 1(c).

Since the material description using state variables $\boldsymbol{\alpha}$ and λ is widely used in various numerical codes of inelastic analysis, let us discuss its possibilities. First, let us write the associated flow rule

$$\dot{\varepsilon}^{\mathrm{p}}_{ij} = \begin{cases} \dfrac{1}{H}\dfrac{\partial f}{\partial \sigma_{ij}}\dfrac{\partial f}{\partial \sigma_{kl}}\dot{\sigma}_{kl} & \text{for } F = 0,\ f_{\dot{\sigma}} = \dfrac{\partial f}{\partial \sigma_{kl}}\dot{\sigma}_{kl} > 0 \\ 0 & \text{for } F < 0 \ \text{ or } \ F = 0,\ f_{\dot{\sigma}} < 0 \end{cases} \tag{2.6}$$

or in vectorial form

$$\dot{\boldsymbol{\varepsilon}}^{\mathrm{p}} = \begin{cases} \dfrac{1}{H}\dfrac{\partial f}{\partial \boldsymbol{\sigma}}\dfrac{\partial f}{\partial \boldsymbol{\sigma}} \cdot \dot{\boldsymbol{\sigma}} = \dfrac{1}{K}\mathbf{n}\dot{\sigma}_{\mathrm{n}} & \text{for } F = 0,\ \dot{\sigma}_{\mathrm{n}} = \dot{\boldsymbol{\sigma}} \cdot \mathbf{n} > 0 \\ 0 & \text{for } F < 0 \ \text{ or } \ F = 0,\ \dot{\sigma}_{\mathrm{n}} < 0 \end{cases} \tag{2.7}$$

where $\mathbf{n}$ denotes the unit vector normal to the yield surface at $\boldsymbol{\sigma}$ and $\dot{\sigma}_{\mathrm{n}} = \dot{\boldsymbol{\sigma}} \cdot \mathbf{n}$ denotes the projection of the stress rate vector on $\mathbf{n}$. The hardening modulus K is

$$K = H\left(\frac{\partial f}{\partial \boldsymbol{\sigma}} \cdot \frac{\partial f}{\partial \boldsymbol{\sigma}}\right)^{-1} = \frac{\dot{\sigma}_{\mathrm{n}}}{|\dot{\boldsymbol{\varepsilon}}^{\mathrm{p}}|} = \frac{\dot{\boldsymbol{\sigma}} \cdot \mathbf{n}}{(\dot{\boldsymbol{\varepsilon}}^{\mathrm{p}} \cdot \dot{\boldsymbol{\varepsilon}}^{\mathrm{p}})^{1/2}} \tag{2.8}$$

and it can be expressed as the ratio of the normal stress rate component to the plastic strain rate modulus. In the uniaxial case the modulus K is proportional to the tangent modulus of the stress–plastic strain curve.

Equations (2.5)–(2.8) should be supplemented by the evolution equations specifying variation of the variables $\boldsymbol{\alpha}$ and λ in the course of plastic flow. Obviously, when there is no plastic flow, these variables should be fixed, thus $\dot{\boldsymbol{\alpha}} = \dot{\lambda} = 0$ when $\dot{\boldsymbol{\varepsilon}}^{\mathrm{p}} = 0$. This condition is satisfied by assuming a homogeneous relationship between the rates of state variables and the plastic strain rate. The simplest assumption is that proposed by Melan (1938) and Prager (1956)

$$\dot{\boldsymbol{\alpha}} = c\dot{\boldsymbol{\varepsilon}}^{\mathrm{p}} = c\mathbf{n}\dot{\lambda} \tag{2.9}$$

and later modified by Ziegler (1959)

$$\dot{\boldsymbol{\alpha}} = c_1(\boldsymbol{\sigma} - \boldsymbol{\alpha})\dot{\lambda}, \qquad \lambda = \lambda_2 \tag{2.10}$$

Here c and c_1 are assumed as constant parameters though they can vary with plastic strain. To investigate the fundamental properties of eqns (2.9) and (2.10), let us calculate the hardening modulus K. As eqn (2.5) is satisfied during the whole deformation process, it is also satisfied in the differential form, thus

$$\frac{\partial f}{\partial \boldsymbol{\sigma}} \cdot \dot{\boldsymbol{\sigma}} - \frac{\partial f}{\partial \boldsymbol{\sigma}} \cdot \dot{\boldsymbol{\alpha}} - \sigma_{\rm p}'(\lambda)\dot{\lambda} = 0 \tag{2.11}$$

where $\sigma_{\rm p}'$ is the derivative of $\sigma_{\rm p}$ with respect to λ. Substituting eqns (2.7) and (2.9) or (2.10) into (2.11), we obtain for the Melan–Prager translational rule

$$K = c + \sigma_{\rm p}'(\lambda)\left(\frac{\partial f}{\partial \boldsymbol{\sigma}} \cdot \frac{\partial f}{\partial \boldsymbol{\sigma}}\right)^{-1/2} \tag{2.12}$$

and for the Ziegler translational rule there is

$$K = c_1(\boldsymbol{\sigma} - \boldsymbol{\alpha}) \cdot \mathbf{n} + \sigma_{\rm p}'(\lambda)\left(\frac{\partial f}{\partial \boldsymbol{\sigma}} \cdot \frac{\partial f}{\partial \boldsymbol{\sigma}}\right)^{-1/2} \tag{2.13}$$

The formulae (2.12) and (2.13) indicate the general property of the model. In fact, when $\sigma_{\rm p}'(\lambda) = 0$, eqn (2.12) predicts $K = c$, i.e. the tangent modulus is constant for any path emanating into the exterior of the elastic domain, and a similar conclusion follows from eqn (2.13). In other words, in eqns (2.12) and (2.13) there is no dependence of K on the direction of previous plastic prestrain.

Let us particularize our discussion to the case of the Huber–Mises yield condition. Instead of eqn (2.5) we now have

$$F(\mathbf{s}, \boldsymbol{\alpha}, \lambda) = \tfrac{3}{2}(\mathbf{s} - \boldsymbol{\alpha}) \cdot (\mathbf{s} - \boldsymbol{\alpha}) - \sigma_{\rm p}^2(\lambda) = 0 \tag{2.14}$$

The flow rule now takes the form

$$\dot{\boldsymbol{\varepsilon}}^{\rm p} = \frac{1}{K}\mathbf{n}\,\dot{\sigma}_{\rm n} = \frac{3}{2K\sigma_{\rm p}^2}(\mathbf{s} - \boldsymbol{\alpha})[(\mathbf{s} - \boldsymbol{\alpha}) \cdot \dot{\mathbf{s}}] \tag{2.15}$$

and

$$\mathbf{n} = \frac{\partial f/\partial \boldsymbol{\sigma}}{|\partial f/\partial \boldsymbol{\sigma}|} = \frac{3(\mathbf{s} - \boldsymbol{\alpha})}{[9(\mathbf{s} - \boldsymbol{\alpha}) \cdot (\mathbf{s} - \boldsymbol{\alpha})]^{1/2}} = \sqrt{\left(\frac{3}{2}\right)}\,\frac{\mathbf{s} - \boldsymbol{\alpha}}{\sigma_{\rm p}} \tag{2.16}$$

In view of eqn (2.12) the expression for K now becomes

$$K = c + \tfrac{2}{3}\sigma_p'(\lambda) \tag{2.17}$$

and a similar expression follows from eqn (2.13). It is evident from eqns (2.14)–(2.17) that the hardening modulus is constant on the yield surface for both loading and reverse loading. Thus the simple kinematic rules (2.9) and (2.10) are unable to describe properly the directional changes in stiffness after initial plastic prestrain. Figure 1(d) illustrates the approximation of the actual reverse loading curve BFG by the simplified hardening rules: kinematic hardening (curve OABB″C″) and isotropic hardening (curve OABB′C′).

Further generalizations of this simple hardening rule were intended to account for non-linearity of the hardening curve and were aimed at simulating not only the reverse loading curve but also the transient material response under cyclically varying stress. Let us discuss briefly some of these generalizations.

Kadashevitch and Novozhilov (1959) related the kinematic hardening rule to the state of residual microstresses depending on the plastic strain tensor and its evolution was assumed to be governed by eqn (2.9). Thus the residual microstress state (back stress) can be identified with the parameter $\boldsymbol{\alpha}$ and assumed to have an elastic or *reversible* character. In particular, it can be removed by reversing the plastic strain. This in turn implies the reversible character of hardening, that is the state of a material can be restored by a proper straining history. Arutunyan and Vakulenko (1965) used the representation (2.4) with the linear translational rule (2.9) and a non-linear function $\sigma_p = \sigma_p(\lambda)$ tending asymptotically to its limiting value, $\sigma_p \to \sigma_u$ for $\lambda \to \infty$. This description has the disadvantage that the tangent moduli during cyclic loading for subsequent hysteresis loops gradually decrease and the loops flatten, the phenomenon usually not observed in experiments.

Eisenberg and Phillips (1968) proposed a non-linear kinematic hardening model by assuming

$$\boldsymbol{\alpha} = c(\lambda)\boldsymbol{\varepsilon}^{p}, \qquad \lambda = \lambda_2 \tag{2.18}$$

and hence there is

$$\dot{\boldsymbol{\alpha}} = c(\lambda)\dot{\boldsymbol{\varepsilon}}^{p} + c'(\lambda)\boldsymbol{\varepsilon}^{p}\dot{\lambda} = [\sqrt{(3/2)}\cdot c(\lambda)\mathbf{n} - c'(\lambda)\boldsymbol{\varepsilon}^{p}]\,\dot{\lambda} \tag{2.19}$$

Thus the translational rule is affected by the plastic strain tensor and is not only governed by the instantaneous irreversible strain rate. A more general hardening model proposed by Mróz *et al.* (1976) started from a similar assumption, namely

$$\dot{\boldsymbol{\alpha}} = F_1(\dot{\boldsymbol{\varepsilon}}^{\mathrm{p}}, \boldsymbol{\varepsilon}^{\mathrm{p}}) \qquad \text{or} \qquad \dot{\boldsymbol{\alpha}} = F_2(\dot{\boldsymbol{\varepsilon}}^{\mathrm{p}}, \boldsymbol{\alpha}) \tag{2.20}$$

or in particular

$$\dot{\boldsymbol{\alpha}} = c(\lambda)\dot{\boldsymbol{\varepsilon}}^{\mathrm{p}} - d(\lambda)\dot{\lambda}\boldsymbol{\varepsilon}^{\mathrm{p}} = [\sqrt{(3/2)} \cdot c(\lambda)\mathbf{n} - d(\lambda)\boldsymbol{\varepsilon}^{\mathrm{p}}]\,\dot{\lambda} \tag{2.21}$$

The difference between eqns (2.19) and (2.21) is that in (2.21) the translational rule depends on two material functions $c = c(\lambda)$ and $d = d(\lambda)$ which determine the hardening properties during loading and reverse loading or in cyclic deformation. Let us discuss these two models since they offer a possibility of more realistic simulation of cyclic deformation of metals.

Starting from the yield condition (2.14) and using eqns (2.15) and (2.16), from the consistency condition (2.11) we obtain

$$K = c(\lambda) - d(\lambda)\frac{(\mathbf{s} - \boldsymbol{\alpha}) \cdot \boldsymbol{\varepsilon}^{\mathrm{p}}}{\sigma_{\mathrm{p}}} + \tfrac{2}{3}\sigma'_{\mathrm{p}}(\lambda) \tag{2.22}$$

The difference between the rules (2.9) and (2.20) follows directly from eqns (2.22) and (2.17). Equation (2.17) predicts the variation of the hardening modulus depending on the relative orientation of the 'effective stress tensor' $\mathbf{s} - \boldsymbol{\alpha}$ and the plastic strain vector $\boldsymbol{\varepsilon}^{\mathrm{p}}$. Figure 2(a) represents the yield surface after the initial stress path OB followed by elastic unloading along BC. Assuming $d(\lambda) > 0$, at the points B and C there is, respectively, $(\mathbf{s} - \boldsymbol{\alpha})_{\mathrm{B}} \cdot \boldsymbol{\varepsilon}^{\mathrm{p}}_{\mathrm{B}} > 0$, $(\mathbf{s} - \boldsymbol{\alpha})_{\mathrm{C}} \cdot \dot{\boldsymbol{\varepsilon}}^{\mathrm{p}}_{\mathrm{C}} < 0$, and from eqn (2.22) it follows that $K(\mathrm{C}) > K(\mathrm{B})$. Figure 2(b) represents the uniaxial stress–strain response during loading and reverse loading predicted by this model, and Fig. 2(c) shows the variation of the tangent modulus E_{t} along these curves. Since for the uniaxial case there is $s_{22} = s_{33} = -\frac{1}{2}s_{11}$, $\varepsilon^{\mathrm{p}}_{22} = \varepsilon^{\mathrm{p}}_{33} = -\frac{1}{2}\varepsilon^{\mathrm{p}}_{11}$, eqn (2.21) provides

$$K = c(\lambda) - \tfrac{3}{2}d(\lambda)\frac{(s_{11} - \alpha_{11}) \cdot \varepsilon^{\mathrm{p}}_{11}}{\sigma_{\mathrm{p}}} + \tfrac{2}{3}\sigma'_{\mathrm{p}}(\lambda) \tag{2.23}$$

and the plastic tangent modulus is expressed as

$$E_{\mathrm{t}} = \frac{\mathrm{d}\sigma_{11}}{\mathrm{d}\varepsilon^{\mathrm{p}}_{11}} = \tfrac{3}{2}K = \tfrac{3}{2}[c(\lambda) \mp d(\lambda)\,\varepsilon^{\mathrm{p}}_{11}] \tag{2.24}$$

where minus and plus refer to tension followed by compression. It is now obvious that the functions $c(\lambda)$ and $d(\lambda)$ can be determined from the uniaxial loading–reverse loading tests. In fact, considering the tangent moduli variation on initial and reverse loading curves, from eqn (2.24) we obtain

$$c_1(\lambda) = \frac{E_t^i - E_t^r}{3\varepsilon_c^p} \tag{2.25}$$

where E_t^i and E_t^r are the tangent moduli on respective curves corresponding to the same value of λ, and ε_c^p is the value of plastic strain at the reversal point.

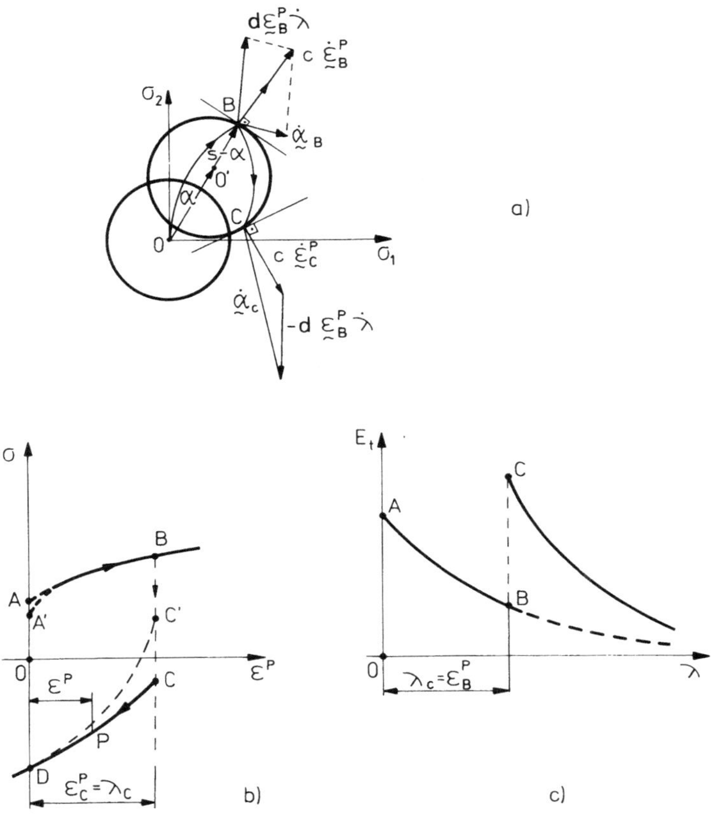

Fig. 2. Non-linear kinematic hardening model with modified translational rule: (a) translation direction of the yield surface; (b) description of the loading–unloading curve; (c) variation of the hardening modulus.

It is thus seen that more sophisticated translation rules depending on plastic strain provide a more realistic description since now the tangent moduli depend on the relative orientation of the stress and plastic strain vectors. A similar translation rule was used by Armstrong and Frederick (1966) and Chaboche (1977), namely

$$\dot{\boldsymbol{\alpha}} = c\dot{\boldsymbol{\varepsilon}}^{\mathrm{p}} - c_1\boldsymbol{\alpha}\dot{\lambda} = [\sqrt{(3/2)} \cdot c\mathbf{n} - c_2\mathbf{m}]\,\dot{\lambda} \tag{2.26}$$

where $c_2 = c_1(\boldsymbol{\alpha} \cdot \boldsymbol{\alpha})^{1/2}$ and $\mathbf{m} = \boldsymbol{\alpha}/(\boldsymbol{\alpha} \cdot \boldsymbol{\alpha})^{1/2}$ is the unit vector coaxial with the vector $\boldsymbol{\alpha}$; c and c_1 are the material parameters being in general functions of plastic strain. The expression (2.26) is similar to the evolution rule (2.21) since both can be interpreted as representing two mechanisms, the first term corresponding to hardening and the other to 'plastic relaxation', since the translation rate $-c_2\mathbf{m}\dot{\lambda}$ tends to reduce the absolute value of the back stress $\boldsymbol{\alpha}$. For monotonic or reverse loading conditions these two hardening rules provide similar results. However, for cyclic loading, especially when ratcheting effects occur, the rule (2.26) is more appropriate since the plastic strain may accumulate and grow incessantly in consecutive cycles whereas the back stress $\boldsymbol{\alpha}$ undergoes only periodic variations.

Khazhinsky (1975), starting from the yield condition (2.14) with σ_{p} = const, assumed

$$\dot{\boldsymbol{\alpha}} = \tfrac{2}{3}\phi(\beta_{\mathrm{e}})\,\dot{\boldsymbol{\varepsilon}}^{\mathrm{p}} \tag{2.27}$$

where

$$\beta_{\mathrm{e}} = [\tfrac{3}{2}(\boldsymbol{\sigma} - c\boldsymbol{\varepsilon}^{\mathrm{p}}) \cdot (\boldsymbol{\sigma} - c\boldsymbol{\varepsilon}^{\mathrm{p}})]^{1/2} = (\tfrac{3}{2}\boldsymbol{\beta} \cdot \boldsymbol{\beta})^{1/2} \tag{2.28}$$

is proportional to the second invariant of the 'active stress tensor' $\boldsymbol{\beta} = \boldsymbol{\sigma} - c\boldsymbol{\varepsilon}^{\mathrm{p}}$ and c is a constant. Assuming the linear relation $\phi(\boldsymbol{\beta}) = A - B\beta_{\mathrm{e}}$, the uniaxial stress–strain curve takes the form

$$\sigma = \sigma_* + c\varepsilon^{\mathrm{p}} - (\sigma_* - \sigma_{\mathrm{p}})\exp\left[-\frac{Bc}{\sigma_* - \sigma_{\mathrm{p}}}\varepsilon^{\mathrm{p}}\right] \tag{2.29}$$

and tends to an asymptote $\sigma = \sigma_* + c\varepsilon^{\mathrm{p}}$. During unloading and reverse loading, the stress–strain curve exhibits different hardening moduli and during cyclic loading a curvilinear loop is created.

A similar effect will be attained by introducing the directional variation of c in the translation rule (2.9). Assume, for instance, that

$$\boldsymbol{\alpha} = \left\{c_2 + c_3\exp\left[-\frac{(\boldsymbol{\sigma} - \boldsymbol{\alpha}) \cdot \boldsymbol{\alpha}}{c_3}\right]\right\}\dot{\boldsymbol{\varepsilon}}^{\mathrm{p}} \tag{2.30}$$

as was proposed by Krieg (1977). The hardening modulus associated with eqn (2.30) varies according to the relation

$$K = c_2 + c_3 \exp\left[-\frac{(\boldsymbol{\sigma}-\boldsymbol{\alpha})\cdot\boldsymbol{\alpha}}{c_3}\right] \tag{2.31}$$

and depends on the relative orientation of the vectors $\boldsymbol{\sigma}-\boldsymbol{\alpha}$ and $\boldsymbol{\alpha}$. Introducing the angle γ between these two vectors, it can be written more generally

$$K = c_2 + c_3 G(|\boldsymbol{\alpha}|, \cos\gamma), \qquad \cos\gamma = \frac{(\boldsymbol{\sigma}-\boldsymbol{\alpha})\cdot\boldsymbol{\alpha}}{\sigma_{\mathrm{p}}|\boldsymbol{\alpha}|} \tag{2.32}$$

where the function $G(|\boldsymbol{\alpha}|, \cos\gamma)$ can be specified experimentally. A similar hardening rule was presented by Eisenberg (1976).

It is thus seen that the generalizations (2.21) and (2.25), (2.27) or (2.30) of the kinematic hardening rule were aimed first of all at including in the model the directional changes of material stiffness due to plastic prestrain. Such a description represents the *Bauschinger anisotropy* occurring for both small and large strain ranges.

To describe the *texture anisotropy* (or *distortional hardening*) induced by large plastic strains, a change of shape of the yield surface and its translation and rotation should be taken into account. A general form of the yield condition for this case

$$F = F(\mathbf{s}, \boldsymbol{\varepsilon}^{\mathrm{p}}, \lambda) = 0 \qquad \text{or} \qquad F = F(\mathbf{s}, \boldsymbol{\alpha}, \lambda) = 0 \tag{2.33}$$

has been discussed in numerous papers. A particular form of eqn (2.33),

$$\tfrac{1}{2}C_{ijkl}(\varepsilon^{\mathrm{p}}_{mn})(s_{ij}-\alpha_{ij})(s_{kl}-\alpha_{kl}) - \sigma_{\mathrm{p}}^2(\lambda) = 0 \tag{2.34}$$

was analysed in several works (cf. Baltov and Sawczuk, 1965; Shrivastava *et al.*, 1973), where the components of the matrix C_{ijkl} were assumed to depend on plastic strains. Let us note that in eqn (2.34) the evolution rule for $\boldsymbol{\alpha}$ can be used in one of the forms discussed in this section. In a recent study of stainless steel anisotropy, Lee and Zaverl (1979) assumed that the matrix C_{ijkl} evolves to its asymptotic state, that is

$$\dot{C}_{ijkl} = \kappa(C^{\mathrm{a}}_{ijkl} - C_{ijkl})\dot{\lambda} \tag{2.35}$$

where κ is a material constant and C^{a}_{ijkl} denotes the stationary anisotropic state reached asymptotically in the course of plastic deformation. The Fourier series approximation of the distorted shape of the yield surface was discussed by Ortiz and Popov (1983).

2.2. Thermodynamic Structure of Constitutive Relations

The present phenomenological description can be supplemented by the thermodynamic background using the concepts of state variables and thermodynamic potentials. In fact, any hardening rule involves one or several state variables, such as back stress $\boldsymbol{\alpha}$, combined with their proper evolution rules.

Assume the state of a material to be represented by its elastic strain $\boldsymbol{\varepsilon}^{\mathrm{e}}$, temperature θ and a set of state variables denoted collectively by $\mathbf{a}$. These variables can be scalars, vectors or tensors. Assume the internal energy u, free energy ϕ, free enthalpy ψ and entropy per unit mass η to be functions of stress, strain and state variables, thus

$$\begin{aligned} u &= u(\boldsymbol{\varepsilon}^{\mathrm{e}}, \mathbf{a}, \theta), \qquad \eta = \eta(\boldsymbol{\varepsilon}^{\mathrm{e}}, \mathbf{a}, \theta) \\ \phi &= u - \theta\eta = \phi(\boldsymbol{\varepsilon}^{\mathrm{e}}, \mathbf{a}, \theta) \\ \rho\psi &= \boldsymbol{\sigma}\cdot\boldsymbol{\varepsilon}^{\mathrm{e}} - \rho\phi, \qquad \psi = \psi(\boldsymbol{\sigma}, \mathbf{a}, \theta) \end{aligned} \tag{2.36}$$

The local energy and entropy balance equations now are expressed as

$$\rho\dot{u} = \rho\dot{r} + \boldsymbol{\sigma}\cdot\dot{\boldsymbol{\varepsilon}} - \operatorname{div}\mathbf{q} \tag{2.37}$$

$$\rho\dot{\eta} = \frac{\rho\dot{r}}{\theta} + \operatorname{div}\left(\frac{\mathbf{q}}{\theta}\right) + \rho\dot{\eta}_{\mathrm{i}} \tag{2.38}$$

where $\mathbf{q}$ is the heat flux and r denotes the rate of heat supply per unit mass. The irreversible entropy production $\dot{\eta}_{\mathrm{i}}$ is specified by

$$\rho\dot{\eta}_{\mathrm{i}} = \mathbf{q}\cdot\mathbf{grad}(1/\theta) + \rho\dot{\eta}_{\mathrm{i}}^{\mathrm{d}} \tag{2.39}$$

where $\dot{\eta}_{\mathrm{i}}^{\mathrm{d}}$ denotes the rate of entropy production due to internal dissipation. For a uniform temperature field, only the second term contributes to entropy production.

Using eqns (2.37) and (2.38), we obtain

$$\begin{aligned} \rho\theta\dot{\eta}_{\mathrm{i}}^{\mathrm{d}} &= -\rho\dot{\phi} - \rho\dot{\theta}\eta + \boldsymbol{\sigma}\cdot\dot{\boldsymbol{\varepsilon}} \\ &= \left(\boldsymbol{\sigma} - \rho\frac{\partial\phi}{\partial\boldsymbol{\varepsilon}^{\mathrm{e}}}\right)\cdot\dot{\boldsymbol{\varepsilon}}^{\mathrm{e}} - \rho\left(\eta + \frac{\partial\phi}{\partial\theta}\right)\cdot\dot{\theta} + \boldsymbol{\sigma}\cdot\dot{\boldsymbol{\varepsilon}}^{\mathrm{p}} - \rho\frac{\partial\phi}{\partial\mathbf{a}}\cdot\dot{\mathbf{a}} \end{aligned} \tag{2.40}$$

or using the free enthalpy ψ,

$$\rho\theta\dot{\eta}_{\mathrm{i}}^{\mathrm{d}} = \rho\dot{\psi} - \boldsymbol{\varepsilon}^{\mathrm{e}}\cdot\dot{\boldsymbol{\sigma}} - \rho\dot{\theta}\eta + \boldsymbol{\sigma}\cdot\dot{\boldsymbol{\varepsilon}}^{\mathrm{p}}$$

$$= \left(\rho \frac{\partial \psi}{\partial \sigma} - \varepsilon^e\right) \cdot \dot{\sigma} + \sigma \cdot \dot{\varepsilon}^p + \rho \frac{\partial \psi}{\partial \mathbf{a}} \cdot \dot{\mathbf{a}} + \rho \left(\frac{\partial \psi}{\partial \theta} - \eta\right) \dot{\theta} \qquad (2.41)$$

Since within the elastic region there is $\dot{\eta}_i^d = 0$, from eqns (2.40) and (2.41) it follows that

$$\sigma = \rho \frac{\partial \phi}{\partial \varepsilon^e}, \quad \eta = -\frac{\partial \phi}{\partial \theta}, \quad \varepsilon^e = \rho \frac{\partial \psi}{\partial \sigma}, \quad \eta = \frac{\partial \psi}{\partial \theta} \qquad (2.42)$$

and the rate of entropy production due to dissipative deformation is

$$\rho \theta \dot{\eta}_i^d = \sigma \cdot \dot{\varepsilon}^p - \mathbf{A} \cdot \dot{\mathbf{a}} > 0 \qquad (2.43)$$

where

$$\mathbf{A} = \rho \frac{\partial \phi}{\partial \mathbf{a}} = -\rho \frac{\partial \psi}{\partial \mathbf{a}} \qquad (2.44)$$

are the internal forces conjugate to $\mathbf{a}$, representing hidden free energy or enthalpy. In fact, setting $\varepsilon^e = \sigma = 0$ in eqn (2.36), we obtain the energy or enthalpy locked within the material $\phi = \phi(0, \mathbf{a}, \theta)$, $\psi = \psi(0, \mathbf{a}, \theta)$ due to non-vanishing state variables $\mathbf{a}$ (e.g. in the form of residual stress energy).

Assume that there exists a one-to-one relationship between state variable rates $\dot{\mathbf{a}}$ and the plastic strain rate $\dot{\varepsilon}^p$, thus

$$\dot{\varepsilon}^p = \mathbf{L}\dot{\mathbf{a}} \qquad (2.45)$$

$$\dot{\mathbf{a}} = \mathbf{M}\dot{\varepsilon}^p, \qquad \mathbf{M} = \mathbf{L}^{-1} \qquad (2.46)$$

where $\mathbf{L}$ and $\mathbf{M}$ are the transformation matrices ($\mathbf{L}\dot{\mathbf{a}}$ denotes the matrix product of $\mathbf{L}$ and $\mathbf{a}$, thus $\dot{\varepsilon}_{ij}^p = L_{ijkl}\dot{a}_{kl}$). The rate of entropy production can thus be represented as

$$\rho \theta \dot{\eta}_i^d = (\sigma - \mathbf{AM}) \cdot \dot{\varepsilon}^p = \sigma^d \cdot \dot{\varepsilon}^p > 0 \qquad (2.47)$$

or

$$\rho \theta \dot{\eta}_i^d = (\sigma \mathbf{L} - \mathbf{A}) \cdot \dot{\mathbf{a}} = \mathbf{A}^d \cdot \dot{\mathbf{a}} > 0 \qquad (2.48)$$

In eqn (2.47), σ^d denotes the *dissipative stress,*

$$\sigma^d = \sigma - \mathbf{AM} = \sigma - \rho \frac{\partial \phi}{\partial \mathbf{a}} \mathbf{M} = \sigma - \sigma^r, \qquad \sigma^r = \mathbf{AM} \qquad (2.49)$$

conjugate to the plastic strain rate $\dot{\varepsilon}^p$, and σ^r is the hidden energy stress. In eqn (2.48), $\mathbf{A}^d$ denotes the *dissipative force* conjugate to the state variable rate $\dot{\mathbf{a}}$. The entropy production can thus be seen in two spaces: in the space of plastic strain with the associated dissipative stress $\boldsymbol{\sigma}^{\mathrm{d}}$ or in the space of state variables with the associated dissipative force $\mathbf{A}^{\mathrm{d}}$. Assuming the behaviour at the level of dissipative forces to be perfectly plastic, we obtain the constitutive equations

$$F(\sigma^{\mathrm{d}}) \leqslant 0 \qquad \text{(yield condition)}$$

$$\text{and when } F(\sigma^{\mathrm{d}}) \equiv 0 \quad \Rightarrow \quad \dot{\varepsilon}^{\mathrm{p}} = \dot{\lambda}\frac{\partial F}{\partial \sigma^{\mathrm{d}}}, \quad \dot{\lambda} \geqslant 0 \qquad \text{(flow rule)} \tag{2.50}$$

or by introducing the specific dissipation function (per unit volume)

$$D = D(\dot{\varepsilon}^{\mathrm{p}}) \tag{2.51}$$

the inverse relations are generated in the form

$$\sigma^{\mathrm{d}} = \frac{\partial D}{\partial \dot{\varepsilon}^{\mathrm{p}}} \tag{2.52}$$

The structure of the relations (2.50)–(2.52) is thus identical to perfect plasticity relations, provided that we use the dissipative stress rather than the total stress. An alternative description is obtained by formulating the yield condition and the flow rule in terms of conjugate variables $\mathbf{A}^{\mathrm{d}}$, $\dot{\mathbf{a}}$, thus

$$F(\mathbf{A}^{\mathrm{d}}) \leqslant 0 \qquad \text{(yield condition)}$$

$$\text{and when } F(\mathbf{A}^{\mathrm{d}}) \equiv 0 \quad \Rightarrow \quad \dot{\mathbf{a}} = \dot{\lambda}\frac{\partial F}{\partial \mathbf{A}^{\mathrm{d}}}, \quad \dot{\lambda} > 0 \qquad \text{(flow rule)} \tag{2.53}$$

and introducing the specific dissipation function $D = D(\dot{\mathbf{a}})$, the inverse relations have the form

$$\mathbf{A}^{\mathrm{d}} = \frac{\partial D(\dot{\mathbf{a}})}{\partial \dot{\mathbf{a}}} \tag{2.54}$$

where $D(\dot{\mathbf{a}})$ and $D(\dot{\varepsilon}^{\mathrm{p}})$ are homogeneous functions of first order of the rates $\dot{a}$ and $\dot{\varepsilon}^{\mathrm{p}}$.

It should be noted that the dissipative forces can be directly obtained by assuming the specific free energy to be a function of the total strain rather than the elastic strain, thus

$$\phi = \phi(\varepsilon, \theta, \mathbf{a}), \qquad \rho\overline{\psi} = \sigma \cdot \varepsilon - \rho\phi = \rho\overline{\psi}(\sigma, \mathbf{a}, \theta) \tag{2.55}$$

Equation (2.40) now takes the form

$$\rho\theta\dot{\eta}_i^d = \left(\sigma - \rho\frac{\partial\phi}{\partial\varepsilon}\right)\cdot\dot{\varepsilon} - \rho\left(\frac{\partial\phi}{\partial\theta} + \eta\right)\dot{\theta} - \rho\frac{\partial\phi}{\partial \mathbf{a}}\cdot\dot{\mathbf{a}} > 0 \qquad (2.56)$$

and since for fixed a there is $\dot{\eta}_i^d = 0$, we have

$$\sigma = \rho\frac{\partial\phi}{\partial\varepsilon}, \qquad \eta = -\frac{\partial\phi}{\partial\theta} \qquad (2.57)$$

and the rate of entropy production is

$$\rho\theta\dot{\eta}_i^d = \mathbf{A}^d\cdot\dot{\mathbf{a}} = -\rho\frac{\partial\phi}{\partial\mathbf{a}}\cdot\dot{\mathbf{a}} > 0 \qquad (2.58)$$

The description (2.55) therefore provides direct specification of dissipative forces $\mathbf{A}^d$, thus

$$\mathbf{A}^d = -\rho\frac{\partial\phi}{\partial\mathbf{a}} = \rho\frac{\partial\bar{\psi}}{\partial\mathbf{a}} \qquad (2.59)$$

Let us note that the free enthalpy $\bar{\psi}$ specified by eqn (2.55) is not equal to ψ specified by eqn (2.36) (in fact, $\rho(\bar{\psi} - \psi) = \boldsymbol{\sigma}\cdot\dot{\boldsymbol{\varepsilon}}^p$). The difference between the two descriptions becomes evident by considering a fictitious 'relaxation' process at fixed $\boldsymbol{\varepsilon}$ and θ. From eqn (2.40) it then follows that

$$\rho\theta\eta_i^d = -\rho\dot{\phi} - \rho\dot{\theta}\eta + \boldsymbol{\sigma}\cdot\dot{\varepsilon} = -\rho\frac{\partial\phi}{\partial\mathbf{a}}\cdot\dot{\mathbf{a}} = \mathbf{A}^d\cdot\dot{\mathbf{a}} > 0 \qquad (2.60)$$

Thus the variation of free energy is totally dissipated in this process. On the other hand, considering a fictitious 'creep' process at fixed $\boldsymbol{\varepsilon}^e$ and θ, from eqn (2.40) we obtain

$$\rho\theta\dot{\eta}_i^d = -\rho\dot{\phi} + \boldsymbol{\sigma}\cdot\dot{\boldsymbol{\varepsilon}}^p = \boldsymbol{\sigma}\cdot\dot{\boldsymbol{\varepsilon}}^p - \rho\frac{\partial\phi}{\partial\mathbf{a}}\cdot\dot{\mathbf{a}} = \boldsymbol{\sigma}\cdot\dot{\boldsymbol{\varepsilon}}^p + \mathbf{A}\cdot\dot{\mathbf{a}} \qquad (2.61)$$

and $\rho\dot{\phi}$ now corresponds to variation of the hidden free energy at fixed $\boldsymbol{\varepsilon}^e$ and θ.

Instead of using the space of dissipative stress $\boldsymbol{\sigma}^d$ or dissipative forces $\mathbf{A}^d$, we can regard $\boldsymbol{\sigma}$ and $\mathbf{A}$ as independent forces, and formulate the yield condition

$$F(\sigma, \mathbf{A}) \leqslant 0 \qquad (2.62)$$

in the $\sigma \times \mathbf{A}$ space. Following eqns (2.50) and (2.53), the generalized flow rule then takes the form

$$\dot{\varepsilon}^{\mathrm{p}} = \dot{\lambda}\frac{\partial F}{\partial \sigma}, \qquad \dot{\mathbf{a}} = \dot{\lambda}\frac{\partial F}{\partial \mathbf{A}} \qquad \dot{\lambda} > 0 \tag{2.63}$$

and the dissipation function $D = D(\dot{\varepsilon}^{\mathrm{p}}, \dot{\mathbf{a}})$ then provides the conjugate forces σ and $\mathbf{A}$, thus

$$\sigma = \frac{\partial D}{\partial \dot{\varepsilon}^{\mathrm{p}}}, \qquad \mathbf{A} = \frac{\partial D}{\partial \dot{\mathbf{a}}} \tag{2.64}$$

As we see from the present discussion, there are several possibilities of representing the dissipative deformation process. First, the concept of the dissipative stress can be used and relations of type (2.50)–(2.52) applied. In this case the evolution rule of the back stress remains arbitrary and does not follow from the flow rule. On the other hand, using the formalism (2.53), (2.54) or (2.62)–(2.64), not only the flow rule for plastic strain rate but also the evolution rules for state variables **a** are specified. To show the equivalence of eqns (2.53) and (2.63), let us write in view of eqn (2.48), with the restrictive hypothesis (2.45),

$$F(\mathbf{A}^{\mathrm{d}}) = F(\mathbf{A} + \sigma\mathbf{L}) \leqslant 0 \tag{2.65}$$

and then we have

$$\dot{\varepsilon}^{\mathrm{p}} = \mathbf{L}\dot{\mathbf{a}} = \dot{\lambda}\mathbf{L}\frac{\partial F}{\partial \mathbf{A}^{\mathrm{d}}} = \dot{\lambda}\frac{\partial F}{\partial \sigma}, \qquad \dot{\mathbf{a}} = \dot{\lambda}\frac{\partial F}{\partial \mathbf{A}^{\mathrm{d}}} = \dot{\lambda}\frac{\partial F}{\partial \mathbf{A}} \tag{2.66}$$

The normality rule for state variables (2.53) was assumed by Rice (1971) whereas the generalized normality relations (2.63) were proposed by Halphen and Nguyen (1975). A less restrictive formulation using the concept of dissipative stress was presented by Vakulenko (1961) (cf. also Mróz, 1972, and Kratochvil and Dillon, 1969). A study by Mandel (1974) of the thermodynamics of plastic deformation for large strains provides a thorough discussion of various approaches to the description of the material state and of the associated evolution rules.

Consider, for instance, the kinematic hardening model (2.5) with σ_p = const. Assume the free energy in the form

$$\rho\phi(\varepsilon, \varepsilon^p) = \phi_1(\boldsymbol{\varepsilon} - \boldsymbol{\varepsilon}^p) + \phi_2(\boldsymbol{\varepsilon}^p) + \phi_3(T) \tag{2.67}$$

and identify the state variable as $\mathbf{a} = \boldsymbol{\varepsilon}^p$. According to eqns (2.55) and (2.59), we have

$$\boldsymbol{\sigma} = \frac{\partial\phi_1}{\partial\boldsymbol{\varepsilon}}, \qquad \mathbf{A}^d = -\rho\frac{\partial\phi}{\partial\mathbf{a}} = \boldsymbol{\sigma} - \rho\frac{\partial\phi_2}{\partial\boldsymbol{\varepsilon}^p} = \boldsymbol{\sigma} - \boldsymbol{\alpha} \tag{2.68}$$

The yield condition expressed in terms of the dissipative stress takes the form

$$F(\mathbf{A}^d) = f(\boldsymbol{\sigma} - \boldsymbol{\alpha}) - \sigma_p = 0 \tag{2.69}$$

In particular, when $\phi_2(\boldsymbol{\varepsilon}^p) = \frac{1}{2}c\boldsymbol{\varepsilon}^p \cdot \boldsymbol{\varepsilon}^p$, the Melan–Prager evolution rule (2.9) is obtained. However, the evolution rules (2.13) or (2.21) do not follow from the representation (2.67).

Concluding this section, it can be stated that the thermodynamic description provides a general unifying structure for constitutive relations especially in cases when at the macro level the body is regarded as perfectly plastic. On the other hand, the generalized normality rules, such as (2.63), may prove too restrictive and requiring too many structural variables. In practical cases the normality is preserved with respect to plastic strain rate whereas the evolution rules are formulated starting from a mechanical concept of the deformation mechanism. In the following section we shall discuss two approaches to the formulation of more complex hardening rules.

3. COMPOSITE MODELS OF HARDENING

The simple hardening rules discussed in the preceding section can provide a sufficiently accurate simulation of deformation histories involving several stress or strain reversals with loading, unloading and reloading paths. However, when we are interested in the study of transient and steady response of metals under multiaxial stress conditions, these hardening rules prove to be insufficient. In particular, the state of hardening was assumed to depend on the 'reversible' state variables such as $\boldsymbol{\varepsilon}^p$ and the irreversible variable λ growing monotonically with the plastic strain. For the steady cyclic state, all functions depending on λ tend to their asymptotic values $\sigma_p(\lambda) \to \sigma_u$, $c_1(\lambda) \to c_{1u}$, $d_1(\lambda) \to d_{1u}$, and then a new steady cyclic state is predicted starting from the first loading cycle for any new amplitude of stress or strain in the uniaxial case or for any subsequent loading programme in a general case. The previous large prestrain or any other loading history has, according to model prediction, no effect on subsequent steady cycles. Thus there is not room for incorporation of the effects of prior monotonic or cyclic loading history on subsequent cyclic behaviour or the possibility to study evolution of transient states toward steady states for different cyclic loading programmes.

It is natural to expect that, if more state variables $\mathbf{a}_r$ are introduced into a model, a more accurate description of the material response should be obtained. To make the constitutive model reasonably simple, only the most important aspects of material response should be included in the model.

3.1. Composite Structure Approach

There are several possibilities for extending the description of hardening rules. The first possibility is to consider a material element as a composition of subelements with simple elastic-plastic properties, e.g. perfectly plastic elements with different yield limits. The stress–strain for the macroelement would then exhibit a hardening effect due to constraint of neighbouring grains and progressing plastic flow of consecutive subelements. This analysis is typical for polycrystals where simple hardening rules for monocrystals and constraint effects of adjacent grains are studied. Let us mention here the classical works by Taylor (1938), Batdorf and Budiansky (1954), and recent studies by Lin (1971) and Asaro (1983).

The simplest mathematical treatment of such composites is obtained by postulating that either stress or strain fields are uniform with the macroelement. For elastic composites these two approaches provide statically or kinematically admissible fields and the resulting lower and upper bounds on average elastic moduli. A similar approximation was used in earlier theories of plasticity of polycrystals. Thus Batdorf and Budiansky (1954) assumed a uniform stress state within the polycrystalline aggregate whereas Taylor (1938) and Lin (1971) assumed a uniform strain field. These two approaches can therefore be regarded as providing lower and upper bounds on the actual stress–strain curve. Recent approaches accounting for inhomogeneity of stress and strain within the macroelement (cf. Hill, 1967), in particular a self-consistent model, provide more accurate analyses but do not change the character of the corner flow rules occurring along the loading path. The occurrence of the corner regime is attributed to the piecewise-linear yield condition with corners for a monocrystal and to the effect of inhomogeneity of deformation.

Let us discuss first the two simplest versions of the composite hardening rules involving series or parallel connections of elastic-plastic subelements. Whereas the series connection implies stress homogeneity, the parallel connection corresponds to strain homogeneity. These two models can be used to provide bounds to the actual hardening modulus of the macroelement.

(i) Assume first a homogeneous stress state within the macroelement. This corresponds to the assumption on the series connection of elastic-plastic subelements (Fig. 3) each characterized by the yield condition

$$F^{(i)} = f^{(i)}(\boldsymbol{\sigma} - \boldsymbol{\alpha}^{(i)}) - \sigma_{\mathrm{p}}^{(i)} = 0, \qquad i = 1, 2, \ldots, n \tag{3.1}$$

where $\sigma_{\mathrm{p}}^{(i)}$ are constant yield limits of particular subelements and $\dot{\boldsymbol{\alpha}}^{(i)} = c^{(i)}\dot{\boldsymbol{\varepsilon}}^{(i)\mathrm{p}}$ according to the Melan–Prager rule (let us note that the assemblage of elastic-perfectly plastic elements would not produce overall hardening). The elastic and plastic strain rates are specified as

$$\begin{aligned} \dot{\boldsymbol{\varepsilon}}^{(i)\mathrm{e}} &= \mathbf{C}^{(i)}\dot{\boldsymbol{\sigma}} \\ \dot{\boldsymbol{\varepsilon}}^{(i)\mathrm{p}} &= \frac{1}{K^{(i)}}\mathbf{n}^{(i)}\dot{\sigma}_{\mathrm{n}}^{(i)} = \frac{1}{H^{(i)}}\frac{\partial f^{(i)}}{\partial \boldsymbol{\sigma}}\left(\frac{\partial f^{(i)}}{\partial \boldsymbol{\sigma}}\cdot\dot{\boldsymbol{\sigma}}\right) \end{aligned} \tag{3.2}$$

where $K^{(i)} = c^{(i)}$ according to eqn (2.12) and $\mathbf{C}^{(i)}$ denotes the elastic compliance matrix of each subelement. The macrostrain is defined as an average value of elastic and plastic strains of particular subelements, thus

$$\begin{aligned} \dot{\boldsymbol{\varepsilon}}^{\mathrm{p}} &= \frac{1}{n}\sum_{i=1}^{n}\dot{\boldsymbol{\varepsilon}}^{(i)\mathrm{p}} = \frac{1}{n}\sum_{i=1}^{n}\frac{1}{H^{(i)}}\frac{\partial f^{(i)}}{\partial \boldsymbol{\sigma}}\left(\frac{\partial f^{(i)}}{\partial \boldsymbol{\sigma}}\cdot\dot{\boldsymbol{\sigma}}\right), \\ \dot{\boldsymbol{\varepsilon}}^{\mathrm{e}} &= \frac{1}{n}\sum_{i=1}^{n}\mathbf{C}^{(i)}\dot{\boldsymbol{\sigma}} = \mathbf{C}\dot{\boldsymbol{\sigma}}, \qquad \mathbf{C} = \frac{1}{n}\sum_{i=1}^{n}\mathbf{C}^{(i)} \end{aligned} \tag{3.3}$$

and

$$\dot{\boldsymbol{\varepsilon}} = \mathbf{C}\dot{\boldsymbol{\sigma}} + \dot{\boldsymbol{\varepsilon}}^{\mathrm{p}} \tag{3.4}$$

Subjecting such an assemblage to proportional loading, the piecewise-linear stress–strain curve is obtained and all active yield surfaces will translate with the stress point (Fig. 3). Upon unloading and reverse loading, the yield surfaces move consecutively backwards and the tangent modulus gradually decreases in the course of deformation. For an infinite number of elements the smooth loading and unloading curves are obtained.

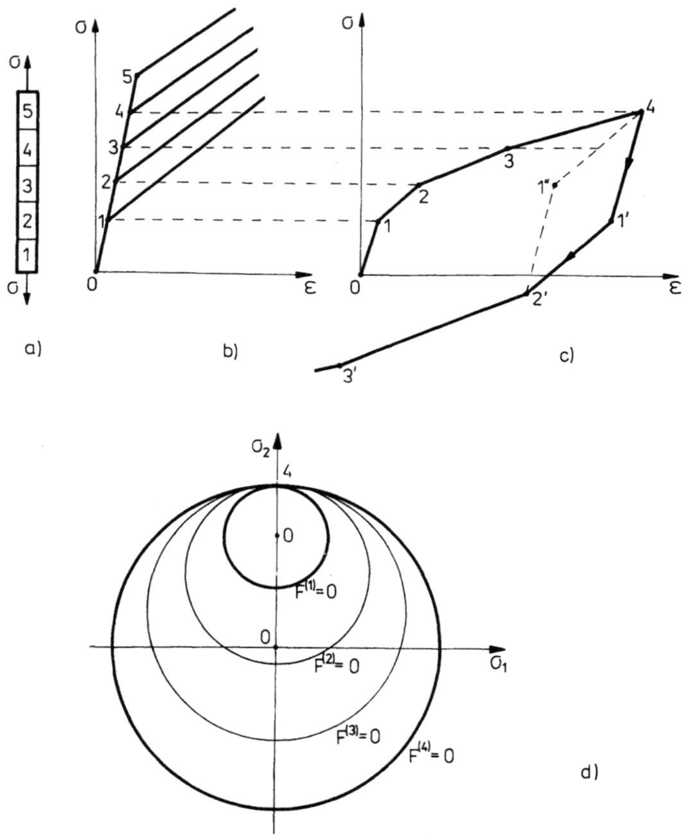

Fig. 3. (a) Composite series connection model; (b) stress–strain curves of subelements; (c) predicted loading–unloading curve; (d) evolution of yield surfaces.

It is easy to establish the relationship between the initial loading curve $\sigma = f(\varepsilon)$ and the reverse loading curve. In fact, if the scale factor 1/2 is used on both coordinate axes, the reverse loading curve is made to coincide with the curve of initial loading. Thus

$$\sigma < \sigma_A, \quad \dot{\sigma} < 0, \quad \frac{\sigma - \sigma_A}{2} = f\left(\frac{\varepsilon - \varepsilon_A}{2}\right) \quad \text{or} \quad \bar{\sigma} = 2f\left(\frac{\bar{\varepsilon}}{2}\right) \tag{3.5}$$

where $\bar{\sigma} = \sigma - \sigma_A$, $\bar{\varepsilon} = \varepsilon - \varepsilon_A$. Equation (3.5) is usually referred to as the Masing relation or hardening rule. When reloading from B to A, the stress–strain curve is identical to eqn (3.5), and

$$\sigma_B < \sigma < \sigma_A, \quad \dot{\sigma} > 0, \quad \frac{\sigma - \sigma_B}{2} = f\left(\frac{\varepsilon - \varepsilon_B}{2}\right) \tag{3.6}$$

and the loop is closed at A; Fig. 3(c). The Masing relation therefore predicts a steady cyclic state after the first stress cycle.

(ii) Assume now that a uniform strain state occurs within the macroelement, $\varepsilon = \bar{\varepsilon}$. This assumption corresponds to a parallel connection of elastic-plastic subelements; Fig. 4(a). The overall hardening is now occurring due to constraint of elastic elements imposed on plastic elements and an assemblage of elastic-perfectly plastic elements exhibits hardening. Denoting elastic and plastic strain components by $\boldsymbol{\varepsilon}^{(i)\mathrm{e}}$ and $\boldsymbol{\varepsilon}^{(i)\mathrm{p}}$, we have

$$\boldsymbol{\sigma}^{(i)} = \mathbf{D}^{(i)}(\boldsymbol{\varepsilon} - \boldsymbol{\varepsilon}^{(i)\mathrm{p}}) \tag{3.7}$$

and

$$f^{(i)}(\boldsymbol{\sigma}^{(i)} - \boldsymbol{\alpha}^{(i)}) - \sigma_{\mathrm{p}}^{(i)} = f^{(i)}[\mathbf{D}^{(i)}\boldsymbol{\varepsilon} - (\mathbf{D}^{(i)} + c^{(i)}\mathbf{I})\boldsymbol{\varepsilon}^{(i)\mathrm{p}}] - \sigma_{\mathrm{p}}^{(i)} = 0 \tag{3.8}$$

The average stress within the macroelement is

$$\boldsymbol{\sigma} = \frac{1}{n}\sum_{i=1}^{n}\boldsymbol{\sigma}^{(i)} = \mathbf{D}\boldsymbol{\varepsilon} - \frac{1}{n}\sum_{i=1}^{n}\mathbf{D}^{(i)}\boldsymbol{\varepsilon}^{(i)\mathrm{p}}, \qquad \mathbf{D} = \frac{1}{n}\sum_{i=1}^{n}\mathbf{D}^{(i)} \tag{3.9}$$

In the case of the Huber–Mises yield condition only deviatoric stresses $\mathbf{s}^{(i)}$ occur in the yield condition. For elastically homogeneous layers there is $\mathbf{s}^{(i)} = 2G(\mathbf{e} - \mathbf{e}^{(i)\mathrm{p}})$ where $\mathbf{e}^{(i)}$ denotes the deviatoric strain and G is the Kirchhoff modulus, and the yield condition has the form

$$\tfrac{3}{2}(\mathbf{s}^{(i)} - \boldsymbol{\alpha}^{(i)}) \cdot (\mathbf{s}^{(i)} - \boldsymbol{\alpha}^{(i)}) - \sigma_{\mathrm{p}}^{(i)2} = 0 \tag{3.10}$$

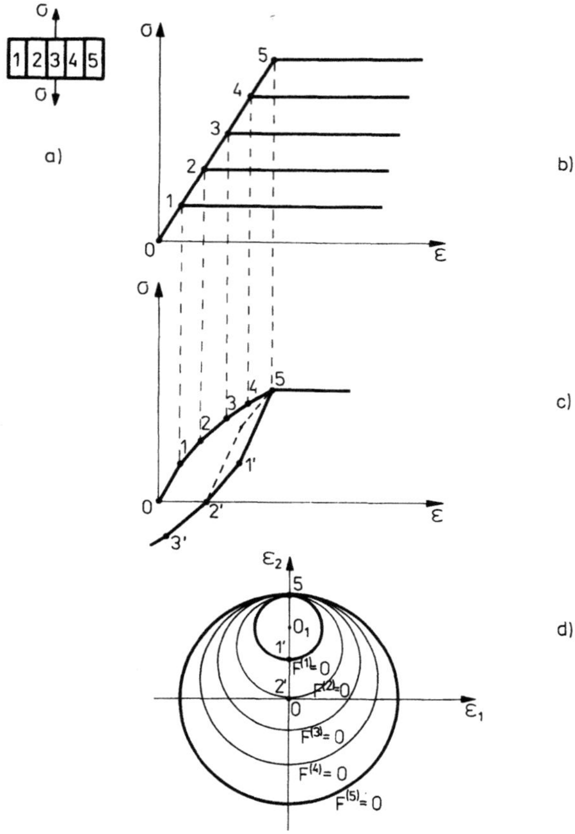

Fig. 4. (a) Composite parallel connection model; (b) stress–strain curves of subelements; (c) predicted loading–unloading curve; (d) evolution of yield surfaces in the strain space.

or in terms of strains

$$(\mathbf{e} - k\,\mathbf{e}^{(i)\mathrm{p}}) \cdot (\mathbf{e} - k\,\mathbf{e}^{(i)\mathrm{p}}) - \frac{\sigma_{\mathrm{p}}^{(i)2}}{2G^2} = 0 \qquad (3.11)$$

where $k = 1 + c/2G$, and $c^{(i)} = c$. In this case it is more convenient to use representation in the strain space: the active yield surfaces translate and consecutively activate subsequent surfaces; cf. Fig. 4(d).

This model was discussed in detail by Besseling (1958) and Wells and Paslay (1969), and applied in the numerical analysis of boundary value problems (cf. Zienkiewicz *et al.*, 1972). The series connection model was discussed by Iwan (1967). An extensive analysis of structural models was also presented in papers by Kafka (1968) and Zarka *et al.* (1980).

It is seen that, for a large number of layers, an accurate description of loading–unloading curves can be obtained. However, this accuracy is accompanied by a large number of material parameters to be identified. Similarly, as for series connection, the Masing relations (3.5) are generated for cyclic loading and closed loops are produced after the first loading cycle.

(iii) The two examples of parallel and series connections can be generalized to apply to any assemblage of elastic and plastic elements. Let us now discuss a general structure of such relations by considering a representative volume element loaded by the surface tractions $t_i^{\mathrm{b}} = \sigma_{ij}^{\mathrm{u}} \nu_j$ or displacements $u_i^{\mathrm{b}} = v_{ij}^{\mathrm{u}} x_j$ where σ_{ij}^{u} denotes the uniform stress tensor on the element boundary and v_{ij} is a constant matrix. The mean stress and strain tensors can now be expressed as follows (cf. Hill, 1967; Mandel, 1971)

$$
\begin{aligned}
V\bar{\sigma}_{ij} &= \tfrac{1}{2}\int (t_i^{\mathrm{b}} x_j + t_j^{\mathrm{b}} x_i)\,\mathrm{d}S \\
V\bar{\varepsilon}_{ij} &= \tfrac{1}{2}\int (u_i^{\mathrm{b}} \nu_j + u_j^{\mathrm{b}} \nu_i)\,\mathrm{d}S
\end{aligned}
\qquad (3.12)
$$

and it can easily be shown that $\bar{\sigma}_{ij} = \sigma_{ij}^{\mathrm{u}}$, $\bar{\varepsilon}_{ij} = \frac{1}{2}(v_{ij}^{\mathrm{u}} + v_{ji}^{\mathrm{u}})$. For the representative volume element there is

$$
\bar{\sigma} \cdot \bar{\varepsilon} = \frac{1}{V}\int \sigma \cdot \varepsilon \,\mathrm{d}V \qquad (3.13)
$$

Let us denote the purely elastic state within the element by σ^0, ε^0, $\mathbf{u}^0$ and assume the usual decomposition

$$
\sigma = \sigma^0 + \sigma^{\mathrm{r}}, \quad \varepsilon = \varepsilon^0 + \varepsilon^{\mathrm{r}} + \varepsilon^{\mathrm{p}} = \varepsilon^{\mathrm{e}} + \varepsilon^{\mathrm{p}}, \quad \varepsilon^{\mathrm{e}} = \varepsilon^0 + \varepsilon^{\mathrm{r}} \qquad (3.14)
$$

where σ^{r}, ε^{r} denote the residual stress and the associated strain fields, $\varepsilon^{\mathrm{r}} = \mathbf{C}\sigma^{\mathrm{r}}$ and ε^{p} denotes the plastic strain. Introduce the elastic stress and strain concentration matrices $\mathbf{A}(\mathbf{x})$ and $\mathbf{B}(\mathbf{x})$, so that

$$
\sigma^0(\mathbf{x}) = \mathbf{A}(\mathbf{x})\bar{\sigma}, \qquad \varepsilon^0(\mathbf{x}) = \mathbf{B}(\mathbf{x})\bar{\varepsilon} \qquad (3.15)
$$

and denoting by $\mathbf{C}_l$, $\mathbf{D}_l$ the local compliance and stiffness matrices, we have

$$
\begin{aligned}
V\bar{\boldsymbol{\sigma}} &= \int \mathbf{D}_l \boldsymbol{\varepsilon}^0 \,\mathrm{d}V = \int \mathbf{D}_l \mathbf{B} \bar{\boldsymbol{\varepsilon}} \,\mathrm{d}V = \mathbf{D}\bar{\boldsymbol{\varepsilon}}, \quad \mathbf{D} = \frac{1}{V}\int \mathbf{D}_l \mathbf{B} \,\mathrm{d}V \\
V\bar{\boldsymbol{\varepsilon}} &= \int \mathbf{C}_l \boldsymbol{\sigma}^0 \,\mathrm{d}V = \int \mathbf{C}_l \mathbf{A} \bar{\boldsymbol{\sigma}} \,\mathrm{d}V = \mathbf{D}\bar{\boldsymbol{\sigma}}, \quad \mathbf{C} = \frac{1}{V}\int \mathbf{C}_l \mathbf{A} \,\mathrm{d}V
\end{aligned}
\tag{3.16}
$$

The virtual work equation for the traction-loaded representative element can be written in the form

$$
\begin{aligned}
V\bar{\boldsymbol{\sigma}} \cdot \bar{\boldsymbol{\varepsilon}}^{\mathrm{p}} &= \int (\boldsymbol{\sigma}^0 + \boldsymbol{\sigma}_{\mathrm{s}}^{\mathrm{r}}) \cdot (\boldsymbol{\varepsilon}_{\mathrm{k}}^{\mathrm{r}} + \boldsymbol{\varepsilon}_{\mathrm{k}}^{\mathrm{p}}) \,\mathrm{d}V = \int \boldsymbol{\sigma}_{\mathrm{s}}^{\mathrm{r}} \cdot \boldsymbol{\varepsilon}_{\mathrm{k}}^{\mathrm{r}} \,\mathrm{d}V \\
&\quad + \int \boldsymbol{\sigma}^0 \cdot \boldsymbol{\varepsilon}_{\mathrm{k}}^{\mathrm{p}} \,\mathrm{d}V + \int \boldsymbol{\sigma}_{\mathrm{s}}^{\mathrm{r}} \cdot \boldsymbol{\varepsilon}_{\mathrm{k}}^{\mathrm{p}} \,\mathrm{d}V
\end{aligned}
\tag{3.17}
$$

where $\boldsymbol{\sigma}_{\mathrm{s}} = \boldsymbol{\sigma}^0 + \boldsymbol{\sigma}_{\mathrm{s}}^{\mathrm{r}}$ and $\boldsymbol{\varepsilon}_{\mathrm{k}} = \boldsymbol{\varepsilon}^0 + \boldsymbol{\varepsilon}_{\mathrm{k}}^{\mathrm{r}} + \boldsymbol{\varepsilon}_{\mathrm{k}}^{\mathrm{p}}$ are statically and kinematically admissible fields for the elements and $\bar{\boldsymbol{\varepsilon}}^{\mathrm{p}} = \bar{\boldsymbol{\varepsilon}} - \bar{\boldsymbol{\varepsilon}}^0$. Setting $\boldsymbol{\sigma}_{\mathrm{s}} = \boldsymbol{\sigma}^0$, $\boldsymbol{\sigma}_{\mathrm{s}}^{\mathrm{r}} = 0$, from eqn (3.17) we obtain

$$
\begin{aligned}
V\bar{\boldsymbol{\sigma}} \cdot \bar{\boldsymbol{\varepsilon}}^{\mathrm{p}} &= \int \boldsymbol{\sigma}^0 \cdot \boldsymbol{\varepsilon}^{\mathrm{p}} \,\mathrm{d}V = \int \mathbf{A}\bar{\boldsymbol{\sigma}} \cdot \boldsymbol{\varepsilon}^{\mathrm{p}} \,\mathrm{d}V = \bar{\boldsymbol{\sigma}} \cdot \int \mathbf{A}^{\mathrm{T}} \boldsymbol{\varepsilon}^{\mathrm{p}} \,\mathrm{d}V \\
V\bar{\boldsymbol{\sigma}} \cdot \bar{\boldsymbol{\varepsilon}}^0 &= \int \boldsymbol{\sigma}^0 \cdot \boldsymbol{\varepsilon}^0 \,\mathrm{d}V = \int \mathbf{A}\bar{\boldsymbol{\sigma}} \cdot \boldsymbol{\varepsilon}^0 \,\mathrm{d}V = \bar{\boldsymbol{\sigma}} \cdot \int \mathbf{A}^{\mathrm{T}} \boldsymbol{\varepsilon}^0 \,\mathrm{d}V
\end{aligned}
\tag{3.18}
$$

and since the equalities (3.18) are valid for any values of $\bar{\boldsymbol{\sigma}}$, there is

$$
\bar{\boldsymbol{\varepsilon}}^{\mathrm{p}} = \frac{1}{V}\int \mathbf{A}^{\mathrm{T}} \boldsymbol{\varepsilon}^{\mathrm{p}} \,\mathrm{d}V, \quad \bar{\boldsymbol{\varepsilon}}^0 = \frac{1}{V}\int \mathbf{A}^{\mathrm{T}} \boldsymbol{\varepsilon}^0 \,\mathrm{d}V, \qquad \bar{\boldsymbol{\varepsilon}} = \bar{\boldsymbol{\varepsilon}}^0 + \bar{\boldsymbol{\varepsilon}}^{\mathrm{p}} \tag{3.19}
$$

Similarly, considering rates of elastic and plastic strain, the virtual work equation (3.17) can be written as follows

$$
V(\bar{\boldsymbol{\sigma}} - \bar{\boldsymbol{\sigma}}_{\mathrm{s}}) \cdot \dot{\bar{\boldsymbol{\varepsilon}}}^{\mathrm{p}} - \int (\boldsymbol{\sigma}^{\mathrm{r}} - \boldsymbol{\sigma}_{\mathrm{s}}^{\mathrm{r}}) \cdot \dot{\boldsymbol{\varepsilon}}^{\mathrm{r}} \,\mathrm{d}V = \int (\boldsymbol{\sigma} - \boldsymbol{\sigma}_{\mathrm{s}}) \cdot \dot{\boldsymbol{\varepsilon}}^{\mathrm{p}} \,\mathrm{d}V \tag{3.20}
$$

where $\boldsymbol{\sigma}_{\mathrm{s}} = \boldsymbol{\sigma}^0 + \boldsymbol{\sigma}_{\mathrm{s}}^{\mathrm{r}}$ denotes any statically admissible stress field corresponding to the mean stress $\bar{\boldsymbol{\sigma}}_{\mathrm{s}}$ and not violating the yield condition, thus

$$
F(\sigma_{\mathrm{s}}) \leqslant 0, \qquad F(\sigma) \leqslant 0 \tag{3.21}
$$

Assuming local normality of the plastic strain rate and convexity of the yield surface, the overall normality and convexity now follows from eqn (3.20) since, by setting $\boldsymbol{\sigma}^{\mathrm{r}} = \boldsymbol{\sigma}_{\mathrm{s}}^{\mathrm{r}}$, we obtain

$$(\bar{\sigma} - \bar{\sigma}_{\mathrm{s}}) \cdot \dot{\bar{\varepsilon}}^{\mathrm{p}} \geqslant 0 \qquad \text{when} \qquad (\sigma - \sigma_{\mathrm{s}}) \cdot \dot{\varepsilon}^{\mathrm{p}} \geqslant 0 \tag{3.22}$$

On the other hand, since the field $\boldsymbol{\varepsilon}^{\mathrm{r}} + \boldsymbol{\varepsilon}^{\mathrm{p}}$ satisfies the compatibility condition, from eqn (3.20) it follows that

$$\int \delta\boldsymbol{\sigma}^{\mathrm{r}} \cdot (\dot{\boldsymbol{\varepsilon}}^{\mathrm{r}} + \dot{\boldsymbol{\varepsilon}}^{\mathrm{p}})\,\mathrm{d}V = 0 \tag{3.23}$$

for any variation of the residual stress field. The equality (3.23) can be interpreted as the evolution rule for $\dot{\boldsymbol{\varepsilon}}^{\mathrm{r}}$. Consider, for instance, the case when the residual stress and strain fields depend *linearly* on a set of stress parameters, R_{l}, such that

$$R_{\mathrm{l}}\dot{r}_{\mathrm{l}} = -\int \boldsymbol{\sigma}^{\mathrm{r}}(\mathbf{R}) \cdot \dot{\varepsilon}^{\mathrm{r}}\,\mathrm{d}V = \int \boldsymbol{\sigma}^{\mathrm{r}}(\mathbf{R}) \cdot \dot{\varepsilon}^{\mathrm{p}}\,\mathrm{d}V \tag{3.24}$$

where r_{l} are conjugate rates to R_{l}. Since there is

$$\delta R_{\mathrm{l}}\dot{r}_{\mathrm{l}} = -\delta R_{\mathrm{l}} \int \frac{\partial \sigma^{\mathrm{r}}}{\partial R_{\mathrm{l}}} \cdot \dot{\varepsilon}^{\mathrm{r}}\,\mathrm{d}V = \delta R_{\mathrm{l}} \int \frac{\partial \sigma^{\mathrm{r}}}{\partial R_{\mathrm{l}}} \cdot \dot{\varepsilon}^{\mathrm{p}}\,\mathrm{d}V \tag{3.25}$$

we have

$$\dot{r}_{\mathrm{l}} = \int \frac{\partial \sigma^{\mathrm{r}}}{\partial R_{\mathrm{l}}} \cdot \dot{\varepsilon}^{\mathrm{p}}\,\mathrm{d}V = \int \dot{\lambda}\frac{\partial \sigma^{\mathrm{r}}}{\partial R_{\mathrm{l}}} \cdot \frac{\partial F}{\partial \sigma}\,\mathrm{d}V = \int \dot{\lambda}\frac{\partial F}{\partial R_{\mathrm{l}}}\,\mathrm{d}V \tag{3.26}$$

Denoting the yield condition for the representative element

$$F(\boldsymbol{\sigma}^0 + \boldsymbol{\sigma}^{\mathrm{r}}) = F[\mathbf{A}\bar{\sigma} + \boldsymbol{\sigma}^{\mathrm{r}}(\mathbf{R})] = G(\bar{\sigma}, \mathbf{R}) = 0 \tag{3.27}$$

we can write

$$\dot{\bar{\varepsilon}}^{\mathrm{p}} = \dot{\Lambda}\frac{\partial G}{\partial \bar{\sigma}}, \qquad \dot{\mathbf{r}} = \dot{\Lambda}\frac{\partial G}{\partial \mathbf{R}} \tag{3.28}$$

that is the generalized normality occurs not only for the macroscopic plastic strain rate but also for the rates $\dot{r}_{\mathrm{l}}$ conjugate to the residual stress parameters R_{l}. The evolution rules (3.28) are identical to (2.63). However, it must be remembered that such rules occur only when the residual stress field is linearly related to the stress parameters R_{l} (which is the case for redundant elastic-plastic frame or beam structures within small deflection theory).

3.2. Representative State Variable Approach

In this sub-section we shall discuss an alternative approach which is similar to the composite structure approach but differs from it in specifying the state variables. Whereas in the previous approach the state variables represented stress or strain of particular subelements of the assemblage, the present method will be related to the concept of integrated states referred to particular levels of progression of plastic strains.

To illustrate the concept, consider a system of blocks interconnected by springs and resting on the frictional plane (Fig. 5). This kind of model has already been discussed by Ivlev (1963), Prager (1966) and Mróz (1972). Let the displacement of point A correspond to total strain, and the displacement of the block 0 represent the macroscopic

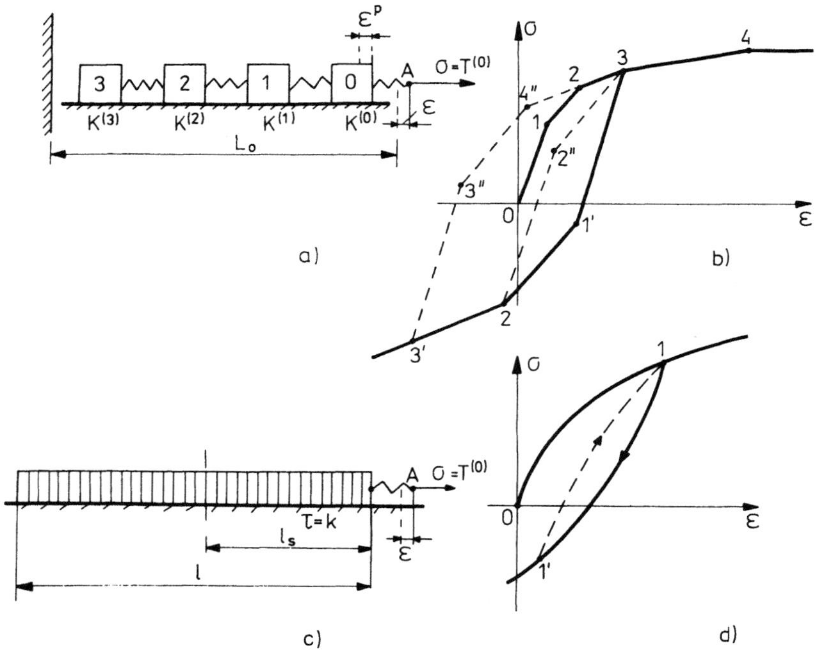

Fig. 5. Frictional models: (a) discrete model composed of a set of blocks; (b) predicted loading–unloading curve; (c) continuous friction model; (d) predicted response curve.

plastic strain, whereas the displacements of blocks $1, 2, 3, \ldots$ and spring forces acting between blocks represent the state variables of the system. Denoting the limiting friction forces of consecutive blocks by $K^{(0)}, K^{(1)}, K^{(2)}, \ldots$ and forces in connecting springs by $T^{(1)}, T^{(2)}, T^{(3)}, \ldots,$ we can write the equilibrium condition of the first block in the state of sliding

$$T^{(0)} - T^{(1)} = K^{(0)} \tag{3.29}$$

where $T^{(0)} = \sigma$ denotes the applied force to the system. When the second block is in the limit state, we have

$$T^{(1)} - T^{(2)} = K^{(1)} \tag{3.30}$$

Combining eqns (3.29) and (3.30), we obtain the equilibrium condition for two blocks

$$T^{(0)} - T^{(2)} = K^{(0)} + K^{(1)} \tag{3.31}$$

Similarly, for k gliding blocks, we obtain

$$T^{(0)} - T^{(k)} = K^{(0)} + K^{(1)} + \ldots + K^{(k-1)} \tag{3.32}$$

i.e. the equilibrium condition for a set of blocks is expressed in terms of the applied force and only one internal force. The forms (3.31) and (3.32) give us the idea of a new set of state variables representing the motion of a specified fraction of blocks. The piecewise-linear stress–strain curve of Fig. 5(b) is characterized by stress values at points 1, 2, 3, 4, and slopes of segments 1–2, 2–3, 3–4, ... representing friction resistance and system stiffness when moving one, two and three consecutive blocks. These quantities can therefore be used as representative for our system. Figure 5(c) represents a continuous system consisting of an elastic bar resting on the frictional foundation with the external force $\sigma = T^{(0)}$ applied at A to the spring connected to the bar. Now, the corresponding stress–strain curve is continuous and the characteristic parameter is the ratio of the slipping portion of the bar to the total bar length, $\eta = l_s/l$; cf. Fig. 5(c). Without going into details of calculation, it can be said that the variable slip fraction η, the friction force $T = kl_s$, and the elastic stiffness of the bar and spring characterize the stress–strain curve. Though the system of Fig. 5(c) can be regarded as the limiting case of the discrete system of Fig. 5(a) with the number of blocks tending to infinity, there is no need to introduce an infinite number of state variables. Using properly specified representative variables, a simple description of the system can be obtained with only several variables of this type.

Let us now discuss the formal structure of the evolution rules using the representative state variable approach; cf. Mróz (1972). Restricting ourselves to isothermal deformation, consider the yield condition in the form

$$F^0(\sigma, \boldsymbol{\alpha}^{(s)}) = 0, \qquad s = 1, 2, \ldots, k \tag{3.33}$$

where $\boldsymbol{\alpha}^{(s)}$ denote collectively scalar or tensor state variables. During deformation in the elastic range, there is no variation of state variables, so $\dot{\boldsymbol{\alpha}}^{(s)} = 0$ when $\dot{\boldsymbol{\varepsilon}}^{\mathrm{p}} = 0$. However, there is no uniform evolution of $\boldsymbol{\alpha}^{(s)}$ during the inelastic deformation process, but particular variables are assumed to vary only for specified states of the material or different evolution rules occur for different internal states. To make a distinction between internal states, let us introduce m additional scalar relations defining the *regions of internal irreversibility,* for instance, in the form

$$F_a^{(q)}(\sigma, \boldsymbol{\alpha}^{(s)}) = 0 \tag{3.34}$$

or

$$F_a^{(q)}(\boldsymbol{\alpha}^{(s)}, \boldsymbol{\varepsilon}^{\mathrm{p}}) = 0 \tag{3.35}$$

Thus, when the equality $F_a^{(q)} = 0$ is satisfied, the corresponding parameter $\boldsymbol{\alpha}^{(q)}$ varies during plastic deformation according to some specified rule. On the other hand, when $F_a^{(q)} < 0$, then $\dot{\boldsymbol{\alpha}}^{(q)} = 0$ whereas the other parameters can vary. It may also be assumed that a single equality of type (3.34) or (3.35) governs the variability of several parameters.

This form of the evolution rules imposes some internal consistency relations. Assume, for instance, the evolution rule for $\boldsymbol{\alpha}^{(q)}$ in the form

$$\begin{aligned} \dot{\boldsymbol{\alpha}}^{(q)} &= \mathbf{b}^{(q)}\dot{\lambda} \qquad \text{for} \qquad F_a^{(q)} = 0 \\ \dot{\boldsymbol{\alpha}}^{(q)} &= 0 \qquad\quad \text{for} \qquad F_a^{(q)} < 0 \end{aligned} \tag{3.36}$$

where $\lambda = \lambda_2$ is specified by eqn (2.2). In view of the flow rule (2.6), there is

$$\dot{\boldsymbol{\alpha}}^{(q)} = \frac{1}{H}\mathbf{b}^{(q)} \left| \frac{\partial F}{\partial \sigma} \right| \left(\frac{\partial F}{\partial \sigma} \cdot \dot{\sigma} \right) = \mathbf{B}^{(q)} \left(\frac{\partial F}{\partial \sigma} \cdot \dot{\sigma} \right) \tag{3.37}$$

On the other hand, the continuity condition for $F_a^{(q)} = 0$ requires that

$$\dot{\boldsymbol{\alpha}}^{(q)} = \chi \mathbf{B}^{(q)} \left(\frac{\partial F_a^{(q)}}{\partial \sigma} \cdot \dot{\sigma} \right) \tag{3.38}$$

where $\mathbf{b}^{(q)}$ and $\mathbf{B}^{(q)} = 1/H|\partial F/\partial \sigma|\mathbf{b}^{(q)}$ are tensors and χ is a scalar constant. Comparing eqns (3.37) and (3.38), we obtain

$$\frac{\partial F}{\partial \sigma} = \chi \frac{\partial F_a^{(q)}}{\partial \sigma} \tag{3.39}$$

i.e. the gradients of the yield and irreversibility surfaces are proportional. This means that in the stress space the regular surfaces $F = 0$ and $F_a^{(q)} = 0$ should be tangential to each other. Further, differentiating eqns (3.34) and (3.35), in view of the rules (3.36) leads to

$$\begin{gathered} H + \sum_{s=1}^{k} \frac{\partial F}{\partial \alpha^{(s)}} \cdot \mathbf{b}^{(s)} \left| \frac{\partial F}{\partial \sigma} \right| = 0 \\ H + \sum_{s=1}^{k} \frac{\partial F_a^{(q)}}{\partial \alpha^{(s)}} \cdot \mathbf{b}^{(s)} \left| \frac{\partial F_a^{(q)}}{\partial \sigma} \right| = 0 \end{gathered} \tag{3.40}$$

Equations (3.40) are the consistency conditions imposing constraints on H and b occurring in the flow rule (2.6) and the evolution rules (3.36). Thus, when the number of constraint equations changes, the hardening modulus H or the evolution tensors $\mathbf{b}^{(s)}$ should also change. Similar consistency relations can be derived when using the irreversibility conditions (3.35).

Now, let us discuss a particular model within such a general structure, based on the concept of the field of hardening moduli; cf. Mróz (1967, 1969). Assume the yield condition in the form

$$F^{(0)} = f(\sigma - \alpha^{(0)}) - \sigma^{(0)2} = 0 \tag{3.41}$$

where f is a homogeneous function of second order, and $\alpha^{(0)}$, $\sigma^{(0)}$ are the back stress and the yield limit. Further, it is postulated that the state of hardening is specified not only by the yield condition but also by a set of nesting surfaces of constant hardening moduli $K_0, K_1, K_2, \ldots, K_n$, defined by eqn (2.8), where K_0 denotes the hardening modulus on the yield surface and K_i are referred to particular surfaces

$$F^{(i)} = f(\sigma - \alpha^{(i)}) - \sigma^{(i)2} = 0 \tag{3.42}$$

The set of equations (3.42) can be regarded as specifying internal irreversibility domains, such as (3.34), and together with the yield condition they constitute a field of hardening moduli in the stress space. From proportional loading and unloading tests this field can easily be identified. In fact, instead of plotting conventional yield surfaces defined by the assumed value of plastic strain, more meaningful information can be produced by specifying the field of hardening moduli since this field indicates the stiffness anisotropy of the material and its variation in the course of plastic deformation; cf. Haythornthwaite (1968).

Consider first the case of proportional loading, for instance, tension and compression. Similarly, as in Fig. 5(b), approximate the stress–strain curve by linear segments 0–1, 1–2, 2–3, 3–4 along which the tangent modulus is constant. In the stress space this approximation can be visualized by introducing surfaces $F^{(0)} = 0$, $F^{(1)} = 0$, $F^{(2)} = 0$, ... defining the regions of constant hardening moduli in the stress space. When the stress point moving from 0 reaches $F^{(0)} = 0$ at 1, the yield surface $F^{(0)} = 0$ starts to translate with the stress point toward a point 2 and from 2 the two surfaces $F^{(0)} = 0$ and $F^{(1)} = 0$ move together until a point 3 is reached. Figure 6(a) shows the field of hardening moduli after reaching the point 3. During unloading and reverse loading the surfaces translate backwards and at 3′ the symmetric configuration with respect to 3 is obtained; Fig. 6(b). It is seen that, when surfaces do not expand but only translate, again the Masing hardening rule is obtained.

For non-proportional loading it is assumed, according to the consistency condition (3.39), that the surfaces $F^{(0)} = 0, F^{(1)} = 0, F^{(2)} = 0$ do not intersect but consecutively contact and push each other. For convex, similar surfaces this is achieved by postulating that the instantaneous motion of the stress point P corresponding to $F^{(l)} = 0$ is directed to the point R on $F^{(l+1)} = 0$ with the same normal as that at P. Similarly, when $F^{(0)} = 0$ and $F^{(1)} = 0$ are tangential to each other at P, the translation of P occurs along PR, where R is an associated point on $F^{(2)} = 0$ corresponding to the same normal. The position of the associated point R is determined from the proportionality relation for any two similar surfaces $F^{(l)} = 0$ and $F^{(l+1)} = 0$,

$$\boldsymbol{\sigma}_{\mathrm{R}}^{(l+1)} - \boldsymbol{\alpha}^{(l+1)} = \frac{\sigma^{(l+1)}}{\sigma^{(l)}}(\boldsymbol{\sigma}_{\mathrm{P}}^{(l)} - \boldsymbol{\alpha}^{(l)}) \tag{3.43}$$

and the vector $\boldsymbol{\beta}$ [Fig. 6(c)] connecting P and R is expressed as

$$\mathbf{PR} = \boldsymbol{\beta} = \frac{1}{\sigma^{(l)}}[(\sigma^{(l+1)} - \sigma^{(l)})\boldsymbol{\sigma}_{\mathrm{R}}^{(l+1)} - \sigma^{(l+1)}\boldsymbol{\alpha}^{(l)} - \sigma^{(l)}\boldsymbol{\alpha}^{(l+1)}] \tag{3.44}$$

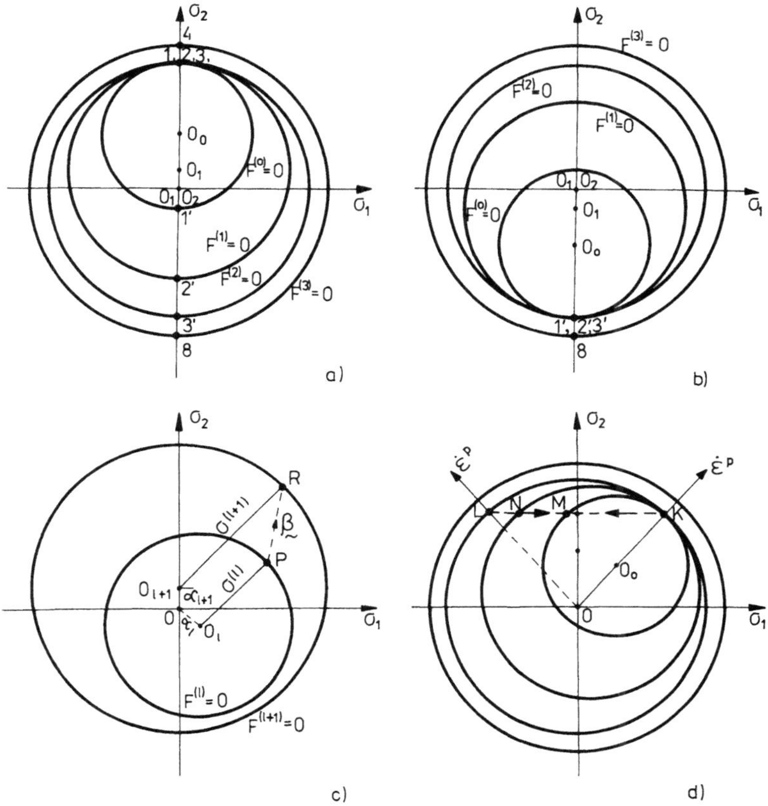

Fig. 6. Evolution of loading surfaces of the multisurface model: (a) after loading to 3; (b) after unloading to 3′; (c) translation rule of two neighbouring surfaces; (d) translation of loading surfaces along KL during cyclic loading.

The relative motion of P with respect to R is assumed to occur along PR, that is

$$\dot{\boldsymbol{\sigma}}_{\mathrm{P}}^{(l)} - \dot{\boldsymbol{\sigma}}_{\mathrm{R}}^{(l+1)} = \boldsymbol{\beta}\dot{\mu} \tag{3.45}$$

where $\dot{\mu}$ is a scalar factor. Since

$$\begin{aligned} \dot{\boldsymbol{\sigma}}_{\mathrm{P}}^{(l)} &= \dot{\boldsymbol{\alpha}}^{(l)} + (\boldsymbol{\sigma}_{\mathrm{P}}^{(l)} - \boldsymbol{\alpha}^{(l)})\frac{\dot{\sigma}^{(l)}}{\sigma^{(l)}} \\ \dot{\boldsymbol{\sigma}}_{\mathrm{R}}^{(l+1)} &= \dot{\boldsymbol{\alpha}}^{(l+1)} + (\boldsymbol{\sigma}_{\mathrm{R}}^{(l+1)} - \boldsymbol{\alpha}^{(l+1)})\frac{\dot{\sigma}^{(l+1)}}{\sigma^{(l+1)}} \end{aligned} \tag{3.46}$$

from eqn (3.45) we obtain

$$\dot{\boldsymbol{\alpha}}^{(l)} = \dot{\boldsymbol{\alpha}}^{(l+1)} + \boldsymbol{\beta}\dot{\mu} + \frac{\dot{\sigma}^{(l+1)} - \dot{\sigma}^{(l)}}{\sigma^{(l)}}(\boldsymbol{\sigma}_{\mathrm{P}}^{(l)} - \boldsymbol{\alpha}^{(l)}) \tag{3.47}$$

and the scalar $\dot{\mu}$ can be determined from the consistency condition (3.40). When all surfaces $F^{(0)} = 0$, $F^{(1)} = 0, \ldots, F^{(l-1)} = 0$ are in contact with the surface $F^{(l)} = 0$ and move with the stress point P, their translation is governed by the motion of P and we have

$$\frac{(\boldsymbol{\sigma}_{\mathrm{P}}^{(l)} - \boldsymbol{\alpha}^{(k)})}{(\boldsymbol{\sigma}_{\mathrm{P}}^{(l)} - \boldsymbol{\alpha}^{(l)})} = \frac{\sigma^{(k)}}{\sigma^{(l)}}, \qquad k = 0, 1, 2, \ldots, l \tag{3.48}$$

Figure 6(d) represents the case when the stress component σ_1 oscillates between the values at K and L with fixed stress component σ_2. Now, the yield and subsequent nesting surfaces will translate with the stress point between K and L, so the steady cyclic state corresponds to progressing accumulation of the plastic strain component $\varepsilon_1^{\mathrm{p}}$ after each stress cycle. The present translation rule is therefore reminiscent of the rules (2.21) and (2.26), where the total translation rate can be decomposed into the component normal to the yield surface and the relaxation component directed to the origin in the stress space.

The field hardening moduli formulation is similar to the series connection model as it is represented by a set of surfaces in the stress space. However, this similarity occurs only for proportional loading paths (cf. Figs 3 and 6). The surfaces $F^{(0)} = 0$, $F^{(1)} = 0, \ldots$ in Fig. 3 represent yield conditions of particular elements and may intersect each other for non-proportional paths, whereas the surfaces $F^{(0)} = 0$, $F^{(1)} = 0, \ldots$ in Fig. 6 represent overall changes in material stiffness due to plastic deformation and remain tangential to each other or separate. These two models are therefore conceptually different and should not be identified (as is sometimes done in the literature). Combined isotropic and kinematic hardening of material within multisurface formulation was treated by Petersson and Popov (1977).

Now let us pass to discussion of particular cases and of some modified versions of this model which may prove useful in practical applications.

Consider first the case when there are only two surfaces $F_0 = 0$ and $F_l = 0$ and the hardening modulus takes two values K_0 and K_l (Fig. 7). In particular, the surface $F_l = 0$ can be identified with the limit surface on which $K_l = 0$. This model was discussed by Mróz (1972) and applied to simulate cyclic behaviour of thin-walled tubes under fixed tension and alternating torsion.

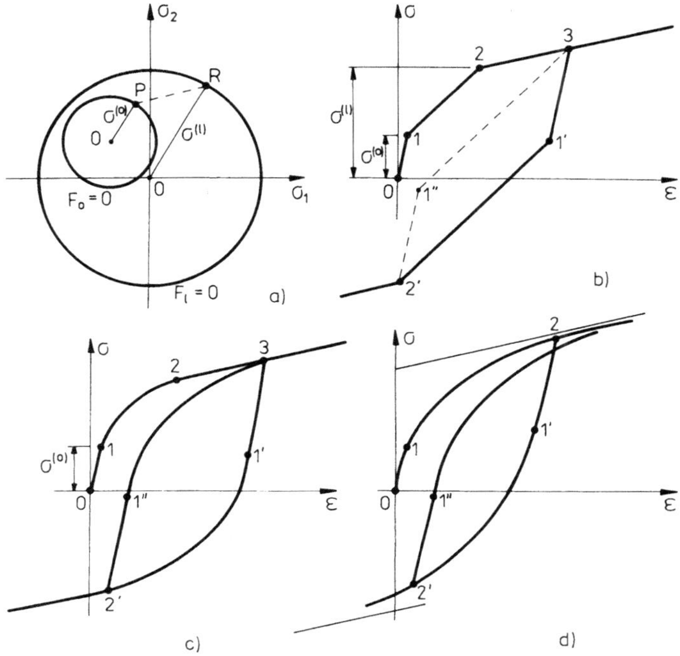

Fig. 7. Two-surface hardening model: (a) configuration of surfaces; (b) piecewise-linear approximation of the stress–strain curve; (c) curvilinear approximation; (d) the case of translation of the bounding surface $F_l = 0$ and the resulting stress–strain curve.

Assume first that the surface $F_l = 0$ remains unchanged before it is contacted by the stress point and next translates with the stress point according to the rule (2.9). The yield surface $F_0 = 0$ translates toward the associated point R in accordance with the rule (3.47) and next it becomes tangential to the exterior surface $F_l = 0$. Figure 7(b) shows the piecewise-linear approximation of the loading and unloading curves corresponding to this model. A next step to generalize this model is to assume that the hardening modulus K varies non-linearly with the distance δ between the surfaces $F_0 = 0$ and $F_l = 0$. Assume that this distance is measured by the length of PR. The maximal value of δ is $\delta_m = \sigma^{(l)} - \sigma^{(0)}$ for the first loading and $\delta_m = 2(\sigma^{(l)} - \sigma^{(0)})$ for subsequent unloading from the surface $F_l = 0$, whereas the minimal value $\delta_{min} = 0$. Assume that the hardening modulus K takes an infinite or very large value for $\delta = \delta_m$ and $K = K_l$ for $\delta = 0$, that is at the contact point between the two surfaces. The variation of K can be, for instance, expressed by the relation

$$K = K_l + a\left(\frac{\delta}{\delta_m - \delta}\right)^m \tag{3.49}$$

where a and m are material parameters and δ_m is the maximal value of δ from the prior deformation history. Figure 7(c) shows the loading, unloading and reloading curves obtained for such a non-linear description. When the stress point reaches the surface $F_l = 0$, the subsequent behaviour is described by the constant hardening modulus K_l.

This non-linear model can further be generalized by assuming that the exterior surface $F_l = 0$ undergoes translation before it is contacted by the stress point. We have then

$$\begin{aligned} F^{(0)} &= f(\boldsymbol{\sigma} - \boldsymbol{\alpha}^{(0)}) - \sigma^{(0)2} = 0 \\ F^{(l)} &= f(\boldsymbol{\sigma} - \boldsymbol{\alpha}^{(l)}) - \sigma^{(l)2} = 0 \end{aligned} \tag{3.50}$$

where

$$\begin{aligned} \dot{\boldsymbol{\alpha}}^{(l)} &= \boldsymbol{\gamma}\dot{\lambda} \\ \dot{\boldsymbol{\alpha}}^{(0)} &= \dot{\boldsymbol{\alpha}}^{(l)} + \boldsymbol{\beta}\dot{\mu} \end{aligned} \tag{3.51}$$

and $\boldsymbol{\beta}$ is specified by eqn (3.44), whereas $\boldsymbol{\gamma}$ is the translation vector of the bounding surface $F_l = 0$. The factor $\dot{\mu}$ is determined from the consistency condition of eqns (3.50). Figure 7(d) represents the corresponding loading, unloading and reloading curves obtained for this last model version. Obviously additional expansion of both surfaces can be incorporated into the model.

The non-linear variation of the hardening modulus for the two-surface model was discussed independently by Dafalias and Popov (1975) and Krieg (1975). The formulation of Dafalias and Popov admits both translation and expansion of the two surfaces. The case when the external surface is a limit surface was applied by Lamba and Sidebottom (1978) in their study of cyclic hardening for non-proportional loading paths.

Let us now discuss the other extreme case of the multisurface hardening rule by assuming that there is an infinite number of nesting surfaces specifying the field of the hardening moduli. This version was discussed by Mróz (1972), by Mróz and Lind (1975) for the case of the Tresca yield condition, and by Koning (1977) for the Mises yield condition. Referring to Fig. 8, assume that in the loading process OA the stress point has moved into the elastic-plastic region and reached the exterior surface $F_m = 0$. At A all surfaces $F^{(k)} = 0$, $k = 0, 1, \ldots, m-1$, are tangential to the surface $F_m = 0$. If the loading process continues, the subsequent incremental response will be specified by the active loading surface $F_m = 0$ and the hardening modulus ascribed to this surface. Therefore all surfaces $F^{(k)} = 0$, $k = 0, 1, \ldots, m-1$, will be the passive surfaces and will not enter the material description. Consider now the unloading process AB. At B the surface $F_{l1} = 0$ tangential to $F_m = 0$ at A is the active loading surface. Let us note that, knowing the position of B,

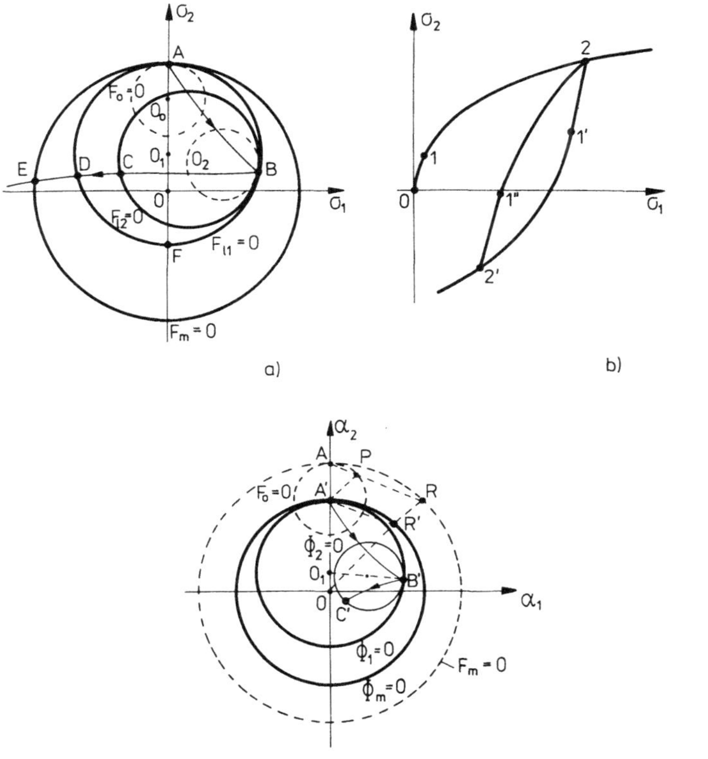

Fig. 8. Continuous description of field of hardening moduli and memory rules: (a) in the stress space; (b) loading–unloading curve; (c) representation in the space of back stress.

the centre and the diameter of the active loading surface $F_{l1} = 0$ are specified. During the next loading event, such as AB for which the stress path is directed into the interior of the domain enclosed by $F_m = 0$, we do not need therefore to trace the positions of all surfaces but only the active loading surface passing through B should be determined. When the loading path at B changes its orientation and follows BCD, a new loading event commences for which the active loading surface $F_{l2} = 0$ is tangential at B to $F_{l1} = 0$ with its centre on O_1B. Thus, at D, after the loading history OABD, only three surfaces characterize the prior history and are stored in the material memory, that is the maximal loading surface $F_m = 0$, first loading surface $F_{l1} = 0$ and second loading surface $F_{l2} = 0$. When the stress point on BCD reaches D, the surface $F_{l2} = 0$ coincides with the surface $F_{l1} = 0$ and the second loading event is erased from the material memory. Similarly, when the stress point reaches E, the loading path ABE is erased from the material memory and the state of the material is represented by the maximal loading surface $F_m = 0$ (which may become modified due to the loading history OABE).

Considering the first loading event and assuming the equations of $F_m = 0$ and $F_{l1} = 0$ in the form

$$\begin{aligned} F_m &= f(\boldsymbol{\sigma} - \boldsymbol{\alpha}^{(m)}) - \sigma^{(m)2} = 0 \\ F_{l1} &= f(\boldsymbol{\sigma} - \boldsymbol{\alpha}^{(l)}) - \sigma^{(l)2} = 0 \end{aligned} \tag{3.52}$$

we have also, for the loading path AB, the relations

$$\boldsymbol{\sigma}^{A} - \boldsymbol{\alpha}^{(l)} = (\boldsymbol{\sigma}^{A} - \boldsymbol{\alpha}^{(m)}) \frac{\sigma^{(l)}}{\sigma^{(m)}} \tag{3.53}$$

from which $\boldsymbol{\alpha}^{(l)}$ and $\sigma^{(l)}$ are determined once the position of the stress point B on $F_{l1} = 0$ is specified. Similarly, for the second loading event, the active loading surface $F_{l2} = 0$ is specified from similar relations.

Assume that the hardening modulus on the active loading surface $F_{li} = 0$ depends on the ratio of the semi-diameters of the maximal loading and active loading surfaces with respect to the diameter of the yield surface, thus

$$K = K(\Delta), \qquad \Delta = \frac{\sigma^{(l)} - \sigma^{(0)}}{\sigma^{(m)} - \sigma^{(0)}} \tag{3.54}$$

It is seen that $\Delta = 0$ when $\sigma^{(l)} = \sigma^{(0)}$ and $\Delta = 1$ when $\sigma^{(l)} = \sigma^{(m)}$.

Similarly to eqn (3.49), the variation of K can, for instance, be described by the relation

$$K = K_m + A\left(\frac{1-\Delta}{\Delta}\right)^p \tag{3.55}$$

where A and p are the material parameters and K_m denotes the value of K on the maximal loading surface.

The plasticity model with distinct yield and loading surfaces, similar to multisurface formulation, was discussed by Eisenberg and Phillips (1971).

A similar concept of the memory of maximal prestress can also be expressed in terms of a set of loading surfaces in the space of back stress $\boldsymbol{\alpha}$; cf. Fig. 8(c). After loading along OA, the α-path is OA. To define the effective scalar measure of $\boldsymbol{\alpha}$, let us introduce the function of $\boldsymbol{\alpha}$, for instance in the form

$$\alpha_e = \phi(\boldsymbol{\alpha}) = (\tfrac{3}{2}\boldsymbol{\alpha}\cdot\boldsymbol{\alpha})^{1/2} > 0 \tag{3.56}$$

$\alpha_e = 0$ for $\boldsymbol{\alpha} = 0$. The maximal value of $\boldsymbol{\alpha}$ reached at A is now represented by the surface

$$\phi_m = \phi(\boldsymbol{\alpha}) - \alpha_m = 0 \tag{3.57}$$

where $\alpha_e = \alpha_m$. The subsequent process will be divided into two classes: those for which $\dot{\alpha}_e = \dot{\alpha}_m > 0$ (α_e loading) and those for which $\alpha_e < \alpha_m$ (α_e unloading). For the α_e unloading process A′B′ lying within the domain $\alpha_e - \alpha_m < 0$, let us introduce another surface within this domain similar to (3.57) and tangential to it at A, that is

$$\phi_1 = [\tfrac{3}{2}(\boldsymbol{\alpha} - \boldsymbol{\alpha}_0^{(1)})\cdot(\boldsymbol{\alpha}_0 - \boldsymbol{\alpha}_0^{(1)})]^{1/2} - \alpha_u^{(1)} = 0 \tag{3.58}$$

where $\boldsymbol{\alpha}_0^{(1)}$ is the centre position of the surface $\phi_1 = 0$ and $\alpha_u^{(1)}$ is proportional to its diameter. Let us note that, when $\boldsymbol{\alpha}'_B$ and $\boldsymbol{\alpha}'_A$ are given, both $\alpha_u^{(1)}$ and $\boldsymbol{\alpha}_0^{(1)}$ can be determined. In fact, since the surface $\phi_1 = 0$ passes through B′, we have

$$\boldsymbol{\alpha}_0^{(1)} = \boldsymbol{\alpha}_{A'} - g^{(1)}\boldsymbol{\nu}_{A'}$$

$$\alpha_u^{(1)} = \sqrt{\left(\frac{3}{2}\right)}\cdot\frac{(\boldsymbol{\alpha}_{B'} - \boldsymbol{\alpha}_{A'})\cdot(\boldsymbol{\alpha}_{B'} - \boldsymbol{\alpha}_{A'})}{2(\boldsymbol{\alpha}_{A'} - \boldsymbol{\alpha}_{B'})\cdot\boldsymbol{\nu}_{A'}} = \sqrt{\left(\frac{3}{2}\right)}\cdot g^{(1)} \tag{3.59}$$

where $\boldsymbol{\nu}_{A'}$ is the unit normal vector along O_1A. The relations (3.59) prescribe the evolution rules for $\boldsymbol{\alpha}_0^{(1)}$ and $\alpha_u^{(1)}$. The first α_e unloading event corresponds to the inequalities $\dot{\alpha}_u^{(1)} > 0$, $\alpha_u^{(1)} < \alpha_m$ and, when $\dot{\alpha}_u^{(1)} < 0$ at B′, a new loading event commences and the subsequent path BC is directed into the interior of $\phi_1 = 0$. Similarly, as previously, we may construct a new surface $\phi_2 = 0$ tangential at B′ to $\phi_1 = 0$ with its diameter growing along the path BC.

It is seen that the definitions of particular loading events expressed in terms of the stress tensor [Fig. 8(a)] and the back stress tensor [Fig. 8(c)] are similar. However, the formulation in terms of stress provides also the evolution rule for $\boldsymbol{\alpha}$. In fact, when the stress point is at P in Fig. 8(c) and P lies within the maximal loading surface $F_m = 0$, then the instantaneous translation of P occurs along PR where R is the associated point on $F_m = 0$. The respective point R on the surface $F_m = 0$ lies on the radial line OR parallel to A′P. Denoting the position of R′ by $\bar{\boldsymbol{\alpha}}$, it may therefore be postulated that the instantaneous evolution of $\boldsymbol{\alpha}$ occurs along A′R′, thus

$$\dot{\boldsymbol{\alpha}} = c(\bar{\boldsymbol{\alpha}} - \boldsymbol{\alpha})\dot{\lambda} \tag{3.60}$$

where c is the material parameter. However, the memory rule formulation in terms of the back stress is more convenient since the evolution rule other than (3.60) can be used in the material description. For a detailed discussion of this model, the reader is referred to Mróz (1981).

An alternative formulation of the hardening rule using two surfaces was proposed by Chaboche *et al.* (1979) and extended by Murakami and Ohno (1982) for the case of creep. Besides the yield condition (2.14), the internal irreversibility surface was introduced in the form

$$F_\alpha = \phi(\boldsymbol{\varepsilon}^{\mathrm{p}} - \boldsymbol{\alpha}) - \rho^2(\boldsymbol{\varepsilon}^{\mathrm{p}}) = 0 \tag{3.61}$$

Thus for states $F_\alpha < 0$ there is $\dot{\boldsymbol{\alpha}} = \dot{\rho} = 0, \dot{\sigma}_{\mathrm{p}} = 0$, whereas the evolution of $\boldsymbol{\alpha}$, ρ, and σ_{p} occurs for states corresponding to the surface $F_\alpha = 0$. Here, β and ρ are new variables whose evolution should be specified. The functional description of evolution of $\boldsymbol{\alpha}$ was presented by Backhaus (1972). This description is closely related to the new formulation of the endochronic model; cf. Valanis and Lee (1982).

4. VISCOPLASTIC HARDENING RULES

In this section we shall discuss the material models in which the irreversible strain component $\boldsymbol{\varepsilon}^{\mathrm{i}}$ is assumed as time dependent. Thus in the usual decomposition

$$\dot{\boldsymbol{\varepsilon}} = \dot{\boldsymbol{\varepsilon}}^{e} + \dot{\boldsymbol{\varepsilon}}^{i}, \qquad \dot{\boldsymbol{\varepsilon}}^{i} = \dot{\boldsymbol{\varepsilon}}^{v} \tag{4.1}$$

the irreversible strain rate $\dot{\boldsymbol{\varepsilon}}^{v}$ varies with time and, for constant stress, the strain rate $\dot{\boldsymbol{\varepsilon}}^{v}$ tends to its asymptotic value. Here we do not decompose the irreversible strain $\boldsymbol{\varepsilon}^{i}$ into the instantaneous (plastic) term $\boldsymbol{\varepsilon}^{p}$ and the time-independent term $\boldsymbol{\varepsilon}^{v}$. This decomposition, sometimes used in constructing the creep models, follows from the assumption that strain developing quickly after load increment is assumed as instantaneous, $\boldsymbol{\varepsilon} = \boldsymbol{\varepsilon}^{p}$, whereas the slowly growing strain is regarded as viscous. Figure 9 shows the familiar decomposition proposed long ago by Odqvist for creep strains developing under applied constant stress. It turns out, however, that transient and steady viscous states can better be described by a proper combination of creep hardening rules.

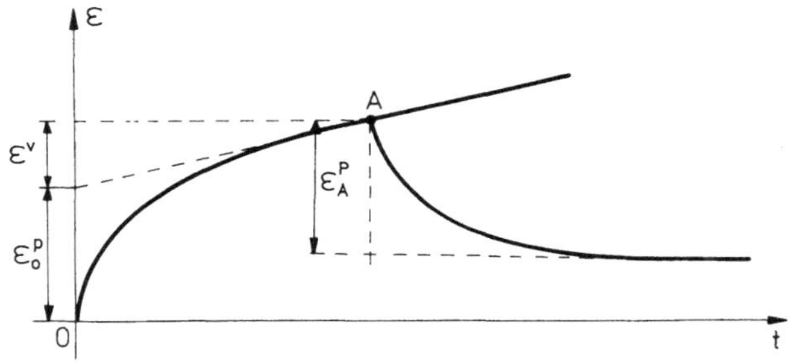

Fig. 9. Decomposition of the transient viscous strain into the instantaneous plastic part ε^{p} and the steady state creep strain ε^{v}.

4.1. Viscous Potential Formulation

Let us first discuss the formulation that is based on a non-linear viscous model. The viscous strain rate is now generated by the viscous potential rather than by the yield condition, thus

$$\dot{\boldsymbol{\varepsilon}}^{v} = \frac{\partial W(\boldsymbol{\sigma}, \boldsymbol{\alpha}, \sigma_{p})}{\partial \boldsymbol{\sigma}} \tag{4.2}$$

with the associated evolution law for the state variables $\boldsymbol{\alpha}$ and σ_{p}. Assume the viscous or viscoplastic potential in the form

$$W(\boldsymbol{\sigma}, \boldsymbol{\alpha}, \sigma_{p}) = W'(\mathrm{s} - \boldsymbol{\alpha}) - \sigma_{p}(\lambda) \tag{4.3}$$

or introducing the invariant

$$\bar{\sigma}_{e} = [\tfrac{3}{2}(\mathrm{s} - \boldsymbol{\alpha}) \cdot (\mathrm{s} - \boldsymbol{\alpha})]^{1/2} \tag{4.4}$$

and assuming the power law for which the viscous stress potential is

$$W' = \frac{\gamma}{n+1}\left(\frac{\bar{\sigma}_e}{\sigma_p} - 1\right)^{n+1} \tag{4.5}$$

the following flow rule is obtained

$$\dot{\boldsymbol{\varepsilon}}^v = \frac{\gamma}{\sigma_p}\left(\frac{\sigma_e}{\sigma_p} - 1\right)^n \tfrac{3}{2}\frac{\mathbf{s} - \boldsymbol{\alpha}}{\bar{\sigma}_e} \qquad \text{for} \quad \bar{\sigma}_e > \sigma_p$$

$$\dot{\boldsymbol{\varepsilon}}^v = 0 \qquad \text{for} \quad \bar{\sigma}_e < \sigma_p \tag{4.6}$$

where γ denotes the material parameter representing the material viscosity. Figure 10(a) shows the contours W = const in the stress plane (plotted schematically as circles). The elastic domain is defined by the inequality

$$F = \bar{\sigma}_e - \sigma_p \leqslant 0 \tag{4.7}$$

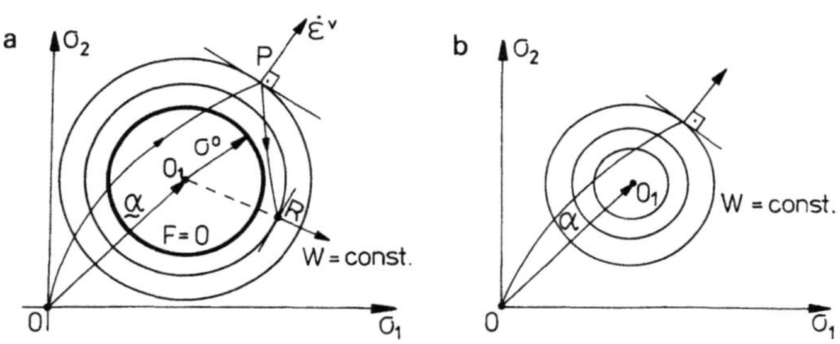

Fig. 10. Viscous potentials in the stress plane: (a) with elastic domain; (b) without elastic domain.

The viscous potential vanishes for $F = 0$, and for $F < 0$ only elastic strains occur. Thus, if the stress reaches the value represented by P and is kept fixed, viscous flow occurs and the yield surface will translate and expand or contract. Figure 10(b) shows the contours W = const for a viscous model with vanishing elastic domain, for which eqns (4.5) and (4.6) become

$$W' = \frac{\gamma}{n+1}\left(\frac{\bar{\sigma}_e}{\sigma_0}\right)^{n+1} = \frac{\bar{\gamma}}{n+1}\bar{\sigma}_e^{n+1}, \qquad \dot{\boldsymbol{\varepsilon}}^v = \tfrac{3}{2}\bar{\gamma}\bar{\sigma}_e^{n-1}(\mathbf{s} - \boldsymbol{\alpha}) \tag{4.8}$$

The flow rule (4.6) and (4.8) should be supplemented by the evolution rules for $\boldsymbol{\alpha}$ and σ_{p}. Let us assume that

$$\begin{aligned} \dot{\boldsymbol{\alpha}} &= A_{\alpha}\dot{\boldsymbol{\varepsilon}}^{\mathrm{v}} - B_{\alpha}\dot{\boldsymbol{\alpha}} \\ \dot{\sigma}_{\mathrm{p}} &= A_{\mathrm{p}}\dot{\lambda} - B_{\mathrm{p}}(\sigma_{\mathrm{p}} - \sigma_{\mathrm{p}0}) \end{aligned} \tag{4.9}$$

where $A_{\alpha}, B_{\alpha}, A_{\mathrm{p}}, B_{\mathrm{p}}$ are four material parameters (or functions) and $\sigma_{\mathrm{p}0}$ denotes the initial value of σ_{p}. It is seen that the first term of eqns (4.9) corresponds to hardening due to viscous flow whereas the second term represents the viscous recovery of the back stress. Here the original idea of Bailey (1926) and Orowan (1946) is incorporated, namely the viscous flow is assumed to be accompanied by simultaneous hardening and recovery processes. This assumption was tested by Mitra and McLean (1966) and the first general theory to account for hardening and recovery was proposed by Malinin and Khadjinsky (1972) and extended by Chaboche (1978). Further elaborations along these lines were developed by numerous authors (cf. review of creep hardening rules by Krieg, 1977).

The present hardening recovery model in its general form thus involves three viscosity parameters, $\gamma, B_{\alpha}, B_{\mathrm{p}}$, one appearing in the flow rule, the others in the evolution rules (4.9). Now let us discuss some particular versions of this model.

Lagneborg (1972) considered only one state variable, namely the dislocation density ρ, and for the uniaxial tension he proposed the following relations

$$\begin{aligned} \dot{\varepsilon}^{\mathrm{v}} &= A(\sigma - B\rho^{1/2})^{n} \\ \dot{\rho} &= C\dot{\varepsilon}^{\mathrm{v}} - D\rho^{2} \end{aligned} \tag{4.10}$$

where A, B, C, D are four material parameters. Let us note that these uniaxial relations much resemble the general relations (4.6) and (4.9). To identify material parameters, the creep tests at various constant stresses and at piecewise constant stress are carried out. Since for the asymptotically reached steady state there is $\dot{\rho} = 0$ and $\dot{\varepsilon}_{\mathrm{v}} = (D/C)\rho^{2}$, the first equation (4.10) provides the creep rate at the steady state

$$\dot{\varepsilon}^{\mathrm{v}} = A\left[\sigma - B\left(\frac{C}{D}\dot{\varepsilon}^{\mathrm{v}}\right)^{1/4}\right]^{n} \tag{4.11}$$

from which two material parameters can be identified. The other two parameters are determined by fitting the model prediction to the transient creep curves. The multidimensional generalization of eqns (4.10) is not directly evident but we may assume that $\rho^2 = \boldsymbol{\rho}|\boldsymbol{\rho}|$ and $\boldsymbol{\alpha} = c(\boldsymbol{\rho}/|\boldsymbol{\rho}|)$, i.e. relate $\boldsymbol{\rho}$ to the back stress $\boldsymbol{\alpha}$. The first equation (4.10) then takes the usual form equivalent to eqn (4.8)

$$\dot{\varepsilon}^{\mathrm{v}} = A\left(\frac{\bar{\sigma}_{\mathrm{e}}}{\sigma_{\mathrm{p}}}\right)^n \frac{3}{2}\frac{\mathbf{s} - \boldsymbol{\alpha}}{\bar{\sigma}_{\mathrm{e}}} \tag{4.12}$$

Following Ponter and Leckie (1976), let us propose the phenomenological constitutive equations which are based on the Bailey–Orowan model. Introducing the viscous potential $W(\sigma, \rho)$, where ρ denotes the state variable, one can postulate that

$$\dot{\varepsilon}^{\mathrm{v}} = \frac{\partial W}{\partial \sigma}, \qquad \dot{\rho} = -h(\rho)\frac{\partial W}{\partial \rho} \tag{4.13}$$

and for the potential $W = F[\phi(\sigma) - g(\rho)] + G(\rho)$ eqns (4.13) yield

$$\dot{\varepsilon}^{\mathrm{v}} = F_{\phi}\frac{\partial \phi}{\partial \sigma}, \qquad \dot{\rho} = -hF_g - hG_{\rho} \tag{4.14}$$

where F_{ϕ}, F_g, G_{ρ} denote derivatives of F and G with respect to ϕ, g, ρ; ϕ denotes the homogeneous function of first order of stress. For the Mises material when $\phi = \sigma_{\mathrm{e}} = [(3/2)\mathbf{s}\cdot\mathbf{s}]^{1/2}$, eqns (4.14) provide the relations analogous to (4.10)

$$\begin{aligned} \dot{\varepsilon}^{\mathrm{v}}_{ij} &= \tfrac{3}{2}A(\sigma_{\mathrm{e}} - B\rho^{1/2})\frac{s_{ij}}{\sigma_{\mathrm{e}}} \\ \dot{\rho} &= C\dot{\lambda} - D\rho^2 \end{aligned} \tag{4.15}$$

and the viscous potential is

$$W = \frac{1}{n+1}A(\sigma_{\mathrm{e}} - B\rho^{1/2})^{n+1} + \frac{2}{5}\frac{BD}{G}\rho^{5/2}, \qquad h = \frac{G\rho^{1/2}}{B} \tag{4.16}$$

It is seen that the mathematical structure of the constitutive relations coincides with the Lagneborg model. Similar constitutive relations were proposed by Hart (1976) who considered, however, a modified recovery relation.

The flow and evolution rules (4.13) can be incorporated into a general formalism discussed in Section 2 using thermodynamic forces and conjugate velocities. Assume generally that the viscous potential depends on stress and the dissipative force **A**, so that $W = W(\sigma, \mathbf{A})$, and

$$\dot{\varepsilon}^{\mathrm{v}} = \frac{\partial W}{\partial \sigma}, \qquad \dot{\mathrm{a}} = -\frac{\partial W}{\partial \mathbf{A}} \tag{4.17}$$

Introducing the dual dissipative potential

$$\Phi(\dot{\varepsilon}^{\mathrm{v}}, \dot{\mathrm{a}}) = \sigma \cdot \dot{\varepsilon}^{\mathrm{v}} - \mathbf{A} \cdot \dot{\mathrm{a}} - W \tag{4.18}$$

the constitutive relations take the form

$$\sigma = \frac{\partial \Phi}{\partial \dot{\varepsilon}^{\mathrm{v}}}, \qquad \mathbf{A} = -\frac{\partial \Phi}{\partial \dot{\mathrm{a}}} \tag{4.19}$$

For partial transformation of conjugate variables, the following potential is obtained

$$\Psi(\dot{\varepsilon}^{\mathrm{v}}, \mathbf{A}) = \sigma \cdot \dot{\varepsilon}^{\mathrm{v}} - W \tag{4.20}$$

and then

$$\sigma = \frac{\partial \Psi}{\partial \dot{\varepsilon}^{\mathrm{v}}}, \qquad \dot{\mathrm{a}} = \frac{\partial \Psi}{\partial \mathbf{A}} \tag{4.21}$$

These relations are similar to (2.62)–(2.64) derived for the case of time-independent plasticity.

To formulate the Bailey–Orowan model within this general formalism, consider the following creep potential

$$W(\sigma, \boldsymbol{\alpha}) = \frac{\gamma_1}{n+1}\bar{\sigma}_{\mathrm{e}}^{n+1} + \frac{\gamma_2}{m+1}\alpha_{\mathrm{e}}^{m+1} \tag{4.22}$$

where $\bar{\sigma}_{\mathrm{e}}$ is defined by eqn (4.4), α_{e} by eqn (3.36), and γ_1, γ_2 are the viscosity parameters. The double normality rule (4.17) now provides the flow and evolution rules

$$\begin{aligned} \dot{\varepsilon}^{\mathrm{v}} &= \frac{\partial W}{\partial \sigma} = \gamma_1 \bar{\sigma}_{\mathrm{e}}^{n-1}\tfrac{3}{2}(\mathrm{s} - \boldsymbol{\alpha}) \\ \dot{\mathrm{a}} &= \frac{\dot{\boldsymbol{\alpha}}}{h} = -\frac{\partial W}{\partial \boldsymbol{\alpha}} = \gamma_1 \bar{\sigma}_{\mathrm{e}}^{n-1}\tfrac{3}{2}(\mathrm{s} - \boldsymbol{\alpha}) - \gamma_2 \alpha_{\mathrm{e}}^{m-1}\tfrac{3}{2}\boldsymbol{\alpha} = \dot{\varepsilon}^{\mathrm{v}} - \gamma_2 \alpha_{\mathrm{e}}^{m-1}\tfrac{3}{2}\boldsymbol{\alpha} \end{aligned} \tag{4.23}$$

The stationary creep rate corresponds to the condition $\dot{\mathbf{a}} = \dot{\alpha} = 0$, from which we have

$$(\dot{\varepsilon}^{v})_{st} = \tfrac{3}{2}\gamma_2 \alpha_e^{m-1}\boldsymbol{\alpha} \tag{4.24}$$

Alternative versions of this model were presented by Chaboche (1977, 1978) and Miller (1976).

4.2. Plasticity Formulation Incorporating the Viscous Character of the Back Stress

Now let us briefly outline the alternative formulation which, though it is based on the Bailey–Orowan hypothesis, uses the familiar plasticity scheme. The creep and recovery processes are accounted for by assuming the viscous character of residual microstresses but neglecting the viscous character of deformation in the flow rule.

Consider first the simpler version and assume the yield condition in the form

$$F = f(\sigma - \alpha) - \sigma_p^2 = 0 \tag{4.25}$$

coupled with the flow rule and evolution rules

$$\begin{aligned} \dot{\varepsilon}^{p} &= \dot{\Lambda}\frac{\partial F}{\partial \sigma} && \text{for } F = \dot{F} = 0 \\ \dot{\varepsilon}^{p} &= 0 && \text{for } F < 0 \text{ or } F = 0,\ \dot{F} < 0 \\ \dot{\alpha} &= C_1(\lambda)\,\dot{\varepsilon}^{p} - C_2(\alpha)\,\alpha \end{aligned} \tag{4.26}$$

where $C_1(\lambda)$ and $C_2(\boldsymbol{\alpha})$ can be material functions and σ_p is the constant yield stress. The consistency condition provides now the expression for the multiplier $\dot{\Lambda}$, namely

$$\dot{\Lambda} = \frac{\dfrac{\partial f}{\partial \sigma}\cdot\dot{\sigma} + C_2\left(\dfrac{\partial f}{\partial \sigma}\cdot\alpha\right)}{C_1\left(\dfrac{\partial f}{\partial \sigma}\cdot\dfrac{\partial f}{\partial \sigma}\right)} \tag{4.27}$$

Let us note that the evolution rule (4.26) is analogous to the rule (4.15) and describes the simultaneous hardening and recovery processes. Using the flow rule in the form (2.6) or (2.7), the hardening moduli H and K can be obtained from eqn (4.27), namely

$$H = \frac{C_1 \left(\frac{\partial f}{\partial \sigma} \cdot \frac{\partial f}{\partial \sigma} \right) \left(\frac{\partial f}{\partial \sigma} \cdot \dot{\sigma} \right)}{\frac{\partial f}{\partial \sigma} \cdot \dot{\sigma} + C_2 \frac{\partial f}{\partial \sigma} \cdot \alpha}, \qquad K = \frac{C_1}{1 + C_2 \frac{\mathbf{n} \cdot \alpha}{\mathbf{n} \cdot \dot{\sigma}}} \tag{4.28}$$

It is seen that the hardening modulus now depends on the stress rate. In particular, for very high stress rates ($\dot{\sigma} \to \infty$), we obtain $K = C_1$, that is the plastic hardening modulus, whereas for slow processes ($\dot{\sigma} \to 0$) we have $K = 0$. In view of eqn (4.27), the loading–unloading condition now takes the form

$$\frac{\partial f}{\partial \sigma} \cdot \dot{\sigma} > -C_2 \left(\frac{\partial f}{\partial \sigma} \cdot \alpha \right) \tag{4.29}$$

For the stress point P on the instantaneous yield surface, this condition specifies the *loading and unloading cones* whose angles depend on the stress rate; Fig. 11(a). For $\dot{\sigma} \to \infty$, the unloading cone becomes a plane tangential to the yield surface at P.

The steady state reached under constant stress ($\dot{\sigma} = 0$) corresponds to the case $\dot{\alpha} = 0$ and then from eqns (4.26) we obtain

$$\dot{\varepsilon}^{\mathrm{p}} = \frac{C_2}{C_1} \alpha \tag{4.30}$$

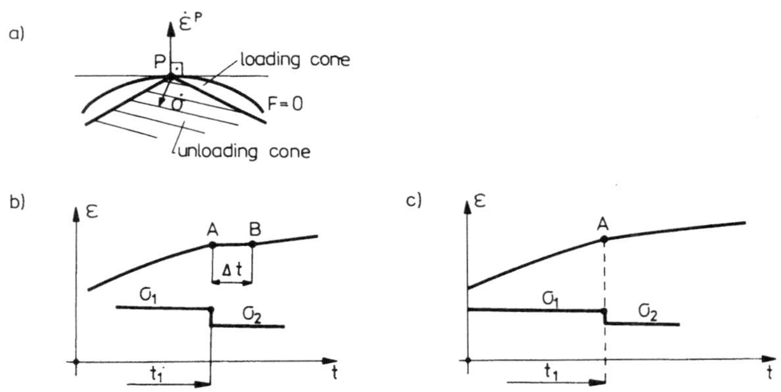

Fig. 11. Plasticity model with viscous recovery of back stress: (a) loading–unloading cones in the stress space; (b) determination of the recovery period Δt following stress reduction; (c) viscous model prediction with no recovery period.

There is an essential difference between the present formulation and the viscous model (4.23). In fact, the time-dependent deformation process now results from the viscous recovery of the back stress, whereas the previous formulation was associated with two viscous processes, one corresponding to viscous flow, the other to viscous recovery of $\boldsymbol{\alpha}$ (hence the material was characterized by two viscosity parameters). This difference can easily be demonstrated by considering a constant stress creep process followed by small reduction of stress from σ_1 to σ_2. Figure 11(b, c) show the differences of predictions of two models. When the plasticity formulation is used such as (4.26), the stress point after stress reduction remains within the elastic domain and there is no plastic flow for some period of time Δt. Next, due to viscous recovery of $\boldsymbol{\alpha}$, the yield surface translates toward the origin and plastic flow starts again at B. On the other hand, when the rules (4.23) are used, the progressive creep occurs after the stress reduction which affects the local slope of the ε–t curve. Mitra and McLean (1966) determined the recovery parameter C_2 for nickel by measuring the periods Δt for different stress reductions. However, for some metals, such a no creep period is not observed and therefore the viscous model is more preferable in such cases.

The plasticity formulation was discussed by Ponter and Leckie (1976) using the normality concept in generating both flow and evolution rules. Kujawski and Mróz (1980) used a similar plasticity formulation but decomposed the back stress into two components $\boldsymbol{\alpha}_1$ and $\boldsymbol{\alpha}_2$, one corresponding to plastic hardening and recovery, the other to viscous hardening and recovery, thus

$$\begin{aligned} \dot{\boldsymbol{\alpha}}_1 &= C_1\dot{\boldsymbol{\varepsilon}}^{\mathrm{p}} - C_2\boldsymbol{\varepsilon}^{\mathrm{p}}\dot{\lambda} \\ \dot{\boldsymbol{\alpha}}_2 &= C_3\dot{\boldsymbol{\varepsilon}}^{\mathrm{p}} - C_4\boldsymbol{\alpha}_2 \end{aligned} \tag{4.31}$$

where C_1, C_2, C_3, C_4 are four material functions. These functions were identified from monotonic and cyclic loading tests.

4.3. Creep Hardening Rule with Memory of Maximal Prestress

Assuming a viscous model of flow (4.6) or (4.7), let us now combine it with the concept of memory of maximal prestress expressed in terms of the maximal loading surface in the space of stress or back stress and of active loading surfaces emanating from the prestress point; cf. Fig. 8. Similarly, as for time-independent deformation, consider a viscoplastic deformation process and introduce the measure of maximal prestress in terms of $\phi(\boldsymbol{\alpha}) = \alpha_e$, that is described by the surface $\phi(\boldsymbol{\alpha}) - \alpha_m = 0$ [eqn (3.57)], so that

$$\alpha_m = \sup_{0<s<t} \alpha_e (t - s) \tag{4.32}$$

During the initial creep process α_e increases from its initial value and then $\alpha_e = \alpha_m$. On the other hand when, after reaching the maximal value $\alpha_e = \alpha_m$ its subsequent value decreases, $\alpha_e < \alpha_m$, the maximal value is kept in the material memory and a new creep process proceeds. The distinction is therefore made between two types of creep processes: the active creep process for which $\dot{\alpha}_e = \dot{\alpha}_m$, $\alpha_e = \alpha_m$, and the creep unloading process for which $\alpha_e < \alpha_m$.

Using the flow rule (4.8) or (4.23), that is neglecting the elastic domain, let us specify the evolution rules for these two processes. For the active creep process, neglect the recovery term and assume that

$$\dot{\boldsymbol{\alpha}}^{(1)} = C_1^{(1)} \boldsymbol{\beta}^{(1)} \dot{\lambda} \qquad \text{for } \alpha_e = \alpha_m, \quad \dot{\alpha}_e > 0 \tag{4.33}$$

where

$$\boldsymbol{\beta}^{(1)} = \frac{\boldsymbol{\sigma} - \boldsymbol{\alpha}}{|\boldsymbol{\sigma} - \boldsymbol{\alpha}|}$$

is the unit vector along the effective stress vector $\boldsymbol{\sigma} - \boldsymbol{\alpha}$. The recovery effect can be incorporated into the function $C_1^{(1)} = C_1^{(1)}(\alpha_e)$, so that the steady creep process will be attained for a specified stress level.

Referring to Fig. 12, assume that the active creep process terminates for the value of $\boldsymbol{\alpha}$ at A_1 when the stress state represented by $\mathbf{P}_1$ is suddenly changed to P_2 and is kept fixed at P_2. The creep unloading process starts when the α-path moves into the interior of the domain enclosed by the maximal loading surface $\phi_m = 0$. As previously, for the creep unloading event we introduce $\phi_1 = 0$ passing through the current α-point B_2 and tangential at A_1 to the surface $\phi_m(\boldsymbol{\alpha}, \alpha_u^{(1)}) = 0$. Equations (3.58) and (3.59) specify the unloading surface. The first creep unloading event is defined by the inequalities $\dot{\alpha}_u^{(1)} > 0$ and $\alpha_u^{(1)} < \alpha_m$ and it terminates when either $\dot{\alpha}_u^{(1)} < 0$ or $\alpha_u^{(1)} = \alpha_m$. The evolution of $\boldsymbol{\alpha}$ during this event is specified by the rule

$$\dot{\boldsymbol{\alpha}}^{(u)} = C_1^{(u)}(\alpha_m, \alpha_u) \boldsymbol{\beta}^{(u)} + \dot{\boldsymbol{\alpha}}^{(1)}, \qquad \alpha_e < \alpha_m, \quad \dot{\alpha}_u > 0 \tag{4.34}$$

where $\boldsymbol{\beta}^{(u)}$ is the unit vector. The first term of eqn (4.34) represents the *unloading recovery* term vanishing when $\alpha^{(u)} = \alpha_m$, whereas the second term represents the continuing hardening. Following previous discussion of the plastic hardening rule, it is assumed that the vector $\boldsymbol{\beta}^{(u)}$ coincides with $\mathbf{s} - \boldsymbol{\alpha}$ for stress states lying inside the surface $\phi_m = 0$ whereas, for states lying in the exterior of this surface, the α-path of the unloading recovery is directed to a point B′ lying on $\phi_m = 0$ and the radial line OP_2 connecting the centre of $\phi_m = 0$ with the stress point. It can therefore be written

$$
\begin{aligned}
\boldsymbol{\beta}^{(u)} &= \frac{\mathbf{s} - \boldsymbol{\alpha}}{|\mathbf{s} - \boldsymbol{\alpha}|} && \text{for} \quad \sigma_e < \alpha_m \\
\boldsymbol{\beta}^{(u)} &= \frac{\bar{\mathbf{s}} - \boldsymbol{\alpha}}{|\mathbf{s} - \boldsymbol{\alpha}|} = \frac{\mathbf{s}\dfrac{\alpha_m}{\sigma_e} - \boldsymbol{\alpha}}{|\mathbf{s} - \boldsymbol{\alpha}|} && \text{for} \quad \sigma_e > \sigma_m
\end{aligned}
\qquad (4.35)
$$

This evolution rule is illustrated in Fig. 12. For the stress point at P_1, the α-path OA_1 follows the radial line OP_1 until the point A_1 is reached. When the stress state is subsequently changed to P_2, the creep unloading process occurs. The first term of eqn (4.34) predicts the α-path following the line A_1B' terminating at B′ whereas the second term predicts the evolution direction along the line connecting P_2 with the instantaneous α-point. As the first term predominates, the evolution of $\boldsymbol{\alpha}$ follows the path A_1B_2B'', departing slightly from the path A_1B_1B'. The function $C_1^{(u)}$ occurring in eqn (4.34) should satisfy the condition $\dot{\boldsymbol{\alpha}}^{(u)} = \dot{\boldsymbol{\alpha}}^{(l)}$ for

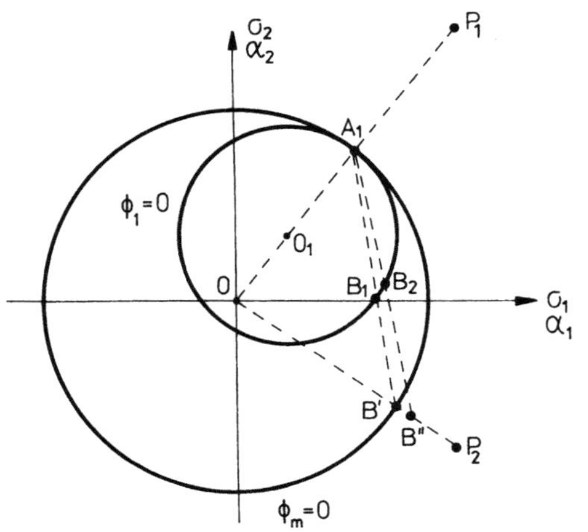

Fig. 12. Creep hardening rule: evolution of the back stress in the α-space.

$\alpha_u = \alpha_m$. This function is therefore expressed in the form

$$C_1^{(u)} = B(\alpha_m - \alpha_u)^l \tag{4.36}$$

where B and l are material parameters.

It is seen that this hardening rule makes a distinction between particular loading events for which the evolution rules of the back stress may differ depending on whether there is creep loading or unloading event. Formulation of this model and its identification and application to cyclic creep deformation was presented by Mróz and Trampczyński (1984). A somewhat similar idea of using creep hardening surface, eqn (3.61), measuring the maximal viscous strain from the past history with respect to the back stress and of different evolution rules was discussed by Murakami and Ohno (1982).

5. CONCLUDING REMARKS

The present chapter provides a review of basic formulations and concepts used in the phenomenological description of plastic deformation and hardening of metals. Whereas the structural approach may offer one way of constructing more accurate models, it is inherently related to the introduction of a number of state variables. Another approach using the concept of representative variables seems more promising. Discretized or continuum description of the field of hardening moduli provides a good example of such an approach. The other important concept is the memory rule which allows a distinction to be made between particular inelastic loading events and to formulate proper evolution rules. It turns out that the evolution rules should depend on the type of loading event and are different for particular events. The two-surface or multisurface formulations provide convenient means to classify loading events, generate memory rules and formulate evolution rules for state variables.

The development of viscoplastic or viscous hardening rules follows similar trends. In fact, the same concepts can be used in both descriptions neglecting or accounting for time-dependent deformation. The creep hardening-recovery models, though useful for simpler loading histories, are unable to describe hardening effects for complex loading paths involving previous prestress histories. Therefore a more successful description would result when the hardening-recovery description is combined with the hardening rule in which the memory of maximal prestress is incorporated.

REFERENCES

Armstrong, P. J. and Frederick, C. O. (1966). A mathematical representation of the multiaxial Bauschinger effect, Centr. Electr. Gen. Board Rep. BN731.

Arutunyan, R. A. and Vakulenko, V. A. (1965). On multiple loading of elastic-plastic solids (in Russian), *Izv. Akad. Nauk SSSR, Mekhanika*, **4**, 53–64.

Asaro, R. J. (1983). Crystal plasticity, *J. Appl. Mech.*, **50**, 921–33.

Backhaus, G. (1972*a*). Zur analytischen Erfassung des allgemeinen Baushinger-effects, *Acta Mech.*, **5**, 31–42.

Backhaus, G. (1972*b*). Fliesspannungen und Fliessbedingung bei zyklischen Verformungen, *Z.A.M.M.*, **56**, 337–48.

Bailey, R. W. (1926). Note on the softening of strain hardening metals and its relation to creep, *J. Inst. Metals*, **35**, 27–40.

Baltov, A. and Sawczuk, A. (1965). A rule of anisotropic hardening, *Acta Mech.*, **1**, 81–91.

Batdorf, B. and Budiansky, B. (1954). Polyaxial stress-strain relations of strain hardening metal, *J. Appl. Mech.*, **21**, 323–6.

Besseling, J. F. (1958). Theory of elastic, plastic, and creep deformation of an initially isotropic material showing anisotropic strain hardening, creep recovery, and secondary creep, *J. Appl. Mech.*, **25**, 529–36.

Chaboche, J. L. (1978). Sur l'utilisation des variables d'état pour la description cyclic and anisotropic behaviour of metals, *Bull. Acad. Sci. Pol.*, **25**, 33–42.

Chaboche, J. L. (1978). Sur l'utilization des variables d'état pour la description du comportement viscoplastique et de la rupture par endommagement, *Proc. Symp. Franco-Polon., Rhéologie et Mécanique*, Pol. Sci. Publ.

Chaboche, J. L., Dang Van, K. and Cordier, G. (1979). Modelization of strain memory effect on the cyclic hardening of 316 stainless steel, *Trans. 5th Conf. Struct. Mech. Reactor Techn.*, Vol. L.

Dafalias, Y. F. and Popov, E. P. (1975). A model of non-linearly hardening materials with complex loading, *Acta Mech.*, **21**, 173–92.

Eisenberg, M. A. (1976). A generalization of plastic flow theory with application to cyclic hardening and softening phenomena, *Trans. ASME, J. Eng. Mat. Techn.*, **97H**, 221–8.

Eisenberg, M. A. and Phillips, A. (1968). On non-linear kinematic hardening, *Acta Mech.*, **5**, 1–13.

Eisenberg, M. and Phillips, A. (1971). A theory of plasticity with non-coincident yield and loading surfaces, *Acta Mech.*, **11**, 247–60.

Halphen, B. and Nguyen, Q. S. (1975). Sur les matériaux standards généralisés, *J. Mécan.*, **14**, 41–60.

Hart, E. W. (1976). Constitutive relations for the non-elastic deformation of metals, *J. Eng. Mat. Techn.*, **98**, 193–203.

Haythornthwaite, R. M. (1968). A more rational approach to strain hardening data. In *Engineering Plasticity*, J. Heyman and F. A. Leckie (Eds), Cambridge Univ. Press.

Hill, R. (1967). The essential structure of constitutive laws for metal composites and polycrystals, *J. Mech. Phys. Solids*, **15**, 79–95.

Ivlev, V. D. (1963). On a theory of composite media (in Russian), *Dokl. Akad. Nauk SSSR*, **8**, 28–30.
Iwan, W. D. (1967). On a class of models for the yielding behaviour of continuous and composite systems, *J. Appl. Mech.*, **34**, 612–17.
Kadashevitch, Yu, I. and Novozhilov, V. V. (1959). Theory of plasticity accounting for residual stresses (in Russian), *Prikl. Math. Mekh.*, **22**, 104–8.
Kafka, V. (1968). The general theory of isothermal elastic-plastic deformation of polycrystals based on an analysis of the microscopic state variables, *Z.A.M.M.*, **48**, 265–82.
Khazhinsky, G. M. (1975). Plastic deformation of metals (in Russian), *Izv. Akad. Nauk SSSR, Mekh. Tv. Tela*, **1**, 175–7.
de Koning, A. U. (1977). A continuum theory of time independent plasticity based on a volume fraction model, National Aerospace Lab. NLR, MP. 77017U, Amsterdam.
Kratochvil, J. and Dillon, O. W. (1969). Thermodynamics of elastic-plastic materials as a theory with internal state variables, *J. Appl. Phys.*, **40**, 3207–18.
Krieg, R. D. (1975). A practical two-surface plasticity theory, *J. Appl. Mech.*, **42**, 641–6.
Krieg, R. D. (1977). Numerical integration of some unified plasticity-creep formulations, Proc. SMiRT-4 Conf., Vol. L, San Francisco.
Kujawski, D. and Mróz, Z. (1980). A viscoplastic material model and its application to cyclic loading, *Acta Mechanica*, **36**, 213–30.
Lagneborg, R. (1972). A modified recovery-creep model and its evaluation, *Met. Sci. J.*, **6**, 127–33.
Lamba, H. S. and Sidebottom, O. M. (1978). Cyclic plasticity for non-proportional paths. Part 1: Cyclic hardening, erasure of memory and subsequent strain hardening experiments, *Trans. ASME, J. Eng. Mat. Techn.*, **100**, 96–103.
Lee, D. and Zaverl, F., Jr (1979). A description of history dependent plastic flow behaviour of anisotropic metals, *Trans. ASME, J. Eng. Mat. Techn.*, **101**, 59–67.
Lin, T. H. (1971). Physical theory of plasticity, *Adv. Appl. Mech.*, **11**, Academic Press.
Malinin, N. N. and Khadjinsky, G. M. (1972). Theory of creep with anisotropic hardening, *Int. J. Mech. Sci.*, **14**, 235–46.
Mandel, J. (1971). *Plasticité et Viscoplasticité*, CISM course nr 97, Springer-Verlag.
Mandel, J. (1974). Thermodynamics and plasticity. In *Foundations of Continuum Thermodynamics*, J. J. Delgado Dominges, N. R. Nina and J. H. Whitelaw, Macmillan Press, pp 283–304.
Melan, E. (1938). Zur Plastizität des raumlichen Kontinuums, *Ing. Arch.*, 116–26.
Miller, A. K. (1976). An inelastic constitutive model for monotonic cyclic, and creep deformation, Parts I and II, *Trans. ASME, J. Eng. Mat. Techn.*, **98**, 47–52.
Mitra, S. K. and McLean, D. (1966). Work hardening and recovery in creep, *Proc. Roy. Soc.*, **A295**, 288–99.
Mróz, Z. (1967). On the description of anisotropic work-hardening, *J. Mech. Phys. Solids*, **15**, 163–75.

Mróz, Z. (1969). An attempt to describe the behaviour of metals under cyclic loading using a more general work-hardening model, *Acta Mech.*, 7, 199–212.
Mróz, Z. (1972). A description of work-hardening of metals with application to variable loading, *Proc. Symp. Foundations of Plasticity*, A. Sawczuk (Ed.), Noordhoff Int. Publ., pp 551–70.
Mróz, Z. (1981). On generalized kinematic hardening rule with memory of maximal prestress, *J. Méc. Appl.*, 5, 241–60.
Mróz, Z. and Lind, N. C. (1975). Simplified theories of cyclic plasticity, *Acta Mech.*, 22, 131–52.
Mróz, Z. and Trampczyński, W. A. (1984). On the creep hardening rule for metals with a memory of maximal prestress, *Int. J. Solids Struct.*, 20, 467–86.
Mróz, Z., Shrivastava, H. P. and Dubey, R. N. (1976). A non-linear hardening model and its application to cyclic loading, 25, 51–61.
Murakami, S. and Ohno, N. (1982). A constitutive equation for creep based on the concept of a creep hardening surface, *Int. J. Solids Struct.*, 18, 597–609.
Orowan, E. (1946). The creep of metals, *J. West. Scot. Iron Steel Inst.*, 54, 45–53.
Ortiz, M. and Popov, E. P. (1983). Distortional hardening rules for metal plasticity, *J. Eng. Mech. Div., Proc. ASCE*, 109 (EM4), 1042–57.
Petersson, H. and Popov, E. P. (1977). Constitutive relations for generalized loading, *J. Eng. Mech. Div., Proc. ASCE*, 103 (EM4), 611–27.
Ponter, A. R. S. and Leckie, F. A. (1976). Constitutive relationship for the time-dependent deformation of metals, *J. Eng. Mat. Techn. Proc. ASME*, 98, 47–52.
Prager, W. (1956). A new method of analysing stress and strain in work-hardening plastic solids, *J. Appl. Mech.*, 23, 493–6.
Prager, W. (1966). Composite stress-strain relations for elastoplastic solids, *Proc. IUTAM Symp. Irreversible Aspects of Continuum Mechanics*, H. Parkus (Ed.), Springer-Verlag.
Rice, J. R. (1971). Inelastic constitutive relations for solids: an internal variable theory and its application to metal elasticity, *J. Mech. Phys. Solids*, 19, 433–55.
Shrivastava, H. P., Mróz, Z. and Dubey, R. N. (1973). Yield criterion and the hardening rule for a plastic solid, *Z.A.M.M.*, 53, 625–33.
Taylor, G. I. (1938). Plastic strain in metals, *J. Inst. Metals*, 62, 306–24.
Vakulenko, A. A. (1961). On relations between stress and strain in inelastic media, *Problems of Elasticity and Plasticity* (in Russian), Vol. 1, Leningrad Univ.
Valanis, K. C. and Lee, C. F. (1982). Some recent developments of the endochronic theory with applications, *Nucl. Eng. Design*, 69, 327–44.
Wells, C. H. and Paslay, P. R. (1969). A small strain plasticity theory for planar slip materials, *J. Appl. Mech.*, 36, 15–21.
Zarka, J., Engel, J. J. and Inglebert, G. (1980). On a simplified inelastic analysis of structures, *Nucl. Eng. Design*, 57, 333–68.
Ziegler, H. (1959). A modification of Prager's hardening rule, *Quart. Appl. Math.*, 17, 55–60.
Zienkiewicz, O. C., Nayak, G. C. and Owen, D. R. J. (1972). Composite and overlay models in numerical analysis of elasto-plastic continua, *Proc. Intern. Symp. Foundations of Plasticity*, A. Sawczuk (Ed.), Noordhoff Int. Publ.

PART IV

SYNTHESIS

CHAPTER 10

Modelling Global Behaviours

J. ZARKA
Laboratoire de Mécanique des Solides, Ecole Polytechnique, Palaiseau, France

ABSTRACT

The purpose of this chapter is to show how it is possible to use the various elements presented in the previous chapters in order to give rational descriptions of the behaviour of metals according to the physical approach or the quasi-physical engineering approach.

The fundamental key is that there are some local microscopical *mechanisms which are responsible for the inelastic strains and which interact between them to give the* global macroscopical *behaviour.*

This chapter is divided into two sections where respectively the two points of view will be presented. These two descriptions are rather old; they were given in Zarka (1968) for the physical approach and Zarka (1979) for the structural approach. But now, as good illustrations of applications of this book, it is hoped that they will be understood better and used more often.

Section 1: A Physical Model for the Behaviour of the Single Crystal

1. THEORETICAL MODEL FOR THE SINGLE CRYSTAL

We know that plastic strains are induced by the motions of defects. For the slip mode, which is the most important one, we see that the crystal structure remains unchanged and that these strains are delayed and often very large. Elastic strains are linked to the mean lattice; they are, on our time scale, instantaneous and generally very small. This fundamental experimental fact allows us to share the elastic strains and plastic strains at any time.

1.1. Geometrical Description of the Crystal

(a) Let us define two families of vectors $\vec{F}_K$ and $\vec{e}_\alpha$ for the volume element (it is always the same that we follow during the strain). These vectors have the following properties (Fig. 1).

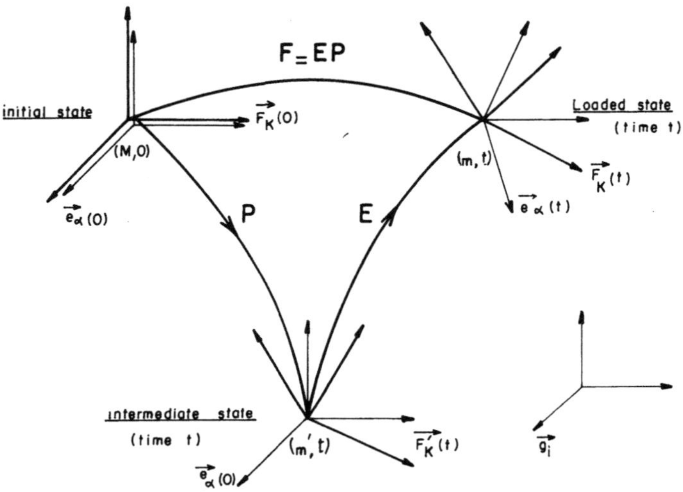

Fig. 1. Representation of the various states of the single crystal.

The $\vec{F}_K$ are three macroscopic vectors; they characterize the total transformation (indexes of quantities which are referred to this frame are noted $K, L, M, \ldots$).

The $\vec{e}_\alpha$ are three microscopic vectors linked to the mean lattice; they characterize the elastic transformation (indexes of quantities in this frame are noted $\alpha, \beta, \gamma, \ldots$).

In the fixed and rectangular frame of reference $\vec{g}_i$ (indexes in this frame i, j, k; in more their positions do not matter):

(i) At $t = 0$ and at the point $X^K \delta^i_K$ they had for components (we choose an initial state which is unloaded and unstrained)

$$\vec{F}_K(0) = \delta^i_K \vec{g}_i \qquad \text{and} \qquad \vec{e}_\alpha(0) = \delta^i_\alpha \vec{g}_i \tag{1.1}$$

where in δ^i_K, δ^i_α, δ^β_α, $\ldots$ δ is the Kronecker symbol.

(ii) At actual time t, and at the point x^i the image of X^K in the total transformation (x^i, t are the Eulerian coordinates; X^K, t are the Lagrangian coordinates), they have for components

$$\vec{F}_K = \frac{\partial x^i}{\partial X^K}\vec{g}_i \qquad \text{and} \qquad \vec{e}_\alpha = x^i_\alpha \vec{g}_i \tag{1.2}$$

$\mathbf{F} = (\partial x^i/\partial X^K)$ is the classical displacement gradient

$\mathbf{E} = (x^i_\alpha)$ defines the elastic transformation

It is useful to introduce their contravariant vectors

$$\vec{F}^K = \frac{\partial X^K}{\partial x^i}\vec{g}_i \qquad \text{and} \qquad \vec{e}^\alpha = \xi^\alpha_i \vec{g}_i \tag{1.3}$$

(By definition, we have $\dfrac{\partial X^K}{\partial x^j}\dfrac{\partial x^j}{\partial X^L} = \delta^K_L \ldots ; x^i_\alpha \xi^\alpha_j = \delta^i_j \ldots .$)

(b) By using the fundamental experimental fact we may say (Fig. 2):
(i) The components of the elastic strain tensor $\mathbf{e}'$

$$\begin{aligned} 2e'_{ij} &= \delta_{ij} - \delta_{\alpha\beta}\xi^\alpha_i \xi^\beta_j \qquad (\text{in the frame } \vec{g}_i) \\ 2e'_{\alpha\beta} &= \delta_{ij}x^i_\alpha x^j_\beta - \delta_{\alpha\beta} \qquad (\text{in the frame } \vec{e}_\alpha) \end{aligned} \tag{1.4}$$

are very small but we know nothing about the elastic rotation.
(ii) The components of the total strain tensor $\mathbf{e}$

$$2e_{ij} = \delta_{ij} - \delta_{KL}\frac{\partial X^K}{\partial x^i}\frac{\partial X^L}{\partial x^j} \qquad (\text{Eulerian tensor}) \tag{1.5}$$

are, as well as the total rotation, generally large.

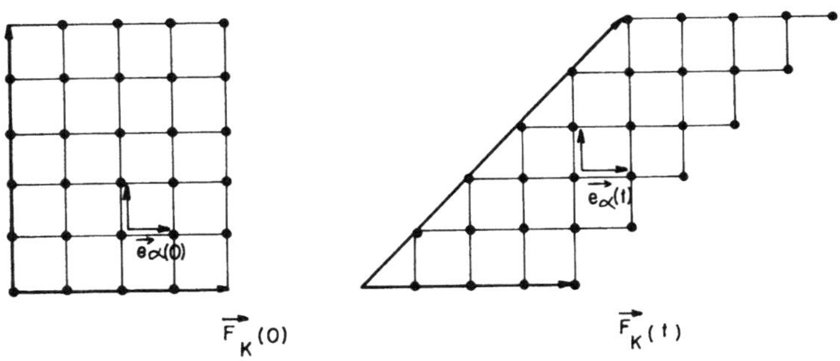

Fig. 2. Physical interpretation of the two families $\vec{F}_K$ and $\vec{e}_\alpha$ during a pure glide transformation.

(iii) It is then necessary to describe the single crystal behaviour in large strains.

(iv) The elastic strains may be instantaneously obtained; in consequence, at the actual time t, by a total unloading in forces and temperature, with the help of a convenient rotation, we could have an intermediate unloaded state in which the microscopic vectors $\vec{e}'$ are equipotent to the $\vec{e}_\alpha(0)$ or $\vec{g}_i$. But in this state there is a new distribution of defects relative to the initial state.

Let us define by $\{y\}$ a family of parameters which describes this distribution.

(c) By classical elasticity theory (as in the mathematical theory of dislocations for example) it is possible to calculate the stress tensor $\mathbf{\Pi}$ when we impose $\mathbf{e}'$ and T for a given distribution $\{y\}$. With its defects, the crystal is only an inhomogeneous medium. The stress tensor $\mathbf{\Pi}$ is referred to the frame $\vec{e}_\alpha$. Conversely we may impose $\mathbf{\Pi}, T$ and calculate $\mathbf{e}'$. The total transformation cannot be characterized so easily. We must follow the crystal step by step and share the elastic part $v''^i_{\,j}$ and the viscoplastic part $v'''^i_{\,j}$ of the Eulerian velocity gradient $v'^i_{\,j} = \partial v^i / \partial x^j$. We need for that some hypothesis. In the same way, in order to determine the temperature field, we must admit some hypothesis on the thermodynamical functions of the volume element.

Before going on, we must recall that the velocity gradient is very important. For example, it allows one to determine the variations of $\vec{F}_K$ and $\vec{F}^K$ (i.e. the material derivatives of the displacement gradients).

$$\dot{\left(\frac{\partial x^i}{\partial X^K}\right)} = v'^i_{\,l}\,\frac{\partial x^l}{\partial X^K}; \qquad \dot{\left(\frac{\partial X^K}{\partial x^i}\right)} = -v'^l_{\,i}\,\frac{\partial X^K}{\partial x^l} \tag{1.6}$$

1.2. Hypothesis on the Viscoplastic Strains

(a) Macroscopically, we saw that plastic strains result from some displacements on some crystallographic planes and along some crystallographic directions. These planes and these directions, since they are crystallographic, are always very well known in the frame $\vec{e}_\alpha$ which represents the mean lattice. They are then easily defined in the frame $\vec{g}_i$.

Consider a displacement direction $\vec{b}$; it is given by a vector which links two atoms, so it will have constant components in the frame $\vec{e}_\alpha$

even if the crystal is strained:

$$\vec{b} = b^a \vec{e}_a = b^a x^i_a \vec{g}_i \equiv b^i \vec{g}_i \tag{1.7}$$

where b^a are the components of this vector for $t = 0$. From it, at any time, we may deduce a unit vector $\vec{h}$ such that

$$\vec{h} = \frac{\vec{b}}{b} \equiv h^i \vec{g}_i \tag{1.8}$$

Consider now $\vec{N}$, a normal vector to a plane which contains atoms; we have, since $\vec{N}$ is a covariant vector,

$$\vec{N} = N_a \vec{e}^a = N_a \xi^a_i \vec{g}_i \equiv N_i \vec{g}_i \tag{1.9}$$

and a unit normal vector

$$\vec{n} = \frac{\vec{N}}{N} \equiv n_i \vec{g}_i \tag{1.10}$$

In Eulerian coordinates, if γ is the relative velocity along $\vec{h}$ between two planes $\vec{n}$ which are unit length apart, the velocity gradient is (Fig. 3)

$$v''^i{}_j = h^i n_j \gamma \tag{1.11}$$

During this transformation, only the $\vec{F}_K$ are transformed; the $\vec{e}_a$ remain fixed. Since $v''^i{}_j$ is not symmetrical, the $\vec{F}_K$ are not only strained but also have a rotation relative to the $\vec{e}_a$ (this is the plastic rotation; it is of course independent of the reference frame $\vec{g}_i$).

(b) Very often there are, during the strain, several different displacements. With the system (r) is associated, by definition, a couple $(\vec{h}^{(r)}, n^{(r)})$. The formula (1.11) is thus changed to

$$v''^i{}_j = \sum_{(r)} h^{i(r)} n^{(r)}_j \gamma^{(r)} \tag{1.12}$$

The summation is extended over all active systems. But we have to know when a system (r) is active and what is the amplitude of the relative velocity $\gamma^{(r)}$.

(i) We assume that the system (r) is active when the inequality

$$f^{(r)}(\boldsymbol{\Sigma}, T, x^i_a, \{y\}) \geqslant 0 \tag{1.13}$$

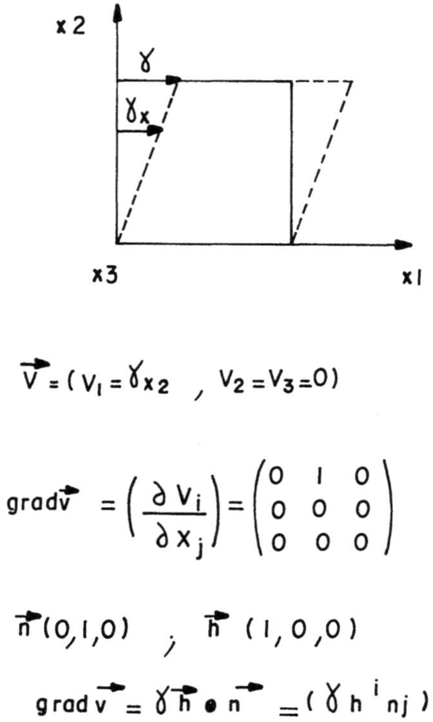

Fig. 3. Geometrical proof of the expression for the velocity gradient.

is satisfied. The classical Cauchy stress tensor $\mathbf{\Sigma}$ is related to $\mathbf{\Pi}$ by

$$\Sigma^{ij} = \frac{1}{\det(x^i_a)} x^i_\gamma x^j_\delta \, \Pi^{\gamma\delta} \tag{1.14}$$

This inequality must be, of course, independent of the frame $\vec{g}_i$ (objectivity principle); the physical basis of plasticity suggests a form to us (Schmid's law):

$$f^{(r)} \equiv h^{(r)}_i n^{(r)}_j \Sigma^{ij} - g^{(r)}(T, \{y\}) \geqslant 0 \tag{1.15}$$

where $h^{(r)}_i n^{(r)}_j \Sigma^{ij} = \tau^{(r)}$ is a scalar; $g^{(r)}$ is the yield limit for this system and may be zero. The intersection in stress space of all the domains $f^{(r)} < 0$ gives the elastic domain of the crystal. Since there is x^i_a, through $\vec{h}^{(r)}$ and $\vec{n}^{(r)}$, the crystal is naturally anisotropic.

(ii) Relative to the fact that $\gamma^{(r)}$ results from delayed motions of the defects (viscoplasticity), we assume that this velocity is only a function of the actual values of $\tau^{(k)}$, T and $\{y\}$, when (1.15) is satisfied:

$$\gamma^{(r)} \equiv \gamma^{(r)}(\tau^{(k)}, T, \{y\})\, Y(f^{(r)}) \tag{1.16}$$

where $Y(x)$ is the Heaviside function (0 if $x \leqslant 0^-$; 1 if $x \geqslant 0^+$). Very often only $\tau^{(r)}$ occurs in eqn (1.16). Some parameters among the family $\{y\}$ may evolute in the same form and we can write

$$\{\dot{y}\} = \sum_{(r)} G^{(r)}(\tau^{(k)}, T, \{y\})\, Y(f^{(r)}) \tag{1.17}$$

We shall see further the meaning of all these hypotheses.

1.3. Hypothesis on the Elastic Strains

Let us consider again the unloaded intermediate state at the actual time t; the volume element has a density ρ' (plastic strain does not generally promote a change of volume; $\rho' = \rho_0$)

$$\rho' = \rho_0 \frac{\det(\partial x^i_\alpha)}{\det(\partial x^I/\partial X^K)} \tag{1.18}$$

The vectors are $\vec{F}'_K$ and $\vec{e}'_\alpha$ $(\equiv \vec{e}_\alpha(0))$; the defects are characterized by $\{y\}$. By definition the defects stay immobile* during the elastic transformation. We may thus calculate exactly the variations that they provoke on the macroscopic elastic properties. Since these variations are generally very weak, in order to evaluate them we prefer to admit hypotheses that are simplifying but reasonable.

(i) We assume that the Voigt model for the elastic aggregate is valid. In this model the strains are the same at any point of the volume element. We form small blocks which contain either no defects at all (perfect crystal) or complete dislocation loops or only voids (Fig. 4). The dislocations have almost no influence on the elastic properties.

(ii) We assume that for the perfect single crystal (index 0 and density ρ_0) the constitutive relations are written in the form

$$\dot{e}'^{0}_{\alpha\beta} = \Lambda(\Pi^0, T)^0_{\alpha\beta\gamma\delta}\, \dot{\Pi}^{0\gamma\delta} + \alpha(\Pi^0, T)^0_{\alpha\beta}\, \dot{T} \tag{1.19}$$

with the integrability conditions

$$\frac{\partial}{\partial T}(\Lambda^0_{\alpha\beta\gamma\delta}) = \frac{\partial}{\partial \Pi^{0\gamma\delta}}(\alpha^0_{\alpha\beta}) \tag{1.20}$$

*There are also reversible motions of defects which provoke reversible strains but these strains are not considered like elastic ones.

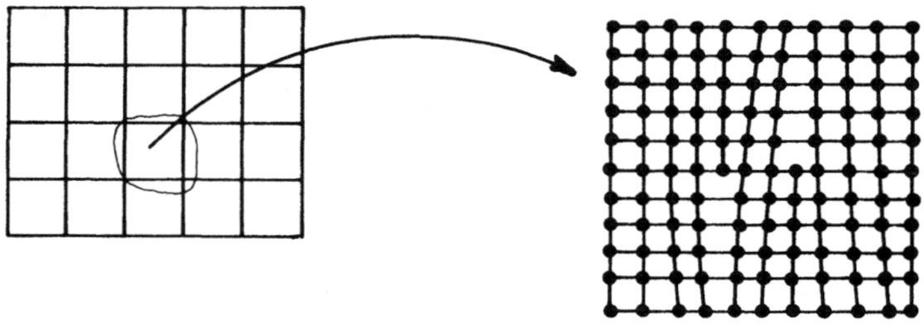

Fig. 4. This single crystal contains two opposite parallel dislocations and is globally similar to a perfect single crystal.

The elastic coefficient matrix Λ^0 and the thermal dilatation matrix α^0 must also satisfy the symmetry conditions of the crystalline structure.

(iii) From the Voigt model we deduce that

$$\dot{e}'_{\alpha\beta} = \dot{e}'^0_{\alpha\beta} \tag{1.21}$$

and by writing

$$\frac{\Pi^{\gamma\delta}\,\dot{e}'_{\gamma\delta}}{\rho'} = \frac{\Pi^{0\gamma\delta}\,\dot{e}'^0_{\gamma\delta}}{\rho_0} \tag{1.22}$$

that

$$\Pi \simeq \frac{\rho'}{\rho_0}\,\Pi^0 \tag{1.23}$$

(The point defects take part in the elastic properties only through the volume change.) Therefore, in the frame $\vec{g}_i$, we obtain for the elastic deformation rate tensor:

$$d'_{ij} \equiv \dot{e}'_{\alpha\beta}\,\xi^\alpha_i\,\xi^\beta_j = \Lambda^0_{\alpha\beta\gamma\delta}\,\xi^\alpha_i\,\xi^\beta_j\,\xi^\gamma_h\,\xi^\delta_k \det\left(\frac{\partial x^l}{\partial X^K}\right)\check{\Sigma}^{hk} + \alpha^0_{\alpha\beta}\,\xi^\alpha_i\,\xi^\beta_j\,\dot{T} \tag{1.24}$$

or

$$d'_{ij} \equiv \Lambda_{ijhk}\,\check{\Sigma}^{hk} + \alpha_{ij}\,\dot{T}$$

By reason of eqn (1.23), the objective derivative of Σ is

$$\check{\Sigma}^{ij} = \frac{x^i_\gamma\,x^j_\delta\,\dot{\Pi}^{\gamma\delta}}{\det(\partial x^l/\partial X^K)} = \dot{\Sigma}^{ij} - v'^i{}_k\,\Sigma^{kj} - v'^j{}_k\,\Sigma^{ki} + v^k{}_k\,\Sigma^{ij} \tag{1.25}$$

We only know the symmetrical part of ${v'}^i{}_j$ (d'_{ij}); it remains the elastic rotation which can be determined thanks to the compatibility equations.

During this elastic transformation, the vectors $\vec{F}_K$ and $\vec{e}_a$ are transformed. The formulae (1.6) give

$$\dot{x}^i_a = {v'}^i{}_j x^j_a; \qquad \dot{\xi}^\beta_i = -{v'}^j{}_i \xi^\beta_j \tag{1.26}$$

The composition of the plastic and elastic strain rates may be seen in Fig. 5.

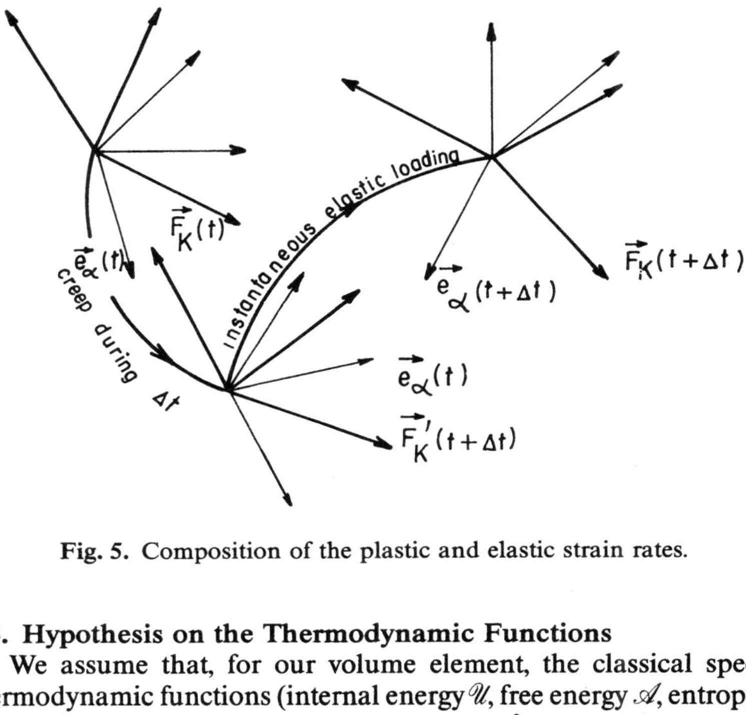

Fig. 5. Composition of the plastic and elastic strain rates.

1.4. Hypothesis on the Thermodynamic Functions

(a) We assume that, for our volume element, the classical specific thermodynamic functions (internal energy $\mathscr{U}$, free energy $\mathscr{A}$, entropy $\mathscr{S}$) exist and that they are regular functions of $\mathbf{\Pi}^0$, T and $\{y\}$:

$$\left.\begin{aligned}\mathscr{A} &\equiv \mathscr{A}(\mathbf{\Pi}^0, T, \{y\})\\ \mathscr{S} &\equiv \mathscr{S}(\mathbf{\Pi}^0, T, \{y\})\end{aligned}\right\} \Rightarrow \mathscr{U} = \mathscr{A} + T\mathscr{S} = \mathscr{U}(\mathbf{\Pi}^0, T, \{y\}) \tag{1.27}$$

(b) The restrictions due to the second law of thermodynamics, together with our hypothesis on the parts of the velocity gradient, imply that the thermodynamic functions are separated into two parts. The classical inequality

$$\frac{\Sigma^{ij} v_{ij}}{\rho} - \dot{\mathscr{A}} - \mathscr{S}\dot{T} \geqslant 0 \tag{1.28}$$

which has to be satisfied for any $\dot{\mathbf{\Pi}}^0$ and $\dot{T}$ leads to

$$\begin{aligned} \mathscr{A} &\equiv \mathscr{A}_1(\mathbf{\Pi}^0, T) + \mathscr{A}_2(T, \{y\}) \\ \mathscr{S} &\equiv \mathscr{S}_1(\mathbf{\Pi}^0, T) + \mathscr{S}_2(T, \{y\}) \end{aligned} \tag{1.29}$$

with

$$\left.\begin{aligned} \Pi^{0\gamma\delta}\, \Lambda^0_{\gamma\delta\alpha\beta} &= \rho_0 \frac{\partial \mathscr{A}_1}{\partial \Pi^{0\alpha\beta}} \\ \Pi^{0\gamma\delta}\, \alpha^0_{\gamma\delta} - \rho_0 \frac{\partial \mathscr{A}_1}{\partial T} &= \rho_0 \mathscr{S}_1 \\ -\frac{\partial \mathscr{A}_2}{\partial T} &= \mathscr{S}_2 \end{aligned}\right\} \tag{1.30}$$

We have therefore

$$\mathscr{U} = \mathscr{U}_1(\mathbf{\Pi}^0, T) + \mathscr{U}_2(T, \{y\}) \equiv \mathscr{A}_1 + T\mathscr{S}_1 + \mathscr{A}_2 + T\mathscr{S}_2 \tag{1.31}$$

In the inequality (1.28) it only remains

$$\Sigma^{ij} \sum_{(r)} h_i^{(r)} n_j^{(r)} \gamma^{(r)} - \sum_{(r)} \rho \frac{\partial \mathscr{A}_2}{\partial y} G^{(r)} \geqslant 0 \tag{1.32}$$

or else

$$\sum_{(r)} \tau^{(r)} \gamma^{(r)} \geqslant \rho \sum_{(r)} \frac{\partial \mathscr{A}_2}{\partial y} G^{(r)} \tag{1.33}$$

This last inequality means that the plastic power is always higher than or equal to the variation of free energy due to the defects.

(c) The first principle gives the equation

$$\sum_{(r)} \tau^{(r)} \gamma^{(r)} - \sum_{(r)} \rho \frac{\partial \mathscr{U}_2}{\partial y} G^{(r)} = \alpha_{ij} T \check{\Sigma}^{ij} + \rho C \dot{T} - \rho r + \frac{\partial q_i}{\partial x_i} \tag{1.34}$$

where $C = T(\partial \mathscr{S} / \partial T)$ is the specific heat; $\vec{q}(q^i)$ is the heat flux by conduction, and r is the heat supplied per unit mass and unit time.

2. PHYSICAL MODEL FOR FACE-CENTRED CUBIC SINGLE CRYSTALS

In the preceding Section 1 we made several hypotheses in order to define our theoretical model. In our opinion, it is fundamental to show that these hypotheses are indeed satisfied, that is to say that it is possible to determine the nature of the parameters $\{y\}$ and to give the explicit forms of the functions $f^{(r)}, \gamma^{(r)}, G^{(r)}$ and $\mathscr{U}_2$. Finally, we have to show that there is qualitative and quantitative agreement between theory and experiment. For that we shall study now the particular case of pure face-centred cubic single crystals. Of course this approach can be generalized to any other crystalline structure and also to impure crystals.

2.1. Description of the Physical Model

We saw in the previous chapters how plastic strains are interpreted by the behaviour of the defects, the most important ones being the dislocations. With the help of these fundamental elements we shall show how it is possible to build a model that is physically admissible, can represent the crystal and fits into the theoretical mathematical framework.

2.1.1. Hypothesis on the Dislocations Distribution

We take into account the shape of dislocation lines (small segments) and their random distribution but in a special form (Fig. 6).

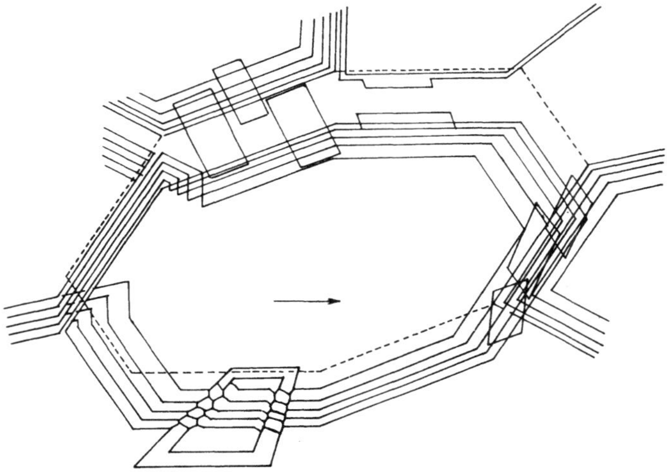

Fig. 6. Schematization of the distribution of the dislocations. (From Steeds (1964).)

(i) We call a mechanism (family) of dislocations (r) the set of small segments of dislocations which are parallel and have the same Burgers vector $\vec{b}^{(r)}$ (their slip planes are thus parallel and have the same unit normal $\vec{n}^{(r)}$).

(ii) We assume that, for each mechanism, it is possible to measure the mean number $N^{(r)}$ per unit volume of the segments and their mean length $l^{(r)}$ (Fig. 7(a)).

(iii) We suppose that, for each mechanism, the centres of the segments are randomly distributed and that the probability law not to find one centre of a segment inside any volume v is

$$P(v) = \exp\{-N^{(r)} v\} \tag{2.1}$$

With this law (Fig. 7(b)) we can define and calculate the shortest mean distance $d^{(r)}$ between two segments of one mechanism (r):

$$d^{(r)} = \frac{0{\cdot}5}{(N^{(r)} l^{(r)})^{1/2}} \tag{2.2}$$

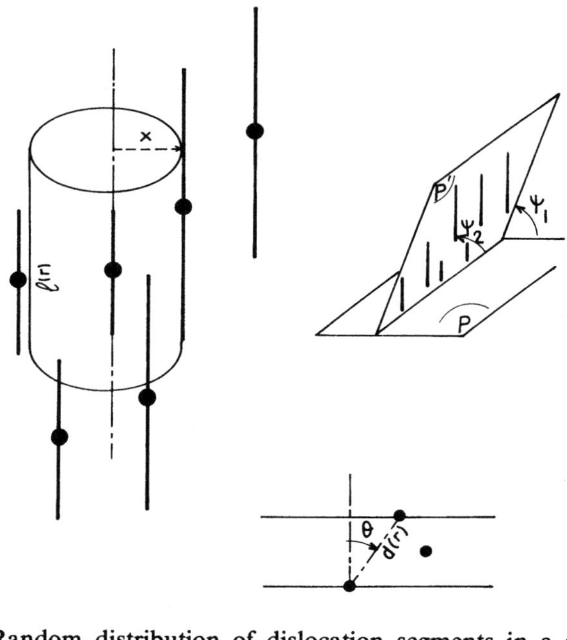

Fig. 7. (a) Random distribution of dislocation segments in a mechanism; (b) shortest mean distance between two segments of a mechanism; (c) geometry for segments belonging to the plane p piercing a plane p'.

We can also calculate the shortest mean distance $D^{(r)}$ between two planes which contain segments of this mechanism:

$$D^{(r)} = \frac{1}{\pi \, (N^{(r)} \, l^{(r)})^{1/2}} \tag{2.3}$$

Finally we can calculate the density per unit surface $n'^{(r)}$ of segments of one mechanism (r) which pierce a particular plane p' (Fig. 7(c)):

$$n'^{(r)} = N^{(r)} \, l^{(r)} \sin \Psi_1 \sin \Psi_2 \tag{2.4}$$

where Ψ_1 is the angle between the planes p and p' and Ψ_2 is the angle between one segment and an intersection line of planes p and p'.

(iv) We suppose that the probability law to find a segment of one mechanism (r) with a length greater or equal to any $\mathscr{L}$ is given by

$$P(\mathscr{L}) = \exp\left\{-\frac{\mathscr{L}}{l^{(r)}}\right\} \tag{2.5}$$

(v) By definition, we call the dislocations distribution the data on the couples $(N^{(r)}, l^{(r)})$ of each mechanism (r).

It is very important to notice that the laws (2.1) and (2.5) are simplifying; there is neither concentration of defects in one part of the crystal (no pile up) nor correlation between the data on the couples $(N^{(r)}, l^{(r)})$.

2.1.2. *Hypothesis on the Interactions between Dislocation Mechanisms*

(a) For a given dislocation distribution we can estimate the force per unit length on a dislocation segment of a particular mechanism (r) which is necessary to allow the following to occur.

(i) A very small displacement in or out of its plane. We determine only the mean actions of dislocations that are parallel to this segment (the others have negligible effects for such a displacement). These first limits occur chiefly in the case of climb or cross-slip.

(ii) A large displacement in its slip plane. Such a displacement is only possible on the slip plane for energy considerations. We determine the mean short-range actions of all the dislocations. These second limits occur chiefly for the glide of the new loops created by the Frank–Read sources.

(b) We carried out the calculations using the classical hypothesis of straight, infinitely long dislocation lines in an infinite isotropic medium, as Taylor did. Nevertheless we weighted their actions by taking into account their finite length when we introduced the quantities $d^{(r)}$ and $D^{(r)}$ [eqns (2.2) and (2.3)]. These elementary approximate calculations are interesting because they lead to explicit formulae; more accurate but not explicit ones can be done in the framework of the mathematical theory of dislocations.

(c) All these different limits can be expressed in the form

$$\mathscr{F}^{(r)} = \mathscr{F}^{(r)}_{\mathrm{PN}} + \mu b^{(r)} \sum_{(k)} \beta^{(r)}_{(k)} b^{(k)} (N^{(k)} l^{(k)})^{1/2} \tag{2.6}$$

β ($\beta^{(r)}_{(k)}$) is a matrix which gives the interactions between the different mechanisms and it may be made fully explicit (see Zarka, 1968).

(d) Let us consider the particular case when only the slip planes $\{111\}$ and the Burgers vectors $\frac{1}{2}\langle 110\rangle$ (modulus of which is b) take part in the deformation.

Let us assume too that the segments are all parallel to the directions $\langle 110\rangle$ and that these directions are equiprobable. Then we may define a slip system by putting together the six mechanisms which have the same slip plane and the same Burgers vector. While the couple $(N^{(r)}, l^{(r)})$ characterizes the mechanism (r), the couple $(\mathscr{N}^{(R)}, l^{(R)})$ characterizes the system (R). Figure 8 shows easily that $\mathscr{N}^{(R)}$, the mean number per unit volume of dislocations of one system (R) (we do not care about their orientation), is equal to $6N^{(r)}$ and that $l^{(R)} = l^{(r)}$.

(i) In this case the mean second limits are given by

$$\mathscr{F}^{(K)} = \mathscr{F}^{(K)}_{\mathrm{PN}} + \mu b^2 \sum_{(R)} D^{(K)}_{(R)} (\mathscr{N}^{(R)} l^{(R)})^{1/2} \tag{2.7}$$

with $D^{(K)}_{(R)} \simeq 1/8$ when (R) and (K) have the same slip plane and parallel Burgers vectors; $D^{(K)}_{(R)} \simeq 1/16$ when (R) and (K) have only the same slip

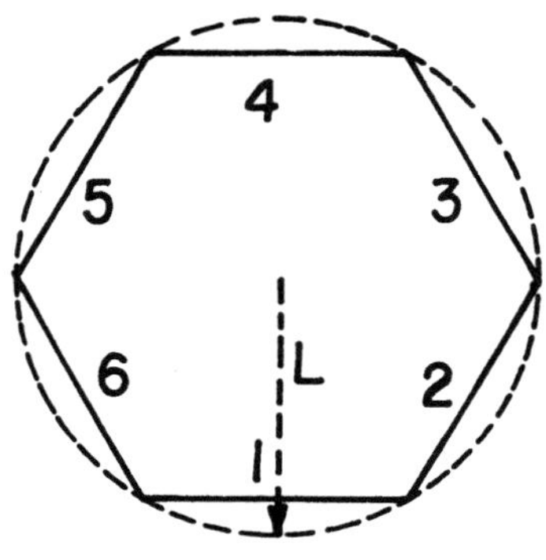

Fig. 8. Segments of the six mechanisms define a system.

plane; $D_{(R)}^{(K)} \simeq 1/12$ when (R) and (K) have neither the same slip plane nor parallel Burgers vectors; $D_{(R)}^{(K)} \simeq 1/20$ when (R) and (K) have not the same slip plane but have parallel Burgers vectors. (The relative values of $D_{(R)}^{(K)}$ show that the latent hardening is smaller than the eigen hardening in the active system; this seems to contradict Zaoui's results in Chapter 7, but it is not possible to adjust our model.)

(ii) Also in this case, from the formula (2.4) we can calculate the mean number per unit surface of dislocations of the system (R) which pierce a plane p' of type $\{111\}$; we take $\Psi_1 \simeq 19°30'$, $\Psi_2 \simeq 30°$ and only 4/6 of segments of the system pierce the plane; we find hence

$$\kappa'^{(R)} \simeq 0{\cdot}54.\, N^{(R)} l^{(R)} \qquad (2.8)$$

(e) Let us come back to the general segments distribution. We assume that the motion of one segment of the mechanism (r) in any plane will be possible when the component of the applied force in this plane $F^{(r)}$ is higher or equal to the limit $\mathscr{F}^{(r)}$. The thermal activation cannot allow the displacement of the whole segment.

2.1.3. *Hypothesis on Some Mean Times of Realization*

(a) Always as functions of the dislocation distribution and the applied forces $F^{(k)}$, we can evaluate some important mean times which are as follows (see Annexe to Chapter 1).

(i) The mean time, t_{GD}, which is necessary for a dislocation to get over another dislocation. We have to create two jogs and, as we saw, that may be thermally activated.

(ii) The mean time, t_1, which is necessary at a Frank–Read source to engender a dislocation loop.

(iii) The mean time, t_{CS}, which is necessary at a split screw dislocation to recombine into a perfect one and to move from its initial plane p into a new one p' by cross-slip.

(iv) The mean time, t_C, which is necessary for the creation and the displacement of one jog along the dislocation; this provokes the progressive climb of the dislocation into the plane p'.

(b) All these elementary times are written in the form

$$t_m = t_m(F^{(r)} - \mathscr{F}^{(r)}, F^{(k)}, T, N^{(k)}, l^{(k)}) \qquad (2.9)$$

They are infinite when $F^{(r)} < \mathscr{F}^{(r)}$; $\mathscr{F}^{(r)}$ is the limit for the corresponding motion.

Under the conditions of the formula (2.7), and when we have a rather low constant temperature ($T \simeq 20\,°C$) and low stresses, only the first times t_{GD} and t_1 are finite; they are approximately

$$t_{GD}^{(R)} \simeq \frac{C_0^{(R)}}{\langle F^{(R)} - \mathscr{F}^{(R)} \rangle} \qquad \text{and} \qquad t_1 \simeq 0 \tag{2.10}$$

where $C_0^{(R)}$ is a physical constant.

2.1.4. *Description of the Physical Model*

(a) We assume the following during the small dt.

(i) For each mechanism (r), some dislocations are sources; they emit $dP^{(r)}$ dislocation loops (per unit volume). These new loops sweep a mean area $\pi L^{(r)2}$ by easy glide before being stopped on the dislocations that pierce their plane $p(\vec{n}^{(r)})$.

(ii) Among the initial immobile segments a number $dp'^{(r)}$ (per unit volume) are moved chiefly by cross-slip and climb into other planes p' at a distance $L'^{(r)}$. Some of these $dp'^{(r)}$ segments will be annihilated or will form a subgrain with other dislocations.

(b) This follows us to write the following.

(i) Since these dislocation motions provoke relative velocities $\gamma^{(r)}$ and $\gamma'^{(r)}$ along $\vec{b}^{(r)}$ between two planes p and two planes p' which are unit length distant,

$$\left.\begin{aligned} \gamma^{(r)}\, dt &= b^{(r)}\, dP^{(r)}\, \pi L^{(r)2} && \text{between } p \\ \gamma'^{(r)}\, dt &= b^{(r)}\, dp'^{(r)}\, l^{(r)}\, L'^{(r)} && \text{between } p' \end{aligned}\right\} \tag{2.11}$$

(ii) And we have for each mechanism (k) the evolution of the couple $(N^{(k)}, l^{(k)})$:

$$dN^{(k)} = \sum_{(r)} dP^{(r)} - \sum_{p'} \omega'^{(k)}\, dp'^{(k)}$$

$$d(N^{(k)}\, l^{(k)}) = \sum_{(r)} dP^{(r)}\, s^{(k)}\, L^{(r)} - \sum_{(p')} \omega'^{(k)}\, dp'^{(k)}\, l^{(k)} \tag{2.12}$$

and thus

$$d(l^{(k)}) = \sum_{(r)} \frac{dP^{(r)}}{N^{(k)}} (s^{(k)}\, L^{(r)} - l^{(k)})$$

The first summation is extended over the mechanisms (r) which have the same Burgers vector and the same plane that the mechanism (k) has; the second summation is extended to the plane p' parallel to the segments of the mechanism (k). The new loops given by (r) are decomposed into segments which belong to different mechanisms (k) (owing to their orientation) and which have a length equal to $s^{(k)} L^{(r)}$. $\omega'^{(k)}\, dp'^{(k)}$ is the number of segments which disappear.

(iii) Climb into the planes p' provokes a change of volume and thus a variation of the number n_p (per unit volume) of point defects:

$$dn_p = \frac{1}{v_p} \sum_{(r)\text{ and } p'} h^{i(r)} n_i'^{(r)} \gamma'^{(r)} \tag{2.13}$$

where v_p is the volume occupied by one defect.

(c) With the aid of the probability laws (2.1) and (2.5), it is possible to evaluate $\pi L^{(r)2}, s^{(k)}, L'^{(r)}, \omega'^{(r)}$ as functions of the dislocation distribution and the applied forces $F^{(k)}$. But here, in order to simplify the equations, we shall use some restrictive hypotheses; when the loading is progressive they are sufficient. We suppose that

(i) $s^{(k)}$ and $\omega'^{(r)}$ are constant
(ii) $\pi L^{(r)2}$ and $L'^{(r)}$ are only functions of the dislocation distribution; e.g. for $\pi L^{(r)2}$ we have

$$\pi L^{(r)2} \simeq \frac{x^{(r)}}{\sum_{(k)} n^{(k)}} \tag{2.14}$$

This summation is extended to all the mechanisms (k) which pierce the plane p; $x^{(r)}$ is the number of jogs on the new loop and may be taken as constant.

(iii) $dP^{(r)}$ and $dp'^{(r)}$ are linked to the dislocation distribution and the forces $F^{(k)}$ thanks to the calculated mean times. For example, for $dP^{(r)}$ we have

$$dP^{(r)} = \omega^{(r)} N^{(r)} \frac{dt}{t_1^{(r)} + x^{(r)} t_{GD}^{(r)}} \tag{2.15}$$

where $\omega^{(r)}$, the fraction of segments that are sources, is assumed to be constant. (Only a few segments are capable of being a source; they have to be long enough, and to have not too many obstacles during their expansion.)

(d) In the particular case of the formulae (2.7), (2.8) and (2.10), and for a constant temperature $T \simeq 20\,^\circ\mathrm{C}$, we can say that for each active system (R) we have only

$$\left.\begin{aligned} \gamma^{(R)}\,\mathrm{d}t &= b\,\mathrm{d}\mathcal{N}^{(R)}\,\pi L^{(R)2} \\ \mathrm{d}(\mathcal{N}^{(R)}\,l^{(R)}) &= \mathrm{d}\mathcal{N}^{(R)}\,2\pi L^{(R)} \\ \pi L^{(R)2} &\simeq \frac{x^{(R)}}{\sum_{(K)} \kappa'^{(K)}} \equiv \frac{x^{(R)}}{\varrho^{(R)}} \\ \mathrm{d}\mathcal{N}^{(R)} &\simeq \omega^{(R)}\mathcal{N}^{(R)}\frac{\mathrm{d}t}{x^{(R)}\,t_{\mathrm{GD}}^{(R)}} \end{aligned}\right\} \tag{2.16}$$

where $\varrho^{(R)} = \sum \kappa'^{(K)}$ is the density per unit surface of all the dislocations which pierce the slip plane of (R); there is no cross-slip or climb.

2.1.5. *Return to the Theoretical Model*

(a) We see now exactly the physical meaning of the parameters $\{y\}$ for the pure single crystal; they are all the couples $(N^{(r)}, l^{(r)})$ and n_{p}. We find immediately the explicit form of $f^{(r)}$ (or $g^{(r)}$), $\gamma^{(r)}$ and $G^{(r)}$ with the help of eqns (2.6), (2.11), (2.12), (2.13), [or (2.7) and (2.16)]. Furthermore, since the self-energy per unit length of a dislocation (r) is approximately given by $\mu b^{(r)2}/C^{(r)}$ (see Chapter 1), we deduce that the specific energy due to the defects, $\mathcal{U}_2(T, \{y\})$, is equal to

$$\rho\,\mathcal{U}_2(T, \{y\}) \simeq \sum_{(r)} \frac{\mu b^{(r)2}}{C^{(r)}} N^{(r)}\,l^{(r)} \tag{2.17}$$

(we neglect the interaction energy between defects).

(b) In order to obtain these expressions we admitted many simplifying hypotheses. For the attention of people who do not agree with them, we indicated how it is possible to achieve more accurate calculations. We do not believe in the utility of such calculations because the dislocations distribution may never be determined exactly by experimentation; this distribution is not fully described by the couples $(N^{(r)}, l^{(r)})$ and it would be necessary to introduce moments up to infinite order, which is not realistic.

We shall see that our approximate calculations are however sufficient to interpret the most important properties of crystal behaviour.

2.2. Viscoplastic and Multiple-plastic Potential of a Crystal

In the restrictive framework of the application of formulae (2.16) it is very easy to prove that there is a viscoplastic potential and, in a particular case when the viscosity may be neglected, a multiple plastic potential. We have to notice that there is no work-softening. (See Chapter 7 for more details.)

2.2.1. *Viscoplastic Potential*

(a) Indeed the formulae (2.10) and (2.16) give for each active system (R):

$$\gamma^{(R)} \simeq b \frac{\omega^{(R)} \mathcal{N}^{(R)}}{C_0^{(R)} \rho^{(R)}} \langle F^{(R)} - \mathcal{F}^{(R)} \rangle \tag{2.18}$$

where $\gamma^{(R)}$, by the intermediacy of $F^{(R)}$, is only a function of the applied shear stress, $\tau^{(R)} = h_i^{(R)} n_j^{(R)} \Sigma^{ij}$, on the isolated system (R). Thus, by putting for each system (R),

$$\mathcal{T}^{(R)}(\Sigma, \mathcal{N}^{(K)}, l^{(K)}) \equiv \frac{\omega^{(R)} \mathcal{N}^{(R)}}{2C_0^{(R)} \rho^{(R)}} \langle F^{(R)} - \mathcal{F}^{(R)} \rangle^2 \tag{2.19}$$

and for the crystal

$$\mathcal{T} = \sum_{(R)} \mathcal{T}^{(R)} \tag{2.20}$$

the formula (1.12) giving the plastic part of the velocity gradient can be written

$$v''^i_j = \left(\frac{\partial \mathcal{T}}{\partial \Sigma^i_j} \right)_{\{y\} = \text{const}} \tag{2.21}$$

The virtual variations of the Cauchy stress tensor Σ are taken in R^9 (and not in R^6).

The proof of eqn (2.21) is evident. Successively, we write

$$\left(\frac{\partial \mathcal{T}}{\partial \Sigma^i_j} \right) = \sum_{(R)} \left(\frac{\partial \mathcal{T}^{(R)}}{\partial \Sigma^i_j} \right) = \sum \left(\frac{\partial \mathcal{T}^{(R)}}{\partial F^{(R)}} \right) \left(\frac{\partial F^{(R)}}{\partial \Sigma^i_j} \right)$$

$$= \sum_{(R)} \frac{\gamma^{(R)}}{b} b_i^{(R)} n_j^{(R)} \equiv \sum_{(R)} h^{i(R)} n_j^{(R)} \gamma^{(R)} \tag{2.22}$$

where $\mathcal{T}$ is Rice's viscoplastic potential of the single crystal.

(b) In generalized stresses space (R^9) (the real stresses are in R^6 since there is the symmetry $\Sigma^{ij} = \Sigma^{ji}$), the equipotential surfaces $\mathscr{T} =$ const > 0 are regular (without any singular point) and the surface $\mathscr{T} = 0$ corresponds to the elastic surface since we have $F^{(R)} \leqslant \mathscr{F}^{(R)}$.

In this space the vector v''^i_j is always normal to the equipotential surface which is defined by the actual value of Σ (Fig. 9). All the surfaces are cylindrical (Σ'^i_i does not occur); eqn (2.20) at once shows that these surfaces are convex and successively fitted in when the const. is increasing.

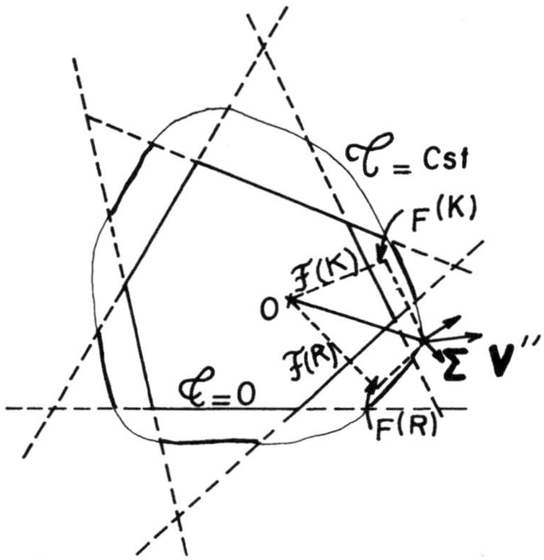

Fig. 9. Viscoplastic potentials for a single crystal.

2.2.2. *Multiple Plastic Potential*

(a) When all the ratios $\omega^{(R)}/C_0^{(R)}$ are very large and the strain rates are rather low, the differences $\langle F^{(R)} - \mathscr{F}^{(R)} \rangle$ for the active systems are thus very small. This means that, at any time, the actual stress state Σ is very near the elastic limit surface and that the viscosity may be neglected. This is classical plasticity.

(b) For each active system (R) between t and $t + \mathrm{d}t$, we have

$$\left.\begin{array}{lcl} F^{(R)}(t) \simeq \mathscr{F}^{(R)}(t) & & F^{(R)}(t) \leqslant \mathscr{F}^{(R)}(t) \\ & \text{or for any } (K) & \\ \dot{F}^{(R)} \simeq \dot{\mathscr{F}}(t) & & F^{(R)}(t+\mathrm{d}t) \leqslant \mathscr{F}^{(R)}(t+\mathrm{d}t) \end{array}\right\} \quad (2.23)$$

with $\overset{\bullet}{\overline{(\mathcal{N}^{(R)} l^{(R)})}} \geqslant 0$. This may be explicitly written in the form

$$\left.\begin{aligned} F^{(R)} &= b h_i^{(R)} n_j^{(R)} \Sigma^{ij} = b\tau^{(R)} \simeq \mathscr{F}_{\mathrm{PN}}^{(R)} + \mu b^2 \sum_{\text{at all }(K)} D_{(K)}^{(R)} (\mathcal{N}^{(K)} l^{(K)})^{1/2} \\ \dot{F}^{(R)} &= b\dot{\tau}^{(R)} = \mu b^2 \sum_{\text{at active }(K)\text{ only}} D_{(K)}^{(R)} \frac{\overset{\bullet}{\overline{(\mathcal{N}^{(K)} l^{(K)})}}}{2(\mathcal{N}^{(K)} l^{(K)})^{1/2}} \end{aligned}\right\} \quad (2.24)$$

From the equations (2.16) we deduce

$$\overset{\bullet}{\overline{(\mathcal{N}^{(K)} l^{(K)})}} = \frac{2\gamma^{(K)}}{bL^{(K)}} \quad (2.25)$$

and by taking this into the second equation (2.24) we obtain

$$\dot{\tau}^{(R)} \simeq \sum_{\text{at active }(K)} H_{(K)}^{(R)} \gamma^{(K)} \qquad (\text{with } \gamma^{(K)} \geqslant 0) \quad (2.26)$$

The matrix **H** is defined by

$$H_{(K)}^{(R)} = \frac{\mu D_{(K)}^{(R)}}{(\mathcal{N}^{(K)} l^{(K)})^{1/2} L^{(K)}} \quad (2.27)$$

(This matrix is not found to be symmetrical; this is linked to the fact that large changes of the dislocation distribution were allowed.) So, as in Mandel's multiple plastic potential theory, we find

$$v''^{i}_{\ j} = \sum_{\text{at active }(K)} \frac{\partial \tau^{(K)}}{\partial \Sigma^{ij}} \gamma^{(K)} = \sum_{\text{at active }(K)} h_i^{(K)} n_j^{(K)} \gamma^{(K)} \quad (2.28)$$

and that the increase of the limit for any system (R) is a linear function of $\gamma^{(K)}$ ($\gamma^{(K)} \geqslant 0$) on the active systems (K).

(c) There is the very important problem of how to determine the active systems (K) and to calculate $\gamma^{(K)}$ when we impose, by the intermediacy of the objective stress tensor derivative, $\dot{\tau}^{(R)}$ on each system (R). In taking as intermediate variables the quantities

$$\dot{X}^{(R)} = \frac{\overset{\bullet}{\overline{(\mathcal{N}^{(R)} l^{(R)})}}}{2(\mathcal{N}^{(R)} l^{(R)})^{1/2}}$$

it is easy to show that, since the matrix **D** ($D_{(K)}^{(R)}$) is symmetrical and is strictly positive for any number of systems lower than 8, we have a quadratic minimization problem with the linear constraints $\dot{X}^{(R)} \geqslant 0$. This may be solved easily with the help of many algorithms.

2.3. Example of Tensile Experiments

We have seen that generally only experiments with tension or compression are performed on the single crystal and that these experiments are very difficult to realize and to interpret chiefly owing to the big changes of geometry. First we shall look at the sort of theoretical curves that are obtained by use of our model and then we shall compare some theoretical and experimental results.

2.3.1. *Position of the Problem*

(i) Now we limit ourselves to classical plasticity and we adopt eqns (2.7), (2.9), (2.10) and (2.16).

(ii) We assume that at any moment we have a simple uniaxial force Q on the crystal and a constant temperature $T \simeq 20\,°\mathrm{C}$.

(iii) Since the elastic strains are very small, we assume that the frame of reference $\vec{g}_i$ is always confused with the frame $\vec{e}_\alpha$. The axes Ox, Oy, Oz are respectively along the directions [000], [010] and [001] (Fig. 10).

(iv) We assume that the matrix of elastic coefficients Λ is constant (independent of the stresses).

Experimentally, during the strain, we can only measure the applied force Q and the crystal length D. Hence we ought to verify these quantities directly if the model gives good results. Generally metallurgists show their tensile curve in the form (τ, Γ) where τ is the applied shear stress and Γ the glide on the principal system. For that they assume that only this system is active, which is not quite correct.

2.3.2. *Equations of the Problem*

(i) Let us denote by S the actual normal section of the crystal; the Cauchy stress tensor is proportional to

$$\sigma = Q/S \tag{2.29}$$

and is given by

$$\Sigma = \sigma \vec{d} \otimes \vec{d} \tag{2.30}$$

where $\vec{d}$ is the unit vector along the tensile axis. This vector is defined by the angles θ, ψ and its components are

$$\left.\begin{aligned} d_x &= \sin\theta \, \sin\psi \\ d_y &= -\sin\theta \, \cos\psi \\ d_z &= \cos\theta \end{aligned}\right\} . \tag{2.31}$$

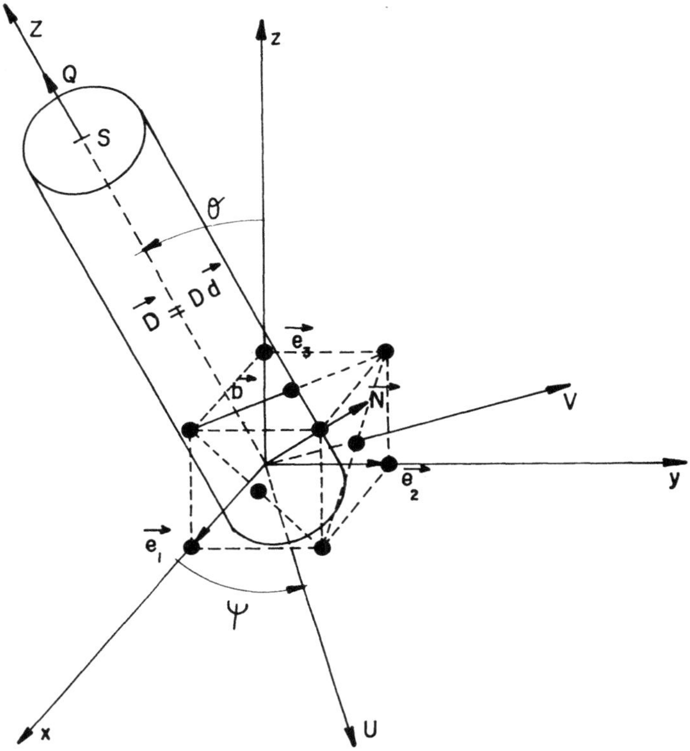

Fig. 10. Geometry of a single crystal specimen during a tensile test.

The vector $\vec{D} = D\vec{d}$ is linked to the tensile axis; it is proportional to the crystal length D and has the same direction. Since elastic strains are small and plastic strains do not cause change of volume, we have

$$S \simeq S_0 D_0 / D \tag{2.32}$$

where S_0 and D_0 are the initial values of S and D. Thus, by using eqn (2.29), we deduce

$$\sigma = QD/S_0 D_0 \tag{2.33}$$

(ii) As we saw, in this simple model, there are 24 slip systems; the components for unit normal and unit glide direction are given in Table 1. The applied force on the system (R) which is given by

TABLE 1

$\vec{n}^{(K)}$: $n_x^{(K)}$ $n_y^{(K)}$ $n_z^{(K)}$	$\vec{h}^{(K)}$: $h_x^{(K)}$ $h_y^{(K)}$ $h_z^{(K)}$			K indices		
$\frac{1}{\sqrt{3}}$ $(11\bar{1})$	$\frac{1}{\sqrt{2}}$ $[101]$	or	$[\bar{1}0\bar{1}]$	1, 1	or	1, 2
	$[0\bar{1}\bar{1}]$	or	$[011]$	2, 1	or	2, 2
	$[\bar{1}10]$	or	$[1\bar{1}0]$	3, 1	or	3, 2
$\frac{1}{\sqrt{3}}$ $(1\bar{1}1)$	$\frac{1}{\sqrt{2}}$ $[\bar{1}01]$			4, 1		
	$[0\bar{1}\bar{1}]$			5, 1		
	$[\bar{1}\bar{1}0]$			6, 1		
$\frac{1}{\sqrt{3}}$ $(1\bar{1}\bar{1})$	$\frac{1}{\sqrt{2}}$ $[\bar{1}\bar{1}0]$			7, 1		
	$[0\bar{1}1]$			8, 1		
	$[\bar{1}0\bar{1}]$	or	$[101]$	9, 1	or	9, 2
$\frac{1}{\sqrt{3}}$ (111)	$\frac{1}{\sqrt{2}}$ $[\bar{1}10]$			10, 1		
	$[01\bar{1}]$			11, 1		
	$[10\bar{1}]$	or	$[\bar{1}01]$	12, 1	or	12, 2

$$F^{(R)} = b_i^{(R)} n_j^{(R)} \Sigma^{ij} = b h_i^{(R)} n_j^{(R)} \Sigma^{ij}$$

is equal to

$$F^{(R)} = b(n_j^{(R)} d_j)(h_i^{(R)} d_i)\sigma \tag{2.34}$$

or by putting

$$m^{(R)} = (h_i^{(R)} d_i)(n_j^{(R)} d_j) \tag{2.35}$$

we obtain

$$F^{(R)} = b m^{(R)} \sigma = b m^{(R)} (QD/S_0 D_0) \tag{2.36}$$

where $m^{(R)}$ is the classical geometrical Schmid coefficient (see Chapter 1). Of course, if $m^{(R)}$ is negative this system cannot take part; instead we must consider the system with the opposite Burgers vector. The limit of large glide for this system (R) is equal to

$$\mathscr{F}^{(R)} = \mathscr{F}_{\text{PN}}^{(R)} + \mu b^2 \sum_{(K)} D_{(K)}^{(R)} (\mathscr{N}^{(K)} l^{(K)})^{1/2} \tag{2.37}$$

This summation is extended to the 24 systems. The system (R) is active when

$$F^{(R)} \geqslant \mathscr{F}^{(R)} \tag{2.38}$$

We have then

$$\left.\begin{aligned}
\gamma^{(R)} &\simeq b\,\frac{\omega^{(R)}\mathcal{N}^{(R)}}{C_0^{(R)}\rho^{(R)}}\langle F^{(R)}-\mathcal{F}^{(R)}\rangle \\
\dot{\mathcal{N}}^{(R)} &= \frac{\gamma^{(R)}}{b\pi L^{(R)2}}; \qquad \overline{(\mathcal{N}^{(R)}l^{(R)})}^{\bullet} = \frac{2\gamma^{(R)}}{bL^{(R)}} \\
\pi L^{(R)2} &= \frac{x^{(R)}}{\rho^{(R)}}
\end{aligned}\right\} \tag{2.39}$$

(iii) In the frame $\vec{g}_i\,(Ox,y,z)$ the derivative of the Cauchy stress tensor, $\dot{\Sigma}^{ij}$, is equal to the objective derivative and there is no elastic rotation since $\vec{e}_a$ and $\vec{g}_i$ are confused. The Eulerian velocity gradient is thus

$$\begin{aligned}
v_{ij} &= \Lambda_{ijhk}\,\dot{\Sigma}^{hk} \quad + \sum_{(R)} h_i^{(R)}\,n_j^{(R)}\,\gamma^{(R)} \\
&\qquad \text{elastic part} + \quad \text{plastic part}
\end{aligned} \tag{2.40}$$

For a cubic crystal, in these particular axes (Ox,y,z), the elastic part can be expanded in the form

$$\left.\begin{aligned}
v'_{xx} &= S_{11}\dot{\Sigma}_{xx} + S_{12}(\dot{\Sigma}_{yy}+\dot{\Sigma}_{zz}) \\
v'_{yy} &= S_{11}\dot{\Sigma}_{yy} + S_{12}(\dot{\Sigma}_{xx}+\dot{\Sigma}_{zz}) \\
v'_{zz} &= S_{11}\dot{\Sigma}_{zz} + S_{12}(\dot{\Sigma}_{xx}+\dot{\Sigma}_{yy}) \\
2v'_{xy} &= S_{44}\dot{\Sigma}_{xy} \\
2v'_{xz} &= S_{44}\dot{\Sigma}_{xz} \\
2v'_{yz} &= S_{44}\dot{\Sigma}_{yz}
\end{aligned}\right\} \tag{2.41}$$

For Al and Cu crystals we have (in the units 10^{-5} mm^2 kgf^{-1})

$$\begin{aligned}
&\text{Al:} \quad S_{11} = 15{\cdot}9; \quad S_{12} = -5{\cdot}8; \quad S_{44} = 35{\cdot}2; \\
&\text{Cu:} \quad S_{11} = 14{\cdot}9; \quad S_{12} = -6{\cdot}2; \quad S_{44} = 13{\cdot}3;
\end{aligned}$$

(iv) The evolution law for $\vec{D}$ is thus

$$\begin{pmatrix}\dot{D}_x\\ \dot{D}_y\\ \dot{D}_z\end{pmatrix} = \begin{pmatrix} v_{xx} & v_{xy} & v_{xz}\\ v_{yx} & v_{yy} & v_{yz}\\ v_{zx} & v_{zy} & v_{zz}\end{pmatrix}\begin{pmatrix}D_x\\ D_y\\ D_z\end{pmatrix} \tag{2.42}$$

From this we deduce that the variation of the crystal length

$$\dot{D} \equiv \left[S_Z \dot{\sigma} + \dot{S}_Z \sigma + \sum_{\text{active }(R)} m^{(R)} \gamma^{(R)} \right] D \tag{2.43}$$

where S_Z, given by

$$S_Z = S_{11} - \{2S_{11} - 2S_{12} - S_{44}\} \{(d_x d_y)^2 + (d_x d_z)^2 + (d_y d_z)^2\} \tag{2.44}$$

is the inverse of the Young's modulus E_Z in the direction OZ of the tensile axis. The unit vector $\vec{d}$ is only submitted to plastic rotation and its evolution is given by

$$\dot{d}_i = v''_{[ij]} d_j \qquad (v''_{[ij]} \text{ antisymmetrical part}) \tag{2.45}$$

(v) Generally the experiments are performed on tensile machines where a constant velocity C of the mobile traverse is given, which means

$$\dot{D} \simeq C \tag{2.46}$$

We have thus to solve for this problem a system of differential equations in σ, ψ, θ and the $\mathcal{N}^{(R)}, \mathcal{N}^{(R)} l^{(R)}$ with as initial conditions for $t = 0$, for example

$$\left.\begin{array}{ll} \sigma(0) = 0; \quad \psi(0) = \psi_0 \quad \text{and} \quad \theta(0) = \theta_0 & \\ \mathcal{N}^{(R)}(0) = \mathcal{N}_0 \quad \text{and} \quad \mathcal{N}^{(R)} l^{(R)}(0) = \mathcal{N}_0 l_0 & \text{for any } (R) \end{array}\right\} \tag{2.47}$$

2.3.3. *Integration of the Equations*

(a) When only the principal system is active, during the whole deformation, the integration can be done easily and explicitly.

(b) When several glide systems are active, the integration has to be performed numerically. It is possible to write two different programs.

(i) The first one is based on the direct integration of the different equations of Section 2.3.2. There are several numerical methods for solving this system. The best one is the Treanor method with variable step. We have several parameters: x_0, ω/C_0, $\mathcal{N}_0$, $\mathcal{N}_0 l_0$, C.

(ii) The second one is based on the multiple plastic potential. Here an implicit scheme is used; we try to have, at any time t

$$F^{(R)}(t) \leqslant \mathcal{F}^{(R)}(t) \qquad \text{for any } (R)$$

and at time $t + \Delta t$

$$F^{(R)}(t + \Delta t) \leqslant \mathcal{F}^{(R)}(t + \Delta t) \quad \text{with} \quad \Delta(\mathcal{N}^{(R)} l^{(R)}) \geqslant 0 \qquad \text{for any } (R)$$

For an active system we have

$$F^{(R)}(t) = \mathscr{F}^{(R)}(t) \quad \text{and} \quad F^{(R)}(t + \Delta t) = \mathscr{F}^{(R)}(t + \Delta t)$$

This is a quadratic minimization problem with linear constraints as we see it. We have here only x_0 and $\mathscr{N}_0 l_0$ as parameters.

Some theoretical curves are drawn in Figs 11 and 12. They are the most typical ones for different orientations of the tensile axis.

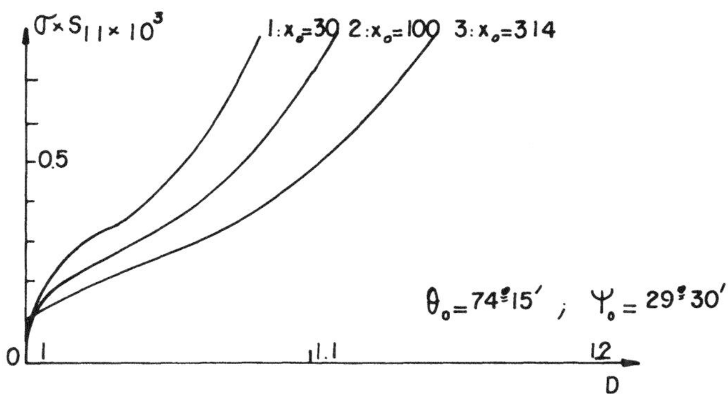

Fig. 11. Influence of the parameter x_0 on the response of the single crystal.

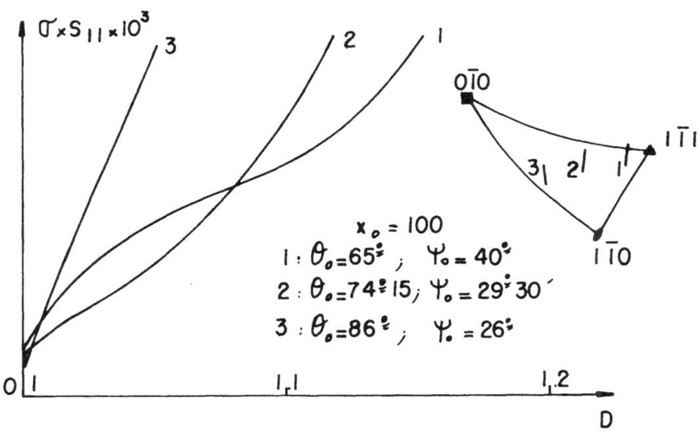

Fig. 12. Influence of the direction of the tensile axis on the response of the single crystal.

2.3.4. *Comparisons with Experimental Results*

(a) It is easy to see from these curves that we interpret not only the two important stages of work-hardening, which are stage I and stage II, but also the transition stages. Of course, as we neglected here the cross-slips and climbs of dislocations and thus the work-softening, stage III is not obtained. Also, we have the dependence with the orientation of the tensile axis.

(b) We understand quickly what happens during the strain.

(i) The transition to stage I and stage II (this is generally a parabola) is linked to the fact that only one system is active (or sometimes two systems but with the same glide plane).

(ii) The transition to stage II shows the activation of one new system, the plane of which cuts the first slip plane.

(iii) During stage II several systems are active.

(iv) For some orientations of the tensile axis, stage I may disappear.

(c) The fitting between theoretical and experimental results may also be quantitative if we have a convenient value of x_0. Of course, owing to the phenomenological nature of this parameter, this fitting is not convincing. Some other measures, where x_0 does not truly take part, show however that the model looks suitable (chiefly in the formula giving the interactions between slip systems; see Zarka, 1972).

2.3.5. *General Remark*

Of course, the main interest of this model is at the fundamental level. We can understand very clearly how all the mechanisms can interact. Naturally such a model cannot be given for engineering applications, not only since it concerns the single crystal but also because too many parameters have to be taken into account.

Section 2: Systematic Structural Modelling of Metals

3. GENERAL FRAMEWORK

3.1. Introduction

To describe inelastic constitutive relations for metals — as was shown in the various chapters of this book and particularly in Chapters 7 and 9 — usually two different approaches are possible.

(i) Phenomenological formulae can be guessed and developed (macroscopical approach).

(ii) The different local processes leading to inelastic strains can be tested and the behaviour can be deduced after some lengthy computations (microscopical approach; see first part of this chapter).

An intermediate approach has been considered in the regime where hardening, softening and creep are possible. It is also based on the fact that the fundamental volume element we have to represent is a polycrystal, an aggregate of crystals or a crystal with several internal structural defects, but a simplification of local behaviours is assumed. Some authors (Besseling, 1953; Mróz, 1972*a*; Zienkiewicz *et al.*, 1973) have introduced it, but indeed it does not differ very much from the microscopical approach. Here we shall give the most important ideas for its understanding as well as some of its powerful applications.

3.2. Global Constitutive Relations

3.2.1. *Experimental Facts and Principle of the Description*

Under applied cyclic stress in the homogeneous uniaxial condition, some important facts are classically underlined (Fig. 13).

(i) The eventual existence of a steady cyclic state for which the stress and strain change periodically; there is *elastic shakedown* (where the sample will sustain a large number of cycles before failure) or *plastic shakedown* (where low cycle fatigue will generally occur).

(ii) Some asymmetrical stress cycles may induce ratcheting (in which progressive plastic strain increments will lead to incremental collapse), even if during some lower mean stresses there may be shakedown.

(iii) Some cycles of loading are necessary to reach the eventual limiting state.

(iv) The eventual ratcheting or the cyclic inelastic strain may be the result of both plastic and viscous phenomena during cyclic tests where varying frequencies or rest-periods are imposed.

The modelling of materials has to take these facts into account.

On the other hand, as was seen all through the other chapters of this book, the material is, at our macroscopical level, a polycrystal, i.e. the fundamental volume element is an assembly of crystals (or grains) which themselves contain various components and some structural defects. Its *global* behaviour will be associated with the *local* behaviour of its different constituents but will possess particular properties too (see Chapter 4 of this book).

Following this picture, we shall just say that the volume element we consider is symbolized by a Discrete Assembly of Local Subelements or

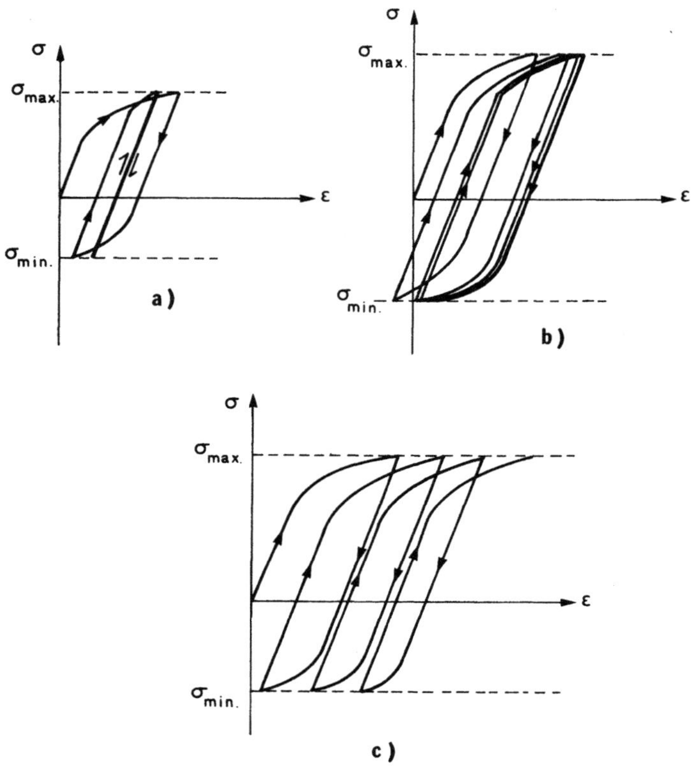

Fig. 13. Cyclic responses for a metal: (a) elastic shakedown; (b) plastic shakedown; (c) ratcheting.

Mechanisms with simple mechanical properties. We shall distinguish four types of such subelements, the first three being similar to the classical rheological models.

(1) Sliders of friction blocks (for instantaneous strains with a threshold, as the glide of one dislocation).
(2) Dashpots (for viscous strains, as the diffusion of defects).
(3) In-parallel sliders–dashpots (for delayed viscous strains with a threshold, as the thermally activated crossing of two dislocations).
(4) Linear springs (for linear elastic strains), forming links between the first three.

We shall take all these subelements as perfect.

As we shall see later, hardening and softening will only be the result of creation and redistribution of residual stresses within the volume element (Mandel, 1977); however, a reasonable description of the behaviour will be obtained.

In order to help in the understanding of the framework, we shall first give some examples of the assembly of uniaxial rheological models. But we do not recommend attempts to find if such an assembly or real dislocations model can be associated with any set of general constitutive relations that we could later easily write down '*a priori*'.

3.2.2. *Some Examples*

(a) *Basic equations for classical uniaxial non-linear rheological models* (Fig. 14).

(1) *Slider.* The applied local stress $\sigma^{(1)}$ must be 'plastically admissible' at any moment, i.e. it has to satisfy

$$-S_0 \leqslant \sigma^{(1)} \leqslant S_0 \tag{3.1}$$

where S_0 is the threshold for gliding. This implies that $\sigma^{(1)}$ belongs to the *immobile convex set* $C_0 \equiv [-S_0, S_0]$. When $\sigma^{(1)}$ is strictly inside C_0 there is no glide, $\dot{\alpha} = 0$; but when $\sigma^{(1)} = +S_0$ we can have any $\dot{\alpha} \geqslant 0$ and when $\sigma^{(1)} = -S_0$ any $\dot{\alpha} \leqslant 0$. These three cases may be symbolically written in the simpler form

$$\dot{\alpha} \in \partial\Psi_{C_0}(\sigma^{(1)}) \tag{3.2}$$

i.e. $\dot{\alpha}$ is an *external* normal to the convex set C_0 in $\sigma^{(1)}$ (see Moreau, 1972, and Chapter 3 of this book).

(2) *Linear dashpot.* $\sigma^{(2)}$, being the applied local stress and η the damping factor, we have

$$\dot{\beta} = \sigma^{(2)}/\eta \tag{3.3}$$

which implies that $\dot{\beta} = 0$ only when $\sigma^{(2)} = 0$.

(3) *In-parallel slider–dashpot.* Owing to the internal kinematical constraint in the group, the displacements of the slider and the dashpot are always identical. If $\sigma^{(3)}$ is the applied stress at the *level* of the group, we have

$$\dot{\gamma} = \frac{1}{\eta}[\langle\sigma^{(3)} - S_0\rangle + \rangle\sigma^{(3)} + S_0\langle] \tag{3.4}$$

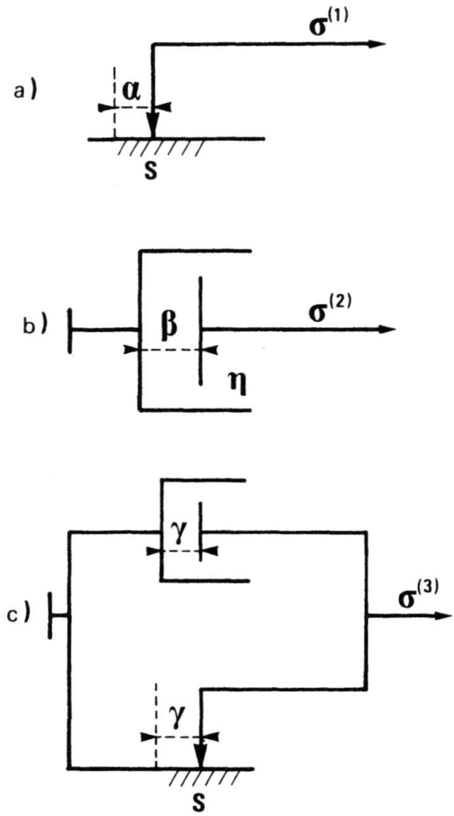

Fig. 14. Rheological models: (a) slider; (b) dashpot; (c) slider and dashpot in parallel.

$$\langle x \rangle \text{ is the positive part of } x \equiv \begin{cases} 0 \text{ if } x \leqslant 0 \\ x \text{ if } x \geqslant 0 \end{cases}$$

$$\rangle x \langle \text{ is the negative part of } x \equiv \begin{cases} x \text{ if } x \leqslant 0 \\ 0 \text{ if } x \geqslant 0 \end{cases}$$

We notice that

$$\dot{\gamma} \equiv 0 \quad \text{when} \quad -S_0 \leqslant \sigma^{(3)} \leqslant S_0 \qquad \text{i.e. when} \quad \sigma^{(3)} \in D_0 \tag{3.5}$$

the *immobile convex set* $[-S_0, S_0]$.

(b) *First example.* Let us consider the elementary assembly which is given in Fig. 15; it is composed of two sliders and three springs.

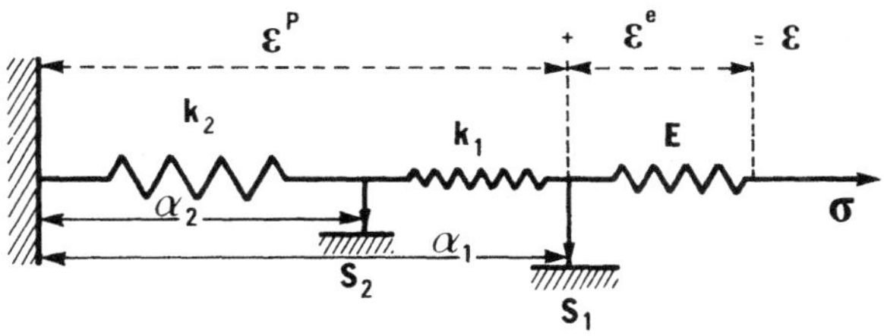

Fig. 15. First assembly of sliders and springs.

(i) The global stress σ is applied to the assembly; the global strain ε is the sum of the global inelastic strain ε^{p} and the global elastic strain ε^{e}. Here we have

$$\left.\begin{aligned} \varepsilon^{\mathrm{e}} &= \sigma/E \\ \varepsilon^{\mathrm{p}} &\equiv \alpha_1 \end{aligned}\right\} \tag{3.6}$$

where $\boldsymbol{\alpha} = \begin{pmatrix}\alpha_1\\ \alpha_2\end{pmatrix}$ are the local inelastic strains.

The knowledge of $\boldsymbol{\alpha}$ is sufficient to describe the internal state of the assembly; they are thus its *internal parameters.*

(ii) At the level of each slider, the local stresses are given by

$$\left.\begin{aligned} \sigma_1^{(1)} &= \sigma - k_1(\alpha_1 - \alpha_2) \\ \sigma_2^{(1)} &= k_1(\alpha_1 - \alpha_2) - k_2\alpha_2 = k_1\alpha_1 - (k_1 + k_2)\alpha_2 \end{aligned}\right\} \tag{3.7}$$

(iii) By definition we shall set

$$\left.\begin{aligned} \sigma_1^{(1)} &\equiv A_1^{(1)}\sigma - \hat{\alpha}_1 \\ \sigma_2^{(1)} &\equiv A_2^{(1)}\sigma - \hat{\alpha}_2 \end{aligned}\right\} \tag{3.8}$$

with here for 'the localization factors' $\mathbf{A}^1 = \begin{pmatrix}A_1^{(1)}\\ A_2^{(1)}\end{pmatrix}$ respectively

$$A_1^{(1)} = 1 \qquad A_2^{(1)} = 0 \tag{3.9}$$

and

$$\hat{\boldsymbol{\alpha}} = \begin{pmatrix}\hat{\alpha}_1\\ \hat{\alpha}_2\end{pmatrix} = \begin{pmatrix}k_1 & -k_1\\ -k_1 & k_1 + k_2\end{pmatrix}\begin{pmatrix}\alpha_1\\ \alpha_2\end{pmatrix} \equiv B^{(1)}\boldsymbol{\alpha} \tag{3.10}$$

We can easily see that the $B^{(1)}$ matrix is a symmetrical strictly positive matrix. $\widehat{\alpha}$, which are *here* linked to $\boldsymbol{\alpha}$ by a one-to-one linear relation, will be called the *transformed internal parameters.*

(iv) With eqns (3.9) the global plastic strain ε^{p} is given by

$$\varepsilon^{\mathrm{p}} = \sum_r A_r^{(1)} \alpha_r \equiv \mathbf{A}^{(1)\mathrm{T}} \boldsymbol{\alpha} \tag{3.11}$$

(v) Since $\sigma_1^{(1)}$ and $\sigma_2^{(1)}$ have to be plastically admissible

$$\sigma_1^{(1)} \in C_1 = [-S_1, S_1] \qquad \text{and} \qquad \sigma_2^{(1)} \in C_2 = [-S_2, S_2]$$

$\widehat{\alpha}_1$ and $\widehat{\alpha}_2$ have respectively to belong to the *mobile convex* sets $\widehat{C}_1(\sigma), \widehat{C}_2(\sigma)$ which are subjected to simple translations

$$\begin{aligned} \widehat{C}_1(\sigma) &\equiv [-S_1 + A_1^{(1)}\sigma,\ S_1 + A_1^{(1)}\sigma] \equiv C_1 + A_1^{(1)}\sigma \\ \widehat{C}_2(\sigma) &\equiv [-S_2 + A_2^{(1)}\sigma,\ S_2 + A_2^{(1)}\sigma] \equiv C_2 + A_2^{(1)}\sigma \end{aligned} \tag{3.12}$$

In the $(\widehat{\alpha}_1, \alpha_2)$ plane, the transformed internal parameters $\widehat{\boldsymbol{\alpha}}$ belong then to the mobile convex set $\widehat{\mathscr{C}}(\sigma)$, the rectangle whose centre is in $A^{(1)}\sigma$ and whose side lengths are $(2S_1, 2S_2)$ (Fig. 16).

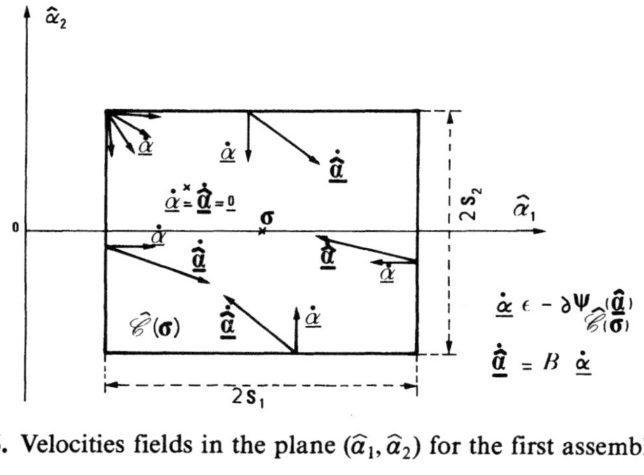

Fig. 16. Velocities fields in the plane $(\widehat{\alpha}_1, \widehat{\alpha}_2)$ for the first assembly.

(vi) For each slider r, the flow rules (3.2) will be written

$$\dot{\alpha}_r \in \partial\Psi_{C_r}(\sigma_r^{(1)}) \quad \text{i.e.} \quad \begin{cases} -S_r < \sigma_r^{(1)} < S_r & \Rightarrow \quad \dot{\alpha}_r \equiv 0 \\ \sigma_r^{(1)} = S_r & \Rightarrow \quad \dot{\alpha}_r \geqslant 0 \\ \sigma_r^{(1)} = -S_r & \Rightarrow \quad \dot{\alpha}_r \leqslant 0 \end{cases}$$

or when introducing (3.8) and assuming known the global stress σ,

$$\left.\begin{array}{lll} -S_r + A_r^{(1)}\sigma < \widehat{\alpha}_r < S_r + A_r^{(1)}\sigma & \Rightarrow & \dot{\alpha}_r \equiv 0 \\ \widehat{\alpha}_r = -S_r + A_r^{(1)}\sigma & \Rightarrow & \dot{\alpha}_r \geqslant 0 \\ \widehat{\alpha}_r = S_r + A_r^{(1)}\sigma & \Rightarrow & \dot{\alpha}_r \leqslant 0 \end{array}\right\} \dot{\alpha}_r \in -\partial\Psi_{\widehat{C}_r(\sigma)}(\widehat{\alpha}_r) \tag{3.13}$$

$\dot{\alpha}_r$ is thus an *internal* normal to the mobile convex set $\widehat{C}_r(\sigma)$ in $\widehat{\alpha}_r$.

(vii) Simultaneously for the two sliders we can write down the flow rules

$$\dot{\boldsymbol{\alpha}} \in -\partial\Psi_{\widehat{\mathscr{C}}(\sigma)}(\widehat{\boldsymbol{\alpha}}) \tag{3.14}$$

and for the velocities of the transformed internal parameters

$$\dot{\widehat{\boldsymbol{\alpha}}} \in -B^1\,\mathrm{d}\Psi_{\widehat{\mathscr{C}}(\sigma)}(\widehat{\boldsymbol{\alpha}}) \tag{3.15}$$

We have shown in Fig. 16 the particular meanings of (3.14) and (3.15).

(c) *Second example.* Let us now consider the assembly which is shown in Fig. 17. We still have two sliders but they are now connected differently. It is very simple to obtain the following properties (see Zarka, 1979):

(i) $\boldsymbol{\alpha} = \begin{pmatrix} \alpha_1 \\ \alpha_2 \end{pmatrix}$ are the two internal parameters of the system.

(ii) The global stress σ being assumed known, the local stresses at the level of each slider are given by

$$\begin{aligned} \sigma_1^{(1)} &= \frac{k_1}{k_1+k_2}\sigma - \frac{k_1 k_2}{k_1+k_2}(\alpha_1 - \alpha_2) \\ \sigma_2^{(1)} &= \frac{k_2}{k_1+k_2}\sigma + \frac{k_1 k_2}{k_1+k_2}(\alpha_1 - \alpha_2) \end{aligned} \tag{3.16}$$

and can be written as in (3.8) with

$$A_1^{(1)} = k_1/(k_1+k_2) \quad (>0); \qquad A_2^{(1)} = k_2/(k_1+k_2) \quad (>0) \tag{3.17}$$

$$\widehat{\boldsymbol{\alpha}} = \begin{pmatrix} \widehat{\alpha}_1 \\ \widehat{\alpha}_2 \end{pmatrix} = \frac{k_1 k_2}{k_1+k_2}\begin{pmatrix} 1 & -1 \\ -1 & 1 \end{pmatrix}\begin{pmatrix} \alpha_1 \\ \alpha_2 \end{pmatrix} = B^{(1)}\boldsymbol{\alpha} \tag{3.18}$$

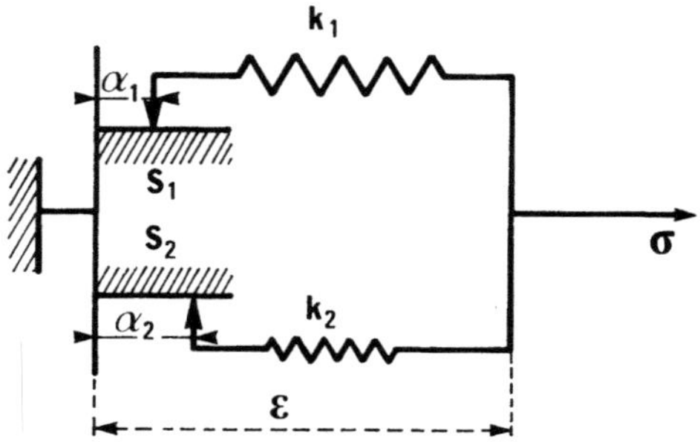

Fig. 17. Second assembly of sliders and springs.

but this time the symmetrical matrix $B^{(1)}$ is singular and we always have the compatibility condition

$$\widehat{\mathscr{L}}\colon\ \widehat{\alpha}_1 + \widehat{\alpha}_2 = 0 \tag{3.19}$$

The transformed internal parameters $\widehat{\boldsymbol{\alpha}}$ are linearly linked to $\widehat{\boldsymbol{\alpha}}$ but the relation is not one-to-one. For a 'compatible' $\widehat{\boldsymbol{\alpha}}$ [satisfying (3.19)] there are an infinite number of solutions $\boldsymbol{\alpha}$ such that (3.18) is verified.

(iii) The equation (3.11) which expresses the global plastic strain is still valid. The global elastic strain is here given by

$$\varepsilon^{e} = \sigma/(k_1 + k_2) \tag{3.20}$$

(iv) The evolution laws for $\boldsymbol{\alpha}$ and $\widehat{\boldsymbol{\alpha}}$ keep the same shapes as in (3.14) and (3.15). The *one* difference from the previous example lies only in the singularity of the $B^{(1)}$ matrix which implies (3.19) (Fig. 18).

(d) *Third example.* In Fig. 19, we have now a more complex assembly of two sliders and three springs. When analysing the local stresses at the level of each slider, we easily obtain

$$A_1^{(1)} = 1 \quad (>0); \qquad A_2^{(1)} = -k_3/(k_2 + k_3) \quad (<0) \tag{3.21}$$

$$B^{(1)} = \begin{pmatrix} k_1 & -k_1 \\ -k_1 & (k_1k_2 + k_2k_3 + k_3k_1)/(k_2 + k_3) \end{pmatrix} \tag{3.22}$$

The $A_r^{(1)}$ have opposite signs; the matrix $B^{(1)}$ is strictly positive. All the other evolution equations are still valid.

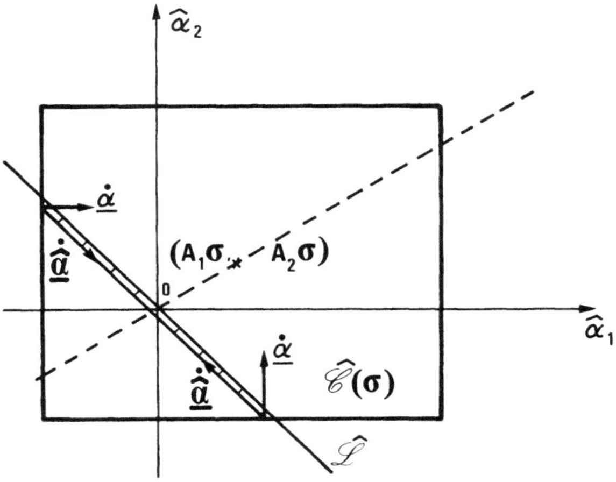

Fig. 18. Velocities fields in the plane $(\widehat{\alpha}_1, \widehat{\alpha}_2)$ for the second assembly.

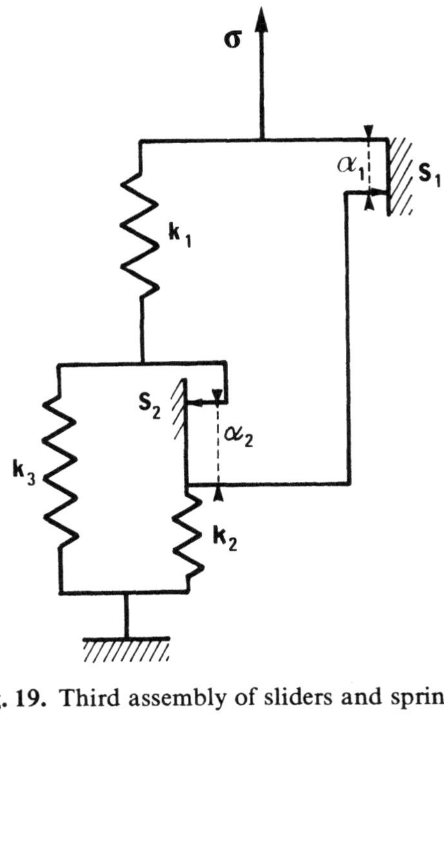

Fig. 19. Third assembly of sliders and springs.

(e) *Fourth example.* We shall consider the assembly of one dashpot and one in-parallel dashpot–slider and one spring as shown in Fig. 20. The spring symbolizes hardening while the dashpot characterizes the softening. The internal parameters of this assembly are $\begin{pmatrix}\beta\\\gamma\end{pmatrix}$, the displacements of the two inelastic models. Always with the same procedure† as in the previous examples, we obtain the following.

(i) The local stresses

$$\sigma^{(2)} = A^{(2)}\sigma - \hat{\beta} \tag{3.23}$$

$$\sigma^{(3)} = A^{(3)}\sigma - \hat{\gamma} \tag{3.24}$$

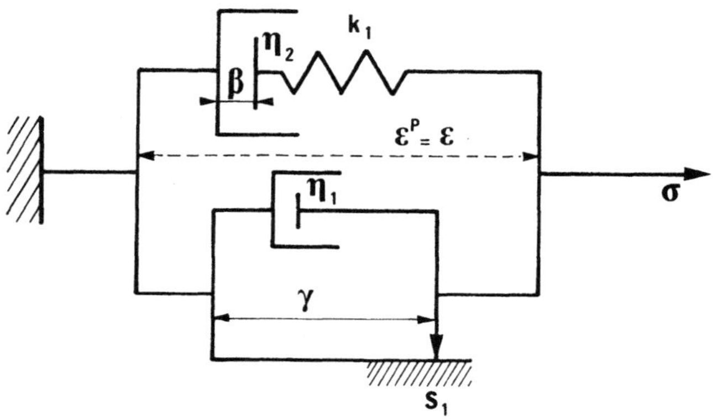

Fig. 20. An example of assembly of a dashpot, an in-parallel slider–dashpot and a spring.

with $A^{(2)} = 0$, $A^{(3)} = 1$; and

$$\begin{pmatrix}\hat{\beta}\\\hat{\gamma}\end{pmatrix} = \begin{pmatrix}k_1 & -k_1\\-k_1 & k_1\end{pmatrix}\begin{pmatrix}\beta\\\gamma\end{pmatrix} \equiv \mathscr{B}\begin{pmatrix}\beta\\\gamma\end{pmatrix} \tag{3.25}$$

The matrix $\mathscr{B}$ is singular and we have the compatibility condition

$$\hat{\mathscr{L}}\colon \hat{\beta} + \hat{\gamma} = 0 \tag{3.26}$$

† We compute the local stresses at the level of each inelastic model. We have two terms; one is linked to the applied stress σ and the other characterizes the residual stresses due to the inelastic strains.

(ii) The evolution laws

$$\dot{\beta} = \sigma^{(2)}/\eta_2 = -\widehat{\beta}/\eta_2 = \widehat{\gamma}/\eta_2 \tag{3.27}$$

$$\dot{\gamma} = \frac{1}{\eta_1}[\langle\sigma^{(3)} - S_1\rangle + \rangle\sigma^{(3)} + S_1\langle] = \frac{1}{\eta_1}[\langle\sigma - \widehat{\gamma} - S_1\rangle + \rangle\sigma - \widehat{\gamma} + S_1\langle] \tag{3.28}$$

(iii) The velocities shown in Fig. 21.

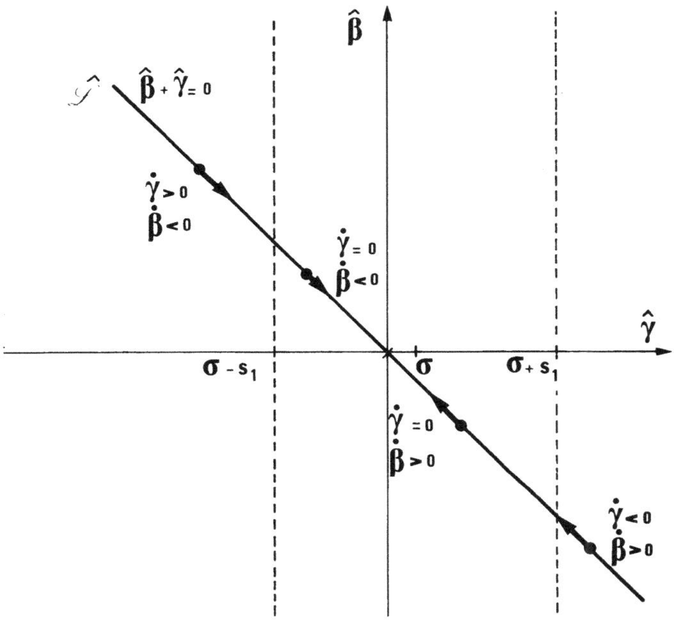

Fig. 21. Associated velocities fields in the plane $(\widehat{\gamma}, \widehat{\beta})$.

3.2.3. *General Formulae* (see Zarka, 1979, for a more precise presentation) Let us now forget the previous uniaxial inelastic models but let us keep in mind only the formalism.

(a) *Internal parameters.* (i) We shall assume that the internal state of our volume element can be characterized by a family **X** of internal parameters (*scalars or tensors*):

$$\mathbf{X} = \begin{pmatrix} \boldsymbol{\alpha} \\ \boldsymbol{\beta} \\ \boldsymbol{\gamma} \end{pmatrix} \tag{3.29}$$

(ii) We shall also assume that these parameters are directly linked to the local inelastic strains of the three different types: $\boldsymbol{\alpha}$ is the vector made of the instantaneous plastic strains with a threshold; $\boldsymbol{\beta}$ is the vector made of the delayed viscous strains without a threshold; $\boldsymbol{\gamma}$ is the vector made of the delayed viscoplastic strains with a threshold. The total number of parameters will be assumed finite and designated by n.

(b) *Local stresses on the inelastic subelements.* At the level of these inelastic subelements or mechanisms, we shall assume that the local stresses are respectively given by

$$\begin{pmatrix} \boldsymbol{\sigma}^{(1)} \\ \boldsymbol{\sigma}^{(2)} \\ \boldsymbol{\sigma}^{(3)} \end{pmatrix} = \begin{pmatrix} \mathbf{A}^{(1)} \\ \mathbf{A}^{(2)} \\ \mathbf{A}^{(3)} \end{pmatrix} \sigma - \begin{pmatrix} \hat{\boldsymbol{\alpha}} \\ \hat{\boldsymbol{\beta}} \\ \hat{\boldsymbol{\gamma}} \end{pmatrix} \equiv \mathbf{A}\sigma - \hat{\mathbf{X}} \tag{3.30}$$

where σ is the global stress tensor applied on the volume element; $\boldsymbol{\sigma}^{(1)}$ is the vector made of all the local stresses (scalars or tensors) for subelements of type 1 (i.e. mechanism with instantaneous plastic strain and a threshold), and so for $\boldsymbol{\sigma}^{(2)}$ and $\boldsymbol{\sigma}^{(3)}$; $\mathbf{A}^{(1)}$ is the vector made of all the localization factors (scalars or tensors) for the subelements of type 1, and so for $\mathbf{A}^{(2)}$ and $\mathbf{A}^{(3)}$.

$\hat{\mathbf{X}} = \begin{pmatrix} \hat{\boldsymbol{\alpha}} \\ \hat{\boldsymbol{\beta}} \\ \hat{\boldsymbol{\gamma}} \end{pmatrix}$, the opposites of residual stresses (scalars or tensors) at the level of inelastic elements, will still be called the transformed internal parameters.

(c) *Transformed internal parameters.* We shall assume that the transformed internal parameters $\hat{\mathbf{X}}$ are linked to the internal parameters $\mathbf{X}$ through a linear application

$$\hat{\mathbf{X}} = \mathscr{B}\mathbf{X} = \begin{pmatrix} B^{(1)} & C^{\mathrm{T}} & D^{\mathrm{T}} \\ C & B^{(2)} & E^{\mathrm{T}} \\ D & E & B^{(3)} \end{pmatrix} \begin{pmatrix} \boldsymbol{\alpha} \\ \boldsymbol{\beta} \\ \boldsymbol{\gamma} \end{pmatrix} \tag{3.31}$$

where $\mathscr{B}$ is a non-negative symmetrical matrix. Two cases will have to be considered:

(i) $\mathscr{B}$ is strictly positive and thus regular

$$\hat{\mathbf{X}} = \mathscr{B}\mathbf{X} \Rightarrow \mathbf{X} = \mathscr{B}^{-1}\hat{\mathbf{X}} \quad \text{(one-to-one relation)}$$

(ii) $\mathscr{B}$ is singular of rank (n-p)

$\hat{\mathbf{X}}$ must belong to the linear subspace of compatibility $\hat{\mathscr{L}}$

To such a compatible $\hat{\mathbf{X}}$ may be associated an infinite numuer of solutions $\mathbf{X}$ such that

$$\mathbf{X} = \mathscr{E}\hat{\mathbf{X}} + \sum_{1,p} \mu_l \mathbf{X}_l \tag{3.32}$$

where $\hat{\mathbf{X}}$ is one particular solution of $\mathscr{B}\,\mathscr{E}\,\hat{\mathbf{X}} = \hat{\mathbf{X}}$; $\mathbf{X}_l$ are p linearly independent solutions of the homogeneous system $\mathscr{B}\,\mathbf{X}_l = (0)$; and μ_l are undetermined factors.

(d) *Evolution laws for the subelements.*

(1) *Instantaneous plasticity.* We shall assume, for each such subelement or mechanism, perfect plasticity behaviour:

(i) C_0, an immobile convex set, defines the threshold* ($\boldsymbol{\sigma}^{(1)}$ must belong to it).

(ii) The associated normality flow rules are satisfied

$$\dot{\boldsymbol{\alpha}} \in \partial\Psi_{C_0}(\boldsymbol{\sigma}^{(1)}) \tag{3.33}$$

i.e. $\dot{\boldsymbol{\alpha}}$ is *one external* normal to the convex set C_0 in $\boldsymbol{\sigma}^{(1)}$.

As we have shown, in the examples, when introducing $\hat{\boldsymbol{\alpha}}$, for all these subelements we can write at any time t:

$$\hat{\boldsymbol{\alpha}} \in \hat{\mathscr{C}}(\sigma) \tag{3.34}$$

with

$$\hat{\mathscr{C}}(\sigma) = \hat{\mathscr{C}}(0) + \mathbf{A}^{(1)}\sigma \tag{3.35}$$

where $\hat{\mathscr{C}}(\sigma)$ is a mobile convex set which is subjected to a simple known translation and which is in fact built locally for each such subelement; and

$$\dot{\boldsymbol{\alpha}} \in -\partial\Psi_{\hat{\mathscr{C}}(\sigma)}(\hat{\boldsymbol{\alpha}}) \tag{3.36}$$

*There might be many inequalities to characterize C_0 which may have corners. The origin must be a centre of symmetry for the convex C_0.

i.e. $\dot{\boldsymbol{\alpha}}$ is *one internal* normal to the convex set $\widehat{\mathscr{C}}(\sigma)$ in $\widehat{\boldsymbol{\alpha}}$.

(2) *Viscosity.* We shall assume that there is for each subelement a potential ω such that

(i) ω is only a convex function of $\sigma^{(2)}$

(ii) $\dot{\boldsymbol{\beta}} = \partial\omega(\sigma^{(2)})/\partial\sigma^{(2)}$ (3.37)

(iii) $\dot{\boldsymbol{\beta}} \equiv 0$ only if $\partial\sigma^{(2)} \equiv 0$ (3.38)

This implies that we have, locally, perfect viscous behaviour without a threshold. In fact we shall later assume linear viscosity as for example in the form

$$\dot{\boldsymbol{\beta}} = \sigma^{(2)}/\eta \tag{3.39}$$

When introducing $\widehat{\boldsymbol{\beta}}$, for all these subelements, we can write

$$\dot{\boldsymbol{\beta}} = -\frac{\partial\omega}{\partial\widehat{\boldsymbol{\beta}}}(\mathbf{A}^{(2)}\sigma - \widehat{\boldsymbol{\beta}}) \tag{3.40}$$

or

$$\dot{\boldsymbol{\beta}} = (\mathbf{A}^{(2)}\sigma - \widehat{\boldsymbol{\beta}})/\eta \tag{3.41}$$

(3) *Viscoplasticity.* For each subelement we shall assume that there exists a convex viscoplastic potential $\boldsymbol{\Omega}$ such that

(i) $\boldsymbol{\Omega}$ is only a convex function of $\sigma^{(3)}$

(ii) $\dot{\boldsymbol{\gamma}} = \partial\boldsymbol{\Omega}(\sigma^{(3)})/\partial\sigma^{(3)}$ (3.42)

(iii) $\dot{\boldsymbol{\gamma}} \equiv 0$ only if $\sigma^{(3)} \in D_0$, an immobile convex set (3.43)

which defines the threshold.

We thus have, locally, perfect viscoplastic behaviour.

Here also, when introducing $\widehat{\boldsymbol{\gamma}}$, we shall obtain for all these subelements

$$\dot{\boldsymbol{\gamma}} = -\frac{\partial\boldsymbol{\Omega}}{\partial\widehat{\boldsymbol{\gamma}}}(\mathbf{A}^{(3)}\sigma - \widehat{\boldsymbol{\gamma}}) \tag{3.44}$$

and

$$\dot{\boldsymbol{\gamma}} \equiv 0 \quad \text{only if} \quad \widehat{\boldsymbol{\gamma}} \in \widehat{\mathscr{D}}(\sigma) \tag{3.45}$$

with the mobile convex set given by

$$\widehat{\mathscr{G}}(\sigma) \equiv \widehat{\mathscr{G}}(0) + \mathbf{A}^{(3)}\sigma \tag{3.46}$$

The properties and the building $\widehat{\mathscr{G}}(\sigma)$ are identical to those of $\widehat{\mathscr{C}}(\sigma)$.

(e) *Global evolution of the volume element.*

(i) Global inelastic strain rates tensor. In order to retain the notion that all the subelements are linked through a linear elastic matrix, as was proved by Mandel (1965), Hill (1967), Bui (1969) and Rice (1970), and as was shown in Chapter 4, we shall take

$$\dot{\varepsilon}^{\mathrm{p}} \equiv \mathbf{A}^{(1)\mathrm{T}}\dot{\boldsymbol{\alpha}} + \mathbf{A}^{(2)\mathrm{T}}\dot{\boldsymbol{\beta}} + \mathbf{A}^{(3)\mathrm{T}}\dot{\boldsymbol{\gamma}} = \mathbf{A}^{\mathrm{T}}\dot{\mathbf{X}} \tag{3.47}$$

Moreover, the local normality and convexity rules induce those for the global behaviour.

(ii) Global inelastic strain rates tensor. We shall assume finally that

$$\dot{\varepsilon}^{\mathrm{e}} \equiv \mathscr{M}\dot{\sigma} \tag{3.48}$$

where $\mathscr{M}$ is the symmetrical positive matrix of elastic coefficients.

(iii) Global strain rates tensor

$$\dot{\varepsilon} = \dot{\varepsilon}^{\mathrm{e}} + \dot{\varepsilon}^{\mathrm{p}} \tag{3.49}$$

(A thermal component could be added too.)

Once given the initial values of $\mathbf{X}$ and $\sigma\,(\mathbf{X}_0, \sigma_0)$, for any loading path in stresses $\sigma(\tau)$ or in strains $\varepsilon(\tau)$ (or mixed components in σ, ε) for $\tau \in [0, T]$, the global behaviour of the volume element can be analysed.

It is worthwhile noting that hardening and softening at this global scale are only the result of stresses redistribution within the volume element since all the plastic and viscoplastic subelements are 'perfect' (Mandel, 1977).

Two fundamental properties of this framework for the inelastic constitutive laws have to be underlined and were proved:

(i) A simple analysis during cyclic loadings will be possible, even if we may think we need to follow step-by-step the volume element and its parameters.

(ii) A similar simple analysis, for any structure which is made of such materials, will also be available.

They are due to the introduction of the transformed internal parameters and their particular linear relation with the internal parameters.

3.3. Global Cyclic Behaviour

3.3.1. *Position*

(i) Many important theoretical papers have been published about shakedown (see Koïter, 1960; König, 1972; Leckie, 1974; Mandel *et al.*, 1972, 1977; Maier, 1969; Mróz, 1972; Nayroles, 1972; Nguyen, 1973; and chiefly Halphen, 1978, or Desbordes, 1978; and there is a review by Nguyen in Chapter 3 of this book.

(ii) Owing to the viscous and viscoplastic subelements we need full knowledge of the cyclic loading path during one period and not only its bounds. Here we shall take a loading path in stresses.

(iii) Subelements of types 2 and 3 are formally similar, the elastic domain for the first ones being only 0. That is the reason why, from now, we shall just keep linearly viscous subelements in 2; the others will be put with 3.

3.3.2. *Purely Linearly Viscous Global Behaviour*

Let us assume, at first, that $\boldsymbol{\alpha}$ and $\boldsymbol{\gamma}$ are all kept at the 0 value; the global behaviour will only be characterized by the linearly viscous subelements. We must solve the system of linear differential equations

$$\dot{\beta}(t) = (\mathbf{A}^{(2)}\sigma(t) - B^{(2)}\beta(t))/\eta \tag{3.50}$$

The submatrix $B^{(2)}$, which was introduced in the matrix $\mathscr{B}$ [eqn (3.32)], is also symmetrical and non-negative. It has non-negative eigenvalues λ_k. The general solution of eqn (3.50) is thus

$$\beta(t) = \beta_v(t) + \sum_k P_k(t) \exp\left\{-\frac{\lambda_k}{\eta_k} t\right\} \tag{3.51}$$

where $\beta_v(t)$ is a periodical particular bounded solution of eqn (3.50) and $P_k(t)$ is a polynomial function of t whose degree is equal to the order of multiplicity of λ_k minus 1.

When $B^{(2)}$ is strictly positive all the λ_k are positive; for t tending to infinity, $\beta(t)$ will tend to $\beta_v(t)$. When $B^{(2)}$ is singular some λ_k are zero; eventually $\beta(t)$ will be unbounded for t tending to infinity.

We shall assume, from now, that $B^{(2)}$ is *always strictly positive*, or otherwise there is no viscous subelement.

3.3.3. *General Response*
(i) For the volume element, let us set

$$\boldsymbol{\beta}(t) \equiv \boldsymbol{\beta}_{\mathrm{v}}(t) + \boldsymbol{\beta}_1(t) \tag{3.52}$$

where $\boldsymbol{\beta}_{\mathrm{v}}(t)$ is the periodical bounded solution of eqn (3.51), and let us define the 'new' transformed parameters $\widehat{\mathbf{X}}_1$ such that

$$\widehat{\mathbf{X}}_1 = \begin{pmatrix} \widehat{\boldsymbol{\alpha}}_1 \\ \widehat{\boldsymbol{\beta}}_1 \\ \widehat{\boldsymbol{\gamma}}_1 \end{pmatrix} \equiv \mathscr{B} \begin{pmatrix} \boldsymbol{\alpha} \\ \boldsymbol{\beta}_1 \\ \boldsymbol{\gamma} \end{pmatrix} \quad \text{and} \quad \widehat{\mathbf{X}} \equiv \widehat{\mathbf{X}}_1 + \begin{pmatrix} C^{\mathrm{T}} & \boldsymbol{\beta}_{\mathrm{v}} \\ B^{(2)} & \boldsymbol{\beta}_{\mathrm{v}} \\ E & \boldsymbol{\beta}_{\mathrm{v}} \end{pmatrix} \tag{3.53}$$

(ii) The evolution laws for the subelements will be changed into

(1) *Subelements*

$$\dot{\boldsymbol{\alpha}} \in -\partial\Psi_{\widehat{\mathscr{C}}_1(\sigma,\boldsymbol{\beta}_{\mathrm{v}})}(\widehat{\boldsymbol{\alpha}}_1) \tag{3.54}$$

with

$$\widehat{\boldsymbol{\alpha}}_1 \in \widehat{\mathscr{C}}_1(\sigma, \boldsymbol{\beta}_{\mathrm{v}}) \tag{3.55}$$

and

$$\widehat{\mathscr{C}}_1(\sigma, \boldsymbol{\beta}_{\mathrm{v}}) \equiv \widehat{\mathscr{C}}(0) + \mathbf{A}^{(1)}\sigma - C^{\mathrm{T}}\boldsymbol{\beta}_{\mathrm{v}} \tag{3.56}$$

(2) *Subelements*

$$\dot{\boldsymbol{\beta}}_1 = -\widehat{\boldsymbol{\beta}}_1/\eta \tag{3.57}$$

and

$$\dot{\boldsymbol{\beta}} = \dot{\boldsymbol{\beta}}_{\mathrm{v}} + \dot{\boldsymbol{\beta}}_1 \tag{3.58}$$

(3) *Subelements*

$$\dot{\boldsymbol{\gamma}} = -\frac{\partial\Omega}{\partial\boldsymbol{\gamma}_1}(\mathbf{A}^{(3)}\sigma - E\boldsymbol{\beta}_{\mathrm{v}} - \widehat{\boldsymbol{\gamma}}_1) \tag{3.59}$$

and

$$\dot{\boldsymbol{\gamma}} \equiv 0 \quad \text{if} \quad \widehat{\boldsymbol{\gamma}}_1 \in \widehat{\mathscr{D}}_1(\sigma, \boldsymbol{\beta}_{\mathrm{v}}) \tag{3.60}$$

with

$$\widehat{\mathscr{D}}_1(\sigma, \boldsymbol{\beta}_{\mathrm{v}}) \equiv \widehat{\mathscr{D}}(0) + \mathbf{A}^{(3)}\sigma - E\boldsymbol{\beta}_{\mathrm{v}} \tag{3.61}$$

3.3.4. *Eventual 'Elastic' Shakedown*

Our definition of 'elastic' shakedown is here that all ($\boldsymbol{\alpha}, \boldsymbol{\beta}_{\mathrm{I}}$ and $\boldsymbol{\gamma}$) will tend towards constant values when t tends to infinity. Two cases must be considered (see Zarka, 1979).

(i) $\mathscr{B}$ is regular. The condition for 'elastic'[†] shakedown is very simple to express:

$$\left.\begin{array}{l}\textit{There is a non-void intersection } \widehat{\mathscr{C}}_l \text{ of convex sets} \\ \widehat{\mathscr{C}}_{\mathrm{I}}\ (\sigma(\tau), \boldsymbol{\beta}_{\mathrm{v}}(\tau)) \text{ in the } \hat{\boldsymbol{\alpha}}_{\mathrm{I}} \text{ space and } \textit{a non-void} \\ \textit{intersection } \widehat{\mathscr{D}}_l \text{ of convex sets } \widehat{\mathscr{D}}_{\mathrm{I}}\ (\sigma(\tau), \boldsymbol{\beta}_{\mathrm{v}}(\tau)) \text{ in} \\ \text{the } \hat{\boldsymbol{\gamma}}_{\mathrm{I}} \text{ space when } \tau \in [0, T]\end{array}\right\} \qquad (3.62)$$

This condition implies indeed that, by taking any $\hat{\boldsymbol{\alpha}}_{\mathrm{I}}^* \in \widehat{\mathscr{C}}_l$, $\hat{\boldsymbol{\beta}}_{\mathrm{I}}^* \equiv 0$ and any $\hat{\boldsymbol{\gamma}}_{\mathrm{I}}^* \in \widehat{\mathscr{D}}_l$, we have found a constant residual stress field $\widehat{\mathbf{X}}_{\mathrm{I}}^* = \begin{pmatrix} \hat{\boldsymbol{\alpha}}_{\mathrm{I}}^* \\ \hat{\boldsymbol{\beta}}_{\mathrm{I}}^* \\ \hat{\boldsymbol{\gamma}}_{\mathrm{I}}^* \end{pmatrix}$ which is plastically admissible, i.e. the condition of the generalized Melan theorem. But, since the convex sets $\widehat{\mathscr{C}}_l$ and $\widehat{\mathscr{D}}_l$ are locally built, here *we have a local condition.*

(ii) $\mathscr{B}$ is singular. The generalized Melan theorem is always true; but this time $\widehat{\mathbf{X}}_{\mathrm{I}}^*$ must also belong to the global subspace of compatibility $\widehat{\mathscr{L}}$ in the $\widehat{\mathbf{X}}_{\mathrm{I}}$ space. The condition for 'elastic' shakedown will now have a *global* character; there is a constant new transformed parameter $\widehat{\mathbf{X}}_{\mathrm{I}}^*$ such that

$$\left.\begin{array}{ll} \widehat{\mathbf{X}}_{\mathrm{I}}^* = \begin{pmatrix} \hat{\boldsymbol{\alpha}}_{\mathrm{I}}^* \\ \hat{\boldsymbol{\beta}}_{\mathrm{I}}^* \\ \hat{\boldsymbol{\gamma}}_{\mathrm{I}}^* \end{pmatrix} \in \widehat{\mathscr{L}} & \\ \text{with } \hat{\boldsymbol{\alpha}}_{\mathrm{I}}^* & \text{strictly included in } \widehat{\mathscr{C}}_l \ \# \ \phi \\ \hat{\boldsymbol{\beta}}_{\mathrm{I}}^* \equiv 0 & \\ \hat{\boldsymbol{\gamma}}_{\mathrm{I}}^* & \text{strictly included in } \widehat{\mathscr{D}}_l \ \# \ \phi \end{array}\right\} \qquad (3.63)$$

For both cases, generally, it is not yet possible to give the number of cycles necessary to reach the limiting state; this number may be infinite.

[†]In fact, the asymptotical global behaviour will be viscoelastic since we must add the particular periodic viscous solution as in (3.53).

3.3.5. *Eventual 'Plastic' Shakedown*

The following two cases have still to be considered.

(i) $\mathscr{B}$ is regular. When the condition (3.62) is not satisfied there will always be plastic shakedown.

(ii) $\mathscr{B}$ is singular. When the condition (3.63) is no longer satisfied it is not yet possible to give *a priori* knowledge of the limiting state. *However when there is no collapse* [i.e. the intersection between $\widehat{\mathscr{S}}$ and $\widehat{\mathscr{C}}_1(\sigma(\tau), \boldsymbol{\beta}_v(\tau))$ for $\tau \in [0, T]$ is non-void which is the necessary condition for the existence of $\widehat{\boldsymbol{\alpha}}_1(t)$] *there will always be a periodic solution* $\widehat{\mathbf{X}}_1(t)$ while $\mathbf{X}_1(t)$ *may be periodic* (plastic shakedown) or *not periodic* (ratcheting).

Here also, in general, the number of cycles to reach the limiting state cannot yet be given.

3.3.6. *Practical Rules During Radial Cyclic Loadings for a Regular $\mathscr{B}$ Matrix*

The cyclic loading will be described as radial if we can write

$$\sigma(\tau) = \lambda(\tau)\,\sigma_{\min} + (1 - \lambda(\tau))\,\sigma_{\max} \tag{3.64}$$

where $\lambda(\tau)$ is a periodic function of time varying monotonically between 0 and 1 and then between 1 and 0, and $\sigma_{\min}$ and $\sigma_{\max}$ are two constant global stress states.

(i) It is evident that two particular loading rates, which will give direct bounds for the real one, may be considered.

A 'static' loading such that $|\dot{\varepsilon}^p|$ will always be very small at any time (T is very 'large'); this will imply that

$$\left.\begin{array}{ll} \widehat{\boldsymbol{\beta}} & \text{will always be near to } 0 \\ \widehat{\boldsymbol{\gamma}} & \text{will always approximately belong to } \widehat{\mathscr{D}}(\sigma) \end{array}\right\} \tag{3.65}$$

In other words, in the volume element, we shall have to consider only the *plastic* subelements 1 and 3 with their internal parameters $(\boldsymbol{\alpha}, \boldsymbol{\gamma})$; the internal parameters $\boldsymbol{\beta}$ of subelements 2 will be known afterwards by the condition $\widehat{\boldsymbol{\beta}} \cong 0$:

$$\boldsymbol{\beta} = B^{(2)^{-1}}(0 - C\boldsymbol{\alpha} - E^{\mathrm{T}}\boldsymbol{\gamma}) \tag{3.66}$$

and by (3.47) we shall deduce ε^p.

A 'dynamic' loading such that $|\dot{\varepsilon}^p|$ will be almost infinite at any time (T is very 'small'); this 'instantaneous' loading will produce $\boldsymbol{\beta} \equiv 0$ and $\boldsymbol{\gamma} \equiv 0$, i.e. only the subelements 1 will take part in the deformation of the volume element.

So, for these two loading rates we need only analyse a volume element with subelements of type 1 and for which only the monotonic variation between $\sigma_{min}, \sigma_{max}$ and then $\sigma_{max}, \sigma_{min}$ is important.

(ii) Zarka and Casier (1979) give some practical rules for obtaining bounds for an elastoplastic structure with kinematic hardening. The procedure will be still similar when the matrix $\mathscr{B}$ is strictly positive (its submatrices $B^{(1)}$ and $\begin{pmatrix} B^{(1)} & D^{\mathrm{T}} \\ D & B^{(3)} \end{pmatrix}$ are also strictly positive). It is evident that for a regular matrix $\mathscr{B}$ only the stress amplitude $\sigma_{max} - \sigma_{min}$ will be important (the mean stress $\sigma_{max} + \sigma_{min}$ will not take part).

4. EXAMPLES OF GLOBAL INELASTIC BEHAVIOURS

4.1. Objectives

In all the following examples we shall write down *a priori* general equations; we shall just keep in mind the mechanisms at the local scale.

From the previous analysis we already know that there may be ratcheting or collapse only when $\mathscr{B}$ is singular. If then we need to represent experimental results according to the eventual observation of the ratcheting or collapse, we shall or shall not take a regular $\mathscr{B}$ matrix.

Now, in the following elementary examples, we shall see some properties which will allow us to interpret more and more complex behaviours. A combination in serial of such simple behaviours may permit interpretation of any of the known cyclic phenomena.

In order to simplify the presentation, here we shall assume that the loading is always biaxial and, if the first component varies periodically with time, the second one will be kept constant at its initial value:

$$\sigma = \begin{pmatrix} \sigma_1(\tau) & 0 & 0 \\ 0 & \sigma_2 & 0 \\ 0 & 0 & 0 \end{pmatrix}$$

This special loading belongs to the radial cyclic loadings (3.64).

4.2. One-dimensional Subelements 1

4.2.1. *One One-dimensional Subelement 1*

We just say that the unique local subelement 1 has a one-dimensional characteristic; this is the case when, for example, we consider the slip of one particular slip mechanism* in a single crystal. In that case we shall write that the most general behaviour within our framework is given by

(i) Local stress

$$\sigma^{(1)} = A_1^{(1)}\sigma_1 + A_2^{(1)}\sigma_2 - \widehat{\alpha} \tag{4.1}$$

(ii) Parameters

$$\widehat{\alpha} = B_{11}^{(1)}\alpha \qquad \text{with} \quad B_{11}^{(1)} \geqslant 0 \tag{4.2}$$

(iii) Flow rules

$$-S \leqslant \sigma^{(1)} \leqslant S \quad \text{or} \quad A_1^{(1)}\sigma_1 - S + A_2^{(1)}\sigma_2 \leqslant \widehat{\alpha} \leqslant A_1^{(1)}\sigma_1 + S + A_2^{(1)}\sigma_2$$
$$\text{or} \quad \widehat{\alpha} \in \widehat{\mathscr{C}}(\sigma_1, \sigma_2); \qquad \dot{\alpha} \in -\partial\Psi_{\widehat{\mathscr{C}}(\sigma_1,\sigma_2)}(\widehat{\alpha}) \tag{4.3}$$

(iv) Component of the global plastic strain tensor along the direction of σ_1

$$\varepsilon_1^{\mathrm{p}} = A_1^{(1)}\alpha \tag{4.4}$$

(There may be other components not only along σ_2 but also in the other directions: $\varepsilon_{ij}^{\mathrm{p}} = A_{ij}^{(1)}\alpha$.)

From (4.2) we can see that only perfect plastic behaviour ($B_{11}^{(1)} = 0$) or kinematic hardening behaviour ($B_{11}^{(1)} > 0$) can be obtained.

The number of cycles of loading to reach the limiting state (shakedown or ratcheting or collapse) is always equal to 1.

4.2.2. *Two One-dimensional Subelements 1*

In Section 3.2.2 we have already shown particular cases of volume elements where two such subelements were put. The most general behaviour will now be given by

*The slip mechanism is, as is well known, defined by a slip plane direction and a slip direction; the corresponding α is in this case the relative glide (scalar) along the slip direction between two planes which are unit distance apart.

(i) Local stresses

$$\begin{aligned} &\text{first subelement} && \sigma_1^{(1)} = A_{11}^{(1)}\sigma_1 + A_{12}^{(1)}\sigma_2 - \hat{\alpha}_1 \\ &\text{second subelement} && \sigma_2^{(1)} = A_{21}^{(1)}\sigma_1 + A_{22}^{(1)}\sigma_2 - \hat{\alpha}_2 \end{aligned} \tag{4.5}$$

(ii) Parameters

$$\hat{\boldsymbol{\alpha}} = \begin{pmatrix} \hat{\alpha}_1 \\ \hat{\alpha}_2 \end{pmatrix} = \begin{pmatrix} B_{11}^{(1)} & B_{12}^{(1)} \\ B_{12}^{(1)} & B_{22}^{(1)} \end{pmatrix} \begin{pmatrix} \alpha_1 \\ \alpha_2 \end{pmatrix} = B^{(1)}\boldsymbol{\alpha} \tag{4.6}$$

with $B^{(1)}$ a non-negative matrix.

(iii) Flow rules

$$\hat{\boldsymbol{\alpha}} \in \hat{\mathscr{C}}(\sigma_1, \sigma_2)$$

the rectangle whose centre is

$$(A_{11}^{(1)}\sigma_1 + A_{12}^{(1)}\sigma_2,\, A_{21}^{(1)}\sigma_1 + A_{22}^{(1)}\sigma_2)$$

and whose sides are (4.7)

$$2S_1, 2S_2$$

and

$$\dot{\boldsymbol{\alpha}} \in -\partial\Psi_{\hat{\mathscr{C}}(\sigma_1,\sigma_2)}(\hat{\boldsymbol{\alpha}})$$

(iv) Component of the global plastic strain tensor along the direction σ_1

$$\varepsilon_1^{\mathrm{p}} = A_{11}^{(1)}\alpha_1 + A_{21}^{(1)}\alpha_2 \tag{4.8}$$

The matrix $B^{(l)}$ is regular

The limiting state will depend only on the stress amplitude $\Delta\sigma_1 = \sigma_{1\,\max} - \sigma_{1\,\min}$.

For increasing values of this stress amplitude we shall obtain elastic shakedown, partial plastic shakedown and total plastic shakedown (Figs 22).

When $A_{11}^{(1)} \geqslant 0$, $A_{21}^{(1)} \geqslant 0$ and $B_{12}^{(1)} \geqslant 0$, the limiting state is always reached after the first cycle (Fig. 22a). On the contrary, when $A_{11}^{(1)}A_{21}^{(1)} < 0$ and $B_{12}^{(1)} < 0$ (or $A_{11}^{(1)}A_{21}^{(1)} > 0$ and $B_{12}^{(1)} > 0$), the elastic shakedown state (when $\hat{\mathscr{C}}_l = \hat{\mathscr{C}}(\sigma_{1\,\min}, \sigma_2) \cap \hat{\mathscr{C}}(\sigma_{1\,\max}, \sigma_2) \# \phi$) is reached whether after one cycle of loading or only after an infinite number of cycles (Fig. 22b) while the plastic shakedown state is reached after any finite number of cycles (Fig. 22c).

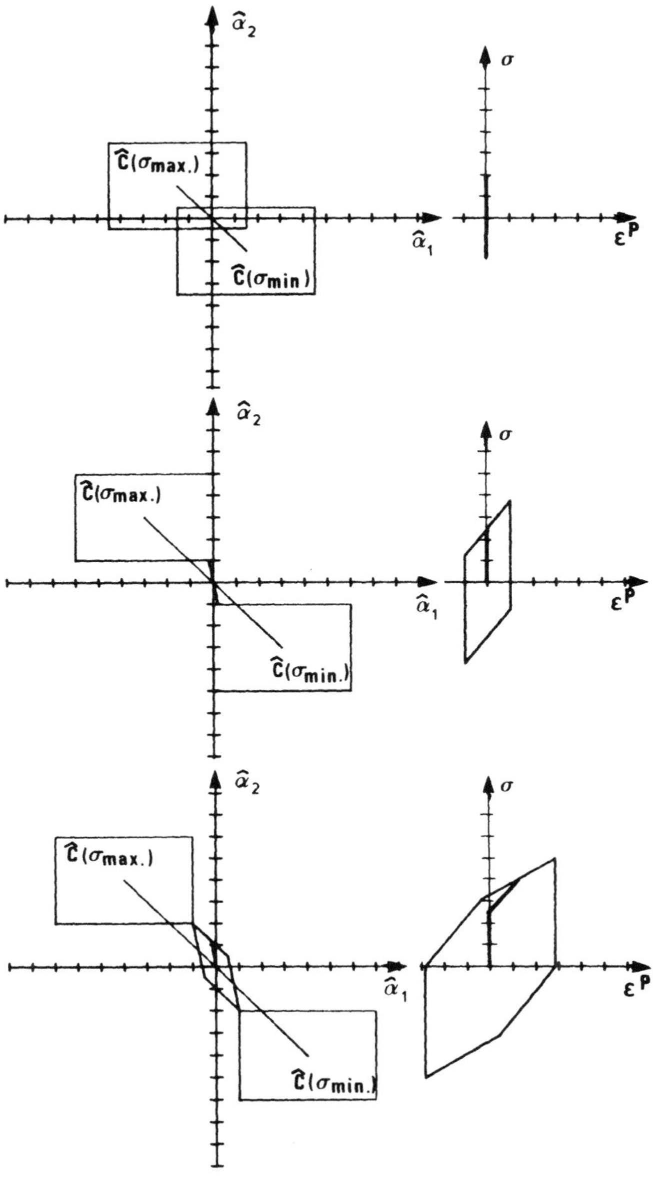

Fig. 22a. Cyclic response for a general assembly made of two sliders and with a regular $B^{(1)}$ matrix: after the first cycle of loading the limiting state is reached.

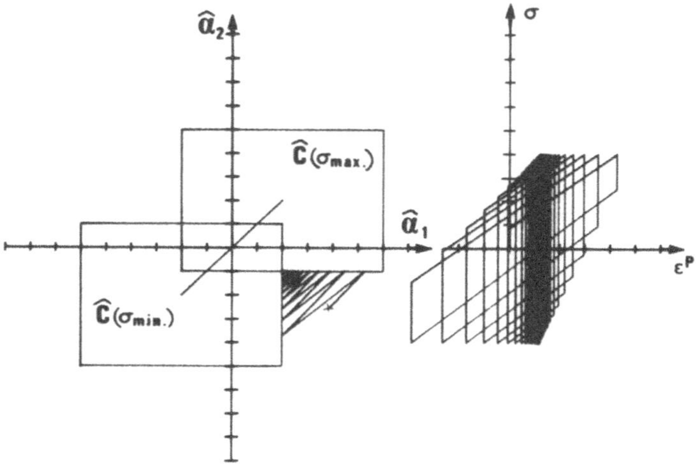

Fig. 22b. Cyclic response for a general assembly made of two sliders and with a regular $B^{(1)}$ matrix: after an infinite number of cycles, elastic shakedown behaviour is reached.

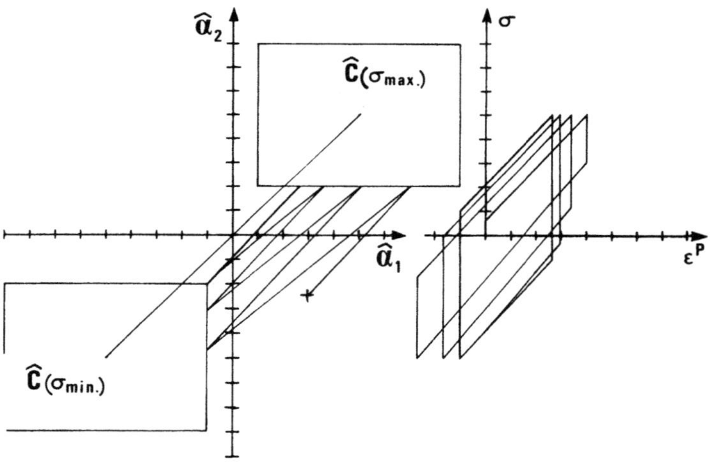

Fig. 22c. Cyclic response for a general assembly made of two sliders and with a regular $B^{(1)}$ matrix: after a finite number of cycles, plastic shakedown behaviour is reached.

The matrix $B^{(1)}$ is singular

This implies that $B_{12}^{(1)} = \partial\sqrt{(B_{11}^{(1)} B_{22}^{(1)})}$ with $\partial = +1$ or -1, and that there is a subspace of compatibility $\widehat{\mathscr{L}}$: $(\partial\sqrt{(B_{22}^{(1)}/B_{11}^{(1)})}\,\widehat{\alpha}_1 - \widehat{\alpha}_2 = 0)$.

There is a solution (no collapse) only when $\widehat{\mathscr{L}} \cap \widehat{\mathscr{C}}(\sigma_1(\tau), \sigma_2) \# \phi$ at any time τ.

The limiting state (elastic or plastic shakedown or ratcheting) is always reached after the first cycle.

The mean stress $\sigma_{1\mathrm{m}} = (\sigma_{1\max} + \sigma_{1\min})$ also takes part; so according to $\sigma_{1\mathrm{m}}$, for a given stress amplitude $\Delta\sigma_1$, it may be elastic shakedown (when $\widehat{\mathscr{L}} \cap \widehat{\mathscr{C}}(\sigma_{1\max}, \sigma_2) \cap \widehat{\mathscr{C}}(\sigma_{1\min}, \sigma_2) \# \phi$), plastic shakedown, ratcheting or collapse (Fig. 23).

4.3. Two-dimensional Subelements 1

4.3.1. *One Two-dimensional Subelement 1*

In order to have a picture of this subelement we may think of a slider which is moving on a plane. The most general behaviour will correspond to

(i) Local stresses

$$\begin{aligned} \sigma_1^{(1)} &= A_{11}^{(1)}\sigma_1 + A_{12}^{(1)}\sigma_2 - \widehat{\alpha}_1 \\ \sigma_2^{(1)} &= A_{21}^{(1)}\sigma_1 + A_{22}^{(1)}\sigma_2 - \widehat{\alpha}_2 \end{aligned} \tag{4.9}$$

(ii) Parameters

$$\widehat{\boldsymbol{\alpha}} = \begin{pmatrix} \widehat{\alpha}_1 \\ \widehat{\alpha}_2 \end{pmatrix} = \begin{pmatrix} B_{11}^{(1)} & B_{12}^{(1)} \\ B_{12}^{(1)} & B_{22}^{(1)} \end{pmatrix} \begin{pmatrix} \alpha_1 \\ \alpha_2 \end{pmatrix} = B^{(1)}\boldsymbol{\alpha} \tag{4.10}$$

(iii) Flow rules. We shall, for example, assume that the two-dimensional threshold C_0 is the circle

$$(\sigma_1^{(1)})^2 + (\sigma_2^{(1)})^2 \leqslant S^2 \tag{4.11}$$

In the $\widehat{\boldsymbol{\alpha}}$ plane, we shall thus find that $\widehat{\mathscr{C}}(\sigma_1, \sigma_2)$ is the circle centred in $(A_{11}^{(1)}\sigma_1 + A_{12}^{(1)}\sigma_2, A_{21}^{(1)}\sigma_1 + A_{22}^{(1)}\sigma_2)$, whose radius is S, and that (Fig. 24)

$$\dot{\boldsymbol{\alpha}} \in -\partial\Psi_{\widehat{\mathscr{C}}(\sigma_1,\sigma_2)}(\widehat{\boldsymbol{\alpha}}) \tag{4.12}$$

(iv) Component of the global plastic strain tensor along the direction σ_1

$$\varepsilon_1^{\mathrm{p}} = A_{11}^{(1)}\alpha_1 + A_{21}^{(1)}\alpha_2 \tag{4.13}$$

The equations (4.9)–(4.13) are similar to (4.5)–(4.8) (only the shapes of the two respective convex sets $\widehat{\mathscr{C}}$ are different). The global behaviours will thus be similar (Fig. 25); the most important difference is in the non-constant hardening modulus during monotonic loading which is induced by the rotation of the internal normal on $\widehat{\mathscr{C}}(\sigma_1, \sigma_2)$.

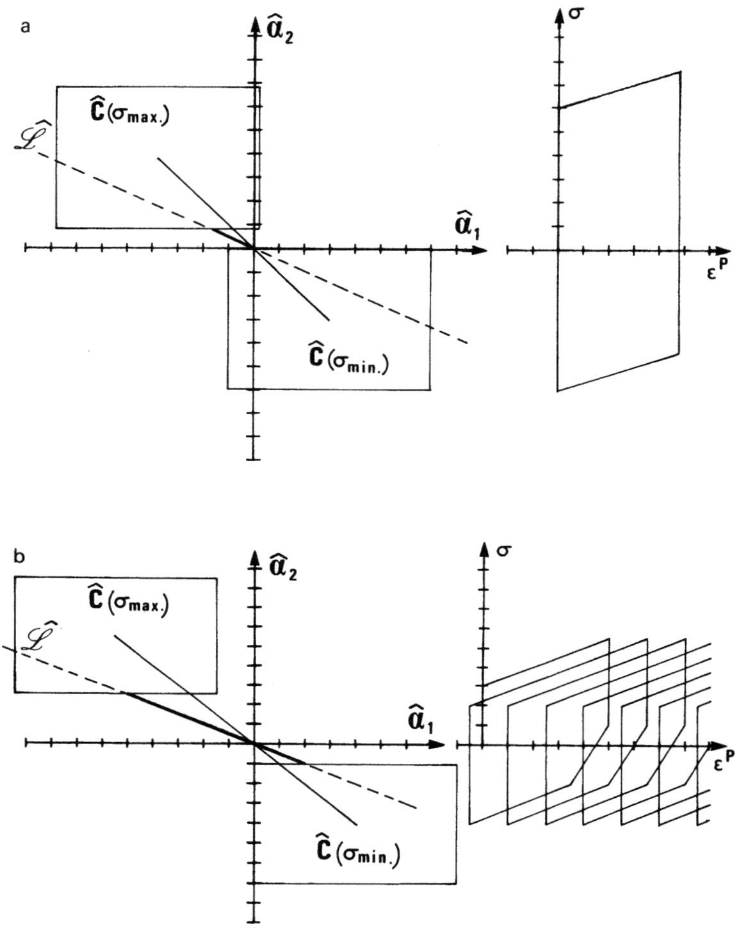

Fig. 23. Cyclic response for a general assembly made of two sliders and with a singular $B^{(1)}$ matrix: (a) elastic or plastic shakedown may be reached; (b) ratcheting may be obtained.

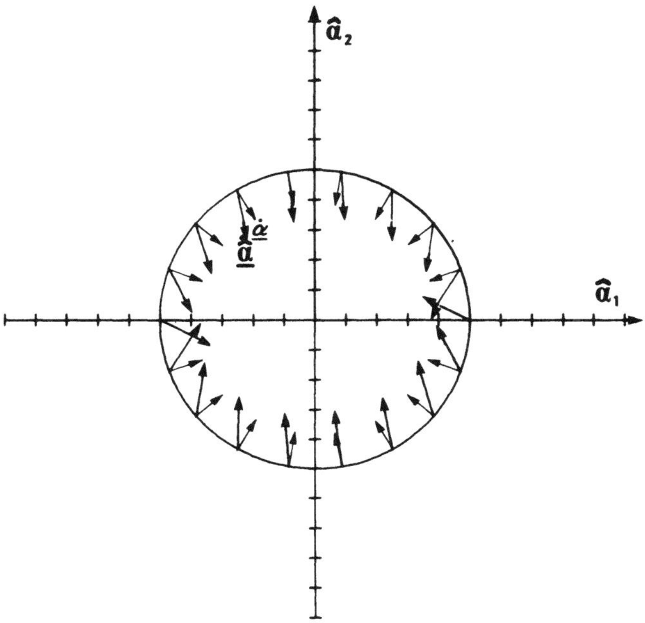

Fig. 24. Velocities fields for a two-dimensional slider in the plane $(\widehat{\alpha}_1, \widehat{\alpha}_2)$.

4.3.2. *Two Two-dimensional Subelements 1*

Now let us assume that we want to find complex behaviours based on two surfaces as introduced, for example, by Mróz (1972) and Dafalias and Popov (1976); however, let us assume that we still want to stay in our framework. We may, for example, take two two-dimensional subelements 1 but with a special $B^{(1)}$ matrix (with this matrix it will be possible to give a plane representation of the different cases of loading):

(i) Local stresses

first subelement
$$\begin{aligned} \sigma_1^{(1)} &= A_{11}^{(1)}\sigma_1 + A_{12}^{(1)}\sigma_2 - \widehat{\alpha}_1 \\ \sigma_2^{(1)} &= A_{21}^{(1)}\sigma_1 + A_{22}^{(1)}\sigma_2 - \widehat{\alpha}_2 \end{aligned} \tag{4.14}$$

second subelement
$$\begin{aligned} \sigma_3^{(1)} &= A_{31}^{(1)}\sigma_1 + A_{32}^{(1)}\sigma_2 - \widehat{\alpha}_3 \\ \sigma_4^{(1)} &= A_{41}^{(1)}\sigma_1 + A_{42}^{(1)}\sigma_2 - \widehat{\alpha}_4 \end{aligned} \tag{4.15}$$

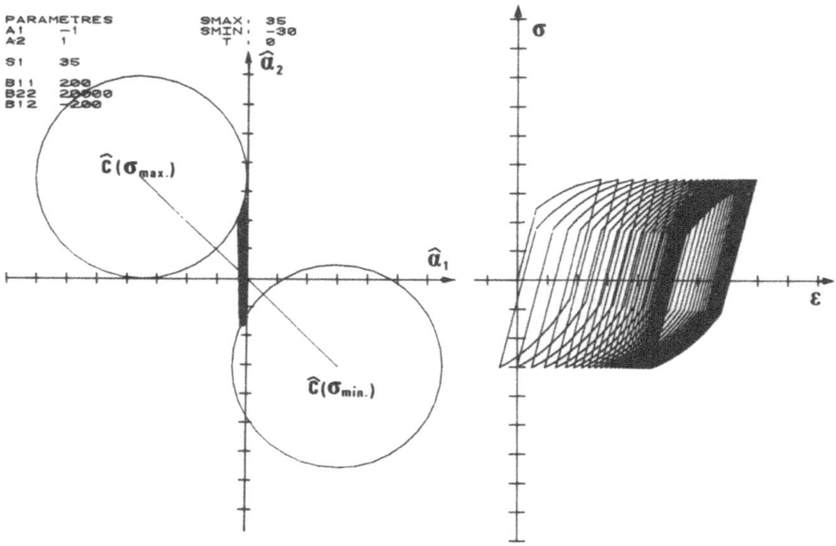

Fig. 25. An example of a plastic shakedown response after a great number of cycles of loading.

(ii) Parameters

$$\hat{\boldsymbol{\alpha}} \equiv \begin{pmatrix} \hat{\alpha}_1 \\ \hat{\alpha}_2 \\ \hat{\alpha}_3 \\ \hat{\alpha}_4 \end{pmatrix} = \begin{pmatrix} B_{11}^{(1)} & B_{12}^{(1)} & hB_{11}^{(1)} & kB_{12}^{(1)} \\ B_{12}^{(1)} & B_{22}^{(1)} & hB_{12}^{(1)} & kB_{22}^{(1)} \\ hB_{11}^{(1)} & kB_{12}^{(1)} & h^2B_{11}^{(1)} & hkB_{12}^{(1)} \\ kB_{12}^{(1)} & kB_{22}^{(1)} & hkB_{12}^{(1)} & k^2B_{22}^{(1)} \end{pmatrix} \begin{pmatrix} \alpha_1 \\ \alpha_2 \\ \alpha_3 \\ \alpha_4 \end{pmatrix} = \boldsymbol{B}^{(1)}\boldsymbol{\alpha} \tag{4.16}$$

with $B_{11}^{(1)}B_{22}^{(1)} - (B_{12}^{(1)})^2 > 0$ and $hk > 0$. The particular $B^{(1)}$ matrix is singular of rank 2. The subspace of compatibility $\widehat{\mathscr{L}}$ is defined by the two linear relations

$$\hat{\alpha}_3 = h\hat{\alpha}_1; \qquad \hat{\alpha}_4 = k\hat{\alpha}_2 \tag{4.17}$$

The limited space $\hat{\boldsymbol{\alpha}}' = \begin{pmatrix} \hat{\alpha}_1 \\ \hat{\alpha}_2 \end{pmatrix}$ will thus be sufficient.

(iii) Flow rules. We shall take for the two-dimensional threshold C_1 and C_2 respectively

$$\left.\begin{aligned} &(\sigma_1^{(1)})^2 + (\sigma_2^{(1)})^2 \leqslant S_1^2 && \text{circle} \\ &(\sigma_3^{(1)}/h)^2 + (\sigma_4^{(1)}/k)^2 \leqslant S_2^2 && \text{ellipse} \end{aligned}\right\} \tag{4.18}$$

These two conditions, expressed for $\widehat{\boldsymbol{\alpha}}' = \begin{pmatrix}\widehat{\alpha}_1 \\ \widehat{\alpha}_2\end{pmatrix}$, show that $\widehat{\boldsymbol{\alpha}}'$ must belong to the intersection of the two circles

$$\widehat{\mathscr{C}}_1(\sigma_1, \sigma_2) \text{ centred in } (A_{11}^{(1)}\sigma_1 + A_{12}^{(1)}\sigma_2, A_{21}^{(1)}\sigma_1 + A_{22}^{(1)}\sigma_2), \text{ radius } S_1 \tag{4.19}$$

and

$$\widehat{\mathscr{C}}_2(\sigma_1, \sigma_2) \text{ centred in } ((A_{31}^{(1)}\sigma_1 + A_{32}^{(1)}\sigma_2)/h, (A_{41}^{(1)}\sigma_1 + A_{42}^{(1)}\sigma_2)/k), \text{ radius } S_2 \tag{4.20}$$

The normality flow rules when using $\widehat{\boldsymbol{\alpha}}'$

$$\begin{pmatrix}\dot{\alpha}_1 \\ \dot{\alpha}_2\end{pmatrix} \in \partial\Psi_{C_1}\begin{pmatrix}\sigma_1^{(1)} \\ \sigma_2^{(1)}\end{pmatrix} \quad ; \quad \begin{pmatrix}\dot{\alpha}_3 \\ \dot{\alpha}_4\end{pmatrix} \in \partial\Psi_{C_2}\begin{pmatrix}\sigma_3^{(1)} \\ \sigma_4^{(1)}\end{pmatrix} \tag{4.21}$$

will be written here

$$\begin{pmatrix}\dot{\alpha}_1 \\ \dot{\alpha}_2\end{pmatrix} \in -\partial\Psi_{\widehat{\mathscr{C}}_1(\sigma_1,\sigma_2)}(\widehat{\boldsymbol{\alpha}}'); \quad \begin{pmatrix}h\dot{\alpha}_3 \\ k\dot{\alpha}_4\end{pmatrix} \in -\partial\Psi_{\widehat{\mathscr{C}}_2(\sigma_1,\sigma_2)}(\widehat{\boldsymbol{\alpha}}') \tag{4.22}$$

From that we shall deduce

$$\dot{\widehat{\boldsymbol{\alpha}}}' = \begin{pmatrix}\dot{\widehat{\alpha}}_1 \\ \dot{\widehat{\alpha}}_2\end{pmatrix} = \begin{pmatrix}B_{11}^{(1)} & B_{12}^{(1)} & B_{11}^{(1)} & B_{12}^{(1)} \\ B_{12}^{(1)} & B_{22}^{(1)} & B_{12}^{(1)} & B_{22}^{(1)}\end{pmatrix} \begin{pmatrix}-\partial\Psi_{\widehat{\mathscr{C}}_1(\sigma_1,\sigma_2)}(\widehat{\boldsymbol{\alpha}}') \\ -\partial\Psi_{\widehat{\mathscr{C}}_2(\sigma_1,\sigma_2)}(\widehat{\boldsymbol{\alpha}}')\end{pmatrix} \tag{4.23}$$

(iv) Components of the global plastic strain tensor in σ_1, σ_2 directions

$$\begin{pmatrix}\varepsilon_1^{\mathrm{p}} \\ \varepsilon_2^{\mathrm{p}}\end{pmatrix} = \begin{pmatrix}A_{11}^{(1)}\alpha_1 + A_{21}^{(1)}\alpha_2 + A_{31}^{(1)}\alpha_3 + A_{41}^{(1)}\alpha_4 \\ A_{12}^{(1)}\alpha_1 + A_{22}^{(1)}\alpha_2 + A_{32}^{(1)}\alpha_3 + A_{42}^{(1)}\alpha_4\end{pmatrix} \tag{4.24}$$

Monotonic loading

(i) In order to have a solution, the intersection of $\widehat{\mathscr{C}}_1(\sigma_1, \sigma_2)$ and $\widehat{\mathscr{C}}_2(\sigma_1, \sigma_2)$ must be non-void. The condition for no collapse, or the 'ultimate' yield surface *in* the (σ_1, σ_2) plane, is thus

$$\begin{aligned} &[(A_{11}^{(1)} - A_{31}^{(1)}/h)\sigma_1 + (A_{12}^{(1)} - A_{32}^{(1)}/h)\sigma_2]^2 + \\ &[(A_{21}^{(1)} - A_{41}^{(1)}/k)\sigma_1 + (A_{22}^{(1)} - A_{42}^{(1)}/k)\sigma_2]^2 \leqslant (S_1 + S_2)^2 \end{aligned} \tag{4.25}$$

(ii) We can follow the evolution of the elastic domain in the (σ_1, σ_2) plane during a monotonic loading (see Fig. 26). This elastic domain changes in shape and size (it is the intersection of two ellipses); it 'seems' to be moreover subjected to a rotation. This evolution is however quite different from that assumed by Mróz (1972*c*).

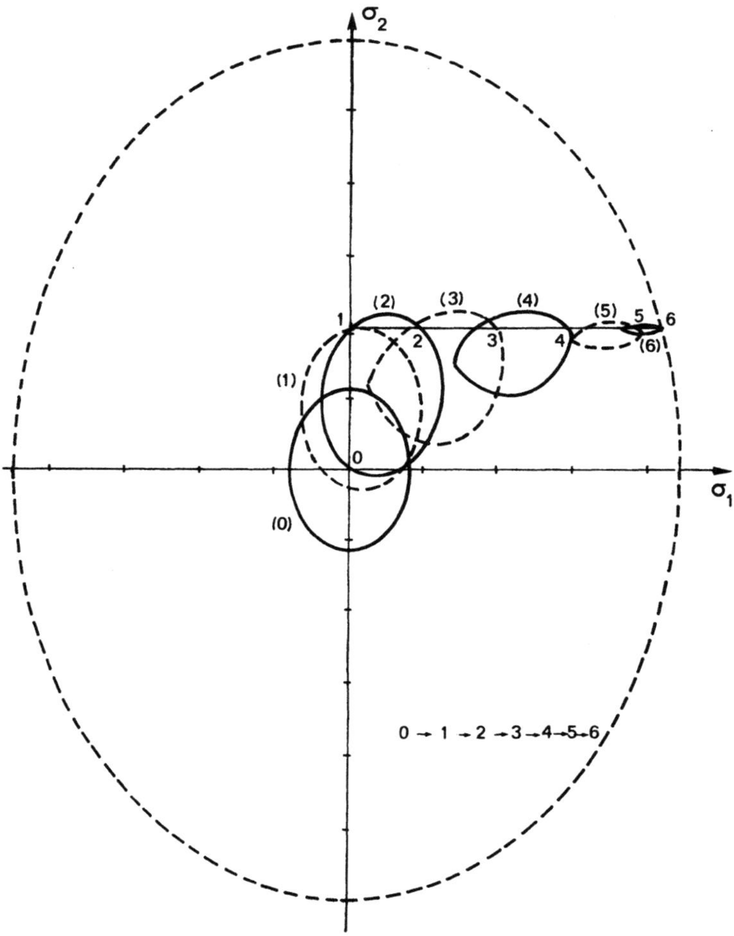

Fig. 26. Evolution of the yield surface inside the loading surface for a particular assembly of two-dimensional sliders during a special monotonic loading.

Cyclic loadings

(i) As long as $\hat{\boldsymbol{\alpha}}'$ stays at any time inside one of the circles $(\hat{\mathscr{C}}_1, \hat{\mathscr{C}}_2)$ and on the boundary of the other one, we find again exactly the behaviour of Section 4.3.1 with only elastic or plastic shakedown (since $B_{11}^{(1)} B_{22}^{(1)} - (B_{12}^{(1)})^2 > 0$).

(ii) As soon as $\hat{\boldsymbol{\alpha}}'$ belongs at some particular time to the intersections of both boundaries of the circles $(\hat{\mathscr{C}}_1, \hat{\mathscr{C}}_2)$, the behaviour will be different and ratcheting may be obtained (Fig. 27).

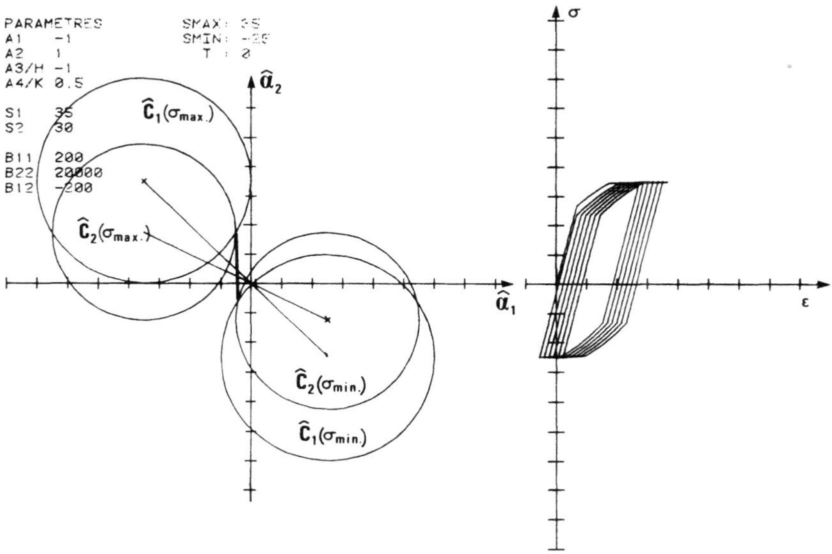

Fig. 27. Ratcheting behaviour for the assembly for a special cyclic loading.

4.4. Effects of Temperature

Without changing anything in our framework, we may take in very easily some influence of temperature if we assume that only the matrices **A**, $\mathscr{B}$ and $\mathscr{M}$ will stay constant. This hypothesis, in fact, implies that the local elastic properties* of the matrix which links the inelastic subelements are independent of temperature. When the temperature range is not too large (a few hundred degrees), this hypothesis is reasonable.

*As in any thermoelastic problem, when there is a coupling effect between temperature and elastic constants, the problem is non-linear and must be solved step by step with a new rigidity matrix.

The local inelastic subelements can have general coupling with temperature:

(i) For subelements 1, we can only have that the size and shape of the thresholds $C_0(\theta)$ is linked to the temperature.

(ii) For subelements 2, we can say that the viscous damping factors $\eta(\theta)$ are functions of temperature.†

(iii) For subelements 3, both the thresholds $D_0(\theta)$ and the viscoplastic potentials $\boldsymbol{\Omega}(\sigma^{(3)}, \theta)$ can be temperature dependent.

The convex sets $\hat{\mathscr{C}}$ and $\hat{\mathscr{D}}$ will then be not only subjected to a translation but also to a change of size; so, for example, instead of eqn (3.35) we shall have

$$\hat{\mathscr{C}}(\sigma, \theta) = \hat{\mathscr{C}}(\theta) + \mathbf{A}^{(1)}\sigma \tag{4.26}$$

However, *their buildings will still be local.*

CONCLUDING REMARKS

We may easily imagine that engineers think that a great work has painfully been performed with only limited results since simple general constitutive laws for metals are not yet available perfectly. However, we are convinced that there is no other way to reach them.

Physical and Quasi-physical Approach

The crystal behaviour seems correctly interpreted when looking at the stress tensor and the strain tensor. However, electron microscope observations are needed in order to show that the predicted calculated values of the internal parameters ($N^{(r)}, l^{(r)}$) are also correct. Perhaps better calculations in the physical model would have been necessary too? A fundamental study of the twinning process has not yet been given.

The polycrystal behaviour could be known after lengthy calculations in each case. This is not realistic. It is as essential for small as for large strains that explicit constitutive relations be given. From the physical studies we understand how we find a kinematic part and an isotropic part in work-hardening (difference between global stress tensor $\boldsymbol{\Sigma}$ and local stress tensor σ). We need now (i) a model for the grain boundary behaviour, (ii) a simple application to the global behaviour of poly-

†The purely viscous analysis in Section 3.3.2 can no more be realized so easily; we reach a non-linear problem.

crystals, (iii) a simple framework for large strains. Perhaps the international CNRS colloquium (Paris, 1985)[†] may have an answer to these problems.

There is still a lot of work to do, but we hope that some elements that we have given here may already be helpful.

Structural Approach

We have described our general framework for solving the problem of global constitutive relations of materials where hardening, softening and creep can occur. Its essential feature consists in the introduction of a family of internal parameters $\mathbf{X}$ (local inelastic strains) which characterizes the local inelastic mechanisms and a family of transformed internal parameters $\hat{\mathbf{X}}$ (associated local residual stresses) which are linearly linked to the previous ones through a symmetrical non-negative matrix $\mathscr{B}$.

Thanks to these $\hat{\mathbf{X}}$, treatment of the plastic or viscoplastic thresholds (local yield conditions) can easily be made; the position of the field $\hat{\mathbf{X}}$ can in fact be known from only the simple purely elastic or linear viscous analysis of the system (i.e. fundamental volume element or structure). The uses of this important property were shown mainly during cyclic loadings: condition for shakedown, bounds for the limiting state etc.

In order to illustrate this framework, we have also given several examples of global behaviours. However, there are other experimental facts which cannot be represented easily within this framework.

(i) An initial deformation of some materials may influence their cyclic curves; this could be the result of the modification of the texture but also of some changes of phases or twinnings.
(ii) Damage is occurring within the material during cyclic loadings. For that, it would have been necessary to assume non-constant $\mathbf{A}, \mathscr{B}, \mathscr{M}$ matrices which would thus imply again a classical step-by-step analysis.

We must also say that we are as yet only in a position to describe the behaviour of materials qualitatively; in order to achieve a quantitative representation of one special material it would be useful to have a systematic analysis of constants identification based on experimental data.

[†]The procedings are the subject of the following volume of this series.

Nevertheless, if we could find such agreement, the analysis of any structure would then be elementary. That is why we are convinced that it is always better, even if we must neglect some aspects of real behaviour, to adopt our proposed framework.

ACKNOWLEDGEMENT

We are thankful to E. D. F.-SEPTEN for supporting this work.

REFERENCES

Besseling, J. F. (1953). Nat. Aera Res. Inst. Report S410.

Bui, H. D. (1969). Evolution de la frontière du domaine élastique des métaux avec écrouissage plastique et comportement élastoplastique d'un agrégat de cristaux cubiques, thesis, Paris.

Dafalias, Y. F. and Popov, E. P. (1976). *J. Appl. Mech.*, **43**, 645.

Desbordes, O. (1978). Adaptation des structures élastiques parfaitement plastiques, Séminaire Matériaux et Structures sous Chargements Cycliques, Palaiseau, Ecole Polytechnique.

Halphen, B. (1978). Problèmes quasistatiques en viscoplasticité, Thesis, Paris.

Hill, R. (1967). The essential structure of constitutive law for metal composites and polycrystals, *J. Mech. Phys. Solids,* **15**, 79.

Koiter, W. T. (1960). General theorems for elastic-plastic solids, *Progress in Solid Mechanics,* Vol. 1, North-Holland, Amsterdam.

König, S. (1972). Shakedown deflections: a finite element approach, *Bulgarian Acad. Sci. Theor. Appl. Mech.,* **III,** 2.

Leckie, F. (1974). A review of bounding techniques in shakedown and ratcheting at elevated temperatures, *WRC Bulletin,* 195.

Maier, G. (1969). Shakedown theory in perfect elastoplasticity, *Meccanica,* **4**(3), 250–60.

Mandel, J. (1965). Généralisation de la théorie de plasticité de W. T. Koiter, *Int. J. Solids Struct.,* **1**, 273–96.

Mandel, J. (1972). Relations de comportement des milieux élastiques plastiques et élastiques-viscoplastiques: Notion de repère directeur, *International Symposium on Foundations of Plasticity* (Varsovie), A. Sawczuk (Ed.), Noordhoff International Publishing.

Mandel, J. (1977). Equations de comportement d'un système élastoviscoplastique dont l'écrouissage est dû à des contraintes résiduelles. *Compt. Rend. Acad. Sci.,* **284.**

Mandel, J., Zarka, J. and Halphen, B. (1977). *Mech. Res. Comm.,*4, 309.

Moreau, J. J. (1972). Rafle par un convexe variable, Séminaire d'Analogie Convexe, Montpellier.

Mróz, Z. (1972*a*). A description of workhardening of metals with application to variable loading. In *Foundations of Plasticity,* A. Sawczuk (Ed.).

Mróz, Z. (1972*b*). On the theory of steady plastic cycles in structures, Proc. SMiRT-1, 6, part L, 489–501.

Mróz (1972*c*). On the description of plastic hardening of materials with application to variable and cyclic loading, *Proc. Symp. Franco–Polonais,* PWN.
Nayroles, B. (1972). Structures algébriques des théories classiques de plasticité. In *Séminaire Plasticité et Viscoplasticité,* D. Radenkovic and J. Salençon (Eds), McGraw-Hill Ediscience.
Nguyen, Q. S. (1973). Contribution à la théorie macroscopique de l'élastoplasticité avec écrouissage, thesis, Paris.
Rice, J. S. (1970). On the structure of stress-strain relations for time-dependent plastic deformation in metals, *Trans. ASME J. Appl. Mech.,* September, 728–37.
Steeds, J. W. (1964). Dislocations arrangement in copper single crystals as a function of strain, *Proc. Roy. Soc.,* **A292**, 343–73.
Zarka, J. (1968). Sur la viscoplasticité des métaux, thèse de doctorat, Paris.
Zarka, J. (1972). Généralisation de la théorie du potentiel plastique multiple en viscoplasticité, *J. Mech. Phys. Solids,* **20**, 179–95.
Zarka, J. (1979). *J. Appl. Mech.,* 3(3), 291–326.
Zarka, J. and Casier, J. (1979). Cyclic loading on an elastoplastic structure; practical rules. In *Mechanics Today,* Vol. 6, J. Nemat-Nasser (Ed.).
Zienkiewicz, O.C., Nayac, G. C. and Owen, D. R. I. (1973). Composite and overlay models in numerical analysis of elastoplastic continua. In *Foundations of Plasticity,* A. Sawczuk, (Ed.), Noordhoff, Groningen.

BIBLIOGRAPHY

Bilby, B. A. (1960). Continuous distributions of dislocations. In *Progress in Solids Mechanics,* I. Sneddon and R. Hill (Eds), Vol. 1, North-Holland, pp. 331–98.
Budiansky, B. and Wu, T. T. (1962). Theoretical prediction of plastic strains of polycrystals, *Proc. 4th U.S. Nat. Congr. Appl. Mech.,* p. 1175.
Campbell, J. D., Simmons, J. A. and Dorn, J. E. (1961). On the dynamic expansion of a Frank-Read source, *Trans. ASME, J. Appl. Mech.,* September.
Cottrell, A. H. (1953). *Dislocations and Plastic Flow in Crystals,* Clarendon Press, Oxford.
Edelglass, S. M., (1966). *Engineering Materials Science,* Structure and mechanical behaviour of solids, Ronald Press.
Friedel, J. (1964). *Dislocations,* Pergamon Press.
Gessinger, H. (1970). Die theoretische Deutung der magnetischen Anisotropie plastich verformter Nickel-Einkristalle, *Phys. Stat. Sol.,* **39,** 581–98.
Gilman, J. J. (1972). The microdynamics of plastic flow, International Symposium on Foundations of Plasticity, Varsone.
Green, A. E. and Naghdi, P. M. (1965). A general theory of an elastic-plastic continuum, *Arch. Rational. Mech. Anal.,* **18,** 251–81.
Havner, K. S. (1972). An analytical model of large deformations effects in crystalline aggregates, *International Symposium on Foundations on Plasticity* (Varsovie), A. Sawczuk (Ed.), Noordhoff International Publishing.
Hill, R. (1963). Elastic properties of reinforced solids, *J. Mech. Phys. Solids,* **11,** 357.
Hutchinson, J. W. (1964). Plastic deformation of fcc polycrystals, *J. Mech. Phys. Solids,* **12,** 11–33.

Jaoul, B. (1965). Etude de la plasticité et applications aux métaux (Dunod).
Kocks, U. F. (1966). A statistical theory of flow stress and work-hardening, *Phil. Mag.*, **13**, 541–66.
Kratochvil, J. (1972). Finite strain theory of inelastic behaviour of crystalline solids, *International Symposium on Foundations of Plasticity* (Varsovie), A. Sawczuk (Ed.), Noordhoff International Publishing.
Kröner, E. (1958). *Kontinuums Theorie der Versetzungen und Eigenspannungen,* Springer-Verlag.
Kröner, E. (1971). Statistical continuum mechanics, Cours du C.I.S.M., Udine.
Kröner, E. and Teodosiu, C. (1972). Lattice defect approach to plasticity and viscoplasticity, *International Symposium on Foundations of Plasticity* (Varsovie), A. Sawczuk (Ed.), Noordhoff International Publishing.
Krönmuller, H. (1967). On the mechanism of work-hardening in f.c.c. metals, *Can. J. Phys.*, **45**, 631–61.
Lee, E. H. and Germain, P. (1972). Elastic-plastic theory at finite strain, *International Symposium on Foundations of Plasticity* (Varsovie), A. Sawczuk (Ed.), Noordhoff International Publishing.
Lin, T. H. (1971). Physical theory of plasticity, *Adv. Appl. Mech.*, **11**, 255–311.
Mandel, J. (1971). Plasticité, Cours du C.I.S.M., Udine.
Mróz, Z. and Zarka, J. (1978). Relations de comportement des métaux sous chargements cycliques, Séminaire Matériaux et Structures sous Chargements Cycliques, Palaiseau, Ecole Polytechnique.
Mughrabi, H. (1971). Elektronenmikroskopiche Untersuchung der Versetzungsanordnung verformter Kupfereinkristall im belasteten Zustand, *Phil. Mag.*, **23**, 869, 897, 931.
Nguyen, Q. S. and Zarka, J. (1972). Quelques méthodes de résolution numérique en élastoplasticité classique et en élastoviscoplasticité, Séminaire Plasticité et Viscoplasticité, Ecole Polytechnique, Paris.
Perzyna, P. (1971). Viscoplasticity: theory and applications, Cours du C.I.S.M., Udine.
Perzyna, P. (1972). Hidden parameter description of plasticity, *International Symposium on Foundations of Plasticity* (Varsovie), A. Sawczuk (Ed.), Noordhoff International Publishing.
Saada, G. V. (1960). Etude théorique de l'écrouissage à partir d'un modèle de dislocations. Publications de l'I.R.S.I.D., Série A, No. 251.
Seeger, A. (1958). Kristallplastizität, *Handbuch der Physik,* Vol. VII/2, Springer.
Sidoroff, F. (1973). The geometrical concept of intermediate configuration and the elastic-plastic finite strain, *Archives of Mechanics,* **25,** 299–308.
Taylor, G. I. (1958). Scientific papers. In *Mechanics of Solids,* Vol. 1, Cambridge, pp. 344–66; 390–8; 586.
Zaoui, A. (1972). Etude de l'influence propre de la desorientation des grains sur le comportement viscoplastique de métaux polycristallins (système c.f.c.), *Mémorial de l'Artillerie Francaise.*
Zarka, J. (1970). Sur la viscoplasticité des métaux, *Mémorial de l'Artillerie Française,* 2ème fascicule, pp. 223–92.

Index

GPSR Compliance

The European Union's (EU) General Product Safety Regulation (GPSR) is a set of rules that requires consumer products to be safe and our obligations to ensure this.

If you have any concerns about our products, you can contact us on ProductSafety@springernature.com

In case Publisher is established outside the EU, the EU authorized representative is:

Springer Nature Customer Service Center GmbH
Europaplatz 3
69115 Heidelberg, Germany

Batch number: 10607212

Printed by Printforce, the Netherlands